bn
15/12/05

)

Molecular Chaperones and Cell Signalling

This book reviews current understanding of the biological roles of extracellular molecular chaperones. It provides an overview of the structure and function of molecular chaperones, their role in the cellular response to stress, and their disposition within the cell. It also questions the basic paradigm of molecular chaperone biology – that these proteins are, first and foremost, protein-folding molecules. The current paradigms of protein secretion are reviewed, and the evolving concept of proteins (such as molecular chaperones) as multi-functional molecules, or "moonlighting proteins," is discussed. The role of exogenous molecular chaperones as cell regulators is examined, and the physiological and pathophysiological roles that molecular chaperones play are described. In the final section, the potential therapeutic use of molecular chaperones is described, and in the final chapter, the crystal ball is brought out and the question – what does the future hold for the extracellular biology of molecular chaperones – is asked.

Brian Henderson is Professor of Cell Biology at the Eastman Dental Institute, University College London, and Head of the Cellular Microbiology Research Group. His major research interests are concerned with bacterial interactions with the host and how such interactions control inflammation and associated tissue destruction. It is through these studies that he identified that molecular chaperones are bacterial virulence factors and started his interest in the direct immunomodulatory actions of cell stress proteins.

A. Graham Pockley is Professor of Immunobiology at the University of Sheffield Medical School and is Head of the Immunobiology Research Unit. Professor Pockley has long-standing interests in the immunobiology of transplant rejection, and his unit is currently focussed on research relating to the biology and immunotherapeutic potential of heat shock proteins, particularly their involvement in the rejection of organ transplants and the development and progression of cardiovascular disease.

Molecular Chaperones and Cell Signalling

Edited by

Brian Henderson
University College London

A. Graham Pockley
University of Sheffield

CAMBRIDGE
UNIVERSITY PRESS

CAMBRIDGE UNIVERSITY PRESS
Cambridge, New York, Melbourne, Madrid, Cape Town, Singapore, São Paulo

Cambridge University Press
40 West 20th Street, New York, NY 10011-4211, USA

www.cambridge.org
Information on this title: www.cambridge.org/9780521836548

First published 2005

Printed in the United States of America

A catalog record for this publication is available from the British Library.

Library of Congress Cataloging in Publication Data
Molecular chaperones and cell signalling / edited by Brian Henderson, A. Graham Pockley.
    p. ; cm.
Includes bibliographical references and index.
ISBN 0-521-83654-9 (hardback)
1. Molecular chaperones.
[DNLM: 1. Molecular chaperones–physiology. QU 55 M716245 2005]
I. Henderson, Brian, PhD. II. Pockley, A. Graham (Alan Graham), PhD. 1960– III. Title.
QP552.M64M64   2005
572′.645 – dc22   2005000664

ISBN-13   978-0-521-83654-8 hardback
ISBN-10   0-521-83654-9 hardback

Contents

prone to aggregation, not because of electrostatic interactions but because they expose highly hydrophobic surfaces to the aqueous environment. For a while the term was restricted to the two proteins that assist the assembly of amphibian nucleosomes and chloroplast rubisco. Its modern usage started when the author suggested that the term could be usefully extended to describe the function of a larger range of proteins that were postulated to assist folding and assembly/disassembly reactions in a wide range of cellular processes [1].

1.3. The general concept of chaperone function

The suggestions made in the first comprehensive description of the chaperone function have so far stood the test of time [8]. Molecular chaperones are defined as being a large and diverse group of proteins that share the property of assisting the non-covalent assembly/disassembly of other macromolecular structures but which are not permanent components of these structures when these are performing their normal biological functions. Assembly is used here in a broad sense and includes several universal intracellular processes: the folding of nascent polypeptide chains both during their synthesis and after release from ribosomes, the unfolding and refolding of polypeptides during their transfer across membranes, and the association of polypeptides with one another and with other macromolecules to form oligomeric complexes.

Molecular chaperones are also involved in macromolecular *dis*assembly processes, such as the partial unfolding and dissociation of subunits when some proteins carry out their normal functions, and the re-solubilisation and/or degradation of proteins partially denatured and/or aggregated by mutation or by exposure to environmental stresses, such as high temperatures and oxidative conditions. Some, but not all, chaperones are also stress or heat shock proteins as the requirement for chaperone function increases under stress conditions that cause proteins to unfold and aggregate. Conversely, some, but not all, stress proteins are molecular chaperones.

It is important to note that this definition is functional, not structural, and it contains no constraints on the mechanisms by which different chaperones may act; this is the reason for the use of the imprecise term 'assist'. Thus, molecular chaperones are defined neither by a common mechanism nor by sequence similarity. Only two criteria need be satisfied to designate a macromolecule a molecular chaperone. Firstly, it must in some sense assist the non-covalent assembly/disassembly of other macromolecular structures, the mechanism being irrelevant, and secondly, it must not be a component of these structures when they are performing their normal biological functions. In all cases studied so far, chaperones bind non-covalently to regions of macromolecules that are inaccessible when these structures are correctly assembled and functioning but that are accessible at other times.

The term non-covalent is used in this definition to exclude those proteins that catalyse co- or post-translational covalent modifications. These are often important for protein assembly, but are distinct from the proteins being considered here. Protein disulphide isomerase may appear to be an exception, but it is not. It is both a covalent modification enzyme and a molecular chaperone, but these activities lie in different parts of the molecule [9] and can be functionally separated by mutation. Other examples include peptidyl-prolyl isomerase, which possesses both enzymatic and chaperone activities in different regions of the molecule, and the α-crystallins, which in the lens of the eye combine two essential functions in the same molecule, contributing to the transparency and refractive index required for vision as well as to the chaperone function, which combats the loss of transparency as the protein chains aggregate with increasing age. The proteasome particle has a chaperone-like activity involved in unfolding proteins prior to their proteolysis. Thus, in principle, there is no reason why molecular chaperones should not possess additional functions, and the possibility that many possess cell-cell signalling functions is the central postulate argued in the other chapters of this volume.

The number of distinct chaperone families continues to rise, and examples occur in all types of cell and in most intracellular compartments. The families are defined on the basis that members within each family have high sequence similarity, whereas members in different families do not. Table 1.1 presents an incomplete list of proteins described as chaperones; however, it must be emphasised that in many cases this description rests on *in vitro* data only and needs confirmation by *in vivo* methods. There is evidence that some chaperones cooperate with each other in defined reaction sequences, but this, along with many other aspects of 'chaperonology', is beyond the scope of this chapter.

1.4. Common misconceptions

As with any new field, misconceptions abound. A common error is to use the term 'chaperonin' synonomously with the term 'chaperone', but it should be noted that the chaperonins are just one particular family of chaperone – i.e., the family that contains GroEL, Hsp60, and tailless complex polypeptide-1 (TCP-1) (see Chapters 5 and 6 for more details on chaperonins). The occasional use of the non-sense term 'molecular chaperonin' in some respectable journals suggests that some people use these terms casually without reference to either their meaning or their history. It should be obvious that the word 'molecular' is used to qualify 'chaperone' because in common usage 'chaperone' refers to a

Table 1.1. Proteins described as molecular chaperones

Family	Proposed roles
Non-steric chaperones	
Nucleoplasmins/nucleophosmins	Nucleosome and ribosome assembly/disassembly
Chaperonins	Folding of newly synthesised and denatured polypeptides
Hsp27/28	Prevention of stress-induced aggregation by adsorbing unfolded chains
Hsp40	Protein folding and transport, oligomer disassembly
Hsp47	Pro-collagen folding in the endoplasmic reticulum (ER)
Hsp70	Protein folding and transport, oligomer disassembly
Hsp90	Cell cycle, hormone activation, signal transduction
Hsp100	Dissolution of insoluble protein aggregates
Calnexin/calreticulin	Folding of glycoproteins in ER
SecB protein	Protein transport in bacteria
Lim protein	Folding of bacterial lipase
Syc protein	Secretion of toxic YOP proteins by bacteria
Protein disulphide isomerase	Prevention of misfolding in ER
ExbB proteins (may be structural rather than chaperones)	Folding of TonB protein in bacteria
Ubiquinated ribosomal proteins	Ribosome assembly in yeast
NAC complex	Folding of nascent proteins
Signal recognition particle	Arrest of translation and targeting to ER membrane
Trigger factor	Folding of nascent polypeptides in bacteria
Prefoldin	Cooperation with chaperonins in folding of newly synthesised polypeptides in *Archaea* and the eukaryotic cytosol
Tim9/Tim10 complex	Prevention of aggregation of hydrophobic proteins during import across mitochondrial intermembrane space
23S Ribosomal RNA	Folding of nascent polypeptides
PrsA protein	Secretion of proteins by *Bacillus subtilis*
Clusterin	Extracellular animal chaperone
Phosphatidylethanolamine	Folding of lactose permease
RNA binding proteins	Folding of RNA
P45	Protection against denaturation in halophilic *Archaea*
Steric chaperones	
PapD proteins	Assembly of bacterial pili
Propeptides (Class I)	Folding of some proteases

person. The term 'molecular chaperonin' is therefore as non-sensical as the term 'molecular immunoglobulin'.

Another common misconception is that molecular chaperones are necessarily promiscuous – i.e., that each assists the assembly of many different types of polypeptide chain. This is true for the Hsp70, Hsp40 and GroE chaperonin families but is not true for Hsp90, PapD, Hsp47, Lim, Syc, ExbB, PrtM/PrsA and prosequences, which are specific for their substrates. Similarly, it is not a universal property of chaperones that they hydrolyse ATP; Hsp100, Hsp90, Hsp70 and the chaperonins hydrolyse ATP, whereas trigger factor, Hsp40, prefoldin, calnexin, protein disulfide isomerase and papD do not. It is not even necessary that the chaperone function resides in molecules separate from their substrates. Thus, some pro-sequences are required for the correct folding of the remainder of the molecule but are then removed [10]. Another example of such intramolecular chaperones is the terminal ubiquitin residues of three ribosomal proteins in yeast; these residues promote the assembly of these proteins into the ribosome but are then removed, thus fulfilling the criteria suggested earlier for the chaperone function [11].

The term 'chemical chaperone' has been proposed to describe small molecules such as glycerol, dimethylsulfoxide and trimethylamine N-oxide that act as protein stabilising agents [12]. This terminology is unfortunate because proteins are also chemicals. However, its usage persists.

Finally, experience suggests that the distinction between molecular chaperones and stress proteins cannot be restated too often. The often-made interpretation, that because a protein accumulates after stress it must be a molecular chaperone, is incorrect, as is the belief that all molecular chaperones are stress proteins. For example, many heat shock proteins are ubiquitin-conjugating enzymes, while the cytosolic chaperonin of eukaryotic cells is not a stress protein.

1.5. Why do molecular chaperones exist?

Given that most denatured proteins that have been examined can refold into their functional conformations on removal of the denaturing agent *in vitro*, the question arises as to why molecular chaperones exist at all. Current evidence suggests that, with two possible exceptions [10], chaperones do not provide steric information for proteins to assemble correctly; rather they either prevent or reverse aggregation processes that would otherwise reduce the yield of functional molecules. Aggregation results because some proteins fold and unfold via intermediate states that expose some interactive surfaces (either charged or hydrophobic) to the environment. In aqueous environments hydrophobic surfaces stick together, while charged surfaces bind to ones bearing the opposite charge, a problem acute in the nucleus where negatively charged nucleic acids are bound to positively charged proteins. Thus, the existence of molecular chaperones does not cast doubt on the validity of the self-assembly principle. Rather, chaperones are required because, to operate efficiently under intracellular conditions, self-assembly needs assistance to avoid unproductive side reactions. This is why the term molecular chaperone is not an example of academic whimsy, but a precise description, because the role of the human chaperone is to improve the efficiency of 'assembly' processes between people without providing the steric information for these processes.

Protein aggregation has long been observed to occur during the *in vitro* refolding of many pure denatured proteins in dilute buffer solutions, but only recently has it been appreciated that the high degree of macromolecular crowding that characterises the intracellular environment makes the aggregation problem much more severe *in vivo*. Although the total concentrations of macromolecules inside cells are in the range 200–400 mg/ml, the properties of the isolated macromolecules are commonly studied *in vitro* at much lower concentrations in uncrowded buffers. The large thermodynamic effects of the high total concentrations of macromolecules inside cells are not generally appreciated and include increasing the association constants of protein aggregation reactions by one to two orders of magnitude [13]. Aggregation is a specific process involving identical or very similar chains and is driven by the interaction of both hydrophobic side chains and main-chain atoms in segments of unstructured backbone that are transiently exposed on the surface of partly folded chains; it is thus a high-order process that increases in rate as the concentration of similar chains or the temperature is raised. Refolding experiments suggest that large multi-domain proteins suffer from aggregation to a greater degree than small single-domain proteins because they fold more slowly via partly folded intermediate states. Thus proteins differ greatly in their propensity to aggregate, and it is likely that chaperones have evolved to combat this tendency of a particular subset of proteins.

These considerations can be reduced to a simple unifying principle. *All cells need a chaperone function to both prevent and reverse incorrect interactions that may occur when potentially interactive surfaces are exposed to the intracellular environment. Such surfaces occur on nascent and newly synthesised unfolded polypeptide chains, on mature proteins unfolded by stress or degradative mechanisms, and on folded proteins in near-native conformation.*

Thus, it is as valid to talk about the chaperone function as it is to talk about the transport function or the defence function of other proteins. We can view the chaperone function as one of the universal mechanisms that enable the crowded state of the cellular interior to be compatible with life.

The best understood chaperones are those involved in the folding of newly synthesised polypeptide chains in *E. coli*. The next section summarises what is known about their mechanisms of action.

1.6. Chaperones involved in *de novo* protein folding

The folding of newly synthesised proteins inside cells differs from the refolding of denatured proteins *in vitro* in two respects [14]. Firstly, protein chains fold inside cells in highly crowded macromolecular environments that favour aggregation. Secondly, protein chains are made vectorially inside cells at a rate slower than the rate of folding. It takes about 20 seconds for a cell of *E. coli* to synthesise a chain of 400 residues at 37 °C; however, *in vitro*, many denatured proteins will refold completely well within this time. Thus there is the possibility that the elongating nascent chain will either misfold because it is incomplete or aggregate with identical elongating chains on the same polysome. It is important to realise that misfolding is conceptually distinct from aggregation. Misfolding can be defined as the chain reaching a partly folded conformation from which it is unable to reach the final functional conformation on a biologically relevant time scale. Misfolded chains may or may not bind to one another to form non-functional aggregates that may be as small as a dimer or large enough to be insoluble. Thus all aggregates are, by definition, misfolded, but to what extent misfolded, but unaggregated, chains occur in cells is unclear.

Molecular chaperones assist the folding of both nascent chains bound to ribosomes and newly synthesised chains released from ribosomes – i.e., in both co-translational and post-translational modes. The chaperones working in these co-translational and post-translational modes are distinct and can be usefully termed small and large chaperones, respectively, because this is a case where size is important for function [15]. Small chaperones are less than 200 kDa in size and include trigger factor, nascent chain-associated complex, prefoldin, the Hsp70 and Hsp40 families and their associated co-chaperones. Co-chaperones are defined as proteins that bind to chaperones to modulate their activity; they may or may not also be chaperones in their own right. Large chaperones are more than 800 kDa in size and include the thermosome in *Archaea*, GroE proteins in *Eubacteria* and the eukaryotic organelles evolutionarily derived from them, and the TCP-1 or TRiC complexes and associated co-chaperones in the cytosol of *Eukarya*. The large chaperones are evolutionarily related and are collectively referred to as the chaperonins. There are no large chaperones in the endoplasmic reticulum lumen of eukaryotic cells, but small chaperones such as BiP (an Hsp70 homologue), calnexin, calreticulin, and protein disulphide isomerase that assist the folding of chains transported into the lumen after synthesis in the cytosol are present. Table 1.2 lists some of the chaperones that assist protein folding.

Table 1.2. Chaperones that assist protein folding

| Family | Other names | | Functions |
	Eukaryotes	Prokaryotes	
Hsp100	Hsp104, 78	ClpA/B/X	Disassembly of oligomers and aggregates
Hsp90	Hsp82, Hsp83, Grp94	HtpG	Regulate assembly of steroid receptors and signal transduction proteins
Hsp70	Hsc70, Ssal-4, Ssb1-2, BiP, Grp75	DnaK, Hsc66, Absent from many *Archaea*	Prevent aggregation of unfolded protein chains
Chaperonins	Hsp60, TRiC, CCT, TCP-1, rubisco subunit binding protein	GroEL, GroES	Sequester partly folded chains inside central cage to allow completion of folding in absence of other folding chains
Hsp40	Ydj1, Sis1, Sec63p, auxilin, zuotin, Hdj2	DnaJ	Stimulate ATPase activity of Hsp70
Prefoldin	GimC	Absent from *Bacteria*, present in *Archaea*	Prevent aggregation of unfolded protein chains
Trigger factor	Absent from *Eukarya*	Present	Bind to nascent chains as they emerge from ribosome
Calnexin, calreticulin	Present	Absent from prokaryotes	Bind to partly folded glycoproteins; located in ER membrane and lumen, respectively
Nascent-chain associated complex (NAC)	Present	Absent from prokaryotes	Bind to nascent chains as they emerge from ribosome
PapD	Absent from *Eukarya*	Present in some	Prevent aggregation of subunits of pili

1.6.1. Small chaperones

Small chaperones bind transiently to small hydrophobic regions (typically seven or eight residues long) on both nascent and completed, newly synthesised chains and thus prevent aggregation both during and after chain elongation by shielding these regions from one another (Figure 1.1) [3]. Trigger factor (48 kDa) is the

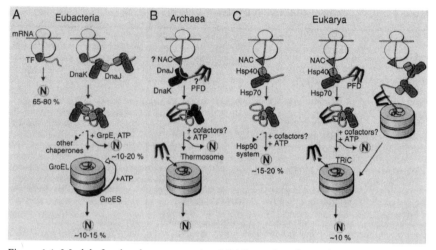

Figure 1.1. Models for the chaperone-assisted folding of newly synthesised polypeptides in the cytosol. (A) *Eubacteria*. TF, trigger factor; N, native protein. Most nascent chains probably interact with TF, and most small proteins (about 65–80% of total chain types) may fold rapidly upon synthesis without further chaperone assistance. Longer chains (10–20% of total chain types) interact subsequently with DnaK and DnaJ and fold after one or several cycles of ATP-dependent binding and release. About 10–15% of total chains fold within the chaperonin GroEL/GroES system. GroEL does not bind to nascent chains and is thus likely to receive its substrates after their release from DnaK. (B) *Archaea*. PFD, prefoldin; NAC, nascent-chain associated complex. Only some archaeal species contain DnaK/DnaJ. The existence of a ribosome-bound NAC homologue and the binding of PFD to nascent chains have not been shown. (C) *Eukarya*. Like TF, NAC probably interacts with many nascent chains. The majority of smaller chains may fold without further chaperone assistance. About 15–20% of chains reach their native states after assistance by Hsp70 and Hsp40, and a specific fraction of these are then transferred to Hsp90. About 10% of chains are passed to the TriC system in a reaction involving PFD. Reprinted from [3] with permission.

first chaperone to bind to nascent chains in prokaryotes because it is associated with the ribosomal large subunit that contains the tunnel from which the chains emerge [16]. A cell of *E. coli* contains about 20,000 copies of this chaperone, sufficient to bind to all nascent chains. Trigger factor shows peptidyl-prolyl isomerase activity and contains a hydrophobic groove which transiently binds to regions of the nascent chain enriched in aromatic residues. It binds to nascent chains as short as 57 residues and dissociates in an ATP-independent manner after the chain is released from the ribosome; this binding does not require prolyl residues in the nascent chain. The isomerase activity may provide a means of keeping nascent chains containing prolyl residues in a flexible state. The eukaryotic cytosol lacks trigger factor; however, its function may be replaced by that of a heterodimeric complex of 33- and 22-kDa subunits, termed the nascent chain–associated complex. Like trigger factor, this complex binds transiently to short nascent chains; however, unlike trigger factor, it does not possess peptidyl-prolyl isomerase activity [17].

Cells lacking trigger factor show no phenotype because its function can be replaced by that of the other major small chaperone, Hsp70 [18, 19]. The Hsp70 family has many 70-kDa proteins distributed between the cytoplasm of *Eubacteria* and some, but not all, *Archaea*, the cytosol of *Eukarya*, and eukaryotic organelles such as the endoplasmic reticulum, mitochondria and chloroplasts. Some, but not all, of these members are also stress proteins. Unlike trigger factor, most of the Hsp70 members do not bind to ribosomes but do bind to short regions of hydrophobic residues exposed on nascent and newly synthesised chains. Such regions occur statistically about every 40 residues and are recognised by a peptide-binding cleft in Hsp70; this recognition involves not just the hydrophobic side chains, but also main-chain atoms in the extended polypeptide backbone of the nascent chain [20].

Most is known about the Hsp70 member in *E. coli*, termed DnaK. Like all Hsp70 chaperones, DnaK contains an ATPase site and occupation of this site by ATP promotes rapid, but reversible, peptide binding. The importance of this ATPase site in cell-cell signalling is described in Chapter 10. ATP hydrolysis then tightens the binding through conformational changes in DnaK. The cycling of ATP between these states is regulated by a 41-kDa co-chaperone of the Hsp40 family, termed DnaJ in *E. coli*, and GrpE, a nucleotide exchange factor that is a co-chaperone, but not a chaperone. DnaJ binds to DnaK through its J domain and increases the rate of ATP hydrolysis, thus facilitating peptide binding. DnaJ, like all the Hsp40 proteins, acts as a chaperone in its own right because it also binds to hydrophobic peptides. Thus DnaK and DnaJ cooperate in binding each other to nascent chains; all Hsp70 chaperones are thought to cooperate with Hsp40 chaperones. The role of GrpE is to stimulate release of ADP from DnaK, allowing the latter to bind another molecule of ATP and so release the peptide. In the eukaryotic cytosol, the role of GrpE is fulfilled by an unrelated co-chaperone called Bag-1 [21]. Some *Archaea* lack Hsp70 proteins; however, their role in protein folding may be replaced by that of an unrelated 90-kDa dimer chaperone called prefoldin.

There is enough DnaK in each *E. coli* cell for one molecule to bind to each nascent chain. DnaK binds to longer chains than trigger factor and so probably binds after trigger factor. When the gene for trigger factor is deleted, the fraction of nascent and newly synthesised chains binding to DnaK increases from about 15% to about 40%. However, removal of the genes for both trigger factor and DnaK in the same cell causes the aggregation of many newly synthesised chains

[18, 19]. The redundancy of important control systems is as good a design principle for cells as it is for passenger planes.

Small chaperones function essentially by reducing the time that potentially interactive surfaces on neighbouring chains are exposed by cycling on and off these chains until they have folded; they do not appear to change the conformation of the chains. Such a simple mechanism can be thought of as analogous to tossing a hot potato from hand to hand until it has cooled enough to be held. However, the other major class of chaperones involved in protein folding function by a much more sophisticated mechanism enabled by their large size.

1.6.2. Large chaperones – the chaperonins

Most is known about GroEL and GroES, the chaperonin and co-chaperonin found in E. coli; however, the general principles of their mechanism (Figure 1.2) are thought to apply also to the thermosome found in *Archaea* and to the TCP-1 complex (also called the TRiC or CCT complex) found in the eukaryotic cytosol. GroEL (800 kDa) consists of two heptameric rings of identical 57-kDa ATPase subunits stacked back to back, containing a cage in each ring [22]. The term cage is used because the walls surrounding each central cavity contain gaps, perhaps to allow entry and exit of nucleotides and water. Each subunit contains three domains. The equatorial domain contains the nucleotide binding site and is connected by a flexible intermediate domain with the apical domain. The latter presents several hydrophobic side chains at the top of the ring orientated towards the cavity of the cage, an arrangement that permits either a partly folded polypeptide chain or a molecule of GroES to bind but prevents binding to another GroEL oligomer.

GroES is a single heptameric ring of 10-kDa subunits that cycles on and off either end of the GroEL in a manner regulated by the ATPase activity of GroEL. At any one time, GroES is bound to only one end of GroEL, leaving the other end free to bind a partly folded polypeptide chain after its release from the ribosome. GroEL does not bind to nascent chains, whereas TCP-1 may do so. The two rings of GroEL are coupled by negative allostery so that only one ring at a time binds nucleotide, but within each ring the binding of nucleotide is cooperative. When either ADP or ATP is bound to one GroEL ring, the GroES sits on top of this ring – now called the *cis* ring. The binding of GroES triggers a large rotation and upward movement of the apical domains, resulting in an enlarged cage and a change in its internal surface properties from hydrophobic to hydrophilic. This enlarged cage can accommodate a single partly folded

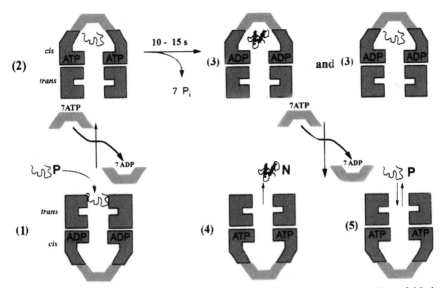

Figure 1.2. Mechanism of action of the GroEL/GroES system in *E. coli*. P, unfolded polypeptide chain; N, native folded chain. Dark shaded illustrates a section through GroEL, light shaded illustrates a section through GroES. For details, see text.

compact polypeptide chain up to about 60 kDa in size, perhaps depending on shape.

The reaction cycle starts with a GroEL-GroES complex containing ADP bound to the *cis* ring (Figure 1.2, step 1). The hydrophobic residues on the apical domains on the other ring, now called *trans*, bind to hydrophobic residues exposed on a partly folded polypeptide chain, presumably after release of small chaperones from this chain. GroES and ATP then bind to this ring, thereby converting it into a new *cis* ring and causing the release of GroES and ADP from the old *cis* ring (Figure 1.2, step 2). This binding of GroES to the *trans* ring displaces the bound polypeptide into the cavity of the cage because some of the hydrophobic residues of the apical domains that bind the polypeptide are the same residues that bind GroES. The displaced chain lying free in the cavity of the cage now has 10–15 seconds to continue folding, a time set by the slow but cooperative ATPase activity of the seven subunits in each ring (Figure 1.2, step 3). The chain thus continues its folding sheltered in a hydrophilic environment containing no other folding chain. Many denatured polypeptide chains will fold completely within 15 seconds in the classic Anfinsen renaturing experiment carried out in a test tube instead of inside GroEL. It is for this reason that I call this mechanism the 'Anfinsen cage model' [23].

The binding of ATP and GroES to the new *trans* ring then triggers the release of GroES and ADP from the *cis* ring containing the polypeptide chain, thereby allowing the latter to diffuse out of the cage into the cytoplasm. If this chain has internalised its hydrophobic residues, it remains free in the cytoplasm (Figure 1.2, step 4). However, any chain that still exposes hydrophobic residues rebinds to the same ring for another round of encapsulation (Figure 1.2, step 5). Rebinding to the same ring rather than the ring of another GroEL oligomer is favoured by the crowding effect created by the high concentration of macromolecules in the cytoplasm and reduces the risk that partly folded chains will meet one another in the cytoplasm in a potentially disastrous encounter [24].

This model was proposed to explain the results of many ingenious *in vitro* experiments; however, recent genetic studies confirm the importance of the Anfinsen cage mechanism in intact cells. Mutants in which the mechanism is prevented by blockage of the entrance to one of the rings of each GroEL oligomer are viable, but the cells form colonies only 10% the size of the wild-type colonies [23, 25]. That these mutants are viable at all suggests that the ring whose entrance is not blocked is acting rather like the small chaperones – reducing aggregation by binding and releasing from the hydrophobic regions on newly synthesised chains.

An unexpected added advantage of the Anfinsen cage mechanism is that, for proteins in a certain size range, encapsulation in the cage increases the rate of folding compared to the rate observed in free solution under conditions where aggregation is not a problem. Thus the rate of folding of bacterial rubisco (50 kDa) is increased four-fold by encapsulation, whereas that of rhodanese (33 kDa) is not affected [26]. This effect can be explained in terms of a type of macromolecular crowding called confinement, in which the proximity of the walls of the confining cage stabilises compact conformations more than extended ones and so enhances the rate of interactions leading to compaction of the folding chain [27]. A possible further advantage of the Anfinsen cage mechanism is that the rotation of the apical domains may cause the bound polypeptide to unfold to some extent and thus destabilises conformations that may have misfolded [28]; this interesting suggestion awaits experimental support.

Current estimates suggest that the fraction of newly synthesised polypeptide chains that bind *in vivo* to either Hsp70 proteins or the chaperonins is in the range 10–20% [3]. Whether the majority of newly synthesised chains bind to other, as yet undiscovered, chaperones or fold unassisted because their sequences have evolved to avoid aggregation is unknown. Nor is it understood what determines that only a defined minority of polypeptides bind to GroEL in the intact cell.

1.7. Other functions of molecular chaperones

The ubiquity and diversity of molecular chaperones is consistent with the possibility that some members have functions additional to the prevention of aggregation. Two examples are well established: the first concerns chaperones that regulate the properties of proteins that have largely folded but are not in their functional native states, whereas the second illustrates a chaperone family that redissolves insoluble protein aggregates.

The Hsp90 family assists in the regulation of signal transduction pathways, the best studied of which are the steroid response pathways mediated by specific receptor proteins [29, 30]. Hsp90 does not act generally in the folding of nascent protein chains as does Hsp70; instead, most of its known substrates are signal transduction proteins folded in a near-native state, ready for interacting with other molecules that trigger their signalling function. Steroid receptors have domains for binding their steroidal ligands, for dimerisation, and for binding regulatory proteins that determine the transcriptional activity of specific sets of genes. In the absence of steroid hormones these receptors are bound to several types of chaperone and co-chaperone, including Hsp70 as well as Hsp90. However, they remain monomeric, do not bind to DNA and are thus functionally inactive as transcription factors. This type of chaperone function involves changes in conformations of the substrate protein and therefore falls within the general area of chaperone-assisted protein folding, but with the aim of regulating function instead of preventing aggregation.

The second example concerns the reversal of protein aggregation. The success of molecular chaperones at preventing aggregation is not perfect, and there is evidence that it declines as organisms age [31, 32]. Protein damage accumulates with time and eventually overloads the repairing ability of chaperones, with Alzheimer's disease being the most dramatic and depressing consequence. The universal heat shock response can be regarded as a mechanism to increase the concentrations of chaperones when cells are subject to environments that cause protein denaturation and subsequent aggregation. The Hsp100 family has the unique ability to redissolve insoluble protein aggregates and acts to rescue or remove those proteins that evade the aggregation-prevention screen provided by other chaperones. The Hsp100 family is a functionally diverse group of oligomeric ATPases that are, in turn, a subfamily of the AAA+ superfamily. All AAA+ proteins form single-ring-shaped complexes reminiscent of the chaperonins. As with the latter, it is thought that the combination of several substrate-binding sites arranged around the ring is crucial to their function [33].

The Hsp100 family member in yeast is called Hsp104; removal of the Hsp104 function by mutation is not lethal, but it does prevent the cell from rescuing the activity of test enzymes inactivated by brief exposure of the cells to 44 °C. Such heat shock causes proteins to aggregate in both the cytoplasm and nucleus of yeast cells; when such cells are placed at 25 °C after the heat shock these aggregates disappear, but this function is lost in cells lacking the Hsp104 function [34]. *In vitro* studies show that, unlike many other chaperones, Hsp104 is unable to prevent the aggregation of denatured proteins but, when mixed with Hsp70 and Hsp40 from yeast, can mediate the recovery of enzymic activity from insoluble aggregates.

1.8. The cell–cell signalling hypothesis

The term cell-cell signalling is conventionally used to encompass all those processes by which cells secrete specific compounds that influence the behaviour of other cells. Given the universality and abundance of molecular chaperones it is plausible to speculate that some of them have such signalling roles, in addition to their known roles in protein folding, and establishing such roles would be an important advance. In my view the current evidence is suggestive rather than conclusive.

Much of the evidence that chaperones occur on cell surfaces and in the extracellular medium relies on the uses of antibodies (see Chapter 2). This evidence needs strengthening by physical characterisation of what the antibodies are detecting – are they detecting proteolytic fragments, subunits, or the oligomeric forms of these chaperones? In the case of the large chaperones such characterisation should be straightforward. It is also not clear in what form exogenous pure chaperones added to cells in culture bind to other cells, given that in some cases (e.g., the mycobacterial chaperonins) it is reported that effects on cell behaviour resist boiling or treatment with proteases (described in Chapter 6). A separate issue that needs further clarification is how these materials appear in the extracellular medium – are they secreted by specific mechanisms by live cells or are they simply released by dead or dying cells? (Partial answers to this question are to be found in Chapters 3, 5 and 12). The ten genes for the mitochondrial chaperonin 60 (Cpn60) of human cells all contain mitochondrial targeting sequences, but there is no established mechanism known for the transport of mitochondrial matrix proteins across the plasma membrane (see Chapter 3). Establishing the existence of such a mechanism would be an important contribution to cell biology.

The minimum hypothesis that is consistent with current evidence is a special form of cell-cell signalling in which cells respond to signals in the form of chaperone fragments released from cells damaged by infection or other stresses. Such damaged cells could include pathogenic bacteria as well as cells of the host. This type of signalling would broadcast the news that some cells in the organism had experienced stresses, including invasion by pathogenic bacteria and viruses, so that ameliorating responses could be triggered. There is convincing evidence that the Cpn60 from *Enterobacter aerogenes* cells found in the saliva of antlion larvae that prey on other insects acts to paralyse the prey, establishing an extracellular role for this particular chaperone [35]. It was shown that the paralytic principle is the intact oligomer of Cpn60 and that it is inactivated by trypsin, but whether it is actively secreted into the saliva or is released from dying bacterial cells was not determined. The Cpn60 from *E. coli* is not toxic to insects but can be converted into one simply by changing the isoleucine at position 100 to valine or changing aspartate at position 338 to glutamate. These remarkable findings are consistent with the possibility that at least some chaperones do have extracellular roles; however, more incisive studies are required to establish this as a widespread phenomenon.

REFERENCES

1. Ellis R J. Proteins as molecular chaperones. Nature 1987, 328: 378–379.
2. Hemmingsen S M, Woolford C, van der Vies S M, Tilly K, Dennis D T, Georgopoulos G C, Hendrix R W and Ellis R J. Homologous plant and bacterial proteins chaperone oligomeric protein assembly. Nature 1988, 333: 330–334.
3. Hartl F U and Hayer-Hartl M. Molecular chaperones in the cytosol: from nascent chain to folded protein. Science 2002, 295: 1852–1858.
4. Ellis R J and Hartl F U. Protein folding and chaperones. *Nature Encyclopedia of the Human Genome*. Macmillan Publishers Ltd 2003, pp 806–810.
5. Laskey R A, Honda B M and Finch J T. Nucleosomes are assembled by an acidic protein that binds histones and transfers them to DNA. Nature 1978, 275: 416–420.
6. Ellis R J. The general concept of molecular chaperones. In Ellis, R. J., Laskey, R. A. and Lorimer, G. H. (Eds.) *Molecular Chaperones*. Chapman and Hall for The Royal Society, London 1993, pp 1–5.
7. Musgrove J E and Ellis R J. The rubisco large subunit binding protein. Phil Trans R Soc Lond B 1986, 313: 419–428.
8. Ellis R J and Hemmingsen S M. Molecular chaperones: proteins essential for the biogenesis of some macromolecular structures. Trends Biochem Sci 1989, 14: 339–342.
9. Puig A and Gilbert H F. Protein disulfide isomerase exhibits chaperone and antichaperone activity in the oxidative folding of lysozyme. J Biol Chem 1994, 269: 7764–7771.
10. Ellis R J. Steric chaperones. Trends Biochem Sci 1998, 23: 43–45.
11. Finley D, Bartel B and Varshavsky A. The tails of ubiquitin precursors are ribosomal proteins whose fusion to ubiquitin facilitates ribosome biogenesis. Nature 1989, 338: 394–401.
12. Welch W J. Role of quality control pathways in human diseases involving protein misfolding. Semin Cell Devel Biol 2004, 15: 31–38.

13. Ellis R J. Macromolecular crowding: obvious but underappreciated. Trends Biochem Sci 2001, 26: 597–603.
14. Hartl F U. Molecular chaperones in cellular protein folding. Nature 1996, 381: 571–579.
15. Ellis R J and Hartl F U. Principles of protein folding in the cellular environment. Cur Opin Struct Biol 1999, 9: 102–110.
16. Schlieker C, Bukau B and Mogk A. Prevention and reversion of protein aggregation by molecular chaperones in the *E. coli* cytosol; implications for their applicability in biotechnology. J Biotech 2002, 96: 13–21.
17. Beatrix B, Sakai H and Wiedmann M. The α and β subunits of the nascent polypeptide complex have distinct functions. J Biol Chem 2000, 275: 37838–37845.
18. Deuerling E, Schulze-Specking A, Tomoyasu A, Mogk A and Bukau B. Trigger factor and DnaK cooperate in folding of newly synthesized proteins. Nature 1999, 400: 693–696.
19. Teter S A, Houry W A, Ang D A, Tradler T, Rockabrand D, Fischer G, Blum P, Georgopoulos C and Hartl F U. Polypeptide flux through bacterial hsp70: DnaK cooperates with trigger factor in chaperoning nascent chains. Cell 1999, 97: 755–765.
20. Bukau B and Horwich A L. The Hsp70 and Hsp60 chaperone machines. Cell 1998, 92: 351–366.
21. Sondermann H, Scheufler C, Schneider C, Hohfeld J, Hartl F U and Moarefi I. Structure of a Bag/hsc70 complex: convergent functional evolution of hsp70 nucleotide exchange factors. Science 2001, 291: 1553–1557.
22. Xu Z, Horwich A L and Sigler P B. The crystal structure of the asymmetric GroEL-GroES- (ADP)$_7$ chaperonin complex. Nature 1997, 388: 741–750.
23. Ellis R J. Protein folding: importance of the Anfinsen cage. Cur Biol 2003, 13: R881–R883.
24. Martin J and Hartl F U. The effect of macromolecular crowding on chaperonin-mediated protein folding. Proc Natl Acad Sci USA 1997, 94: 1107–1112.
25. Farr G W, Fenton W A, Rospert S and Horwich A L. Folding with and without encapsulation by cis and trans-only GroEL-GroES complexes. EMBO J 2001, 22: 3220–3230.
26. Brinker A, Pfeifer G, Kerner M J, Naylor D J, Hartl F U and Hayer-Hartl M. Dual function of protein confinement in chaperonin-assisted protein folding. Cell 2001, 107: 223–233.
27. Takagi F, Koga N and Takada S. How protein thermodynamics and folding are altered by the chaperonin cage: molecular simulations. Proc Natl Acad Sci USA 2003, 100: 11367–11372.
28. Thirumalai D and Lorimer G H. Chaperonin-mediated protein folding. Ann Rev Biophys Biomol Struct 2001, 30: 245–269.
29. Young J C, Moarefi I and Hartl F U. Hsp90: a specialized but essential protein-folding tool. J Cell Biol 2001, 154: 267–273.
30. Smith D F. Chaperones in progesterone receptor complexes. Semin Cell Devel Biol 2000, 11: 45–52.
31. Csermley P. Chaperone overload is a possible contributor to 'civilization diseases'. Trends Genetics 2001, 17: 701–704.
32. Soti C and Csermley P. Molecular chaperones and the aging process. Biogerontology 2000, 1: 225–233.

33. Glover J R and Tkach J M. Crowbars and rachets: hsp100 chaperones as tools in reversing aggregation. Biochem Cell Biol 2001, 79: 557–568.
34. Parsell D A, Kowal A S, Singer M A and Lindquist S. Protein disaggregation by heat shock protein hsp104. Nature 1994, 372: 475–477.
35. Yoshida N, Oeda K, Watanabe E, Mikami T, Fukita Y, Nishimura K, Komai K and Matsuda K. Protein function. Chaperonin turned insect toxin. Nature 2001, 411: 44.

2

Intracellular Disposition of Mitochondrial Molecular Chaperones: Hsp60, mHsp70, Cpn10 and TRAP-1

Radhey S. Gupta, Timothy Bowes, Skanda Sadacharan and Bhag Singh

2.1. Introduction

This chapter reviews work on the intracellular disposition of a number of molecular chaperones that are generally believed to be localised and function mainly within the mitochondria of eukaryotic cells. However, in recent years, compelling evidence has accumulated from many lines of investigation indicating that several of these mitochondrial (m-) chaperones are also localised and perform important functions at a variety of other sites/compartments within cells (see [1, 2]). The four chaperone proteins that are the subjects of this chapter include the following: (i) the 60-kDa heat shock chaperonin protein (Hsp60, also known as chaperonin 60, Cpn60), which is a major protein in both stressed and unstressed cells and plays an essential role in the proper folding and assembly into oligomeric complexes of other proteins [3–6]; (ii) the 10-kDa heat shock chaperonin (Hsp10 or Cpn10), which is a co-chaperone for Hsp60 in the protein folding process [7]; (iii) the mitochondrial homologue of the major 70-kDa heat shock protein (mHsp70), which plays a central role in the import of various proteins into mitochondria and their proper folding [4, 6]; and (iv) the mitochondrial Hsp90 protein, which was originally identified in mammalian cells as the tumour necrosis factor receptor-associated protein-1 (TRAP-1) [8, 9] and is commonly referred to by this latter name.

All of these proteins are encoded by nuclear genes, and, after translation of their transcripts in the cytosol, their protein products are then imported into mitochondria. The import of protein into mitochondria is generally a highly efficient process which occurs very rapidly and generally to completion [10, 11]. Further, once imported into mitochondria, the proteins are not known to exit under normal physiological conditions. Hence, their presence at extra-mitochondrial sites raises important questions regarding the possible mechanisms by which they have reached these locations [1, 2]. The presence of these proteins at other sub-cellular locations also greatly broadens the range of functions with which they are likely to be involved within the cell. This chapter provides a brief review of the cellular distributions of these chaperone proteins and the significance of their distributions on their cellular functions. Other aspects of these proteins are covered in various reviews [3–5, 11, 12] and elsewhere in this volume. The reader is referred to Chapter 3 for a discussion of non-classical pathways of protein export in eukaryotic cells and to Chapter 12 for a review of molecular chaperone release from cells and of molecular chaperones in the circulation.

2.2. Sub-cellular localisation of the mitochondrial molecular chaperones

2.2.1. Hsp60/chaperonin 60

Hsp60 or chaperonin 60 (Cpn60), which is the eukaryotic homologue of the bacterial GroEL protein, constitutes one of the major and most characterised molecular chaperone proteins in both stressed and unstressed cells [4, 13]. In eukaryotic organisms, this protein is primarily found in organelles such as mitochondria and chloroplasts, which have originated from bacteria belonging to the proteobacteria and cyanobacteria groups, respectively [14, 15]. Of these, the mitochondrial Hsp60 has been extensively studied. It is encoded by nuclear DNA and synthesised as a larger precursor containing an N-terminal mitochondrial targeting sequence (MTS), which is cleaved during import of the precursor protein into the matrix compartment [16].

The mature form of the Hsp60 found in mitochondria and various other compartments lacks the MTS sequence [17–19]. Hsp60 was initially discovered in mammalian cells as a protein (P_1) that was specifically altered in Chinese hamster ovary (CHO) cell mutants resistant to the microtubule inhibitor podophyllotoxin [20–22]. Cellular sub-fractionation and immunofluorescence studies indicated that this protein was primarily localised in the mitochondrial matrix compartment [23, 24]. The matrix localisation of Hsp60 (P_1) was totally unexpected in view of the earlier genetic and biochemical studies that strongly indicated that this protein interacted with tubulin, which is not found in mitochondria [25]. Although earlier cell fractionation and immunofluorescence studies suggested that Hsp60 was exclusively a mitochondrial protein, subsequent work reviewed below provides strong evidence that it is also present outside of mitochondria, including on the cell surface (see [1, 2]).

One of the earliest observations pointing to the presence of Hsp60 on the cell surface came from studies on murine and human T cells that recognised the mycobacterial Hsp60 (GroEL). These cells were also found to be stimulated by a protein present on the surface of stressed macrophages and certain tumour cells [26, 27], and this stimulation was blocked by both polyclonal and monoclonal antibodies to Hsp60 [28], indicating that such cells were expressing Hsp60 on their cell surface.

More definitive evidence for the presence of Hsp60 on the surface of cells was provided by its immunoprecipitation from surface-iodinated or surface-biotinylated proteins by polyclonal and monoclonal antibodies specific for Hsp60 [28, 29]. Hsp60 has also been identified on the cell surface by chemical cross-linking of live cells, where it was found associated with the plasma membrane resident p21^{ras} protein [17], suggesting its possible involvement in signal transduction events. Hsp60 in the plasma membrane is also found to be concomitantly enhanced in CHO cell mutants exhibiting an increase in the A system of amino acid transport, suggesting its possible association with the corresponding amino acid transporter [18].

In another study, the plasma membrane–associated Hsp60 was found to be specifically phosphorylated upon activation of Type I protein kinase A [19]. Interestingly, this study also found that histone 2B formed a complex with plasma membrane–associated Hsp60. Phosphorylation of both Hsp60 and histone 2B by Type I protein kinase A disrupted their association, leading to expulsion of histone 2B, but not Hsp60, from the membrane. Hsp60 in the plasma membrane has also been shown to bind to the high-density lipoprotein [30] and has been indicated to play a role in the peptide presentation process [31, 32]. The presence of both Hsp60 and tubulin in the plasma membrane also provides a plausible explanation for the puzzling observation that led to the discovery of Hsp60 in mammalian cells [21, 22] – that mutational changes in this protein cause resistance to anti-mitotic drugs that bind to tubulin. In our earlier studies, in which Hsp60 (P_1) was identified as a tubulin-associated protein [20, 33], we suggested that tubulin in the plasma membrane is associated with Hsp60, such that mutational changes in Hsp60 can alter drug binding to tubulin [2, 34]. All of the aforementioned studies strongly suggest that Hsp60 in the plasma membrane functions as a membrane chaperone, which enables other soluble proteins to exhibit membrane association.

A number of studies have reported an increased cell surface expression of Hsp60 under stressed or apoptotic conditions. In aortic endothelial cells exposed to cytokines or high temperature, increased expression of Hsp60 has been detected on the cell surface by fluorescence imaging [35], and this has been shown to make such cells susceptible to complement-dependent lysis by Hsp60-specific antibodies [36]. Increased expression of Hsp60 and Hsp70 on the cell surface has also been observed in T cells undergoing apoptosis [37]. In a recent study, Hsp60 was found to interact with Bax in the cytoplasm of cardiomyocytes [38].

However, during hypoxia, Hsp60 is re-localised to the plasma membrane, allowing Bax to translocate to mitochondria to induce apoptosis. Although certain types of cells or conditions might enhance or induce cell-surface Hsp60 expression, it is important to recognise that surface expression of Hsp60 is a common characteristic of eukaryotic cells and is not limited to stressed or apoptotic cells [1, 2].

Extensive work has been carried out on the sub-cellular localisation of Hsp60 in different cultured cell lines as well as tissues by means of immunogold labelling (or immuno-electron microscopy (Immuno-EM)), employing monoclonal and polyclonal antibodies. Although immuno-EM labelling of cultured mammalian cells (CHO, BSC-1 kidney cells, PC12 neuronal, Daudi Burkitt's lymphoma and human diploid fibroblasts) has demonstrated the majority of Hsp60 labelling to be found within mitochondria [39], 15–20% of the reactivity is consistently observed at discrete extra-mitochondrial sites, including unidentified cytoplasmic vesicles and granules, sites on endoplasmic reticulum, and at the cell surface (Figure 2.1A) [29, 39]. Using backscattered electron imaging of intact cells, the cell surface copy number of Hsp60 has been estimated to be approximately 200–2000 molecules per cell, in CHO and CEM-SS human T lymphocyte cell lines, which appears to represent about 1–10% of the total cellular Hsp60 [29].

The sub-cellular distribution of Hsp60 has also been examined in different mammalian tissues using a high-resolution immuno-EM technique [40–42]. In some tissues, such as heart, kidney (proximal and distal tubules), skeletal muscle, adrenal gland and spleen, reactivity to Hsp60 antibody was primarily restricted to mitochondria [41]. However, in a number of other tissues, strong and specific labelling due to Hsp60 antibody has been observed in a number of other compartments in addition to mitochondria. In pancreatic β-cells, strong reactivity with Hsp60 antibodies is also seen in mature insulin secretory granules (ISGs) (Figure 2.1B) [40]. In this instance, the Hsp60 antibodies specifically labelled the central core of the mature ISGs, but no labelling was seen in immature secretory granules [40]. In rat liver, specific labelling with Hsp60 antibodies has been observed in mitochondria and peroxisomes [39, 41, 43]. The Hsp60 reactivity in peroxisomes is primarily associated with the urate oxidase crystalline core, which is a distinguishing characteristic of rat liver peroxisomes. In pancreatic acinar cells and pituitary, strong labelling with Hsp60 antibodies has been observed in zymogen granules (ZGs) (Figure 2.1C) and growth hormone granules (GHGs), respectively [41]. The labelling of these compartments with Hsp60 antibodies is completely abolished upon pre-adsorption of the antibodies with recombinant Hsp60, thereby providing evidence that it is specific for Hsp60 [41].

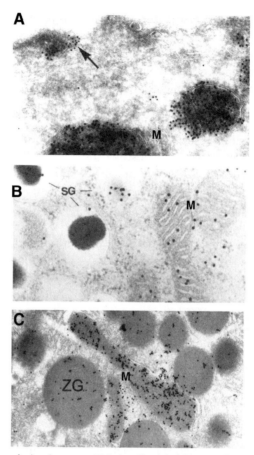

Figure 2.1. High-resolution immuno-EM visualisation of Hsp60 distribution in different cells and tissue using Hsp60-specific antibodies. (A) Immunogold labelling of cryosections of CHO cells; labelling is clearly seen in both mitochondria (M) and on the cell surface. (B) In mouse pancreatic beta cells, labelling is seen both within mitochondria and in the dense core of mature insulin secretory granules (SG). (C) In pancreatic acinar cell sections, strong labelling is observed both within mitochondria and zymogen granules (ZG). In all of the preceding cases, very little labelling is seen in the ER and Golgi compartments. From Brudzynski et al. [40]; Soltys and Gupta [39]; Cechetto et al. [41].

It is of much interest that, although strong labelling due to Hsp60 antibodies is observed in ZGs, GHGs and ISGs, there is negligible (i.e., close to background) labelling seen in immature ISGs, as well as in the endoplasmic reticulum (ER) and Golgi compartments [41]. These results are in marked contrast to those obtained with antibodies to other proteins such as insulin and amylase, which are targeted to the above compartments and in which strong labelling of the entire

ER–Golgi pathway is observed [44, 45]. The absence of significant Hsp60 labelling in the cytoplasm, as well as the ER–Golgi compartments, using different monoclonal and polyclonal antibodies, suggests that Hsp60 could be reaching these granules via a novel mechanism that is different from that which uses the classical ER–Golgi pathway [44, 46]. The presence of Hsp60 in secretory granules suggests that certain cell types should secrete Hsp60. Indeed, the secretion of an Hsp60-like protein has been reported for cultured neuroglial cells and a neuroblastoma cell line [47]. Velez-Granell and colleagues, employing a polyclonal antibody against the bacterial GroEL (from *Chromatium vinosum*), have observed considerable Hsp60 reactivity along the ER–Golgi secretory pathway [42]. However, this antibody also exhibited very high background labelling [42], hence the significance of these results is not clear.

The possible physiological function of Hsp60 in these compartments is presently not known. Hsp60 is associated with the central core of the mature ISGs, but it is not present in immature ISGs. The main difference between these two types of granules is that, during transition from immature ISGs to mature ISGs, pro-insulin is enzymatically cleaved to form insulin, which then, by a poorly understood process, is extensively condensed to form the highly compacted core of the mature granules [48]. The central core of these mature ISGs thus represents a highly organised, supra-molecular structure, the main function of which appears to be to maintain insulin at a very high concentration in a functional form that is ready to be secreted. The Hsp60 in other types of granules such as ZGs and GHGs, may be playing an analogous role. In a similar manner, in peroxisomes, Hsp60 is associated with the urate oxidase crystalline cores, which also constitute a higher order supra-molecular structure which likely requires a chaperone for its assembly. In accordance with its established functions in the formation of oligomeric protein complexes, and in protein secretion in bacteria [4, 49], we have suggested that Hsp60 in different types of granules (ZG, ISGs, GHGs) and peroxisomes also plays a chaperone role in the condensation of proteins within these compartments, and in maintaining the highly compacted proteins in functional forms required for their biological actions [39–41].

2.2.2. Hsp10 (Cpn10)

Hsp10, or Cpn10, is the eukaryotic homologue of the bacterial GroES protein, which serves as a co-chaperone for Hsp60 (GroEL) in the protein folding and assembly processes [3, 12]. Similar to Hsp60, this protein is present in eukaryotic cells in organelles such as mitochondria and chloroplasts [7, 15, 45]. Unlike most other mitochondrial matrix proteins, Hsp10 does not contain an N-terminal cleavable MTS; rather its N-terminal sequence has the ability to form an amphipathic alpha helix which possibly enables it to cross the mitochondrial membrane [50]. Surprisingly, Hsp10 has also been shown to be identical to a protein previously identified as early pregnancy factor (EPF), which appears in maternal serum within 24 hours after fertilisation [51, 52]. The evidence that EPF from human platelets is identical to the Hsp10 protein is provided by several observations: (i) The amino acid sequences of three different fragments covering more than 70% of the EPF show complete identity with the human Hsp10 protein [52]. (ii) Purified rat Hsp10 is found to be as active in the EPF bioassay as the platelet-derived EPF, and this activity can be neutralised by a monoclonal antibody to EPF. In contrast to the mammalian Hsp10, bacterial GroES is not active in the bioassay, providing evidence of specificity [53]. (iii) In the presence of ATP, EPF (similar to Hsp10) forms a stable complex with Hsp60 which co-elutes from a gel filtration column. Further, immobilised Hsp60, in the presence of ATP, removes all EPF activity from pregnancy serum, providing evidence of a specific interaction between these proteins [52, 53]. Fletcher and colleagues have indicated that EPF in mouse cells may be encoded by an intronless gene, the pattern of expression of which is similar to that of EPF activity [54].

The sub-cellular distribution of Hsp10 in rat tissues has been examined in detail using the high-resolution immuno-EM technique employing polyclonal antibodies raised against different regions of human Hsp10 [45]. In all rat tissues examined including liver, heart, pancreas, kidney, anterior pituitary, salivary gland, thyroid and adrenal gland, antibodies to Hsp10 strongly labelled mitochondria. However, in a number of tissues, in addition to mitochondria, strong and specific labelling with the Hsp10 antibodies is also observed in several extra-mitochondrial compartments. These sites included ZGs in pancreatic acinar cells (Figure 2.2A), GHGs in anterior pituitary (Figure 2.2B) and pancreatic polypeptide granules in islet cells. These granules likely provide the pathway for the secretion of this protein into the blood stream, in which it can serve the function of EPF. The Hsp10 labelling in these compartments is at least comparable to (if not higher than) that seen in mitochondria and it has been shown to be specific by different means [45]. In contrast to these secretory granules, the labelling in cytoplasm, nucleus and ER is generally very weak and in most cases at, or near, background levels [45]. These observations are very similar to those attained using Hsp60 antibodies (Figure 2.1) and indicate that the Hsp10 is reaching these compartments by a novel pathway [45].

In addition to these granules, specific reactivity of the Hsp10 antibodies has also been observed within mature erythrocytes (Figure 2.2C) [45]. This observation is surprising because erythrocytes are believed to be devoid of

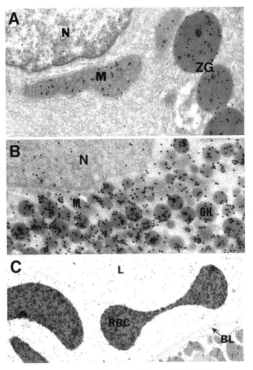

Figure 2.2. Immuno-EM localisation of Hsp10 (Cpn10) in different rat tissues using specific antibodies. (A) Labelling of mitochondria and zymogen granules (ZG) in pancreatic acinar cells. (B) Labelling of GHGs in anterior pituitary sections. (C) Immunogold labelling of red blood cells (RBC). BL, basal lamina; L, lumen. From Sadacharan et al. [45].

mitochondria and various other organelles [55, 56]. However, it is possible that, although mature erythrocytes extrude all mitochondria, they retain specific mitochondrial proteins that may be required for particular functions. Our studies indicate that, similar to Hsp10, cross-reactive proteins to Hsp60 antibodies are also present in erythrocytes (unpublished results). The possible roles that Hsp60 and Hsp10 may play in erythrocytes is presently unclear; however, based on their established chaperone function, it is possible that these proteins are involved in either the assembly or functioning of haemoglobin, which is their primary constituent [55, 56].

2.2.3. mHsp70/DnaK chaperone

Mitochondrial Hsp70 (mHsp70) is the mitochondria-targeted member of the highly conserved Hsp70/DnaK family of proteins. This protein has been independently identified as mortalin, as the 74-kDa peptide binding protein (PBP74) [57], as the 75-kDa glucose regulating protein (Grp75) [58], and as mHsp70 [59, 60]. Similar to Hsp60, it contains an N-terminal MTS that is responsible for its targeting to the mitochondrial matrix [59, 60]. Within mitochondria, it functions as a monomeric ATPase that binds to exposed hydrophobic amino acid residues in proteins to prevent their aggregation or misfolding [4, 12]. Additionally, mHsp70 plays a central role in the mitochondrial import of proteins by binding to, and pulling in, unfolded polypeptide chains entering through the translocase of the inner (outer) membrane translocon [61].

As with the other DnaK homologues, mHsp70 works in conjunction with its co-chaperones, mDnaJ (hTid-1/Hsp40 homologue) and GrpE, which modulate its ATP exchange and ATPase activity [62]. In an earlier study, in which mHsp70 was identified as the PBP74, a protein involved in the processing of antigens, it was shown to be localised in a number of sites including the endocytic vesicles of B cells containing internalised antigen, the ER, and the plasma membrane [63]. mHsp70 has also been shown to interact with exogenously added fibroblast growth factor type-1 as well as with interleukin receptor type-1 and thought to play a role in their internalisation [58, 64].

Studies on mHsp70 localisation by high-resolution immuno-EM demonstrate that, in addition to its expected localisation within mitochondria, this protein is also present at discrete sub-cellular locations including the plasma membrane, endocytic vesicles, and unidentified cytoplasmic granules in both CHO and BSC-1 cells (Figure 2.3A) [60]. Our recent studies show that, similar to Hsp60 and Hsp10, antibodies to mHsp70 also show strong reactivity towards a number of different types of granules including ZGs (Figure 2.3B). Further, as in the case of Hsp60 and Hsp10, very little reactivity can be observed in the cytoplasm and the ER–Golgi compartments, indicating that these proteins are reaching these granules via some novel, yet to be discovered, pathway (see Chapter 3).

Wadhwa and Kaul's groups have independently identified mHsp70 as 'mortalin', a protein that is implicated in conferring the senescent phenotype on cultured mammalian cells [65, 66]. Immunofluorescence studies with antibodies to mortalin in normal fibroblasts generally show a pancytosolic labelling, whereas its localisation varies from a fibrous peri-nuclear to granular staining of the juxtanuclear cap in different immortal cell lines [66]. Based on the distribution pattern of this protein in normal and transformed cells, a number of different complementation groups have been identified [65]. Although these findings are of much interest for understanding the cellular function of mHsp70 in cellular senescence, due to the limited resolution of confocal microscopy it is difficult to determine whether the observed differences in mHsp70 distribution are simply due to altered mitochondrial distributions

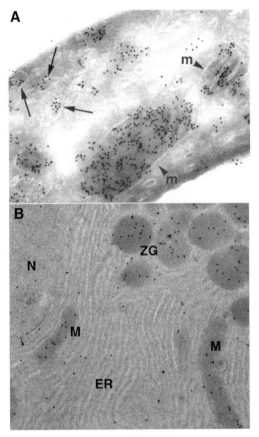

Figure 2.3. Immuno-EM localisation of mHsp70 in (A) cryosections of BSC-1 cells, and (B) in tissue sections of pancreatic acinar cells. In panel (B), the labelling is specific to mitochondria and ZGs and only background labelling is seen in the ER and the nucleus. From Singh et al. [60] and unpublished results.

(or morphology) in different cells, or whether mHsp70 is present at certain discreet extra-mitochondrial sites/compartments in any of these cells [65].

2.2.4. Hsp90 or TRAP-1 protein

Due to their endosymbiotic origin from bacteria, mitochondria are known to contain distinct homologues of various heat shock chaperone proteins. The mitochondrial homologues of Hsp60, Hsp70, Hsp10, as well DnaJ (hTid-1/Hsp40) and GrpE have all been well characterised [67, 68]. In contrast, until recently, no homologue of the major 90-kDa heat shock protein (Hsp90) was identified in mitochondria. However, sequence homology studies have led to the realisation that a protein, TRAP-1, corresponds to the mitochondrial homologue of the Hsp90 protein [8, 9]. TRAP-1 shows greater sequence homology to the Hsp90 homologue from Gram negative bacteria (HtpG), from which mitochondria have originated, than to either the cytosolic or ER-resident forms of human Hsp90 [8, 69]. Additionally, both TRAP-1 and HtpG sequences lack a charged region that is present in all eukaryotic nucleocytosolic Hsp90s [69, 70].

The evidence that TRAP-1 is primarily a mitochondrial protein is provided by several lines of investigations: (i) In immunofluorescence studies, different monoclonal and polyclonal antibodies to TRAP-1 protein are all found to specifically stain mitochondria [8, 69]. (ii) Sub-cellular fractionation of rat liver mitochondria indicates that TRAP-1 is primarily present in mitochondria. Within mitochondria, TRAP-1 reactivity is primarily observed in the matrix and the outer membrane fractions [69]. (iii) TRAP-1 in both human and *Drosophila* is synthesised as a longer precursor protein containing an N-terminal targeting sequence bearing various characteristics of a typical mitochondrial matrix targeting sequence [8]. Similar to other mitochondrial proteins, this sequence is not present in the mature protein and is likely cleaved off during mitochondrial import. (iv) Immunogold labelling of different tissue sections shows strong and specific labelling of mitochondria in all cases [69].

In a number of tissues such as liver and spleen, labelling with TRAP-1 antibody was exclusively seen in mitochondria. However, in several other tissues, labelling at additional discrete locations has also been observed (discussed later). Although TRAP-1 is clearly a mitochondrial protein, and also an Hsp90 homologue, its function within mitochondria is presently not known. Similar to other Hsp90 homologues, TRAP-1 binds ATP and shows ATPase activity which can be blocked by the Hsp90 inhibitors, geldanamycin and radicicol [8]. However, its lack of binding to the Hsp90 co-chaperones p23 and Hop (p60) and its inability to substitute for Hsp90 in other functions [8] indicate that its cellular function has diverged from other well-characterised Hsp90 homologues.

Although the function of TRAP-1 (mHsp90) within mitochondria is not known, it interacts with a variety of proteins involved in diverse functions outside of mitochondria. TRAP-1 was first identified on the basis of its interaction with the intracellular domain of the type I tumour necrosis factor receptor (TNFR-1) in the yeast two-hybrid system [9]. TNFR-1 is a cell surface receptor involved in a variety of cellular events, including cytotoxicity, fibroblast proliferation and antiviral responses (see subsequent chapters that deal with pro-inflammatory cytokine induction by various chaperones). In other studies,

TRAP-1 has been shown to interact with the retinoblastoma protein [71] as well as two proteins involved in hereditary multiple exostoses, EXT1 and EXT2 [72]. The retinoblastoma protein is a nuclear protein involved in cell cycle progression and differentiation [73]. The functions of EXT1 and EXT2 however, are not definitively known, although indirect evidence indicates that these proteins, found primarily in the ER, are involved in glycosaminoglycan synthesis and act as tumour suppressors [72].

In a recent study the *Dictyostelium* homologue of TRAP-1 has been found to display differential sub-cellular localisation in response to nutrient starvation [74]. This protein is normally localised to mitochondria and the cortical membrane region of the cells. However, after 6 hours in media devoid of nutrients, TRAP-1 was found only within mitochondria. TRAP-1 translocates back to the mitochondria from the cortical membrane during starvation, and this translocation could be induced by an as yet uncharacterised secreted factor(s) present in the media of starved cells [74]. In all cases, TRAP-1 was determined to be in its mature form by Western blotting. These results are interesting because they suggest that TRAP-1 is able to cross back into mitochondria without the need for its MTS. In another interesting study, exogenous tumour necrosis factor (TNF) was found to be delivered to mitochondria [75]. Although it remains unclear how exogenously added TNF is directed to mitochondria, TNF binding to TRAP-1 and the subsequent translocation of the complex to the mitochondria is a distinct possibility.

In accordance with its interaction with many extra-mitochondrial proteins, immuno-EM studies on TRAP-1 show that, in addition to mitochondria, strong and significant reactivity to TRAP-1 antibodies is seen in a number of tissues at discrete extra-mitochondrial sites [69]. In pancreatic acinar cells, TRAP-1 is present in both ZGs (Figure 2.4A) as well as glucagon granules [69]. Lower, but significant, reactivity of TRAP-1 has also been seen in the nuclei of pancreatic acinar cells [69]. In endothelial cells lining blood vessels, strong reactivity to TRAP-1 antibody has been observed on the apical cell surface (Figure 2.4B). The labelling is concentrated in foci, which could represent sites where TNFR-1 is localised, or areas in which vesicles containing this protein are either fusing or pinching off from the cell membrane. Significant reactivity of TRAP-1 has also been observed in cardiac sarcomeres, the functional significance of which is presently unknown [69].

2.3. Conclusions and future directions

The presence of various mitochondrial heat shock proteins (viz. Hsp60, Hsp10, mHsp70 and TRAP-1) outside of the mitochondrion, and their involvement in

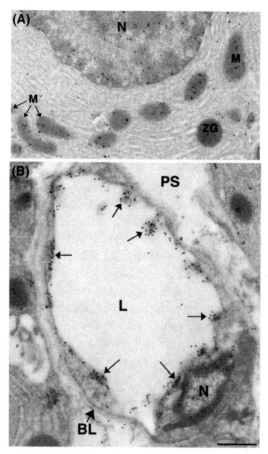

Figure 2.4. Immuno-EM localisation of TRAP-1 (mHsp90) in rat tissue sections using specific antibodies. (A) Localisation of TRAP-1 in mitochondria, ZGs and nucleus in pancreatic acinar cells. (B) Presence of TRAP-1 on the surface of endothelial cells lining a blood vessel. BL, basal lamina; PS, peri-capillary space. From Cechetto and Gupta [69].

important cellular functions, is widely recognised. However, the question as to how these proteins arrive at these locations is generally ignored and has not received due attention. The primary reason for this is that most mitochondrial proteins are encoded by nuclear DNA and translated in the cytosol before being targeted to mitochondria. Hence, it is generally assumed that these proteins can easily move from the cytosol to various other cellular destinations directly, and their presence at extra-mitochondrial location presents no conceptual problem [10, 76]. However, the available facts do not support this simplistic assumption. In this context, it is important to recognise that most of these proteins are synthesised as larger precursor molecules containing an N-terminal leader sequence that targets the proteins to mitochondria. Most of these proteins are also encoded by single copy genes and there is no evidence for the occurrence of alternate processing at either the transcription or translation levels.

The targeting sequences of these proteins are specifically cleaved by a mitochondrial matrix resident protease and these sequences are not present in the mature protein found within mitochondria [10, 76]. Extensive evidence now indicates that various proteins discussed in this review (viz. Hsp60, mHsp70, TRAP-1), as well as other mitochondrial proteins that have been identified at extra-mitochondrial locations, all comprise mature forms of the protein lacking the targeting sequence. Further, for Hsp60, mHsp70 and several other mitochondrial proteins, it has been demonstrated that the conversion of the precursor into the mature form requires entry of the precursor (or at least its targeting sequence) into mitochondria. In cells treated with the potassium ionophores such as nonactin or valinomycin, which inhibit mitochondrial import of the precursor proteins, no conversion of the precursor forms into the mature proteins is observed [39, 60]. Further, under normal conditions the amounts of the precursor proteins that are present in cells are too low to be detected, indicating that the mitochondrial targeting is highly efficient and leading to no significant accumulation of the precursor proteins in the cytosol, or their mis-targeting to other compartments.

In addition to the proteins discussed in this chapter, most other mitochondrial proteins whose sub-cellular distributions in different cells and tissues have been examined in detail by means of high-resolution techniques are found to be present, besides mitochondria, at other specific sites in cells. Examples of such proteins include mitochondrial aspartate aminotransferase, which also functions as a fatty acid transporter on the cell surface [77, 78], and the p32 protein, which is involved in nuclear functions and also acts as a cell surface receptor for complement C1q [79, 80]. Interestingly, cytochrome C, the exit of which from mitochondria is believed to play a central role in apoptosis [81], is also present in ZGs and GHGs in normal rat tissues [82]. Additionally, our recent work indicates that a number of proteins that are encoded by mitochondrial DNA and that are transcribed and translated within mitochondria are also present outside of mitochondria at specific locations (unpublished results). The presence of these proteins at extra-mitochondrial locations challenges the widely held notion that mitochondria constitute a dead-end compartment in cells, from which proteins cannot exit under normal physiological conditions [1, 2, 10, 83, 84].

The preceding observations have led us to postulate the existence of specific mechanisms for protein transport from mitochondria to other cellular compartments [1, 2]. The existence of specific transport mechanisms from mitochondria should not be surprising in view of their origin from Gram negative bacteria, which possess an ever-growing number of mechanisms for protein export/secretion across the cell membrane [2]. Although such mechanisms remain to be characterised in eukaryotic cells, the presence of various mitochondrial chaperones (and other proteins) at extra-mitochondrial sites indicates that their cellular functions are not restricted to mitochondria, and many of these proteins likely play important roles in diverse cellular processes (see [1, 2]).

Acknowledgements

This work has been supported by a research grant from the Canadian Institute of Health Research to R.S.G.

REFERENCES

1. Soltys B J and Gupta R S. Mitochondrial-matrix proteins at unexpected locations: are they exported? Trends Biochem Sci 1999, 24: 174–177.
2. Soltys B J and Gupta R S. Mitochondrial proteins at unexpected cellular locations: export of proteins from mitochondria from an evolutionary perspective. Int Rev Cytol 2000, 194: 133–196.
3. Bukau B and Horwich A L. The Hsp70 and Hsp60 chaperone machines. Cell 1998, 92: 351–366.
4. Craig E A, Gambill B D and Nelson R J. Heat shock proteins: molecular chaperones of protein biogenesis. Microbiol Rev 1993, 57: 402–414.
5. Ellis R J and Hartl F U. Protein folding in the cell: competing models of chaperonin function. FASEB J 1996, 10: 20–26.
6. Hartl F U, Martin J and Neupert W. Protein folding in the cell: the role of molecular chaperones Hsp70 and Hsp60. Ann Rev Biophys Biomol Struct 1992, 21: 293–322.
7. Lubben T H, Gatenby A A, Donaldson G K, Lorimer G H and Viitanen P V. Identification of a groES-like chaperonin in mitochondria that facilitates protein folding. Proc Natl Acad Sci USA 1990, 87: 7683–7687.
8. Felts S J, Owen B A, Nguyen P, Trepel J, Donner D B and Toft D O. The hsp90-related protein TRAP1 is a mitochondrial protein with distinct functional properties. J Biol Chem 2000, 275: 3305–3312.
9. Song H Y, Dunbar J D, Zhang Y X, Guo D and Donner D B. Identification of a protein with homology to hsp90 that binds the type 1 tumor necrosis factor receptor. J Biol Chem 1995, 270: 3574–3581.
10. Herrmann J M and Neupert W. Protein transport into mitochondria. Cur Opin Microbiol 2000, 3: 210–214.
11. Pfanner N and Neupert W. The mitochondrial protein import apparatus. Ann Rev Biochem 1990, 59: 331–353.
12. Hendrick J P and Hartl F U. Molecular chaperone functions of heat-shock proteins. Ann Rev Biochem 1993, 62: 349–384.

13. Hartl F U. Molecular chaperones in cellular protein folding. Nature 1996, 381: 571–579.

14. Gray M W. Evolution of organellar genomes. Cur Opin Genetics Develop 1999, 9: 678–687.

15. Gupta R S. Evolution of the chaperonin families (Hsp60, Hsp10 and Tcp-1) of proteins and the origin of eukaryotic cells. Mol Microbiol 1995, 15: 1–11.

16. Singh B, Patel H V, Ridley R G, Freeman K B and Gupta R S. Mitochondrial import of the human chaperonin (HSP60) protein. Biochem Biophys Res Commun 1990, 169: 391–396.

17. Ikawa S and Weinberg R A. An interaction between p21ras and heat shock protein hsp60, a chaperonin. Proc Natl Acad Sci USA 1992, 89: 2012–2016.

18. Jones M, Gupta R S and Englesberg E. Enhancement in amount of P1 (hsp60) in mutants of Chinese hamster ovary (CHO-K1) cells exhibiting increases in the A system of amino acid transport. Proc Natl Acad Sci USA 1994, 91: 858–862.

19. Khan I U, Wallin R, Gupta R S and Kammer G M. Protein kinase A-catalyzed phosphorylation of heat shock protein 60 chaperone regulates its attachment to histone 2B in the T lymphocyte plasma membrane. Proc Natl Acad Sci USA 1998, 95: 10425–10430.

20. Gupta R S, Ho T K, Moffat M R and Gupta R. Podophyllotoxin-resistant mutants of Chinese hamster ovary cells. Alteration in a microtubule-associated protein. J Biol Chem 1982, 257: 1071–1078.

21. Jindal S, Dudani A K, Singh B, Harley C B and Gupta R S. Primary structure of a human mitochondrial protein homologous to the bacterial and plant chaperonins and to the 65-kilodalton mycobacterial antigen. Mol Cell Biol 1989, 9: 2279–2283.

22. Picketts D J, Mayanil C S and Gupta R S. Molecular cloning of a Chinese hamster mitochondrial protein related to the 'chaperonin' family of bacterial and plant proteins. J Biol Chem 1989, 264: 12001–12008.

23. Gupta R S and Austin R C. Mitochondrial matrix localization of a protein altered in mutants resistant to the microtubule inhibitor podophyllotoxin. Eur J Cell Biol 1987, 45: 170–176.

24. Gupta R S and Dudani A K. Mitochondrial binding of a protein affected in mutants resistant to the microtubule inhibitor podophyllotoxin. Eur J Cell Biol 1987, 44: 278–285.

25. Gupta R S. Mitochondria, molecular chaperone proteins and the *in vivo* assembly of microtubules. Trends Biochem Sci 1990, 15: 415–418.

26. Fisch P, Malkovsky M, Kovats S, Sturm E, Braakman E, Klein B S, Voss S D, Morrissey L W, DeMars R, Welch W J, Bolhuis R L H and Sondel P M. Recognition by human Vγ9/Vδ2 T cells of a GroEL homolog on Daudi Burkitt's lymphoma cells. Science 1990, 250: 1269–1273.

27. Koga T, Wand-Wurttenberger A, DeBruyn J, Munk M E, Schoel B and Kaufmann S H E. T cells against a bacterial heat shock protein recognize stressed macrophages. Science 1989, 245: 1112–1115.

28. Kaur I, Voss S D, Gupta R S, Schell K, Fisch P and Sondel P M. Human peripheral γδ T cells recognize hsp60 molecules on Daudi Burkitt's lymphoma cells. J Immunol 1993, 150: 2046–2055.

29. Soltys B J and Gupta R S. Cell surface localization of the 60 kDa heat shock chaperonin protein (hsp60) in mammalian cells. Cell Biol Int 1997, 21: 315–320.

30. Bocharov A V, Vishnyakova T G, Baranova I N, Remaley A T, Patterson A P and Eggerman T L. Heat shock protein 60 is a high-affinity high-density lipoprotein binding protein. Biochem Biophys Res Commun 2000, 277: 228–235.

31. Lukacs K V, Lowrie D B, Stokes R W and Colston M J. Tumor cells transfected with a bacterial heat-shock gene lose tumorigenicity and induce protection against tumors. J Exp Med 1993, 178: 343–348.

32. Wells A D and Malkovsky M. Heat shock proteins, tumor immunogenicity and antigen presentation: an integrated view. Immunol Today 2000, 21: 129–132.

33. Gupta R S and Gupta R. Mutants of chinese hamster ovary cells affected in two different microtubule-associated proteins. Genetic and biochemical studies. J Biol Chem 1984, 259: 1882–1890.

34. Soltys B J and Gupta R S. Mitochondrial molecular chaperones Hsp60 and mHsp70: Are their roles restricted to mitochondria? In Abe, H. and Latchman, D. S. (Eds.) *Handbook of Experimental Pharmacology: Heat Shock Proteins.* Springer-Verlag New York, Inc., New York: 1998, pp 69–100.

35. Xu Q, Schett G, Seitz C S, Hu Y, Gupta R S and Wick G. Surface staining and cytotoxic activity of heat-shock protein 60 in stressed aortic endothelial cells. Circ Res 1994, 75: 1078–1085.

36. Schett G, Metzler B, Mayr M, Amberger A, Niederwieser D, Gupta R S, Mizzen L, Xu Q and Wick G. Macrophage-lysis mediated by autoantibodies to heat shock protein 65/60. Atherosclerosis 1997, 128: 27–38.

37. Poccia F, Piselli P, Vendetti S, Bach S, Amendola A, Placido R and Colizzi V. Heat-shock protein expression on the membrane of T cells undergoing apoptosis. Immunology 1996, 88: 6–12.

38. Gupta S and Knowlton A A. Cytosolic heat shock protein 60, hypoxia, and apoptosis. Circulation 2002, 106: 2727–2733.

39. Soltys B J and Gupta R S. Immunoelectron microscopic localization of the 60-kDa heat shock chaperonin protein (Hsp60) in mammalian cells. Exp Cell Res 1996, 222: 16–27.

40. Brudzynski K, Martinez V and Gupta R S. Immunocytochemical localization of heat-shock protein 60-related protein in beta-cell secretory granules and its altered distribution in non-obese diabetic mice. Diabetologia 1992, 35: 316–324.

41. Cechetto J D, Soltys B J and Gupta R S. Localization of mitochondrial 60-kD heat shock chaperonin protein (Hsp60) in pituitary growth hormone secretory granules and pancreatic zymogen granules. J Histochem Cytochem 2000, 48: 45–56.

42. Velez-Granell C S, Arias A E, Torres-Ruiz J A and Bendayan M. Molecular chaperones in pancreatic tissue: the presence of cpn10, cpn60 and hsp70 in distinct compartments along the secretory pathway of the acinar cells. J Cell Sci 1994, 107: 539–549.

43. Velez-Granell C S, Arias A E, Torres-Ruiz J A and Bendayan M. Presence of Chromatium vinosum chaperonins 10 and 60 in mitochondria and peroxisomes of rat hepatocytes. Biol Cell 1995, 85: 67–75.

44. Rothman S S. Protein transport by the pancreas. Science 1975, 190: 747–753.

45. Sadacharan S K, Cavanagh A C and Gupta R S. Immunoelectron microscopy provides evidence for the presence of mitochondrial heat shock 10-kDa protein (chaperonin 10) in red blood cells and a variety of secretory granules. Histochem Cell Biol 2001, 116: 507–517.

46. Jamieson J D and Palade G E. Intracellular transport of secretory proteins in the pancreatic exocrine cell. II. Transport to condensing vacuoles and zymogen granules. J Cell Biol 1967, 34: 597–615.

47. Bassan M, Zamostiano R, Giladi E, Davidson A, Wollman Y, Pitman J, Hauser J, Brenneman D E and Gozes I. The identification of secreted heat shock 60-like protein from rat glial cells and a human neuroblastoma cell line. Neurosci Lett 1998, 250: 37–40.

48. Orci L, Vassalli J-D and Perrelet A. The insulin factory. Scientific American 1988, 259: 85–94.

49. Hendrick J P and Hartl F U. The role of molecular chaperones in protein folding. FASEB J 1995, 9: 1559–1569.

50. Jarvis J A, Ryan M T, Hoogenraad N J, Craik D J and Hoj P B. Solution structure of the acetylated and noncleavable mitochondrial targeting signal of rat chaperonin 10. J Biol Chem 1995, 270: 1323–1331.

51. Cavanagh A C and Morton H. The purification of early-pregnancy factor to homogeneity from human platelets and identification as chaperonin 10. Eur J Biochem 1994, 222: 551–560.

52. Quinn K A, Cavanagh A C, Hillyard N C, McKay D A and Morton H. Early pregnancy factor in liver regeneration after partial hepatectomy in rats: relationship with chaperonin 10. Hepatology 1994, 20: 1294–1302.

53. Cavanagh A C. Identification of early pregnancy factor as chaperonin 10: implications for understanding its role. Rev Reprod 1996, 1: 28–32.

54. Fletcher B H, Cassady A I, Summers K M and Cavanagh A C. The murine chaperonin 10 gene family contains an intronless, putative gene for early pregnancy factor, Cpn10-rs1. Mamm Genome 2001, 12: 133–140.

55. Alberts B, Johnson A, Lewis J, Raff M, Roberts K and Walter P. *Molecular Biology of the Cell*. Garland Publishing, Inc., New York: 2002.

56. Weiss L. *Cell and Tissue Biology: A Textbook of Histology*. Urban and Schwarzenberg, Baltimore: 1988.

57. Domanico S Z, DeNagel D C, Dahlseid J N, Green J M and Pierce S K. Cloning of the gene encoding peptide-binding protein 74 shows that it is a new member of the heat shock protein 70 family. Mol Cell Biol 1993, 13: 3598–3610.

58. Mizukoshi E, Suzuki M, Loupatov A, Uruno T, Hayashi H, Misono T, Kaul S C, Wadhwa R and Imamura T. Fibroblast growth factor-1 interacts with the glucose-regulated protein GRP75/mortalin. Biochem J 1999, 343: 461–466.

59. Bhattacharyya T, Karnezis A N, Murphy S P, Hoang T, Freeman B C, Phillips B and Morimoto R I. Cloning and subcellular localization of human mitochondrial hsp70. J Biol Chem 1995, 270: 1705–1710.

60. Singh B, Soltys B J, Wu Z C, Patel H V, Freeman K B and Gupta R S. Cloning and some novel characteristics of mitochondrial Hsp70 from Chinese hamster cells. Exp Cell Res 1997, 234: 205–216.

61. Ungermann C, Neupert W and Cyr D M. The role of Hsp70 in conferring unidirectionality on protein translocation into mitochondria. Science 1994, 266: 1250–1253.

62. Caplan A J, Cyr D M and Douglas M G. Eukaryotic homologues of *Escherichia coli* DnaJ: a diverse protein family that functions with hsp70 stress proteins. Mol Biol Cell 1993, 4: 555–563.

63. VanBuskirk A M, DeNagel D C, Guagliardi L E, Brodsky F M and Pierce S K. Cellular and subcellular distribution of PBP72/74, a peptide-binding protein that plays a role in antigen processing. J Immunol 1991, 146: 500–506.
64. Sacht G, Brigelius-Flohe R, Kiess M, Sztajer H and Flohe L. ATP-sensitive association of mortalin with the IL-1 receptor type I. Biofactors 1999, 9: 49–60.
65. Kaul S C, Taira K, Pereira-Smith O M and Wadhwa R. Mortalin: present and prospective. Exp Gerontol 2002, 37: 1157–1164.
66. Wadhwa R, Kaul S C, Ikawa Y and Sugimoto Y. Identification of a novel member of mouse hsp70 family. Its association with cellular mortal phenotype. J Biol Chem 1993, 268: 6615–6621.
67. Choglay A A, Chapple J P, Blatch G L and Cheetham M E. Identification and characterization of a human mitochondrial homologue of the bacterial co-chaperone GrpE. Gene 2001, 267: 125–134.
68. Syken J, Macian F, Agarwal S, Rao A and Münger K. TID1, a mammalian homologue of the drosophila tumor suppressor lethal(2) tumorous imaginal discs, regulates activation-induced cell death in Th2 cells. Oncogene 2003, 22: 4636–4641.
69. Cechetto J D and Gupta R S. Immunoelectron microscopy provides evidence that tumor necrosis factor receptor-associated protein 1 (TRAP-1) is a mitochondrial protein which also localizes at specific extramitochondrial sites. Exp Cell Res 2000, 260: 30–39.
70. Gupta R S. Phylogenetic analysis of the 90 kD heat shock family of protein sequences and an examination of the relationship among animals, plants, and fungi species. Mol Biol Evol 1995, 12: 1063–1073.
71. Chen C F, Chen Y, Dai K, Chen P L, Riley D J and Lee W H. A new member of the hsp90 family of molecular chaperones interacts with the retinoblastoma protein during mitosis and after heat shock. Mol Cell Biol 1996, 16: 4691–4699.
72. Simmons A D, Musy M M, Lopes C S, Hwang L Y, Yang Y P and Lovett M. A direct interaction between EXT proteins and glycosyltransferases is defective in hereditary multiple exostoses. Hum Mol Genet 1999, 8: 2155–2164.
73. Herwig S and Strauss M. The retinoblastoma protein: a master regulator of cell cycle, differentiation and apoptosis. Eur J Biochem 1997, 246: 581–601.
74. Morita T, Amagai A and Maeda Y. Unique behavior of a dictyostelium homologue of TRAP-1, coupling with differentiation of D. discoideum cells. Exp Cell Res 2002, 280: 45–54.
75. Ledgerwood E C, Prins J B, Bright N A, Johnson D R, Wolfreys K, Pober J S, O'Rahilly S and Bradley J R. Tumor necrosis factor is delivered to mitochondria where a tumor necrosis factor-binding protein is localized. Lab Invest 1998, 78: 1583–1589.
76. Glick B and Schatz G. Import of proteins into mitochondria. Ann Rev Genet 1991, 25: 21–44.
77. Bradbury M W and Berk P D. Mitochondrial aspartate aminotransferase: direction of a single protein with two distinct functions to two subcellular sites does not require alternative splicing of the mRNA. Biochem J 2000, 345: 423–427.
78. Cechetto J D, Sadacharan S K, Berk P D and Gupta R S. Immunogold localization of mitochondrial aspartate aminotransferase in mitochondria and on the cell surface in normal rat tissues. Histol Histopathol 2002, 17: 353–364.
79. Ghebrehiwet B and Peerschke E I. Structure and function of gC1q-R: a multiligand binding cellular protein. Immunobiology 1998, 199: 225–238.

80. Soltys B J, Kang D and Gupta R S. Localization of P32 protein (gC1q-R) in mitochondria and at specific extramitochondrial locations in normal tissues. Histochem Cell Biol 2000, 114: 245–255.

81. Green D R and Reed J C. Mitochondria and apoptosis. Science 1998, 281: 1309–1312.

82. Soltys B J, Andrews D A, Jemmerson R and Gupta R S. Cytochrome c localizes in secretory granules in pancreas and anterior pituitary. Cell Biol Int 2001, 25: 331–338.

83. Poyton R O, Duhl D M J and Clarkson G H D. Protein export from the mitochondrial matrix. Trends Cell Biol 1992, 2: 369–375.

84. Smalheiser N R. Proteins in unexpected locations. Mol Biol Cell 1996, 7: 1003–1014.

Changing Paradigms of Protein Trafficking and Protein Function

3

Novel Pathways of Protein Secretion

Giovanna Chimini and Anna Rubartelli

3.1. Introduction

Intercellular communications are fundamental for many of the biological processes that are involved in the survival of living organisms, and secretory proteins are among the most important messengers in this network of information. Proteins destined for this function are endowed with a hydrophobic signal peptide which targets them to the endoplasmic reticulum (ER) and are released in the extracellular environment by a 'classical' pathway of constitutive or regulated secretion. However, in the early 1990s it became evident that non-classical mechanisms must exist for the secretion of some proteins which, despite their extracellular localisation and function, lack a signal peptide. Indeed, the family of these leaderless secretory proteins continues to grow and comprises proteins that, although apparently unrelated, share both structural and functional features. This chapter will review current hypotheses on the mechanisms underlying non-classical secretion and discuss their implications in the regulation of the inflammatory and immune response. The relevance of non-classical secretion pathways to molecular chaperone biology is also discussed in Chapters 2 and 12.

3.2. Leaderless secretory proteins

Secretory mechanisms that are discrete to the classical pathways appear early in evolution. Gram negative bacteria are endowed with many (up to six) types of secretion mechanisms that are, at least in part, independent of the general secretory pathway, the prototype being the haemolysin secretion system [1]. In addition, two pathways of secretion that avoid the ER exist in yeast. Whereas one of these seems to be activated only as a detoxifying tool [2], the other is essential because it allows the release of the a-factor, a key mating factor [3]. In higher eukaryotes, leaderless secretory proteins display some common structural features, such as a relatively low molecular mass (12–45 kDa with few exceptions), the absence of N-linked glycosylation, even if potential sites are present, and the presence of free cysteines that are not engaged in disulphide bridges [4]. These characteristics suggest that leaderless secretory proteins avoid the ER, in which post-translational modifications such as N-linked glycosylation and formation of disulphide bridges take place. Moreover, the pharmacological evidence that brefeldin A, a drug that blocks secretion of classical secretory proteins, does not affect the release of leaderless proteins [5, 6] further supports the hypothesis that their secretion must follow non-classical export routes.

Due to the relatively large cytosolic accumulation of these proteins, it was orig-inally proposed that their presence outside the cell could be the consequence of a passive release of cytoplasmic content, as might occur in the case of cell death. This possibility was ruled out by observations that the presence of leader-less proteins in the extracellular environment is selective and does not correlate with the presence of other cytoplasmic proteins such as lactate dehydrogenase, a marker for cell lysis [4]. Leaderless secretion has also been shown to be energy- or temperature-dependent and can be blocked by a number of treatments or drugs. On the basis of this evidence, it is now currently accepted that the secre-tion of leaderless proteins is controlled by an active mechanism that excludes the ER–Golgi apparatus.

Table 3.1 lists the most studied leaderless secretory proteins. Some of these belong to the interleukin family (IL-1, IL-16, IL-18), others are cytosolic enzy-mes or derive from proteins that have a well-defined intracellular function (thioredoxin, thioredoxin reductase, phosphoglucose isomerase/AMF, EMAP II, caspase I, transglutaminase), and others are nuclear proteins which may be readdressed to the extracellular compartment (HMGB-1, engrailed-2). An unex-pected nuclear localisation has also been reported. Although in some cases this localisation is associated with a function (galectin-3 is involved in pre-mRNA splicing [33] and macrophage migration inhibitory factor (MIF) might work as a transcription factor modulator [34]), in the case of fibroblast growth factor (FGF)-1 and -2 and IL-1α a nuclear function is still debated.

The list of leaderless secretory proteins continues to grow. However, some proteins that have previously been considered to belong to this class have turned out to function mostly, if not exclusively, intracellularly. For instance, pro- and para-thymosine alpha, previously regarded as thymic hormones, now appear to be essential nuclear factors [35] and their extracellular role is debated [36]; platelet-derived endothelial cell growth factor turned out to be identical to the well-known enzyme thymidine phosphorylase and to play an angiogenic role as a result of its intracellular enzymatic activity [37].

Table 3.1. Leaderless secretory proteins

	Extracellular function	Intracellular function	Key reference
IL-1α	Pro-inflammatory immune mediator	Activator of transcription	[7, 8]
IL-1β	Pro-inflammatory immune mediator	–	[7]
IL-18	Pro-inflammatory immune mediator	–	[9]
caspase I (ICE)	?	IL-1/IL-18 converting enzyme	[10, 11]
IL-16	Pro-inflammatory immune mediator, T-lymphocyte chemoattractant	–	[12]
FGF-1	Growth, angiogenic and motility factor	Growth regulator (nuclear)?	[13]
FGF-2	Growth, angiogenic and motility factor	Growth regulator (nuclear)?	[14]
FGF-9	Neurotrophin, Growth survival factor	–	[15]
CNTF	Neurotrophin	–	[16]
MIF	Pro-inflammatory mediator, shock factor	Transcription factor modulator	[17]
EMAP II	Pro-inflammatory immune mediator	Apoptosis mediator	[18, 19]
Annexin I	Anti-inflammatory immune mediator	Vesicular traffic	[20–22]
AMF	Growth motility factor	Glycolytic enzyme	[23]
TRX/ADF	Immune mediator, chemotactic factor, involved in implantation and establishment of pregnancy	Major cellular disulfide reductase	[24]
TRX reductase	Enzyme reducing oxidised TRX?	Enzyme reducing oxidised TRX	[24]
Galectines	Pro/anti-inflammatory factors	Phagocytosis	[25–27]
Transglutaminase	Protein cross-linking	Apoptosis (intracellular protein cross-linking)	[28]
Coagulation factor XIII	Coagulation	–	[29]
HMGB1	Pro-inflammatory, shock factor	Chromatin component	[30]
Engrailed-2	Transcription factor by intercellular transfer?	Homeodomain transcription factor	[31, 32]

Finally, it is worth stressing that, from a functional point of view, most of these proteins display functions related to the regulation of inflammatory processes [37]. The finding that molecular chaperones also act to control inflammation is, therefore, suggestive.

3.3. Mechanisms of leaderless secretion

3.3.1. Targeting motifs

Although a considerable amount of effort has been dedicated to the understanding of leaderless secretion, the molecular mechanism is still only partially defined. It is likely that this mechanism involves a sequential series of events. First of all, recognition is required and a given leaderless secretory protein must be selected from amongst a myriad of cytosolic macromolecules. A common sorting motif has not yet been identified, although in a number of instances a requirement for a specific primary structure has been determined. In two cases, a single cysteine is crucial for secretion and the mutation of Cys 30 in FGF-1 [38, 39] and Cys 277 in tissue transglutaminase [40] is sufficient to impair their externalisation. Short sequences are required for secretion of chick ciliary neurotrophic factor (CNTF) and galectin-3, and in both proteins these lie in the N-terminal domain. Whereas for CNTF this sequence (AA 46–53) contains six hydrophobic aminoacids [41], in the case of galectin-3 two proline residues (Pro 90 and Pro 93) in the context of AA 89–96 seem crucial [42, 43].

Post-translational modifications play an important role in the secretion of two nuclear factors: acetylation for the chromatin component high mobility group box 1 protein (HMGB-1) [44] and phosphorylation for the homeoprotein engrailed-2 [31]. In both cases, the protein shuttles continuously from the nucleus to the cytoplasm due to nuclear export signals, but the equilibrium is almost completely shifted towards nuclear accumulation. Hyperacetylation of HMGB-1 and de-phosphorylation of engrailed-2 relocate the proteins to the cytoplasm. In turn, cytoplasmic availability allows secretion. Phosphorylation also appears to be required for the secretion of the glycolytic enzyme phosphohexose isomerase/autocrine motility factor (AMF). In this case, however, the modification induces secretion, whereas the non-phosphorylated enzyme remains within the cell [45]. Phosphohexose isomerase/AMF is the archetypal moonlighting protein and is discussed again in Chapters 4 and 5.

Other post-translational modifications such as mirystoylation or farnesylation might positively modulate the secretion of certain leaderless secretory proteins such as IL-1α [46] and galectin-3 [42] by enhancing their recruitment to cell membranes. Nevertheless, these modifications do not represent a bona fide secretory signal, because other cytosolic proteins undergo lipid modification but are not secreted.

3.3.2. Reaching the extracellular space: vesiculation or translocation?

Once recognised, the proteins must be externalised. This can be accomplished by two different mechanisms, namely vesiculation or translocation.

3.3.2.1. Vesiculation

Vesicles may result from either outward or inward membrane bending. In the first case, evagination of the membrane leads to bleb formation followed by the release of extracellular vesicles. Their membrane instability allows the rapid solubilisation of the protein. This mechanism, which has been first documented for the lectin L-14/galectin-1 [47], implies a prior concentration of the protein in patches beneath the plasma membrane. Interestingly, bleb formation can be restricted to the apical membrane in epithelial cells, resulting in polarised secretion [48]. This is also the case of the apocrine secretion described in glandular cells of the male reproductive system which leads to the apical release of a number of leaderless proteins [49] including tissue transglutaminase [50] and MIF [51]. Moreover, microvesicle shedding has been proposed to mediate IL-1β release from a monocytic cell line [52].

In the second case, repeated invaginations of the plasma membrane generate multivesicular bodies. These are multilamellar complexes containing vesicles called exosomes that are released upon the fusion of the multivesicular bodies with the plasma membrane [53]. Exosomes have been observed primarily in haemopoietic cells in which they can serve a variety of functions, including shedding of transferrin receptor during reticulocyte maturation [54] or antigen presentation [55]. In addition, exosomes may mediate externalisation of leaderless proteins. Indeed, galectin-3 and several annexins have been detected by proteomic analysis in exosomes from dendritic cells (DCs) [56], suggesting that they are selectively recruited in this compartment. The potential role of exosomes in the release of molecular chaperones is discussed in Chapter 12.

3.3.2.2. Translocation

Leaderless proteins that are not secreted by vesiculation must cross a cell membrane to reach the extracellular space. Membrane translocation has to meet a number of requirements. Firstly, as a general rule, proteins must unfold in order to acquire a 'translocation competent conformation.' Unfolding is usually assisted by cytoplasmic chaperones [57]. This has been clearly demonstrated for secretion of leaderless proteins in prokaryotes [58], whereas it has not been directly addressed in mammalian leaderless secretion. Interestingly, in the case of FGF-1, unfolding does not seem crucial for translocation. Indeed, Prudovsky et al. [59]

suggest that FGF-1 is secreted as a dimer assembled in a multi-molecular complex with other cytosolic proteins. However, this apparent inconsistency is not the only exception to the rule because peroxisomal proteins may be imported into peroxisomes as heterotrimers [60]. Secondly, protein translocation implies the presence of dedicated membrane transporters. Several specific transporters have been identified as mediators of membrane crossing in most intracellular organelles, from the ER to mitochondria and peroxisomes. Also, the lysosomal membrane is equipped with a transporter (Lamp-2A) which is able to import up to 30% of cytosolic proteins under stress conditions [61].

In yeast and bacteria, the secretion of leaderless proteins [1, 3] is, in most cases, dependent on membrane proteins belonging to the family of ATP-binding cassette (ABC) transporters. This family of transporters is conserved in mammals, in which 48 transporters organised into 7 structural classes (A to G) have been identified [62]. They include, amongst others, the multi-drug resistance protein, responsible for tumour resistance to chemotherapeutic drugs; the cystic fibrosis gene product; the ER proteins Tap-1 and Tap-2, involved in translocation of antigenic peptides; and ABCA1, a crucial regulator of clearance of apoptotic cells also implicated in lipid homeostasis.

By analogy with yeast and bacteria, the implication of ABC transporters in leaderless secretion has also been investigated in mammals. Although a direct demonstration has not been provided, pharmacological evidence strongly suggests that member(s) of the ABCA class mediate translocation of a number of leaderless proteins. Specifically, glybenclamide, a known inhibitor of the activity of ABCA1, has been reported to impair the secretion of IL-1β [63], Annexin I [64], MIF [65] and HMGB-1 (Rubartelli, unpublished observation).

In principle, translocation of a leaderless protein may occur at the plasma membrane or at the membrane of any intracellular organelle able to undergo exocytosis. Despite extensive study, direct evidence for protein translocation at the plasma membrane was lacking until recently, when *in vitro* translocation of FGF-2 and galectin-1 across inside-out membrane vesicles was demonstrated [66]. In contrast, a large body of evidence suggests that leaderless proteins may be imported into cytoplasmic organelles associated with the lysosomal compartment. Immunohistochemical studies have shown the presence of Annexin I in eosinophil granules [67], of chicken CNTF in an endosomal compartment of transfected cells [41], of FGF-2 in mast cell secretory granules [68], and of MIF within secretory granules of pituitary cells [69]. Morphological and biochemical evidence has reported the localisation of engrailed-2 [70] and HMGB-1 [71] in intracellular vesicles belonging to the early endocytic pathway and of IL-1β [72], IL-18 [73], caspase-I and Annexin I (Rubartelli, unpublished results) in the endolysosomes of activated monocytes.

Taken together, these results suggest that the lysosomal compartment is involved in leaderless secretion. Because secretory lysosomes are particularly abundant in haemopoietic cells, in which they are implicated in immune inflammatory processes [74], lysosome-mediated secretion of leaderless proteins is consistent with the role played by these proteins in the modulation of inflammation. Interestingly, the involvement of acidic vesicles in the export of leaderless proteins is evolutionarily conserved; as in *Dictyostelium discoideum*, translocation into exocytic contractile vacuoles of DdCAD-1, a leaderless adhesion protein, is necessary for its externalisation [75].

As indicated earlier, the ABCA inhibitor glybenclamide blocks secretion of several leaderless proteins: more specifically, glybenclamide prevents the appearance in secretory lysosomes of IL-1β [72] and other leaderless proteins (Rubartelli, unpublished results). This implies a lysosomal localisation of the putative ABCA protein responsible for translocation, in line with the presumption that all intracellular membranes may be endowed with transporters of the ABC family [76].

It is to be noted that, depending on the cell system, the same leaderless protein seems to use different pathways of secretion. For example, IL-1β has been found to be released by a myelomonocytic cell line via vesiculation [52], whereas in primary monocytes it undergoes lysosome-mediated exocytosis [72]. Galectin-1, one of the first examples of secretion by vesicle shedding [47], has been found to be capable of translocating directly at the plasma membrane [66]. Similarly, FGF-2 is able to cross plasma membrane vesicles [66], despite the fact that it was previously reported to accumulate in mast cell granules and be secreted after degranulation [68]. Whether these discrepancies are due to the experimental model or reflect physiological modulation remains to be elucidated.

3.4. Lysosome-mediated polarised secretion

In general, lysosome-mediated secretion is a regulated process in that a triggering signal is required to induce exocytosis [74]. In the case of IL-1β, two steps are needed for secretion by monocytes. Firstly, an inflammatory stimulus such as lipopolysaccharide (LPS) induces synthesis which results in cytosolic accumulation and lysosomal translocation; then a second extracellular signal triggers exocytosis resulting in IL-1β release [72, 77]. A similar two-step mechanism seems to account for the regulated secretion of other pro-inflammatory leaderless cytokines such as IL-18 [73] and HMGB-1 [71]. In all these cases, the signal triggering secretion is generated during the process of inflammation. ATP, promoting IL-1β and IL-18 secretion [78], is released by monocytes themselves and by other cells involved in inflammation (i.e., platelets) soon after

LPS stimulation [79]. In contrast, active phospholipids such as phosphatidyl-choline, which are responsible for secretion of HMGB-1, appear later in the inflammatory microenvironment [71].

Interestingly, in addition to inflammatory cells such as monocytes, DCs, the professional antigen presenting cells, also express inflammatory leaderless cytokines upon activation by maturational stimuli such as LPS or the engagement of CD40. However, in these cells, soluble signals seem unable to drive secretion; rather secretion occurs following interaction of DCs with antigen-specific T cells [73, 80]. Morphological approaches have demonstrated that the interaction between DCs and CD8$^+$ T cells is associated with the recruitment of IL-1β– or IL-18–containing secretory lysosomes in the areas of contact among the cells. This results in a polarisation of these organelles and evidence of lysosome exocytosis at the intercellular space – the so-called 'immunological synapse' [81]. These findings warrant two considerations.

On the one hand, they underlie the existence of a bi-directional cross-talk between T cells and DCs, in which a T cell induces the functional polarisation of a DC and the DC responds by a degranulation that is orientated towards the same T cell, with obvious relevance for the control of the immune response. On the other hand, the different ways by which monocytes and DCs regulate secretion may account for the different function of IL-1β and IL-18 in inflammation and the immune response (Fig. 3.1). Monocytes respond to soluble signals with generalised exocytosis, thus allowing the spreading of inflammatory cytokines in the microenvironment, whereas DCs respond to the localised signal provided by the interacting T cell. This restricts the area of release to the immunological synapse and allows the activation of target cells without a wider distribution of the cytokine, thus controlling inflammation. Thus, lysosome-mediated secretion of inflammatory leaderless proteins allows polarised secretion in non-polarised cells.

3.5. Advantages of a leaderless secretory pathway

Leaderless secretion is generally inefficient and of a relatively 'short range' compared to the classical pathway, and this questions the rationale of its conservation. We can envisage several explanations for this. First of all, the vast majority of leaderless secretory proteins are cytokines, as characterised by a high biological activity and their involvement in the regulation of inflammatory processes, both as inducers and as silencers. The cytokine network needs to be perfectly balanced in order to control the onset, progression and resolution of inflammation; indeed, its dysregulation might lead to pathological conditions such as chronic inflammation and autoimmune diseases [7]. In this respect, slow and inefficient

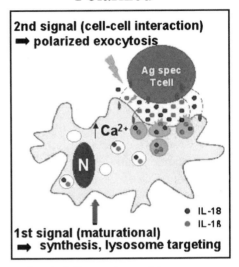

Figure 3.1. Lysosome-mediated leaderless secretion in inflammation and immune response. *Non-polarised:* A first inflammatory stimulus (e.g., LPS) induces monocytes to actively synthesise IL-1ß [7] and to hyperacetylate HMGB-1 that is readdressed from the nucleus to the cytoplasm [44]. Both proteins accumulate into the cytosol and in part into secretory lysosomes [71, 72]. A second extracellular soluble signal (e.g., ATP for IL-1β [79], lysophosphatydylcholine for HMGB-1 [71]) triggers generalised lysosome exocytosis. Although IL-18 is constitutively expressed by monocytes, both signals (LPS and ATP) are needed for secretion to occur [78]. *Polarised:* In DCs, a first maturational stimulus, soluble (LPS, TNF-α) or cell-mediated (CD40 triggering by CD40 ligand expressing activated CD4$^+$ T cells), induces IL-1β synthesis. The second signal is provided by antigen-specific T cells, which induces a $[Ca^{2+}]_i$ rise, followed by the recruitment of IL-1β–containing secretory lysosomes towards the interacting T cell, and exocytosis which is restricted to the intercellular space (immunological synapse) [81]. A similar mechanism undergoes IL-18 secretion [73, 80].

secretion may become an advantage in that it self-limits the protein activity in a restricted environment and allows the control of potentially toxic effects. In addition, many leaderless proteins require reducing conditions to maintain their function.

Mutation assays have demonstrated the requirement of one or more free cysteines for IL-1β [82], IL-18 [83] and FGF-1 [84]. Galectins are sensitive to oxidation if not bound to their appropriate glycoconjugates [85], and at least some of the cytokine functions of thioredoxin (TRX) [24] and MIF [86] require the active enzymatic redox site CXXC. These observations suggest that in principle these proteins should rapidly lose their activity in the oxidising extracellular milieu. On the one hand, this can be a further way of regulating the potentially dangerous hyperactivity of these proteins; however, on the other hand, it is tempting to speculate that the parallel secretion of oxidation-sensitive cytokines and of TRX and TRX reductase in the local microenvironment of the immune response serves to lengthen their survival span and, for instance, to prolong immunostimulation [87].

Another advantage may be the prevention of intracellular autocriny via the compartmentalisation of receptor and ligand. Indeed, in physiological situations, co-expression of leaderless cytokines and their specific receptors is a common event. Thus, during their transit to the extracellular milieu, leaderless proteins should avoid the interaction with their own receptors, because this might lead to undue early activation with unwanted consequences, such as uncontrolled proliferation or even cell transformation [4]. In the case of the leaderless protein FGF-2, the insertion of a secretory leader sequence at the N terminus has been shown to induce cell transformation in the cell line expressing the FGF-2 receptor [88]. Exclusion from the classical secretory pathway may also prevent misfolding of some leaderless proteins.

As mentioned before, many of these proteins bear free sulphhydryl groups which must be maintained in the reduced state in order to guarantee folding and bioactivity. The ER lumen is highly oxidising and favours the formations of disulphide bridges. For proteins that present free thiols, the transit through the ER might thus result in either retention or secretion in a non-functional folding. Similarly, avoiding the Golgi compartment can be an advantage for galectins, because sugars, which are highly abundant in the Golgi lumen, may trap these proteins and thereby impair their transport.

For some proteins, leaderless secretion occurs only under non-physiological conditions. For instance, in yeast, over-expression of some endogenous proteins, the intracellular accumulation of which may be toxic, induces their secretion through a non-classical export system [2]. A similar mechanism also seems to exist in mammals, an example being the mitochondrial sulphotransferase rhodanese, which in physiological conditions accumulates in mitochondria, but when over-expressed is also efficiently secreted [89]. Similarly, heterologous expression of green fluorescent protein (GFP) leads to its cytosolic accumulation but also activates the secretion of improperly folded molecules [90]. Non-classical secretion may thus act as a safety valve, maintaining cellular homeostasis when the cytoplasmic degradative pathways are overloaded. In this context, it is worth stressing that results obtained from studies in which chimaeric proteins, bearing putative sequences for leaderless secretion, are over-expressed must be interpreted with caution, because secretion might result from misfolding rather than from true recognition.

Finally, it must be remembered that many leaderless secretory proteins also have an intracellular function. Some of these are cytosolic enzymes: TRX and MIF are oxide reductases [24, 91]; AMF is the ubiquitous glycolytic enzyme phosphohexose isomerase [23] and tissue transglutaminase catalises the cross-linking of intracellular proteins [28]. Others such as HMGB-1 and engrailed-2 are nuclear factors [30, 32], and Annexin I regulates vesicular traffic [20]. It is possible that, on the basis of its physiological and developmental state, a cell addresses a cytosolic protein towards an additional extracellular function – the non-classical secretory route would guarantee the likelihood of this double function.

REFERENCES

1. de Lima Pimenta A, Blight M A, Chervaux C and Holland I B. Protein secretion in Gram negative bacteria. In Kuchler, K., Rubartelli, A. and Holland, B. (Eds.) *Unusual Secretory Pathways: From Bacteria to Man.* Chapman & Hall Landes Bioscience, New York/Austin, TX 1997, pp 1–48.
2. Cleves A E, Cooper D N, Barondes S H and Kelly R B. A new pathway for protein export in *Saccharomyces cerevisiae.* J Cell Biol 1996, 133: 1017–1026.
3. Kuchler K and Egner R. Unusual protein secretion and translocation pathways in yeast: implication of ABC transporters. In Kuchler, K., Rubartelli, A. and Holland, B. (Eds.) *Unusual Secretory Pathways: From Bacteria to Man.* Chapman & Hall Landes Bioscience, New York/Austin, TX 1997, pp 49–86.
4. Rubartelli A and Sitia R. Secretion of mammalian proteins that lack a signal sequence. In Kuchler, K., Rubartelli, A. and Holland, B. (Eds.) *Unusual Secretory Pathways: From Bacteria to Man.* Chapman & Hall Landes Bioscience, New York/Austin, TX 1997, pp 87–115.
5. Rubartelli A, Cozzolino F, Talio M and Sitia R. A novel secretory pathway for interleukin-1 beta, a protein lacking a signal sequence. EMBO J 1990, 9: 1503–1510.
6. Rubartelli A, Bajetto A, Allavena G, Wollman E and Sitia R. Secretion of thioredoxin by normal and neoplastic cells through a leaderless secretory pathway. J Biol Chem 1992, 267: 24161–24164.
7. Dinarello C A. Proinflammatory cytokines. Chest 2000, 118: 503–508.
8. Werman A, Werman-Venkent R, White R, Lee J K, Werman B, Krelin Y, Voronov E, Dinarello C A, and Apte R N. The precursor form of IL1alpha is an intracrine proinflammatory activator of transcription. Proc Natl Acad Sci U S A, 2004, 101: 2434–2439.
9. Dinarello C A and Fantuzzi G. Interleukin-18 and host defense against infection. J Infect Dis 2003, 187: S370–S384.
10. Fantuzzi G and Dinarello C A. Interleukin-18 and interleukin-1β: two cytokine substrates for ICE (caspase-1). J Clin Immunol 1999, 19: 1–11.
11. Laliberte R E, Eggler J and Gabel C A. ATP treatment of human monocytes promotes caspase-1 maturation and externalization. J Biol Chem 1999, 274: 36944–36951.
12. Cruikshank W W, Kornfeld H and Center D M. Interleukin-16. J Leuk Biol 2000, 67: 757–766.

13. Christofori G. The role of fibroblast growth factors in tumour progression and angiogenesis. In Bicknell, R., Lewis, C. E. and Ferrara, N. (Eds.) *Tumour Angiogenesis*. Oxford University Press, Oxford, UK 1997, pp 201–238.

14. Bikfalvi A, Savona C, Perollet C and Javerzat S. New insights in the biology of fibroblast growth factor-2. Angiogenesis 1998, 1: 155–173.

15. Tsai S J, Wu M H, Chen H M, Chuang P C and Wing L Y. Fibroblast growth factor-9 is an endometrial stromal growth factor. Endocrinology 2002, 143: 2715–2721.

16. Sleeman M W, Anderson K D, Lambert P D, Yancopoulos G D and Wiegand S J. The ciliary neurotrophic factor and its receptor, CNTFR alpha. Pharm Acta Helv 2000, 74: 265–272.

17. Calandra T. Macrophage migration inhibitory factor and host innate immune responses to microbes. Scand J Infect Dis 2003, 35: 573–576.

18. Ko Y G, Park H, Kim T, Lee J W, Park S G, Seol W, Kim J E, Lee W H, Kim S H, Park J E and Kim S. A cofactor of tRNA synthetase, p43, is secreted to up-regulate proinflammatory genes. J Biol Chem 2001, 276: 23028–23033.

19. Berger A C, Alexander H R, Tang G, Wu P S, Hewitt S M, Turner E, Kruger E, Figg W D, Grove A, Kohn E, Stern D and Libutti S K. Endothelial monocyte activating polypeptide II induces endothelial cell apoptosis and may inhibit tumor angiogenesis. Microvasc Res 2000, 60: 70–80.

20. Rescher U, Zobiack N and Gerke V. Intact Ca^{2+}-binding sites are required for targeting of annexin 1 to endosomal membranes in living HeLa cells. J Cell Sci 2000, 113: 3931–3938.

21. Solito E, Nuti S and Parente L. Dexamethasone-induced translocation of lipocortin (annexin) 1 to the cell membrane of U-937 cells. Br J Pharmacol 1994, 112: 347–348.

22. Christmas P, Callaway J, Fallon J, Jones J and Haigler H T. Selective secretion of annexin 1, a protein without a signal sequence, by the human prostate gland. J Biol Chem 1991, 266: 2499–2507.

23. Tsutsumi S, Yanagawa T, Shimura T, Fukumori T, Hogan V, Kuwano H and Raz A. Regulation of cell proliferation by autocrine motility factor/phosphoglucose isomerase signaling. J Biol Chem 2003, 278: 32165–32172.

24. Arner E S and Holmgren A. Physiological functions of thioredoxin and thioredoxin reductase. Eur J Biochem 2000, 267: 6102–6109.

25. Rabinovich G A, Baum L G, Tinari N, Paganelli R, Natoli C, Liu F T and Iacobelli S. Galectins and their ligands: amplifiers, silencers or tuners of the inflammatory response? Trends Immunol 2002, 23: 313–320.

26. Rabinovich G A, Rubinstein N and Toscano M A. Role of galectins in inflammatory and immunomodulatory processes. Biochim Biophys Acta 2002, 1572: 274–284.

27. Sano H, Hsu D K, Apgar J R, Yu L, Sharma B B, Kuwabara I, Izui S and Liu F T. Critical role of galectin-3 in phagocytosis by macrophages. J Clin Invest 2003, 112: 389–397.

28. Griffin M, Casadio R and Bergamini C M. Transglutaminases: Nature's biological glues. Biochem J 2002, 368: 377–396.

29. Grundmann U, Amann E, Zettlmeissl G and Kupper H A. Characterization of cDNA coding for human factor XIIIa. Proc Natl Acad Sci USA 1986, 83: 8024–8028.

30. Wang H, Bloom O, Zhang M, Vishnubhakat J M, Ombrellino M, Che J, Frazier A, Yang H, Ivanova S, Borovikova L, Manogue K R, Faist E, Abraham E, Andersson J,

Andersson U, Molina P E, Abumrad N N, Sama A and Tracey K J. HMG-1 as a late mediator of endotoxin lethality in mice. Science 1999, 285: 248–251.

31. Maizel A, Tassetto M, Filhol O, Cochet C, Prochiantz A and Joliot A. Engrailed homeoprotein secretion is a regulated process. Development 2002, 129: 3545–3553.

32. Maizel A, Bensaude O, Prochiantz A and Joliot A. A short region of its homeodomain is necessary for engrailed nuclear export and secretion. Development 1999, 126: 3183–3190.

33. Dagher S F, Wang J L and Patterson R J. Identification of galectin-3 as a factor in pre-mRNA splicing. Proc Natl Acad Sci USA 1995, 92: 1213–1217.

34. Kleemann R, Hausser A, Geiger G, Mischke R, Burger-Kentischer A, Flieger O, Johannes F J, Roger T, Calandra T, Kapurniotu A, Grell M, Finkelmeier D, Brunner H and Bernhagen J. Intracellular action of the cytokine MIF to modulate AP-1 activity and the cell cycle through Jab1. Nature 2000, 408: 211–216.

35. Vareli K, Frangou-Lazaridis M, van der Kraan I, Tsolas O and van Driel R. Nuclear distribution of prothymosin alpha and parathymosin: evidence that prothymosin alpha is associated with RNA synthesis processing and parathymosin with early DNA replication. Exp Cell Res 2000, 257: 152–161.

36. Hannappel E and Huff T. The thymosins. Prothymosin alpha, parathymosin, and beta-thymosins: structure and function. Vit Horm 2003, 66: 257–296.

37. Focher F and Spadari S. Thymidine phosphorylase: a two-face Janus in anticancer chemotherapy. Cur Cancer Drug Targets 2001, 1: 141–153.

38. Jackson A, Tarantini F, Gamble S, Friedman S and Maciag T. The release of fibroblast growth factor-1 from NIH 3T3 cells in response to temperature involves the function of cysteine residues. J Biol Chem 1995, 270: 33–36.

39. Tarantini F, Gamble S, Jackson A and Maciag T. The cysteine residue responsible for the release of fibroblast growth factor-1 residues in a domain independent of the domain for phosphatidylserine binding. J Biol Chem 1995, 270: 29039–29042.

40. Balklava Z, Verderio E, Collighan R, Gross S, Adams J and Griffin M. Analysis of tissue transglutaminase function in the migration of Swiss 3T3 fibroblasts: the active-state conformation of the enzyme does not affect cell motility but is important for its secretion. J Biol Chem 2002, 277: 16567–16575.

41. Reiness C G, Seppa M J, Dion D M, Sweeney S, Foster D N and Nishi R. Chick ciliary neurotrophic factor is secreted via a nonclassical pathway. Mol Cell Neurosci 2001, 17: 931–944.

42. Menon R P and Hughes R C. Determinants in the N-terminal domains of galectin-3 for secretion by a novel pathway circumventing the endoplasmic reticulum-Golgi complex. Eur J Biochem 1999, 264: 569–576.

43. Gong H C, Honjo Y, Nangia-Makker P, Hogan V, Mazurak N, Bresalier R S and Raz A. The NH2 terminus of galectin-3 governs cellular compartmentalization and functions in cancer cells. Cancer Res 1999, 59: 6239–6245.

44. Bonaldi T, Talamo F, Scaffidi P, Ferrera D, Porto A, Bachi A, Rubartelli A, Agresti A and Bianchi M E. Monocytic cells hyperacetylate chromatin protein HMGB1 to redirect it towards secretion. EMBO J 2003, 22: 5551–5560.

45. Haga A, Niinaka Y and Raz A. Phosphohexose isomerase/autocrine motility factor/neuroleukin/maturation factor is a multifunctional phosphoprotein. Biochim Biophys Acta 2000, 1480: 235–244.

46. Stevenson F T, Bursten S L, Fanton C, Locksley R M and Lovett D H. The 31-kDa precursor of interleukin 1 alpha is myristoylated on specific lysines within the 16-kDa N-terminal propiece. Proc Natl Acad Sci USA 1993, 90: 7245–7249.

47. Cooper D N and Barondes S H. Evidence for export of a muscle lectin from cytosol to extracellular matrix and for a novel secretory mechanism. J Cell Biol 1990, 110: 1681–1691.

48. Lindstedt R, Apodaca G, Barondes S H, Mostov K E and Leffler H. Apical secretion of a cytosolic protein by Madin-Darby canine kidney cells. Evidence for polarized release of an endogenous lectin by a nonclassical secretory pathway. J Biol Chem 1993, 268: 11750–11757.

49. Hermo L and Jacks D. Nature's ingenuity: bypassing the classical secretory route via apocrine secretion. Mol Reprod Dev 2002, 63: 394–410.

50. Steinhoff M, Eicheler W, Holterhus P M, Rausch U, Seitz J and Aumuller G. Hormonally induced changes in apocrine secretion of transglutaminase in the rat dorsal prostate and coagulating gland. Eur J Cell Biol 1994, 65: 49–59.

51. Eickhoff R, Wilhelm B, Renneberg H, Wennemuth G, Bacher M, Linder D, Bucala R, Seitz J and Meinhardt A. Purification and characterization of macrophage migration inhibitory factor as a secretory protein from rat epididymis: evidences for alternative release and transfer to spermatozoa. Mol Med 2001, 7: 27–35.

52. MacKenzie A, Wilson H L, Kiss-Toth E, Dower S K, North R A and Surprenant A. Rapid secretion of interleukin-1β by microvesicle shedding. Immunity 2001, 15: 825–835.

53. Murk J L, Stoorvogel W, Kleijmeer M J and Geuze H J. The plasticity of multivesicular bodies and the regulation of antigen presentation. Sem Cell Dev Biol 2002, 13: 303–311.

54. Johnstone R M, Adam M, Hammond J R, Orr L and Turbide C. Vesicle formation during reticulocyte maturation. Association of plasma membrane activities with released vesicles (exosomes). J Biol Chem 1987, 262: 9412–9420.

55. Théry C, Zitvogel L and Amigorena S. Exosomes: composition, biogenesis and function. Nat Rev Immunol 2002, 2: 569–579.

56. Théry C, Boussac M, Véron P, Ricciardi-Castagnoli P, Raposo G, Garin G and Amigorena S. Proteomic analysis of dendritic cell-derived exosomes: A secreted subcellular compartment distinct from apoptotic vesicles. J Immunol 2001, 166: 7309–7318.

57. Fink A L. Chaperone-mediated protein folding. Physiol Rev 1999, 79: 425–449.

58. Holland I B, Benhabdelhak H, Young J, de Lima Pimenta A, Schmitt L and Blight M A. Bacterial ABC transporters involved in protein translocation. In Holland, I. B. (Ed.) *ABC Proteins: from Bacteria to Man*. Academic Press, London/San Diego 2003, pp 209–242.

59. Prudovsky I, Bagala C, Tarantini F, Mandinova A, Soldi R, Bellum S and Maciag T. The intracellular translocation of the components of the fibroblast growth factor 1 release complex precedes their assembly prior to export. J Cell Biol 2002, 158: 201–208.

60. McNew J A and Goodman J M. An oligomeric protein is imported into peroxisomes *in vivo*. J Cell Biol 1994, 127: 1245–1257.

61. Cuervo A M, Mann L, Bonten E J, d'Azzo A and Dice J F. Cathepsin A regulates chaperone-mediated autophagy through cleavage of the lysosomal receptor. EMBO J 2003, 22: 47–59.

62. Dean M, Hamon Y and Chimini G. The human ATP-binding cassette (ABC) transporter superfamily. J Lipid Res 2001, 42: 1007–1017.

63. Hamon Y, Luciani M F, Becq F, Verrier B, Rubartelli A and Chimini G. Interleukin-1β secretion is impaired by inhibitors of the ATP binding cassette transporter, ABC1. Blood 1997, 90: 2911–2915.

64. Chapman L P, Epton M J, Buckingham J C, Morris J F and Christian H C. Evidence for a role of the adenosine 5′-triphosphate-binding cassette transporter A1 in the externalization of annexin I from pituitary folliculo-stellate cells. Endocrinology 2003, 144: 1062–1073.

65. Flieger O, Engling A, Bucala R, Lue H, Nickel W and Bernhagen J. Regulated secretion of macrophage migration inhibitory factor is mediated by a non-classical pathway involving an ABC transporter. FEBS Lett 2003, 551: 78–86.

66. Schafer T, Zentgraf H, Zehe C, Brugger B, Bernhagen J and Nickel W. Unconventional protein secretion: direct translocation of fibroblast growth factor 2 across the plasma membrane of mammalian cells. J Biol Chem 2003, 279: 6244–6251.

67. Oliani S M, Damazo A S and Perretti M. Annexin 1 localisation in tissue eosinophils as detected by electron microscopy. Med Inflamm 2002, 11: 287–292.

68. Qu Z, Kayton R J, Ahmadi P, Liebler J M, Powers M R, Planck S R and Rosenbaum J T. Ultrastructural immunolocalization of basic fibroblast growth factor in mast cell secretory granules. Morphological evidence for bfgf release through degranulation. J Histochem Cytochem 1998, 46: 1119–1128.

69. Nishino T, Bernhagen J, Shiiki H, Calandra T, Dohi K and Bucala R. Localization of macrophage migration inhibitory factor (MIF) to secretory granules within the corticotrophic and thyrotrophic cells of the pituitary gland. Mol Med 1995, 1: 781–788.

70. Joliot A, Maizel A, Rosenberg D, Trembleau A, Dupas S, Volovitch M and Prochiantz A. Identification of a signal sequence necessary for the unconventional secretion of Engrailed homeoprotein. Cur Biol 1998, 8: 856–863.

71. Gardella S, Andrei C, Ferrera D, Lotti L V, Torrisi M R, Bianchi M E and Rubartelli A. The nuclear protein HMGB1 is secreted by monocytes via a non-classical, vesicle-mediated secretory pathway. EMBO Rep 2002, 3: 995–1001.

72. Andrei C, Dazzi C, Lotti L, Torrisi M R, Chimini G and Rubartelli A. The secretory route of the leaderless protein interleukin 1beta involves exocytosis of endolysosome-related vesicles. Mol Biol Cell 1999, 10: 1463–1475.

73. Gardella S, Andrei C, Poggi A, Zocchi M R and Rubartelli A. Control of interleukin-18 secretion by dendritic cells: role of calcium influxes. FEBS Lett 2000, 481: 245–248.

74. Blott E J and Griffiths G M. Secretory lysosomes. Nat Rev Mol Cell Biol 2002, 3: 122–131.

75. Sesaki H, Wong E F and Siu C H. The cell adhesion molecule DdCAD-1 in *Dictyostelium* is targeted to the cell surface by a nonclassical transport pathway involving contractile vacuoles. J Cell Biol 1997, 138: 939–951.

76. Holland I B. *ABC Proteins: from Bacteria to Man*. Academic Press, London/San Diego: 2003.

77. Andrei C, Margiocco P, Poggi A, Lotti L V, Torrisi M R, Rubartelli A. Phospholipases C and A_2 control lysosome – mediated IL-1beta secretion: Implications for inflammatory processes. Proc Natl Acad Sci USA 2004, 101: 9745–9750.

78. Perregaux D G, McNiff P, Laliberte R, Conklyn M and Gabel C A. ATP acts as an agonist to promote stimulus-induced secretion of IL-1β and IL-18 in human blood. J Immunol 2000, 165: 4615–4623.

79. Di Virgilio F, Chiozzi P, Ferrari D, Falzoni S, Sanz J M, Morelli A, Torboli M, Bolognesi G and Baricordi O R. Nucleotide receptors: an emerging family of regulatory molecules in blood cells. Blood 2001, 97: 587–600.

80. Gardella S, Andrei C, Costigliolo S, Poggi A, Zocchi M R and Rubartelli A. Interleukin-18 synthesis and secretion by dendritic cells are modulated by interaction with antigen-specific T cells. J Leuk Biol 1999, 66: 237–241.

81. Gardella S, Andrei C, Lotti L V, Poggi A, Torrisi M R, Zocchi M R and Rubartelli A. CD8$^+$ T lymphocytes induce polarized exocytosis of secretory lysosomes by dendritic cells with release of interleukin-1β and cathepsin D. Blood 2001, 98: 2152–2159.

82. Kamogashira T, Masui Y, Ohmoto Y, Hirato T, Nagamura K, Mizuno K, Hong Y M, Kikumoto Y, Nakai S and Hirai Y. Site-specific mutagenesis of the human interleukin-1β gene: structure-function analysis of the cysteine residues. Biochem Biophys Res Comm 1988, 150: 1106–1114.

83. Pei D S, Fu Y, Sun Y F and Zhao H R. Site-directed mutagenesis of the cysteines of human IL-18 and its effect on IL-18 activity. Sheng Wu Hua Xue Yu Sheng Wu Wu Li Xue Bao (Shanghai) 2002, 34: 57–61.

84. Ortega S, Schaeffer M T, Soderman D, DiSalvo J, Linemeyer D L, Gimenez-Gallego G and Thomas K A. Conversion of cysteine to serine residues alters the activity, stability, and heparin dependence of acidic fibroblast growth factor. J Biol Chem 1991, 266: 5842–5846.

85. Cho M and Cummings R D. Galectin-1, a β-galactoside-binding lectin in Chinese hamster ovary cells. I. Physical and chemical characterization. J Biol Chem 1995, 270: 5198–5206.

86. Kleemann R, Kapurniotu A, Mischke R, Held J and Bernhagen J. Characterization of catalytic centre mutants of macrophage migration inhibitory factor (MIF) and comparison to Cys81Ser MIF. Eur J Biochem 1999, 261: 753–766.

87. Angelini G, Gardella S, Ardy M, Ciriolo M R, Filomeni G, Di Trapani G, Clarke F, Sitia R and Rubartelli A. Antigen-presenting dendritic cells provide the reducing extracellular microenvironment required for T lymphocyte activation. Proc Nat Acad Sci USA 2002, 99: 1491–1496.

88. Rogelj S, Weinberg R A, Fanning P and Klagsbrun M. Basic fibroblast growth factor fused to a signal peptide transforms cells. Nature 1988, 331: 173–175.

89. Sloan I S, Horowitz P M and Chirgwin J M. Rapid secretion by a nonclassical pathway of overexpressed mammalian mitochondrial rhodanese. J Biol Chem 1994, 269: 27625–27630.

90. Tanudji M, Hevi S and Chuck S L. Improperly folded green fluorescent protein is secreted via a non-classical pathway. J Cell Sci 2002, 115: 3849–3857.

91. Lue H, Kleemann R, Calandra T, Roger T and Bernhagen J. Macrophage migration inhibitory factor (MIF): mechanisms of action and role in disease. Microbes Infect 2002, 4: 449–460.

4

Moonlighting Proteins: Proteins with Multiple Functions

Constance J. Jeffery

4.1. Introduction

Moonlighting proteins, also referred to as 'gene sharing', refer to a subset of multifunctional proteins in which two or more different functions are performed by one polypeptide chain, and the multiple functions are not a result of splice variants, gene fusions, or multiple isoforms [1]. In addition, they do not include proteins with the same function in multiple locations or protein families in which different members have different functions, if each individual member has only one function. A single protein with multiple functions may seem surprising, but there are actually many cases of proteins that 'moonlight'.

4.2. Examples and mechanisms of combining two functions in one protein

The current examples of moonlighting proteins include enzymes, DNA binding proteins, receptors, transmembrane channels, chaperones and ribosomal proteins (Table 4.1). In general, there are several different methods by which a moonlighting protein can combine two functions within one polypeptide chain. A single protein can have a second function when it moves to a different cellular location; when it is expressed in a different cell type; when it binds a substrate, product, or cofactor; when it interacts with another protein to form a multimer, or when it interacts with a large multiprotein complex. In addition, a few enzymes have two active sites for different substrates (Figure 4.1). The methods are not mutually exclusive and sometimes a combination of methods is employed.

Cellular location: Several cytosolic or nuclear enzymes have a second function outside of the cell. Phosphoglycerate kinase and phosphoglucose isomerase

Table 4.1. Moonlighting proteins

One function	Another function	Reference
Plasmin reductase	Phosphoglycerate kinase	[2]
Phosphoglucose isomerase	Neuroleukin, autocrine motility factor, differentiation and maturation mediator	[3–6]
Thymidine phosphorylase	Platelet-derived endothelial cell growth factor	[7]
Thymosin β4 (sequester actin)	Secreted chemotaxis ligand	[8]
SMC3 (sister chromatin cohesion)	Basement membrane bamacam	[9, 10]
Histone H1	Thyroglobulin receptor	[11]
Neuropilin (VEGF receptor)	Receptor for semaphorin III (nerve axons)	[12]
Thymidylate synthase	Translation inhibitor	[13]
birA biotin sythetase	bio operon repressor	[14]
PutA proline dehydrogenase	Transcriptional repressor	[15]
Aconitase	Iron responsive element binding protein (IRE-BP)	[16]
Paramyxovirus hemaglutinin	Neuraminidase	[17]
4a-Carbinolamine dehydratase	Dimerization cofactor (DCoH)	[18]
δ-Aminolevulinic acid dehydratase	Proteasome inhibitory subunit CF-2	[19]
Ribosomal proteins	DNA repair, translational regulators, etc.	[20]
Clf1p pre-mRNA splicing factor	Initiation of DNA replication	[21–24]
Proteasome base complex	RNA pollll transcription	[25]
Cyclooxygenase-1	Heme-dependent peroxidase	[26]
Lysyl hydroxylase isoform 3	Collagen glucosyltransferase	[27]
CFTR chloride channel	Regulator of other epithelial anion channels	[28]
Mitochondrial Lon protease	Chaperone	[29]
Bacterial FtsH chaperone	Metalloprotease	[29]
Lens crystallins	Heat shock proteins, lactate dehydrogenase, argininosuccinate lyase, retinaldehyde dehydrogenase, enolase, quinone oxidoreductase, glyceraldehyde-3-phosphate dehydrogenase	[30]
PHGPx (glutathione peroxidase)	Sperm structural protein	[31]
E. coli thioredoxin	Subunit of T7 DNA polymerase	[38]
PMS2 mismatch repair enzyme	Hypermutation of antibody variable chains	[39]
Leukotriene A4 hydrolase	Aminopeptidase	[40]
1-cys peroxiredoxin (peroxidase)	Phospholipase aiPLA2	[41]
Tetrahymena citrate synthase	14-nm cytoskeletal protein	[42]
Transferrin receptor	Glyceraldehyde-3-phosphate dehydrogenase	[43]
Lactose synthetase	Galactosyltransferase	[44]
Homing endonuclease	Intron splicing factor	[45]
N. crassa tyrosyl tRNA synthetase	Promotes folding of group I introns	[46]
Cytochrome c (electron transport)	Apoptosis	[47]

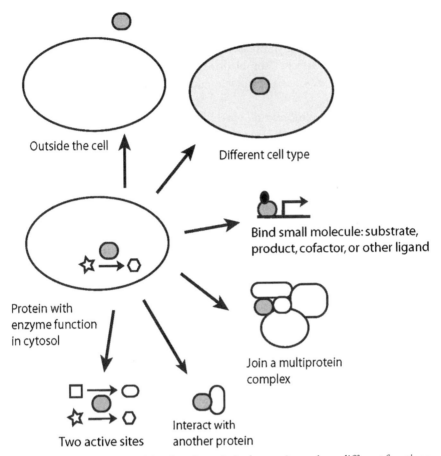

Figure 4.1. Methods of switching functions. A single protein can have different functions in different cellular locations, when expressed in different cell types; when it binds a ligand, substrate, product, or cofactor; when it interacts with another protein to form a multimer; when it interacts with a large multiprotein complex; or by having two binding sites for different substrates or ligands.

catalyse the seventh and second steps, respectively, in glycolysis in the cytosol of most cells. Both are also found to have a second function outside of the cell. Phosphoglycerate kinase is a disulphide reductase that reduces plasmin [2]. The reduced plasmin undergoes proteolysis to produce angiostatin, an angiogenesis inhibitor. Phosphoglucose isomerase binds to cell surface receptors on target cells and causes a variety of effects, including differentiation of pre-B cells to antibody secreting cells, an increase in motility of some tumour cells, and differentiation of HL-60 leukaemia cells to monocytes [3–6].

Thymidine phosphorylase, which is also called platelet-derived endothelial cell growth factor, removes the phosphoryl group from thymidine and deoxyuridine in the cytoplasm and stimulates chemotaxis of endothelial cells outside of the cell [7]. Thymosin beta 4 sulphoxide is an inhibitor of actin polymerisation in the cytosol and serves as a negative modulator of the inflammatory response outside the cell [8].

Whereas the preceding examples include cytosolic proteins that serve as soluble growth factors, enzymes, or cytokines outside of the cell, other cytoplasmic or nuclear proteins have an extracellular second function in which they are not soluble. The mouse SMC3 protein (structural maintenance of chromosome 3), also known as bamacam, functions in sister chromatid cohesion in the nucleus and is also a component of the basement membrane [9, 10]. Histone H1 is another nuclear protein with a function outside the cell, but it remains attached to the extracellular surface of the cell membrane and serves as a receptor for thyroglobulin [11].

Different cell types: Expression by multiple cell types can also result in a protein having multiple functions. For example, neuropilin is a cell surface receptor in neurons and endothelial cells [12]. When expressed in neurons, neuropilin binds semaphorin III and plays a role in axonal guidance. When expressed in endothelial cells, it binds vascular endothelial growth factor (VEGF) and helps signal the need for new blood cells.

Binding substrate, product, cofactor, or other ligand: Binding to a substrate, product, cofactor, or other ligand can cause a change in the function of a protein. The enzymes thymidylate synthase [13], biotin synthetase (*Escherichia coli birA*) [14], PutA proline dehydrogenase [15], and aconitase [16] (also called iron responsive binding protein, IRE-BP) are three cytosolic or membrane-bound enzymes that detect changes in the cellular concentration of a ligand and then bind to DNA or RNA and regulate transcription or translation.

Paramyxovirus hemagglutinin-neuraminidase responds to changes in the pH of its environment by changing conformation of several amino acid side chains and a loop in the active site. These movements may enable a switch between the sialic acid binding and hydrolysis functions of the protein [17].

Forming a complex with other proteins: Entering into multiprotein complexes is another method by which a protein can exhibit a moonlighting function. In some cases, the moonlighting protein interacts with only one or a few other proteins. 4α-carbinolamine dehydratase (also called DCoH) is an enzyme in liver cells. It also binds to the transcription factor HNF1 δ (hepatic nuclear factor

1δ). By influencing the dimerisation of HNF1 δ, DCoH regulates the binding of the transcription factor to DNA [18].

In other cases a protein becomes part of a large multiprotein complex, such as the proteasome or the ribosome, which is composed of many different polypeptide chains. Delta-aminolevulinic acid dehydratase, and enzyme in heme biosynthesis, is the same protein as the 240-kDA inhibitory component of the proteasome [19]. Several other cytoplasmic or nuclear enzymes have been found to be identical to proteins in the ribosome (reviewed in [20]). In addition, there are a few examples of moonlighting proteins that participate in multiple multiprotein complexes, changing roles with the different polypeptide partners. *Saccharomyces cerevisiae* Clf1p apparently performs different functions by interacting with different proteins in two multiprotein complexes for pre-mRNA splicing and the initiation of DNA replication. It interacts with the U5 and U6 small nuclear ribonucleoprotein particles, pre-mRNA, and other components of pre-mRNA splicing reactions. It also interacts with the DNA replication initiation protein Orc2p in the origin of replication complex [21–24].

Sug1/Rpt6 and Sug2/Rpt4 are AAA proteins (an ATP-dependent protein superfamily including molecular chaperones involved in protein assembly/disassembly) that form part of the base complex of the proteasome. The base complex and lid complex, which make up the 19S particle, join with the 20S proteolytic complex to catalyse proteolysis. However, the base complex also plays a role in RNA polIII transcription, without the lid complex or the 20S particle. In response to galactose induction, the base complex moves to the GAL1-10 promoter, and the Sug1/Rpt6 and Sug2/Rpt4 proteins interact directly with the Gal4 transactivator to alter transcription levels from the GAL1-10 promoter [25].

Multiple binding sites for different substrates or ligands: Other proteins do not necessarily have a switch mechanism to change functions; they simply have multiple binding sites or active site pockets for different ligands or substrates. The enzyme prostaglandin H2 synthase-1 has two active sites: a heme-dependent peroxidase active site and a cyclooxygenase active site. The two active site pockets are found near each other in the enzyme structure [26], and both catalyse reactions in the synthesis of prostaglandin H2. Similarly, the enzyme lysyl hydroxylase 3 catalyses two steps in collagen biosynthesis [27].

Overall, there are a number of ways in which a protein can combine two functions within one polypeptide chain. The methods are not mutually exclusive, and, in some cases, a combination of factors contribute to switching between functions.

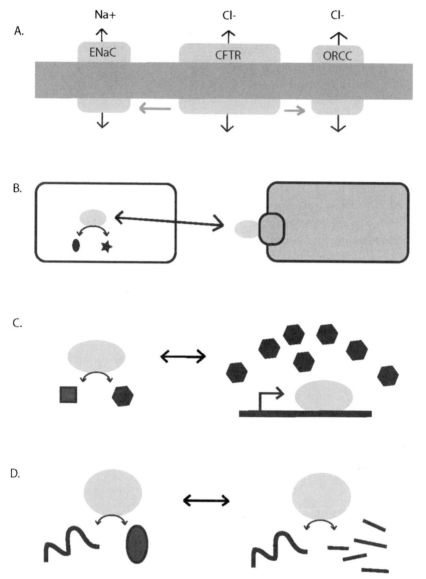

Figure 4.2. Examples of benefits provided by proteins that moonlight: (A) Coordination of functions within a cell. The CFTR is a chloride channel and it regulates (horizontal grey arrows) the activity of the EnaC sodium channel and the ORCC chloride channel. (B) Coordination between different cell types. Some proteins with catalytic activity in the cytoplasm of one cell type (black oval to star) also have an extracellular function in which they bind to a receptor on other cells types. (C) Feedback. Some enzymes that catalyse a chemical reaction (square to hexagon) can sense when the accumulation of product (hexagon) is high and bind to DNA to inhibit the synthesis of more copies of the enzyme. (D) Switch between pathways. Some chaperones that help a protein fold (curved line to black oval) also have a protease activity that can degrade a protein (curved line to short line fragments), depending on cellular conditions.

4.3. Why have moonlighting proteins?

The wide variety of moonlighting proteins and combinations of functions suggests that moonlighting evolved independently many times. This suggests that there are benefits to having moonlighting proteins, or that it is relatively easy for a second function to evolve. Analysis of the examples in Table 4.1 suggests that both are true. In fact, moonlighting proteins appear to provide several kinds of benefits to the organism. Also, there are two proposed mechanisms for moonlighting proteins to have evolved that make use of general physical properties of many protein structures.

4.3.1. Benefits to the organism

Having moonlighting proteins can provide a means to coordinate different biochemical pathways, a means to respond to stress or changes in the environment, and as a feedback mechanism (Figure 4.2).

Coordination: As the complex modern cell evolved, a need arose for methods to coordinate the many intracellular biochemical pathways for signalling, transport, biosynthesis and other functions, and moonlighting proteins provide one such mechanism. For example, combining the two enzymatic functions of lysyl hydroxylase 3, described earlier, within one protein might help coordinate two steps involved in collagen maturation. The cystic fibrosis transmembrane conductance regulator (CFTR) is a chloride channel and also regulates the activity of the outwardly rectifying chloride channel (ORCC) and a sodium channel (ENaC) (Figure 4.2) [28]. The ability of one transmembrane channel to coordinate the activity of several kinds of channels helps to maintain ion homeostasis within epithelial cells. In addition, as multicellular organisms developed, the need for coordination of activities between different cells, cell types, and organs arose, which might be one reason there are multiple examples of intracellular enzymes with a second function as a cytokine or growth factor (Figure 4.2).

Switch between pathways: The combination of two alternative functions within one protein might also provide an efficient method to switch between two pathways in response to changing conditions in the environment, such as changes in food supply or the introduction of a stress (Figure 4.2). Changes in cellular iron concentrations cause a decrease in the catalytic activity of aconitase, a cytosolic enzyme. Aconitase, also known as the iron-responsive element binding protein, then binds to DNA to cause changes in transcription of proteins involved in iron accumulation [16]. The two functions of delta-aminolevulinic acid dehydratase, as an enzyme in the heme biosynthesis pathway and as the proteasome inhibitor CF-2, might provide a method to switch between protein degradation and heme biosynthesis [19]. Mitochondrial Lon protease, which is both a protease and a chaperone (reviewed in [29]), provides a more general switch between protein degradation and protein biosynthesis.

Feedback: Some enzymes that catalyse one step of a biochemical pathway moonlight as a sensor for the overall level of activity of the pathway by measuring the concentration of substrates or products and then binding to DNA or RNA in order to regulate transcription or translation of enzymes within the pathway (Figure 4.2). This combination of functions provides a feedback mechanism to regulate the activity of the pathway. Thymidylate synthase 3, biotin synthetase (birA), and PutA proline dehydrogenase are three examples of enzymes that also bind DNA or RNA to regulate transcription or translation in response to changing levels of substrate, product, or cofactor [13–15].

No clear benefit: Although in many cases it appears there is indeed a benefit to having two functions within one polypeptide chain, it is not always clear if there is a connection between the two functions in some moonlighting proteins. As described earlier, phosphoglucose isomerase is both an enzyme in glycolysis in the cell cytosol and an extracellular cytokine [3–6]. It is not clear why an organism would make use of a glycolytic enzyme in this way. One possibility is that after the second function evolved both functions benefit the organism independently and there was no selective pressure to remove it.

4.4. Models for the evolution of moonlighting proteins

But how can a protein develop a second function within the same polypeptide chain? From consideration of the examples listed in Table 4.1, it appears that there are two general mechanisms for a protein to evolve a second function (Figure 4.3).

Recruitment of a protein without significant change in protein structure: The first method involves recruitment of the protein for a new function, perhaps as a new organ or cell type evolves, without major changes in protein structure (Figure 4.3). The crystallins, members of the small heat shock protein family, are a classic example of this method. Several crystallins are ubiquitous, soluble, cytosolic enzymes that were recruited for a second function in the lens when

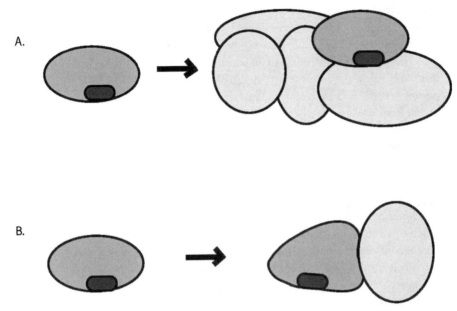

Figure 4.3. Methods of evolution. A protein can evolve a second function (A) by recruitment without a significant change in protein structure or (B) by modification of 'unused' solvent exposed surface area. In both cases, the original function, as represented by the black oval 'active site' remains.

the eye evolved [30]. Two other proteins that appear to have evolved a second function by recruitment into a multiprotein complex are PHGPx (glutathione peroxidase), a soluble enzyme that is also a sperm structural protein [31], and *E. coli* thioredoxin, which was adopted by the T7 phage as a subunit of its DNA polymerase [32].

Modification of solvent-exposed surface area: The other general method for a protein to develop a moonlighting function is based on the observation that several ubiquitous enzymes appear to have evolved a protein-binding site on the protein surface in addition to their catalytic sites (Figure 4.3). One example is phosphoglucose isomerase (PGI). PGI is found in almost all species and so must have evolved over three billion years ago. Throughout all that time, the enzyme active site has been conserved, but the protein surface has undergone many changes. As is seen in many enzymes, the active site pocket of PGI is actually rather small compared to the total surface area of the protein. In this relatively large protein – a dimer with 557 amino acids in each subunit – there is a large amount of apparently unused solvent exposed surface area that might not be under tight evolutionary pressure. The random accumulation of mutations on the surface of PGI could provide the material and opportunity that might have resulted in the evolution of an additional binding site. The protein surface might be able to accommodate these changes without adversely affecting the protein's first function: catalysis. In fact, a comparison between the structures of bacterial and mammalian PGI indicates that several alpha helices, surface loops, and surface pockets have undergone considerable change through evolution, whereas the active site pockets of the two enzymes have remained almost identical [33].

Both the recruitment and modification of surface area methods of evolution of a moonlighting function make use of general features of protein structure and could apply to many proteins. Whichever way that a protein evolves a new function, as long as the new function does not adversely affect the original function of the protein, then the second function might provide some of the benefits just described. It is therefore possible that there would be an advantage to retain both functions during further evolution, or at least no selective pressure to eliminate either.

4.5. How many proteins moonlight?

The two proposed methods of evolving moonlighting functions could have happened to many proteins. In addition, the examples of moonlighting proteins identified to date include many types of proteins, including enzymes, transcription factors, channels, and receptors, and proteins from many diverse organisms and cell types. This wide variety of proteins and combinations of functions suggest that many types of proteins can moonlight. While we cannot put a distinct value on the number of proteins that moonlight, it is possible that moonlighting proteins might be common, and we could find that many other proteins also have additional functions that have not yet been found.

4.6. Identifying more moonlighting proteins

Although it is not yet clear how many proteins might moonlight, moonlighting proteins that have different functions in different locations or within different multiprotein complexes might be especially common because they could provide the key benefits listed earlier: coordinating cellular activities, response to changes in the environment, and feedback mechanisms. The current interest in large-scale proteomics studies that apply biochemical or genetic methods to characterise the locations, protein–protein interactions, or expression levels of thousands of proteins is likely to lead to the identification of more examples of moonlighting proteins.

4.6.1. Methods employing protein locations

Mass spectrometry and two-dimensional gel electrophoresis can be used to identify proteins in complex mixtures such as whole cells, organelles, or multiprotein complexes. Similarly, analysis of RNA levels using micro-array methods can suggest which proteins are being expressed in a chosen cell type or tissue. The resulting protein expression profiles can be repeated for different cell types, at different times in development, before and after application of a signal, or in diseased and healthy tissues. The differences in protein expression patterns are then used to help deduce the function of that protein. For example, in general, groups of proteins that function together in a biochemical pathway, multiprotein complex or signalling pathway are often expressed in the same cell types and under the same growth conditions, whereas in other cell types none of the proteins in those complexes or pathways are expressed – in other words, an 'all or none' pattern.

However, a moonlighting protein might have an unusual pattern of expression which is inconsistent with its function in a single biochemical pathway or protein complex. For example, it might be expressed along with the other enzymes in a multiprotein complex or multienzyme pathway in some cell types or under some conditions, but it might also be expressed in other cell types without the other proteins, in which it performs its second function. When protein expression profiles identify a protein in an unexpected cell type, organelle, or multiprotein complex, it is possible that the protein might have a moonlighting function.

4.6.2. Interactions with different binding partners

Identifying binding partners, whether other proteins or small molecules, might also help to identify moonlighting proteins, because the development of a new binding site is one way by which moonlighting proteins are proposed to evolve. Micro-array technology can be used to measure the interaction of a protein with other proteins, small molecules, antibodies, or peptides *in vitro*. The yeast two-hybrid method is used to determine *in vivo* physical interactions between proteins; for example, in protein complexes or in signalling networks. In fact, yeast two-hybrid screens are notorious for identifying many 'false positive', protein–protein interactions that do not seem to be involved in the protein complex or signalling network being studied. Moonlighting proteins can provide a possible explanation for some of the apparent 'false positives' that are observed in yeast two-hybrid experiments.

4.6.3. Unexpected protein expression levels

An unusually high expression level of a particular protein might also suggest the protein has a second function. In general, many proteins in biochemical and signalling pathways are not needed at high concentrations in the cell because they can be used repeatedly, function in a highly specialised pathway, or are involved in a cascade that amplifies a signal. However, a second function, perhaps as part of a protein complex, might require much higher levels of protein expression. Delta-aminolevulinic acid dehydratase is expressed at levels far more than is needed to catalyse a step in heme biosynthesis (up to 1% of total soluble protein). However, the surprisingly high level of protein expression makes more sense when we consider that the protein is also a proteasome inhibitory subunit.

4.7. Moonlighting proteins in disease and rational drug design

Whereas the preceding discussion included many examples of proteins in basic physiological pathways, moonlighting proteins can also be important in studies of disease. Already, moonlighting proteins have been found to be involved in tumour cell motility, angiogenesis, DNA synthesis or repair, chromatin and cytoskeleton structure, and cystic fibrosis (reviewed in [34]). The ability of a protein to moonlight can complicate the elucidation of molecular mechanisms of disease, the identification of biomarkers of disease progression, and the development of novel therapeutics.

The presence of moonlighting proteins can complicate understanding of the molecular mechanisms of disease development. Even in the case where the key proteins involved have been identified, the observed role of a particular protein in disease progression might be difficult to explain if only one function of a moonlighting protein is known. In fact, even if both functions are known, the identification of a molecular mechanism by which the proteins cause the observed symptoms can be complicated, and it might not be clear which function (or both) is responsible for the observed symptoms.

Even in the case of a genetic disease caused by altered levels or activity of a single protein, the effects of mutations on one function of a protein – for example its enzymatic activity – might not be sufficient to explain the disease symptoms. Instead, the mutation might affect a second function, such as interactions with another protein. Even if all the functions of a protein are known, the function (or both) that is affected by a disease-causing mutation might be unclear. Similarly, moonlighting proteins can complicate the understanding of mutant phenotypes of model organisms developed from experimental methods that alter the level of expression of a protein, such as gene knockouts, RNA interference, anti-sense RNA or protein over-expression.

Some specific proteins whose expression levels differ between healthy and diseased cells can serve as biomarkers for diagnosis or for following the progression of the disease. However, if a protein moonlights, the presence or absence of a disease state might be only one of many factors that affect its expression level. In that case, the lack of a direct correlation between expression levels and disease state would prevent a moonlighting protein from being a good biomarker for the disease.

The ability of a protein to moonlight can also complicate selection of potential drug targets and the development of novel therapeutics to treat disease. It is important that a drug that alters a protein activity alters the correct activity. Modifying other protein activities not involved in the disease can result in increased toxicity and side effects. A review by Searls describes in more detail the importance of considering many potential mechanisms in the evolution of protein functions, including moonlighting, gene redundancy, orthology, paralogy, and crosstalk, in selecting a suitable target for drug development [35].

4.8. Word of caution

Although the preceding discussion emphasises that moonlighting might be quite common, it is important to consider two points of caution.

Although the presence of a protein in multiple cellular locations, multiple cell types, or multiprotein complexes, or the observation of unexpected results, can suggest that a protein is moonlighting, it is not a guarantee that the protein has multiple functions. Some single-function proteins are found in multiple locations or cell types because a single function might be used in both places, for example a kinase activity. A protein with the same function in two different locations is not a moonlighting function. Evidence that the protein truly does two different things in the two locations is needed, and so it is important that the initial observations are complemented with further biochemical characterisation or other studies before a protein is determined to be moonlighting. Combining the results of multiple experimental methods, such as biochemical assays of catalytic activity, yeast two-hybrid data, and mass spectrometric analysis of cellular location, would provide more evidence of multiple functions.

Another area of caution is in regards to assigning functions to proteins based on amino acid sequence homology. If one protein is a moonlighting protein, its homologues or other isoforms might have one, the other, or both functions. For example, the *E. coli* aspartate receptor is also the receptor for maltose binding protein; however, the homologous aspartate receptor from a related bacterium, *Salmonella typhimurium*, does not bind to maltose binding protein [36, 37]. Similarly, within one organism, one isoform of a protein might have multiple functions, whereas other isoforms might each have only one function. For example, three isoforms of lysyl hydroxylase (in collagen synthesis) share approximately 60% overall amino acid sequence identity, and all three have lysyl hydroxylase catalytic activity. However, only isoform 3 (LH3) also contains galactosylhydroxylysyl glucosyltransferase catalytic activity.

4.9. Conclusions

A variety of different proteins has been found to moonlight, with different functions, mechanisms to switch between functions, ways in which they can benefit an organism, and methods by which they might have evolved, and this variety suggests that many more proteins might moonlight. In general, identifying one function of a protein is not always followed by a search for additional functions of a protein, and although there are several types of experiments that can suggest that a protein moonlights, there is no general method to identify which additional proteins moonlight. However, moonlighting proteins might be a common mechanism of communication and cooperation between the many different functions and pathways within a complex modern cell or between different cell types within an organism, and they might help explain complex disease symptoms or unexpected phenotypes from gene knockout experiments in model organisms. Perhaps the identification of more moonlighting proteins might also help explain why the human genome encodes only approximately twice as many proteins as *S. cerevisiae*, a single-celled yeast.

4.10. Acknowledgements

Research on moonlighting proteins in the Jeffery laboratory is supported by grants from the American Cancer Society.

REFERENCES

1. Jeffery C J. Moonlighting proteins. Trends Biochem Sci 1999, 24: 8–11.
2. Lay A J, Jiang X-M, Kisker O, Flynn E, Underwood A, Condron R and Hogg P J. Phosphoglycerase kinase acts in tumour angiogenesis as a disulphide reductase. Nature 2000, 408: 869–873.

3. Xu W, Seiter K, Feldman E, Ahmed T and Chiao J W. The differentiation and maturation mediator for human myeloid leukemia cells shares homology with neuroleukin or phosphoglucose isomerase. Blood 1996, 87: 4502–4506.

4. Watanabe H, Takehana K, Date M, Shinozaki T and Raz A. Tumor cell autocrine motility factor is the neuroleukin/phosphohexose isomerase polypeptide. Cancer Res 1996, 56: 2960–2963.

5. Gurney M E, Apatoff B R, Spear G T, Baumel M J, Antel J P, Bania M B and Reder A T. Neuroleukin: a lymphokine product of lectin-stimulated T cells. Science 1986, 234: 574–581.

6. Gurney M E, Heinrich S P, Lee M R and Yin H S. Molecular cloning and expression of neuroleukin, neurotrophic factor for spinal and sensory neurons. Science 1986, 234: 566–574.

7. Furukawa T, Yoshimura A, Sumizawa T, Haraguchi M and Akiyama S-I. Angiogenic factor. Nature 1992, 356: 668.

8. Young J D, Lawrence A J, Maclean A G, Leung B P, McInnes I B, Canas B, Pappin D J C and Stevenson R D. Thymosin beta 4 sulfoxide is an anti-inflammatory agent generated by monocytes in the presence of glucocorticoids. Nat Med 1999, 5: 1424–1427.

9. Wu R R and Couchman J R. cDNA cloning of the basement membrane chondroitin sulfate proteoglycan core protein, bamacan: a five domain structure including coiled-coil motifs. J Cell Biol 1997, 136: 433–444.

10. Darwiche N, Freeman L A and Strunnikov A. Characterization of the components of the putative mammalian sister chromatid cohesion complex. Gene 1999, 233: 39–47.

11. Brix K, Summa W, Lottspeich F and Herzog V. Extracellularly occuring histone H1 mediates the binding of thyroglobulin to the cell surface of mouse macrophages. J Clin Invest 1998, 102: 283–293.

12. Soker S, Takashim S, Miao H Q, Neufeld G and Klagsbrun M. Neuropilin-1 is expressed by endothelial and tumor cells as an isoform-specific receptor for vascular endothelial growth factor. Cell 1998, 92: 735–745.

13. Chu E, Koeller D M, Casey J L, Drake J C, Chabner B A, Elwood P C, Zinn S and Allegra C J. Autoregulation of human thymidylate synthase messenger RNA translation by thymidylate synthase. Proc Natl Acad Sci USA 1991, 88: 8977–8981.

14. Barker D F and Campbell A M. Genetic and biochemical characterization of the birA gene and its product: evidence for a direct role of biotin holoenzyme synthetase in repression of the biotin operon in *Escherichia coli*. J Mol Biol 1981, 146: 469–492.

15. Ostrovsky de Spicer P and Maloy S. PutA protein, a membrane-associated flavin dehydrogenase, acts as a redox-dependent transcriptional regulator. Proc Natl Acad Sci USA 1993, 90: 4295–4298.

16. Kennedy M C, Mende-Mueller L, Blondin G A and Beiner H. Purification and characterization of cytosolic aconitase from beef liver and its relationship to the iron-responsive element binding protein. Proc Natl Acad Sci USA 1992, 89: 11730–11734.

17. Crennell S, Takimoto T, Portner A and Taylor G. Crystal structure of the multifunctional paramyxovirus hemagglutinin-neuraminidase. Nat Struct Biol 2000, 7: 1068–1074.

18. Citron B A, Davis M D, Milstien S, Gutierrez J, Mendel D B, Crabtree G R and Kaufman S. Identity of 4α-carbinolamine dehydratase, a component of the

phenylalanine hydroxylation system, and DCoH, a transregulator of homeodomain proteins. Proc Natl Acad Sci USA 1992, 89: 11891–11894.

19. Guo G G, Gu M and Etlinger J D. 240-kDa proteasome inhibitor (CF-2) is identical to delta aminolevulinic acid dehydratase. J Biol Chem 1994, 269: 12399–12402.

20. Wool I G. Extraribosomal functions of ribosomal proteins. Trends Biochem Sci 1996, 21: 164–165.

21. Zhu W, Rainville I R, Ding M, Bolus M, Heintz N H and Pederson D S. Evidence that the pre-mRNA splicing factor Clf1p plays a role in DNA replication in *Saccharomyces cerevisiae*. Genetics 2002, 160: 1319–1333.

22. Russell C S, Ben-Yehuda S, Dix I, Kupiec M and Beggs J D. Functional analyses of interacting factors involved in both pre-mRNA splicing and cell cycle progression in *Saccharomyces cerevisiae*. RNA 2000, 6: 1565–1572.

23. Ben-Yehuda S, Dix I, Russell C S, McGarvey M, Beggs J D and Kupiec M. Genetic and physical interactions between factors involved in both cell cycle progression and pre-mRNA splicing in *Saccharomyces cerevisiae*. Genetics 2000, 156: 1503–1517.

24. Chung S, McLean M R and Rymond B C. Yeast ortholog of the *Drosophila* crooked neck protein promotes spliceosome assembly through stable U4/U6.U5 snRNP addition. RNA 1999, 5: 1042–1054.

25. Gonzalez F, Delahodde A, Kodadek T and Johnstom S A. Recruitment of a 19S proteasome subcomplex to an activated promoter. Science 2002, 296: 548–550.

26. Picot D, Loll P J and Garavito R M. The X-ray crystal structure of the membrane protein prostaglandin H2 synthase-1. Nature 1994, 367: 243–249.

27. Heikkinen J, Risteli M, Wang C, Latvala J, Rossi M, Valtavaara M and Myllyla R. Lysyl hydroxylase 3 is a multifunctional protein possessing collagen glucosyltransferase activity. J Biol Chem 2000, 275: 36158–36163.

28. Stutts M J, Canessa C M, Olsen J C, Hamrick M, Cohn J A, Rossier B C and Boucher R C. CFTR as a cAMP-dependent regulator of sodium channels. Science 1995, 269: 847–850.

29. Suzuki C K, Rep M, van Dijl J M, Suda K, Grivell L A and Schatz G. ATP-dependent proteases that also chaperone protein biogenesis. Trends Biochem Sci 1997, 22: 118–123.

30. Piatigorsky J. Multifunctional lens crystallins and corneal enzymes. More than meets the eye. Ann NY Acad Sci 1998, 842: 7–15.

31. Ursini F, Heim S, Kiess M, Maiorino M, Roveri A, Wissing J and Flohe L. Dual function of the selenoprotein PHGPx during sperm maturation. Science 1999, 285: 1393–1396.

32. Mark D F and Richardson C C. *Escherichia coli* thioredoxin: a subunit of bacteriophage T7 DNA polymerase. Proc Natl Acad Sci USA 1976, 73: 780–784.

33. Jeffery C J, Bahnson B J, Chien W, Ringe D and Petsko G A. Crystal structure of rabbit phosphoglucose isomerase, a glycolytic enzyme that moonlights as neuroleukin, autocrine motility factor, and differentiation mediator. Biochemistry 1999, 39: 955–964.

34. Jeffery C J. Multifunctional proteins: examples of gene sharing. Ann Med 2003, 35: 28–35.

35. Searls D B. Pharmacophylogenomics: Genes, evolution and drug targets. Nature Rev Drug Discovery 2003, 2: 613–623.

36. Wolff C and Parkinson J S. Aspartate taxis mutants of the *Escherichia coli* tar chemoreceptor. J Bacteriol 1988, 170: 4509–4515.

37. Mowbray S L and Koshland D E J. Mutations in the aspartate receptor of *Escherichia coli* which affect aspartate binding. J Biol Chem 1990, 265: 15638–15643.

38. Tabor S, Huber H E and Richardson C C. *Escherichia coli* thioredoxin confers processivity on the DNA polymerase activity of the gene 5 protein of Bacteriophage T7. J Biol Chem 1987, 262: 16212–16223.

39. Cascalho M, Wong J, Steinberg C and Wabl M. Mismatch repair co-opted by hypermutation. Science 1998, 279: 1207–1210.

40. Thunnissen M M G M, Nordlunch P and Heggstrom J Z. Crystal structure of human leukotriene A4 hydrolase, a bifunctional enzyme in inflammation. Nat Struct Biol 2001, 8: 131–135.

41. Chen J-W, Dodia C, Feinstein S I, Jain M K and Fisher A B. 1-Cys peroxiredoxin, a bifunctional enzyme with glutathione peroxidase and phospholipase A2 activities. J Biol Chem 2000, 275: 28421–28427.

42. Numata O. Multifunctional proteins in Tetrahymena: 14-nm filament protein/citrate synthase and translation elongation factor-1 alpha. Int Rev Cytol 1996, 164: 1–35.

43. Modun B, Morrissey J and Williams P. The staphylococcal transferrin receptor: a glycolytic enzyme with novel functions. Trends Microbiol 2000, 8: 231–237.

44. Brew K, Vanaman T C and Hill R L. The role of alpha-lactalbumin and the A protein in lactose synthetase: a unique mechanism for the control of a biological reaction. Proc Natl Acad Sci USA 1968, 59: 491–497.

45. Bolduc J M, Spiegel P C, Chatterjee P, Brady K L, Downing M E, Caprara M C, Waring R B and Stoddard B L. Structural and biochemical analysis of DNA and RNA binding by a bifunctional homing endonuclease and group I intron splicing factor. Genes Develop 2003, 17: 2875–2888.

46. Caprara M G, Mohr G and Lambowitz A M. A tyrosyl-tRNA synthetase protein induces tertiary folding of the group I intron catalytic core. J Mol Biol 1996, 257: 512–531.

47. Lim M L, Lum M G, Hansen T M, Roucou X and Nagley P. On the release of cytochrome c from mitochondria during cell death signaling. J Biomed Sci 2002, 9: 488–506.

5

Molecular Chaperones: The Unorthodox View

Brian Henderson and Alireza Shamaei-Tousi

5.1. Introduction

Like a Brian Rix farce, in which the characters' identities are continuously changing, the functions of the class of protein known as molecular chaperones has been unfolding continuously over the past decade resulting in substantial confusion. However, like such farces, we are confident that the dénouement will be a complete surprise and will provide a new world picture of the processes with which molecular chaperones are involved. This short chapter aims to introduce the reader to the rapidly changing world of molecular chaperones as an aid to the reading of the rest of the chapters in this volume.

5.2. Molecular chaperones are protein folders

Our story starts with a huff and a puff with the study of the response of the polytene chromosomes of *Drosophila* to various stressors. This revealed novel patterns of specific chromosomal puffs, in response to heat, and a variety of other environmental stresses, representing the transcription of selected genes [1, 2]. The behaviour of cells exposed to various stresses became known as the heat shock response or the cell stress response and we now appreciate the very large number of environmental factors to which cells will respond in this stereotypical manner. The 'molecularisation' of the cell stress response occurred in the late 1980s with the pioneering work of Ellis and colleagues [3], who introduced both the concept of protein chaperoning and the term molecular chaperone. The enormous amount of work currently being carried out on the structural biology and molecular and cellular mechanisms of molecular chaperones has its genesis in this paper. The reader is referred to Chapter 1 in which John Ellis reviews the protein chaperoning function of molecular chaperones and warns of the pitfalls of incorrect definitions in relation to molecular chaperones and stress proteins.

5.3. Molecular chaperones are potent immunogens

Contemporaneously with the discovery of molecular chaperones as protein-folding 'machines' was the realisation that these proteins were potent immunogens involved in immune responses to infection [4] and also in autoimmunity [5]. The possibility of a connection between these two events was suggested by Irun Cohen [5], who continues this argument in Chapter 16. The discovery of the immune response to molecular chaperones was surprising because these proteins are highly conserved. Furthermore, human molecular chaperones such as chaperonin (Cpn) 10 and Cpn60 proteins can be considered to be bacterial molecules because the mitochondrion evolved from an α-proteobacterium [6]. The unexpected immunogenicity of molecular chaperones is presumably related to the capacity that these proteins have to activate myeloid cells. The interactions of molecular chaperones, ranging in mass from 8 kDa (ubiquitin) to 90 kDa (Hsp90), with myeloid and other cell types is detailed in many of the later chapters in this volume.

5.4. Molecular chaperones as moonlighting proteins

By the late 1980s and early 1990s the paradigm of molecular chaperones as intracellular proteins acting as 'catalysts' of protein folding was being forged [7]. Although this is clearly a major function for these proteins, evidence began to emerge that they have other intracellular and extracellular functions. The first line of evidence for this non-orthodox view of molecular chaperones was the finding of their presence on the surfaces of cells. For example, Cpn60 was identified on the cell surface of γδ T cells [8]. The cellular disposition of molecular chaperones is detailed in Chapter 2. The binding of molecular chaperones to membranes is a continuing motif in the literature. For example, GroESL oligomers have been shown to stabilise artificial membranes [9]. A report suggests that type II chaperonins in *Archaea* function primarily to stabilise cellular membranes [10]. One experimental finding that has not been followed up was the report that Cpn60 induced pores in membranes [11]. Perhaps the most interesting association of molecular chaperones with cell membranes is the finding that the receptor for the potent pro-inflammatory Gram-negative bacterial component, lipopolysaccharide (LPS), contains the heat shock proteins Hsp70 and Hsp90 [12, 13].

LPS is a major issue for those working on the non-folding functions of molecular chaperones (e.g., [14]) because many of the proteins being used are recombinant proteins made in *Escherichia coli*. In the past few years a number of papers have appeared suggesting that all of the actions of molecular chaperones are due to LPS contamination [15–17]. This problem is dealt with by a number of the authors and there can be few fields of study in which more care is taken with LPS contamination of recombinant proteins. The recent finding that the *Helicobacter pylori* Cpn60 protein activates macrophages by a mechanism that does not involve the LPS (TLR4) or bacterial lipopeptide (TLR2) receptors reveals that non-proteinaceous bacterial contaminants are unlikely to account for the biological activity of molecular chaperones [18]. However, the watchword has to be vigilance.

These reports certainly begin to suggest that molecular chaperones may have functions in addition to their protein-folding actions. Thus it is obvious that molecular chaperones can also be grouped into the widening pool of proteins with multiple functions and now known as moonlighting proteins. The concept of moonlighting proteins has been reviewed in Chapter 4.

5.4.1. The unfolding moonlighting functions of molecular chaperones

In the past decade a surprisingly large number of apparent non-folding functions have been ascribed to one or another of the molecular chaperones (Table 5.1) and some of these proteins have a number of different biological actions. These results are still controversial and the molecular chaperone field is divided into those that believe that the non-folding actions of molecular chaperones are artefactual and those that hold that they are part of the systems biology of the cell stress response. The classic example of the former position is the belief that the cytokine-inducing actions of molecular chaperones are due to contamination of these proteins with LPS. Some of the criticism (in this case balanced criticism) about the extracurricular actions of molecular chaperones is voiced in Chapter 1. It is assumed that similar criticisms were levelled at the findings that most of the glycolytic enzymes have moonlighting functions. The protein currently holding the prize for most extracurricular activity is phosphoglucoisomerase (PGI). Over the past 20 years this protein, which has a CXXC motif identical to that found in the molecular chaperone thioredoxin and in certain chemokines, has been independently identified as three different cytokines and an implantation factor. Thus this protein is also neuroleukin [41], autocrine motility factor [42], differentiation and maturation mediator [43] and an implantation factor [44]. Any criticism of these findings seems to have dissipated

Table 5.1. Non-folding actions of molecular chaperones

Proteins	Molecular mass (kDa)	Additional functions
Ubiquitin	8	Antibacterial activity [19]
Thioredoxin	12	ADF – a T cell cytokine [20]
		A novel chemoattractant [21]
		Chemokine inhibitor [22]
		Modulates glucocorticoid action [23]
Chaperonin 10	10 (oligomer)	Early pregnancy factor [24]
		Osteolytic factor [25]
α-Crystallin	18–20	Activates microglia [26]
Cyclophilins	∼ 20	Secretory pro-inflammatory macrophage product [27]
		Chemotactic activity [28]
		Parasite inducer of IL-12 [29]
Hsp27	27	Induces IL-10; anti-inflammatory [30]
Hsp60/Cpn60	60	Modulates myeloid cell and vascular endothelial cell function
Hsp70	70	Cytokine inducer (various chapters in volume) or inhibitor [31]
		Receptor for LPS [11, 12]
Bip	70	Negative regulator of inflammation [32, 33]
Hsp90	90	Immunomodulator acting to present peptides to T lymphocytes
Grp94/Gp96	96	As above
		A cell surface receptor for Gram-negative bacteria [34, 35]
		A receptor for bacterial invasion [34, 35]
		A factor involved in surface TLR expression [36]
		A direct ligand for activating cells [37–40]

and they are now part of the mainstream of the biochemistry and cellular biology of glycolysis.

5.4.2. Moonlighting actions of individual molecular chaperones

The following discussion will briefly deal with the reported moonlighting actions of molecular chaperones and will deal with them in terms of increasing molecular mass (Table 5.1). Much of this information is dealt with more extensively elsewhere in this volume.

Ubiquitin: This is an 8.5-kDa intracellular protein involved in the controlled degradation of proteins and thus just comes under the remit of molecular chaperone. It has recently been discovered that this protein has antibacterial actions [19] and it therefore joins the multitude of proteins and peptides that function to defend us against bacteria.

Thioredoxin: This is a 12-kDa redox protein with a CXXC motif, which acts intracellularly as a hydrogen donor to ribonucleotide reductase. It has been known since the late 1980s that thioredoxin is a secreted cytokine [20], termed adult T cell-leukaemia-derived factor, with autocrine growth properties on T lymphocytes. Thioredoxin is found in the serum in normal individuals [45]. In addition to acting on T cells, thioredoxin is a unique chemoattractant with a different mechanism of action to the large family of chemotactic cytokines known as the chemokines [21]. Surprisingly, in spite of being identified as a chemoattractant, thioredoxin can also block cellular responses to LPS by suppressing the activity of known chemokines [22]. Indeed, circulating levels of thioredoxin appear to be important in AIDS, and it has been proposed that high levels of this molecular chaperone in the blood of HIV-infected individuals with low CD4$^+$ T cell counts directly impair survival by blocking pathogen-induced chemotaxis and thus prevent myeloid cell defences crucial for survival [46]. This chemotaxis suppressing activity of thioredoxin is likely to have a therapeutic effect and one study has shown that this molecular chaperone can block experimental inflammatory or fibrotic lung injury [47]. Finally, thioredoxin has also been shown to be involved in control of glucocorticoid action at the level of glucocorticoid-inducible gene expression. This interaction reveals a link between cellular and physiological stress responses [23]. The authors suggest that the homeostatic control of the multicellular organisms must require a link between the cellular stress responses and the physiological (organismal) stress response.

Chaperonin 10/Hsp10/early pregnancy factor: The fetus is equivalent to an allograft because the mother and fetus will generally express a different profile of major histocompatibility antigens. Thus an obvious question is how is fetal rejection controlled? Almost 30 years ago an immunosuppressive factor was identified in the sera of pregnant mothers and was termed early pregnancy factor (EPF) [48]. Significant efforts were made to identify EPF; however, it took until 1991 for the suggestion to be made that EPF was actually thioredoxin [49]. A second group identified EPF as chaperonin (Cpn) 10 (Hsp10) [50], a 10-kDa protein which forms a heptameric structure that interacts with Cpn60 to promote protein folding (Cpn10 is a co-chaperone). This finding raised an enormous amount of interest and criticism [51]. However, the recent cloning and expression of human EPF (Cpn10/Hsp10) in eukaryotic cells and in *E. coli* has revealed that the recombinant Cpn10 has EPF activity both *in vitro* and *in* *vivo* and that activity depended upon the presence of appropriate N-terminal modification [52]. The studies that suggested EPF was thioredoxin have not been repeated.

Another strand of this story comes from the work of Coates and colleagues, which is described to a limited extent in Chapter 6. Coates was the first to clone and express the *cpn10* gene of *Mycobacterium tuberculosis* [53] and had shown that this protein inhibited inflammation in both adjuvant arthritis in the rat [54] and experimental allergic asthma in the mouse [55]. In the former study, the activity of the whole molecule could be replicated by synthetic N-terminal peptides that are free of LPS.

Tuberculosis of the bone causes major damage and the *M. tuberculosis* Cpn10 was also found to be a potent inducer of bone resorption and the major osteolytic component of this organism [25]. Using synthetic peptides the active site in *M. tuberculosis* Cpn10 has been identified as the mobile loop [25]. The group that had identified EPF as Cpn10 has subsequently shown that human Cpn10 is able to inhibit inflammation in animals with experimental allergic encephalomyelitis, a much-used model of autoimmunity [56, 57]. Here is the first evidence of a molecular chaperone acting as a secreted hormone or cytokine and able to inhibit immune/inflammatory responses in a key process (pregnancy) [24]. Much more information is required before we can fully understand the role played by Cpn10 in the control of the early phase of pregnancy.

α-**Crystallin:** This is a member of the small heat shock protein family – proteins of approximate molecular mass of 20 kDa that form extremely large aggregates (see Chapter 1). It is reported that this protein activates microglial cells [26], which are the myeloid cell population in the brain with major roles in brain defences against infection.

Cyclophilins: A family of proteins (>30 genes in the human genome) with peptidyl-prolyl isomerase activity and a capacity to bind the cyclic peptide immunosuppressants such as cyclosporine. It has been reported that cyclophilin is secreted by LPS-activated macrophages and has chemotactic activity [27, 28]. Members of this protein family have been found in biological fluids including human milk [58, 59] and blood [59]. Elevated levels are also found in the synovial fluid of patients with rheumatoid arthritis [60] and in patients with sepsis [61]. *Toxoplasma gondii* is a protozoan responsible for toxoplasmosis in humans. It has been reported that this eukaryotic parasite releases a potent IL-12–stimulating protein which has recently been identified as C-18 cyclophilin. This protein activates dendritic cells (DCs) by binding to the CCR5 receptor [29]. These findings emphasise the diversity of the interactions that can occur between molecular chaperones and the chemokine system of cytokines.

Hsp27: The literature on this molecular chaperone is reviewed by Miller-Graziano in Chapter 13. The key observation is that exposure of human monocytes to Hsp27 induces the production of the anti-inflammatory cytokine, IL-10 [30]. This suggests that Hsp27 may have anti-inflammatory functions. The ability of extracellular molecular chaperones to act as inhibitors of immunity and inflammation appears to be a theme. Such anti-inflammatory actions of molecular chaperones rules out the possibility that the functions being described are due to LPS contamination.

Cpn60: There is now very good evidence that this molecule is a stimulator of a range of cells, including myeloid cells, vascular endothelial cells and epithelial cells, and this literature has been reviewed by Coates in Chapter 6 and the topic is also touched on in other chapters. Two issues will be addressed in this brief section. The first is the nature of the receptor for this protein. There appears to be a range to choose from, including CD14, TLR2 and TLR4 [62, 63]. However, a number of the Cpn60 proteins tested do not appear to bind to any of these receptors (e.g., [18]). The simplest explanation for this is that cells can discriminate between Cpn60 proteins from different species. The most striking demonstration of the ability to recognise differences in Cpn60 proteins is the finding that the salivary symbiont (*Enterobacter aerogenes*) of the insect predator known as the antlion produces a neurotoxin used by the insect in catching its prey. This neurotoxin turns out to be the Cpn60 protein of this bacterium. Strikingly, single-residue changes in the *E. coli* equivalent protein, GroEL, turn this best-studied of molecular chaperones into a potent insect neurotoxin [64]. The possible consequences of this will be discussed at the end of the chapter.

More recently it has been shown that eukaryotic Cpn60 interacts with Bax and Bak. These two are pro-apoptotic cytolysis proteins that stimulate the release of cytochrome c and apoptosis. Binding to Cpn60 may regulate the activity of these two pro-apoptotic proteins by preventing them from oligomerising and inserting into the mitochondrial membrane [65, 66]. This action resembles the role that Hsp90 has in the normal cell [67, 68]. The second issue is the presence of Cpn60 (Hsp60) in the blood of humans. The protein, which is N-terminally recognised and processed in mitochondria, can also be found on the surface of endothelial cells and macrophages [69]. The authors have established that the levels of Hsp60 in the blood of a population of normal individuals (healthy British civil servants) are stratified into three groups: (i) those below assay detection; (ii) those with measurable, but low levels and (iii) those with extremely high, biologically active, circulating levels. The latter can be in the hundreds of micrograms per millilitre range. There is no explanation for this stratification of plasma Hsp60 concentrations and these results imply a mechanism of production and/or release, or a mechanism of removal of Hsp60 that differs enormously within the normal human population. Preliminary data fail to support the hypothesis that the difference in levels is due to differences in transcriptional rates [70]. However, it has been shown that elevated levels of Hsp60 in the blood of healthy individuals could be associated with an unfavourable lipid profile, high TNF-α levels and low socioeconomic status. TNF-α plays an important role in atherogenesis and the development of acute coronary syndromes [71], and low socioeconomic status and social isolation have been related to chronic heart diseases [72]. This will be dealt with in more detail in Chapter 12. Some questions and speculations about human Cpn60 are detailed in Figure 5.1.

Hsp70: The human genome sequence has revealed that *Homo sapiens* is in possession of 13 hsp70 genes and thus one has to be careful when reviewing the literature on Hsp70 and its biological actions and receptors that one compares apples with apples and not with oranges. Other chapters in this volume (Chapters 7, 8, 9, 10 and 14) deal with aspects of a number of Hsp70 family proteins, including Bip. There is controversy in the literature about the receptor(s) used by exogenous Hsp70 to activate cells (see Chapter 10). There is also controversy about the nature of the signal induced by peptide-free Hsp70. Most studies of the human or mycobacterial Hsp70 show an activation of myeloid cell cytokine synthesis (see Chapters 7–10). However, in a recent report, *M. tuberculosis* Hsp70 has been shown to induce the production of the anti-inflammatory cytokine IL-10 and reduce the production of TNF-α [31].

Bip: This is another member of the Hsp70 protein family which is located in the lumen of the endoplasmic reticulum and was originally identified as an immunoglobulin heavy chain-binding protein. Transcription of Bip (also known as glucose-regulated protein (Grp) 78) is enhanced when the glucose concentration is lowered. The transcription of a number of molecular chaperones is regulated by environmental glucose levels and it is interesting to speculate why this evolved. For example, hypoglycaemia is one response to infection and to the key Gram-negative inflammogen, LPS [73]. As Corrigal and Panayi review in Chapter 14, Bip is now known to be a potent negative regulator of inflammation acting, like Hsp27, as an inducer of the anti-inflammatory cytokine IL-10 [32]. Bip also induces the production of soluble TNF receptor II and IL-1 receptor antagonist, inhibits the recall antigen response by peripheral blood mononuclear cells (PBMCs) to tuberculin purified protein derivative

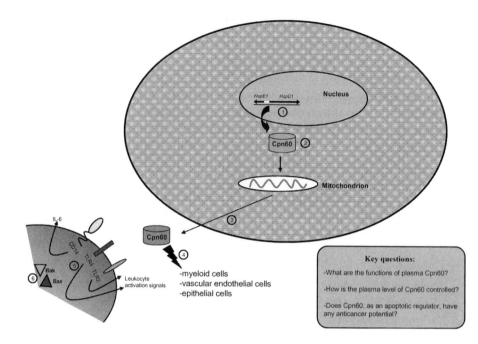

① The *Hsp60* gene, also called *HSPD1*, is linked head to head with the *Hsp10* (*HSPE1*) gene on chromosome 2. These genes are separated by a bidirectional promoter with two-fold greater transcriptional activity in the Hsp60 direction compared to the Hsp10 direction.

② Hsp60 is synthesized as a precursor with an N-terminal mitochondrial targeting sequence, or mitochondrial import peptide (MIP).

③ While Cpn60 is recognised to be mainly a mitochondrial protein, some proportion can also be found at discrete extramitochondrial sites such cell surfaces and circulation.

④ Cpn60 is known to activate myeloid cells, vascular endothelial cells and epithelial cells.

⑤ Soluble Cpn60 activates leukocytes, releasing TNF-α, nitric oxide and IL-6 via TLR2, TLR4 or CD14 receptors.

⑥ By binding to pro-apoptotic Bax and Bak, Cpn60 may have a regulatory role in apoptotic process in the normal cell.

Figure 5.1. Model for the actions of Cpn60 inside and outside of cells: some questions and speculations about mitochondrial Cpn60 (Hsp60), a protein that has been shown to have an increasing range of biological actions.

and down-regulates CD86 and HLA-DR expression of human PBMCs [33]. An obvious hypothesis is that inflammation results in lowered circulating glucose levels and that this triggers the transcription of a number of molecular chaperones, at least one of which, Bip, has anti-inflammatory properties. Is this a homeostatic regulatory network? Further work is required to determine if glucose-regulated molecular chaperones play a role in regulating immune responses.

Hsp90: This is a major cytoplasmic heat shock protein family with a growing range of key cellular functions – notably the ability to render cryptic protein gene polymorphisms that damage protein function [74]. Much of the current excitement about Hsp90 centres on its ability to bind peptides and present them to cytotoxic T cells, and it is believed that this activity will be the basis of a novel and very effective treatment for cancer [75]. In addition, Hsp90 is part of the LPS receptor complex and thus participates in the innate immune response to a key pathogen-associated molecular pattern [12, 18]. Related to this is the finding that the Hsp90 homologue, gp96 (also known as glucose-regulated protein (Grp) 94) is involved in the folding of Toll-like receptors (TLR).

Inactivation of the gene encoding gp96 results, not as one would expect – in cell death under stressed conditions – but in the inability to express Toll-like receptors on the cell surface. Thus the description of gp96 as a cell stress protein may be inaccurate [36]. While putatively a protein of the endoplasmic reticulum, gp96 is also found on cell surfaces throughout the vertebrate phylum and this surface expression is up-regulated by exposure of cells to bacteria and their constituents [76]. This cell surface targeting of gp96 is involved in the maturation of DCs [77]. Indeed, it turns out that gp96 is a receptor for one of the outer membrane proteins (OmpA) of *E. coli* and is involved in the ability of this bacterium to invade eukaryotic cells [34, 35]. In addition to being an infection-related receptor, soluble gp96 is also able to activate leukocytes through binding to receptors including the TLRs [37–39] and gp96 appears to act as a Th2-specific co-stimulatory molecule [40]. The binding of gp96 (with associated peptides) to cells utilises the cell surface 'receptor' CD91 [78]. However, there appears to be a price to pay for the surface expression of gp96 in terms of the induction of autoimmunity [79].

This quick trawl through the literature reveals that many molecular chaperones have a range of actions in addition to the proposed main function as protein-folding proteins. A good example of the changing paradigm in molecular chaperone biology is the finding that the intracellular pathogenic bacteria known as *Chlamydia* contain three *cpn60* genes. This is not unusual, as certain members of the Rhizobia can contain up to six *cpn60* genes. What is striking is that two of the Chlamydial genes encode proteins that have lost the ATPase domain required for the protein-folding ability of these molecules [80]. Confirmation of this loss of protein-folding activity is shown by the inability of the genes encoding these two aberrant proteins to complement an *E. coli* GroEL mutant [80]. These are not some form of pseudo-gene, because it has been shown that all three proteins are independently expressed by *Chlamydia* [81]. Thus, here is a situation in which cells have evolved proteins that look like Cpn60 molecules but which have no capacity to fold proteins. Can these be referred to as non-folding molecular chaperones? The key question is – what biological role do these non-folding chaperones have? Perhaps these proteins are not moonlighting and have completely different roles to play from the assumed protein-folding role.

5.5. A physiological role for molecular chaperones

As reviewed in Chapter 1, molecular chaperones, through their protein-folding and chaperoning actions, are vital for homeostatic cell function. The new paradigm that is the subject of most of this book is that some, if not all, molecular chaperones have additional non-folding actions and that these contribute to cell–cell signalling involved in homeostatic regulation of the organism. The necessary supporting evidence for this paradigm is that molecular chaperones must be found at the surface of cells and/or in the extracellular fluid in order to be able to transmit a signal from cell to cell. There is now substantial evidence that many of the molecular chaperones are found in the blood and there is emerging evidence that these proteins are found in other body fluids such as saliva [82] and seminal fluid [83]. Indeed, it must also be remembered that one molecular chaperone, clusterin (apolipoprotein J) is normally present in body fluids [84]. The extracellular disposition of molecular chaperones and the relationship that this might have to disease processes is described in detail in Chapter 12.

If molecular chaperones are found in the extracellular milieu, then this raises the thorny point about how they get out of the cell. Many have dismissed the idea that molecular chaperones could function as intercellular signalling molecules because there is no known mechanism to account for their secretion from cells. This is actually a poor argument, because for many years we had no idea how many key signalling proteins such as IL-1, IL-16, IL-18, fibroblast growth factor and annexin, to name but a few, were released from cells. As described in Chapter 3, a pathway that has been elucidated over the past decade or more – the leaderless secretory protein pathway – is responsible for the release of the aforementioned proteins and also of the molecular chaperone thioredoxin. It is now known that the glycolytic moonlighting protein, PGI, is secreted via this pathway. Other novel pathways of protein secretion also exist (e.g., [85]) and hence the absence of a mechanism for secreting molecular chaperones cannot be used as an argument that they cannot be released from cells. Discovering the mechanisms of molecular chaperone secretion should be a priority for those interested in the biology of these proteins.

So if we accept that molecular chaperones are released from cells, what do they do? It is now appreciated that we have no shortage of receptors for molecular chaperones such as Cpn60, Hsp70, Hsp90 and cyclophilin with experimental evidence identifying CD14, TLR2, TLR4, LOX1, CD40, CD91 and CCR5 as receptors for these various molecular chaperones. Of course, we do not have identifiable receptors for the other molecular chaperones. However, it is now clear that the various molecular chaperones described can bind to and activate a wide range of cells.

The authors propose that there are four major functions of extracellular molecular chaperones.

1. The first is as signals warning the multicellular organism that certain of its constituents are under stress and modulating the function of nearby cells in case the stress continues or expands. The biological response of cells of innate immunity to stress proteins may overlap with the concept of proteinaceous danger signals as postulated by Matzinger [86]. This warning signal function was termed stress broadcasting in a recent review [87]. It is not clear whether molecular chaperones in the circulation are part of this stress broadcasting mechanism or are involved in linking cellular stress to higher order systems control.

2. The second is as a physiological input signal to the immune system in the form of pro- and anti-inflammatory molecular chaperones which may better be called *stress* cytokines. The pro- and anti-inflammatory actions of molecular chaperones have been identified in earlier sections and will not be discussed further; they are also discussed elsewhere in this volume. This is a novel but testable hypothesis that we hope will be explored in the near future.

3. Increasing evidence exists that another function of molecular chaperones is to provide adjuvant-like signals through the ability of molecules like Hsp70 and Hsp90 to present peptides to antigen-presenting cells. This activity may be a key feature of Matzinger's danger model and is an obvious foundation for a novel therapy for cancer. This is reviewed in detail in Chapters 17 and 18.

4. This is extremely speculative and springs from the findings of the last decade or so that vertebrates live with a very large number of bacterial species. It is estimated that *H. sapiens* have 2–3,000 bacterial species as their constant companions. London Zoo only keeps about 6–700 species of animal. Contrast this with the 40 or 50 bacteria that cause human disease [88]. How do we discriminate between these friendly bacteria and the ones that mean us harm? We propose that one set of signals is the molecular chaperones. There is little evidence for this hypothesis as yet, other than the finding that many receptors recognise molecular chaperones and the work on *E. aerogenes*, which reveals that single nucleotide changes in GroEL can dramatically alter the biological actions of this protein [64].

As these speculations were being put onto paper, another idea emerged from the literature. It is now suggested that Cpn60 is involved in sperm capacitation, a key event required for fertilisation [89].

5.6. Conclusions

John Ellis, a pioneer in the study of molecular chaperones, has thrown down a gauntlet in Chapter 1 with the statement 'This view [that molecular chaperones have non-folding roles] has not found general acceptance, partly because it is novel [as was the concept of proteins folding proteins [90]] and partly because of the paucity of high-quality evidence compared with that available in support of the protein folding paradigm'. The authors would argue that there is now a large amount of 'high-quality' evidence in terms of papers in *Nature*, the *Journal of Experimental Medicine*, the *Journal of Clinical Investigation*, the *Journal of Immunology* and *International Immunology* to name but a few, which supports the hypothesis of the non-folding actions of molecular chaperones. As Sherlock Holmes was want to say 'when you have excluded the possible then the impossible must be true'.

Acknowledgements

The authors are grateful to the Arthritis Research Campaign (programme grant HO600) and to the British Heart Foundation (PG/03/029) for financial support.

REFERENCES

1. Ritossa F A. A new puffing pattern induced by temperature shock and DNP in *Drosophila*. Experientia 1962, 18: 571–573.
2. Ashburner M. Pattern of puffing activity in the salivary gland chromosomes of *Drosophila*. V. Response to environmental treatments. Chromosoma 1970, 31: 356–376.
3. Hemmingsen S M, Woolford C, van der Vies S M, Tilly K, Dennis D T, Georgopoulos G C, Hendrix R W and Ellis R J. Homologous plant and bacterial proteins chaperone oligomeric protein assembly. Nature 1988, 333: 330–334.
4. Young D B, Ivanyi J, Cox J H and Lamb J R. The 65kDa antigen of mycobacteria – a common bacterial protein? Immunol Today 1987, 8: 215–219.
5. Cohen I R and Young D B. Autoimmunity, microbial immunity and the immunological homunculus. Immunol Today 1991, 12: 105–109.
6. Horner D S, Hirt R P, Kilvington S, Lloyd D and Embley T M. Molecular data suggest an early acquisition of the mitochondrion endosymbiont. Proc Royal Soc London B Biol Sci 1996, 263: 1053–1059.
7. Hartl F U and Hayer-Hartl M. Molecular chaperones in the cytosol: from nascent chain to folded protein. Science 2002, 295: 1852–1858.

8. Fisch P, Malkovsky M, Kovats S, Sturm E, Braakman E, Klein B S, Voss S D, Morrissey L W, DeMars R, Welch W J, Bolhuis R L H and Sondel P M. Recognition by human Vγ9/Vδ2 T cells of a GroEL homolog on Daudi Burkitt's lymphoma cells. Science 1990, 250: 1269–1273.

9. Torok Z, Horvath I, Goloubinoff P, Kovacs E, Glatz A, Balogh G and Vigh L. Evidence for a lipochaperonin: association of active protein-folding GroESL oligomers with lipids can stabilize membranes under heat shock conditions. Proc Natl Acad Sci USA 1997, 94: 2192–2197.

10. Trent J D, Kagawa H K, Paavola C D, McMillan R A, Howard J, Jahnke L, Lavin C, Embaye T and Henze C E. Intracellular localization of a group II chaperonin indicates a membrane-related function. Proc Natl Acad Sci USA 2003, 100: 15589–15594.

11. Alder G M, Austen B M, Bashford C L, Mehkert A and Pasternak C A. Heat shock proteins induce pores in membranes. Biosci Reports 1990, 10: 509–518.

12. Triantafilou K, Triantafilou M and Dedrick R L. A CD14-independent LPS receptor cluster. Nat Immunol 2001, 2: 338–344.

13. Triantafilou K, Triantafilou M, Ladha S, Mackie A, Dedrick R L, Fernandez N and Cherry R. Fluorescence recovery after photobleaching reveals that LPS rapidly transfers from CD14 to Hsp70 and Hsp90 on the cell membrane. J Cell Sci 2001, 114: 2535–2545.

14. Bausinger H, Lipsker D, Ziylan U, Manie S, Briand J P, Cazenave J P, Muller S, Haeuw J F, Ravanat C, de la Salle H and Hanau D. Endotoxin-free heat-shock protein 70 fails to induce APC activation. Eur J Immunol 2002, 32: 3708–3713.

15. Gao B and Tsan M F. Induction of cytokines by heat shock proteins and endotoxin in murine macrophages. Biochem Biophys Res Commun 2004, 317: 1149–1154.

16. Gao B and Tsan M F. Endotoxin contamination in recombinant human Hsp70 preparation is responsible for the induction of TNFα release by murine macrophages. J Biol Chem 2003, 278: 174–179.

17. Gao B and Tsan M F. Recombinant human heat shock protein 60 does not induce the release of tumor necrosis factor α from murine macrophages. J Biol Chem 2003, 278: 22523–22529.

18. Gobert A P, Bambou J C, Werts C, Balloy V, Chignard M, Moran A P and Ferrero R L. Helicobacter pylori heat shock protein 60 mediates interleukin-6 production by macrophages via a Toll-like receptor (TLR)-2-, TLR-4-, and myeloid differentiation factor 88-independent mechanism. J Biol Chem 2004, 279: 245–250.

19. Metz-Boutigue M H, Kieffer A E, Goumon Y and Aunis D. Innate immunity: involvement of new neuropeptides. Trends Microbiol 2003, 11: 585–592.

20. Tagaya Y, Maeda Y, Mitsui A, Kondo N, Matsui H, Hamuro J, Brown N, Arai K, Yokota T and Wakasugi H. ATL-derived factor (ADF), an IL-2 receptor/Tac inducer homologous to thioredoxin; possible involvement of dithiol-reduction in the IL-2 receptor induction. EMBO J 1989, 8: 757–764.

21. Bertini R, Howard O M, Dong H F, Oppenheim J J, Bizzarri C, Sergi R, Caselli G, Pagliei S, Romines B, Wilshire J A, Mengozzi M, Nakamura H, Yodoi J, Pekkari K, Gurunath R, Holmgren A, Herzenberg L A, Herzenberg L A and Ghezzi P. Thioredoxin, a redox enzyme released in infection and inflammation, is a unique chemoattractant for neutrophils, monocytes, and T cells. J Exp Med 1999, 189: 1783–1789.

22. Nakamura H, Herzenberg L A, Bai J, Araya S, Kondo N, Nishinaka Y, Herzenberg L A and Yodoi J. Circulating thioredoxin suppresses lipopolysaccharide-induced neutrophil chemotaxis. Proc Natl Acad Sci USA 2001, 98: 15143–15148.

23. Makino Y, Okamoto K, Yoshikawa N, Aoshima M, Hirota K, Yodoi J, Umesono K, Makino I and Tanaka H. Thioredoxin: a redox-regulating cellular cofactor for glucocorticoid hormone action. Cross talk between endocrine control of stress response and cellular antioxidant defense system. J Clin Invest 1996, 98: 2469–2477.

24. Morton H. Early pregnancy factor: an extracellular chaperonin 10 homologue. Immunol Cell Biol 1998, 76: 483–496.

25. Meghji S, White P, Nair S P, Reddi K, Heron K, Henderson B, Zaliani A, Fossati G, Mascagni P, Hunt J F, Roberts M M and Coates A R. *Mycobacterium tuberculosis* chaperonin 10 stimulates bone resorption: a potential contributory factor in Pott's disease. J Exp Med 1997, 186: 1241–1246.

26. Bhat N R and Sharma K K. Microglial activation by the small heat shock protein, α-crystallin. Neuroreport 1999, 10: 2869–2873.

27. Sherry N, Yarlett A, Strupp A and Cerami A. Identification of cyclophilin as a proinflammatory secretory product of lipopolysaccharide-activated macrophages. Proc Natl Acad Sci USA 1992, 89: 3511–3515.

28. Xu Q, Lefeva M C, Fischkoff S A, Handschumacher R E and Lyttle C R. Leukocyte chemotactic activity of cyclophilin. J Biol Chem 1992, 267: 11968–11971.

29. Aliberti J, Valenzuela J G, Carruthers V B, Hieny S, Andersen J, Charest H, Reis e Sousa C, Fairlamb A, Ribeiro J M and Sher A. Molecular mimicry of a CCR5 binding-domain in the microbial activation of dendritic cells. Nat Immunol 2003, 4: 485–490.

30. De A K, Kodys K M, Yeh B S and Miller-Graziano C. Exaggerated human monocyte IL-10 concomitant to minimal TNF-α induction by heat-shock protein 27 (Hsp27) suggests Hsp27 is primarily an anti-inflammatory stimulus. J Immunol 2000, 165: 3951–3958.

31. Detanico T, Rodrigues L, Sabritto A C, Keisermann M, Bauer M E, Zwickey H and Bonorino C. Mycobacterial heat shock protein 70 induces interleukin-10 production: immunomodulation of synovial cell cytokine profile and dendritic cell maturation. Clin Exp Immunol 2004, 135: 336–342.

32. Corrigall V M, Bodman-Smith M D, Fife M S, Canas B, Myers L K, Wooley P, Soh C, Staines N A, Pappin D J, Berlo S E, van Eden W, van der Zee R, Lanchbury J S and Panayi G S. The human endoplasmic reticulum molecular chaperone BiP is an autoantigen for rheumatoid arthritis and prevents the induction of experimental arthritis. J Immunol 2001, 166: 1492–1498.

33. Corrigall V M, Bodman-Smith M D, Brunst M, Cornell H and Panayi G S. Inhibition of antigen-presenting cell function and stimulation of human peripheral blood mononuclear cells to express an anti-inflammatory cytokine profile by the stress protein BiP: Relevance to the treatment of inflammatory arthritis. Arthritis Rheum 2004, 50: 1164–1171.

34. Prasadarao N V, Srivastava P K, Rudrabhatla R S, Kim K S, Huang S H and Sukumaran S K. Cloning and expression of the *Escherichia coli* K1 outer membrane protein A receptor, a gp96 homologue. Infect Immun 2003, 71: 1680–1688.

35. Khan N A, Shin S, Chung J W, Kim K J, Elliott S, Wang Y and Kim K S. Outer membrane protein A and cytotoxic necrotizing factor-1 use diverse signaling mechanisms

for *Escherichia coli* K1 invasion of human brain microvascular endothelial cells. Microb Pathogenesis 2003, 35: 35–42.

36. Randow F and Seed B. Endoplasmic reticulum chaperone gp96 is required for innate immunity but not cell viability. Nat Cell Biol 2001, 3: 891–896.

37. Panjwani N N, Popova L and Srivastava P K. Heat shock proteins gp96 and hsp70 activate the release of nitric oxide by APCs. J Immunol 2002, 168: 2997–3003.

38. Vabulas R M, Braedel S, Hilf N, Singh-Jasuja H, Herter S, Ahmad-Nejad P, Kirschning C J, Da Costa C, Rammensee H G, Wagner H and Schild H. The endoplasmic reticulum-resident heat shock protein Gp96 activates dendritic cells via the Toll-like receptor 2/4 pathway. J Biol Chem 2002, 277: 20847–20853.

39. Radsak M P, Hilf N, Singh-Jasuja H, Braedel S, Brossart P, Rammensee H G and Schild H. The heat shock protein Gp96 binds to human neutrophils and monocytes and stimulates effector functions. Blood 2003, 101: 2810–2815.

40. Banerjee P P, Vinay D S, Mathew A, Raje M, Parekh V, Prasad D V, Kumar A, Mitra D and Mishra G C. Evidence that glycoprotein 96 (B2), a stress protein, functions as a Th2-specific costimulatory molecule. J Immunol 2002, 169: 3507–3518.

41. Chaput M, Claes V, Portetelle D, Cludts I, Cravador A, Burny A, Gras H and Tartar A. The neurotrophic factor neuroleukin is 90% homologous with phosphohexose isomerase. Nature 1988, 332: 454–455.

42. Watanabe H, Takehana K, Date M, Shinozaki T and Raz A. Tumor cell autocrine motility factor is the neuroleukin/phosphohexose isomerase polypeptide. Cancer Res 1996, 56: 2960–2963.

43. Xu W, Seiter K, Feldman E, Ahmed T and Chiao J W. The differentiation and maturation mediator for human myeloid leukemia cells shares homology with neuroleukin or phosphoglucose isomerase. Blood 1996, 87: 4502–4506.

44. Schulz L C and Bahr J M. Glucose-6-phosphate isomerase is necessary for embryo implantation in the domestic ferret. Proc Natl Acad Sci USA 2003, 100: 8561–8566.

45. Kogaki H, Fujiwara Y, Yoshiki A, Kitajima S, Tanimoto T, Mitsui A, Shimamura T, Hamuro J and Ashihara Y. Sensitive enzyme-linked immunosorbent assay for adult T-cell leukemia-derived factor and normal value measurement. J Clin Lab Anal 1996, 10: 257–261.

46. Nakamura H, De Rosa S C, Yodoi J, Holmgren A, Ghezzi P, Herzenberg L A and Herzenberg L A. Chronic elevation of plasma thioredoxin: inhibition of chemotaxis and curtailment of life expectancy in AIDS. Proc Natl Acad Sci USA 2001, 98: 2688–2693.

47. Hoshino T, Nakamura H, Okamoto M, Kato S, Araya S, Nomiyama K, Oizumi K, Young H A, Aizawa H and Yodoi J. Redox-active protein thioredoxin prevents proinflammatory cytokine- or bleomycin-induced lung injury. Am J Respir Crit Care Med 2003, 168: 1075–1083.

48. Morton H, Rolfe B and Clunie G J. An early pregnancy factor detected in human serum by the rosette inhibition test. Lancet 1977, 1(8008): 394–397.

49. Clarke F M, Orozco C, Perkins A V, Cock I, Tonissen K F, Robins A J and Wells J R. Identification of molecules involved in the 'early pregnancy factor' phenomenon. J Reprod Fertil 1991, 93: 525–539.

50. Cavanagh A C and Morton H. The purification of early-pregnancy factor to homogeneity from human platelets and identification as chaperonin 10. Eur J Biochem 1994, 222: 551–560.

51. Lash G E and Legge M. Early pregnancy factor: an unresolved molecule. J Assist Reprod Genet 1997, 14: 495–496.

52. Somodevilla-Torres M J, Morton H, Zhang B, Reid S and Cavanagh A C. Purification and characterisation of functional early pregnancy factor expressed in Sf9 insect cells and in *Escherichia coli*. Protein Expr Purif 2003, 32: 276–287.

53. Atkins D, al Ghusein H, Prehaud C and Coates A R M. Overproduction and purification of *Mycobacterium tuberculosis* chaperonin 10. Gene 1994, 150: 145–148.

54. Ragno S, Winrow V R, Mascagni P, Lucietto P, Di Pierro F, Morris C J and Blake D R. A synthetic 10-kD heat shock protein (hsp10) from *Mycobacterium tuberculosis* modulates adjuvant arthritis. Clin Exp Immunol 1996, 103: 384–390.

55. Riffo-Vasquez Y, Spina D, Page C, Tormay P, Singh M, Henderson B and Coates A R M. Effect of *Mycobacterium tuberculosis* chaperonins on bronchial eosinophilia and hyperresponsiveness in a murine model of allergic inflammation. Clin Exp Allergy 2004, 34: 712–719.

56. Zhang B, Walsh M D, Nguyen K B, Hillyard N C, Cavanagh A C, McCombe P A and Morton H. Early pregnancy factor treatment suppresses the inflammatory response and adhesion molecule expression in the spinal cord of SJL/J mice with experimental autoimmune encephalomyelitis and the delayed-type hypersensitivity reaction to trinitrochlorobenzene in normal BALB/c mice. J Neurol Sci 2003, 212: 37–46.

57. Athanasas-Platsis S, Zhang B, Hillyard N C, Cavanagh A C, Csurhes P A, Morton H and McCombe P A. Early pregnancy factor suppresses the infiltration of lymphocytes and macrophages in the spinal cord of rats during experimental autoimmune encephalomyelitis but has no effect on apoptosis. J Neurol Sci 2003, 214: 27–36.

58. Spik G, Haendler B, Delmas O, Mariller C, Chamoux M, Maes P, Tartar A, Montreuil J, Stedman K, Kocher H P, Roland Kellers R, Hiestand P C and Movva N R. A novel secreted cyclophilin-like protein (SCYLP). J Biol Chem 1991, 266: 10735–10738.

59. Allain F, Boutillon C, Mariller C and Spik G. Selective assay for CypA and CypB in human blood using highly specific anti-peptide antibodies. J Immunol Methods 1995, 178: 113–120.

60. Billich A, Winkler G, Aschauer H, Rot A and Peichl P. Presence of cyclophilin A in synovial fluids of patients with rheumatoid arthritis. J Exp Med 1997, 185: 975–980.

61. Tegeder I, Schumacher A, John S, Geiger H, Geisslinger G, Bang H and Brune K. Elevated serum cyclophilin levels in patients with severe sepsis. J Clin Immunol 1997, 17: 380–386.

62. Kol A, Lichtman A H, Finberg R W, Libby P and Kurt-Jones E A. Heat shock protein (HSP) 60 activates the innate immune response: CD14 is an essential receptor for HSP60 activation of mononuclear cells. J Immunol 2000, 164: 13–17.

63. Vabulas R M, Ahmad-Nejad P, da Costa C, Miethke T, Kirschning C J, Hacker H and Wagner H. Endocytosed HSP60s use toll-like receptor 2 (TLR2) and TLR4 to activate the toll/interleukin-1 receptor signaling pathway in innate immune cells. J Biol Chem 2001, 276: 31332–31339.

64. Yoshida N, Oeda K, Watanabe E, Mikami T, Fukita Y, Nishimura K, Komai K and Matsuda K. Protein function. Chaperonin turned insect toxin. Nature 2001, 411: 44.

65. Gupta S and Knowlton A A. Cytosolic heat shock protein 60, hypoxia, and apoptosis. Circulation 2002, 106: 2727–2733.

66. Kirchhoff S R, Gupta S and Knowlton A A. Cytosolic heat shock protein 60, apoptosis, and myocardial injury. Circulation 2002, 105: 2899–2904.

67. Knowlton A A and Sun L. Heat-shock factor-1, steroid hormones, and regulation of heat-shock protein expression in the heart. Am J Physiol Heart Circ Physiol 2001, 280: H455–464.

68. Pratt W B. The hsp90-based chaperone system: involvement in signal transduction from a variety of hormone and growth factor receptors. Proc Soc Exp Biol Med 1998, 217: 420–434.

69. Xu Q, Luef G, Weimann S, Gupta R S, Wolf H and Wick G. Staining of endothelial cells and macrophages in atherosclerotic lesions with human heat shock protein-reactive antisera. Arteriosclerosis Thrombosis 1993, 13: 1763–1769.

70. Shamaei-Tousi A, Steptoe A, Coates A and Henderson B. Circulating chaperonin 60 in the plasma of British civil servants. Biochem Soc Trans 2004, 'abstr 13'.

71. Libby P. Current concepts of the pathogenesis of the acute coronary syndromes. Circulation 2001, 104: 365–372.

72. Hemingway H and Marmot M. Evidence based cardiology: psychosocial factors in the aetiology and prognosis of coronary heart disease. Systematic review of prospective cohort studies. Brit Med J 1999, 318: 1460–1467.

73. Olson N C, Hellyer P W and Dodam J R. Mediators and vascular effects in response to endotoxin. Brit Vet J 1995, 151: 489–522.

74. Sangster T A, Lindquist S and Queitsch C. Under cover: causes, effects and implications of Hsp90-mediated genetic capacitance. Bioessays 2004, 26: 348–362.

75. Srivastava P. Roles of heat-shock proteins in innate and adaptive immunity. Nat Rev Immunol 2002, 2: 185–194.

76. Morales H, Muharemagic A, Gantress J, Cohen N and Robert J. Bacterial stimulation upregulates the surface expression of the stress protein gp96 on B cells in the frog Xenopus. Cell Stress Chaperones 2003, 8: 265–271.

77. Singh-Jasuja H, Scherer H U, Hilf N, Arnold-Schild D, Rammensee H-G, Toes R E M and Schild H. The heat shock protein gp96 induces maturation of dendritic cells and down-regulation of its receptor. Eur J Immunol 2000, 30: 2211–2215.

78. Binder R J and Srivastava P K. Essential role of CD91 in re-presentation of gp96-chaperoned peptides. Proc Natl Acad Sci USA 2004, 101: 6128–6133.

79. Liu B, Dai J, Zheng H, Stoilova D, Sun S and Li Z. Cell surface expression of an endoplasmic reticulum resident heat shock protein gp96 triggers MyD88-dependent systemic autoimmune diseases. Proc Nat Acad Sci USA 2003, 100: 15824–15829.

80. Karunakaran K P, Noguchi Y, Read T D, Cherkasov A, Kwee J, Shen C, Nelson C C and Brunham R C. Molecular analysis of the multiple GroEL proteins of Chlamydiae. J Bacteriol 2003, 185: 1958–1966.

81. Gerard H C, Whittum-Hudson J A, Schumacher H R and Hudson A P. Differential expression of three *Chlamydia trachomatis* hsp60-encoding genes in active vs persistent infections. Microb Pathogenesis 2004, 36: 35–39.

82. Fabian T K, Gaspar J, Fejerdy L, Kaan B, Balint M, Csermely P and Fejerdy P. Hsp70 is present in human saliva. Med Sci Monit 2003, 9: 62–65.

83. Utleg A G, Yi E C, Xie T, Shannon P, White J T, Goodlett D R, Hood L and Lin B. Proteomic analysis of human prostasomes. Prostate 2003, 56: 150–161.

84. Trougakos I P and Gonos E S. Clusterin/apolipoprotein J in human aging and cancer. Int J Biochem Cell Biol 2002, 34: 1430–1448.

85. Rammes A, Roth J, Goebeler M, Klempt M, Hartmann M and Sorg C. Myeloid-related protein (MRP) 8 and MRP14, calcium-binding proteins of the S100 family, are secreted by activated monocytes via a novel, tubulin-dependent pathway. J Biol Chem 1997, 272: 9496–9502.

86. Matzinger P. The Danger Model: A renewed sense of self. Science 2002, 296: 301–305.

87. Maguire M, Coates A R M and Henderson B. Chaperonin 60 unfolds its secrets of cellular communication. Cell Stress Chaperones 2002, 7: 317–329.

88. Wilson M, McNab R and Henderson B. *Bacterial Disease Mechanisms: An Introduction to Cellular Microbiology.* Cambridge University Press: 2002.

89. Asquith K L, Baleato R M, McLaughlin E A, Nixon B and Aitken R J. Tyrosine phosphorylation activates surface chaperones facilitating sperm-zona recognition. J Cell Sci 2004, 117: 3645–3657.

90. Ellis R J. Proteins as molecular chaperones. Nature 1987, 328: 378–379.

Extracellular Biology of Molecular Chaperones: Molecular Chaperones as Cell Regulators

6

Cell-Cell Signalling Properties of Chaperonins

Anthony Coates and Peter Tormay

6.1. Introduction

In the beginning, bacteria evolved chaperonins (Cpns) to help in the folding of other proteins. Then, about one and a half billion years ago, bacteria began to live with eukaryotic cells. The career of the chaperonin began to expand beyond the protein folding area, and they developed new functions in order to adapt to the eukaryotic evolutionary niche. As the eukaryotes became ever more complex, the chaperonins evolved into cell-cell signalling molecules. The sophistication of this new role has only recently begun to emerge.

The chaperonins that are the subject of this chapter belong to the 60- and 10-kDa classes and are called Cpn60 and Cpn10, respectively. The folding actions of these proteins have been described in detail in Chapter 1. If there is more than one Cpn60 or Cpn10 in any one species, they are called Cpn60.1, Cpn60.2 and so on [1]. Cpn60 proteins are also called heat shock protein (Hsp) 60 or 65, and Cpn10s are also named Hsp10. For example, the *Mycobacterium tuberculosis* genome contains two *cpn60* genes. One of these, termed *cpn60.1* [2], appears to form an operon with the *cpn10* gene. This is the usual relationship in most bacteria. The second *cpn60* gene encodes Cpn60.2, the well-known Hsp65 protein of *M. tuberculosis*, and is found elsewhere in the genome.

6.2. What is a cell-cell signalling molecule?

A cell-cell signalling molecule is one that directly communicates a message from one cell to another. The signalling molecule may be attached to the outside of the broadcasting cell or it may be released from it and may attach to audience cells. The broadcasting cells may be bacterial or eukaryotic, because chaperonins exist in both eukaryotes and prokaryotes. The audience cells respond in a variety of ways, such as cytokine release.

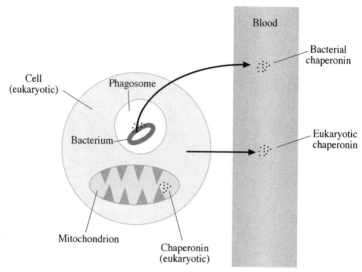

Figure 6.1. Chaperonins are produced by bacteria such as *M. tuberculosis* inside phago-somes. The chaperonin migrates into the phagosome and can be detected in tissues and in the blood. Eukaryotic chaperonins are found in the mitochondrion and can also be detected in the blood.

6.3. Are chaperonins released from cells?

Traditionally, chaperonins are regarded as intracellular, because they help to fold other proteins inside cells, and so, if this is their only function, why bother to wander into the extracellular space? The first suggestion that chaperonins might occur extracellularly was the observation that early pregnancy factor, sub-sequently reported to be Cpn10, could be detected in the serum of animals and of humans [3]. Since then, numerous reports have described chaperonins, both eukaryotic and prokaryotic, in extracellular situations [4–13]. For example, bacterial chaperonins have been detected on the cell surface [10], in the extracel-lular matrix [12] and in the serum of patients with tuberculosis [14] (Figure 6.1). Mammalian chaperonins are also detected on the cell surface [15], are secreted from glial cells and neuroblastoma cells [8] and are found in the serum of hu-mans [5, 16]. It is not known for certain whether these chaperonins are secreted from live cells or whether they are released from dying ones. However, there is evidence that supports the idea that some chaperonins are secreted.

For instance, inside the macrophage phagosome, ingested *M. tuberculosis* Cpn10 accumulates in the wall of the bacterium and in the matrix of the phagosomes [17] (Figure 6.1). The chaperonins dissociate into partially helical monomers which interact with acidic lipids. This may represent two important steps in the mechanism of secretion of the protein into the external environment. So it seems that some chaperonins, although lacking signal peptides, interact with membranes in a similar way to such peptides. The cell location and secretion of these proteins is dealt with in detail in Chapter 2 and to lesser extents in other chapters in this volume.

Interestingly, human Cpn10 behaves in a similar way to the mycobacterial protein and is found extracellularly; however, the Cpn10 of *Escherichia coli* is different in that it adopts a dimeric β-sheet which is not found outside the cell. These data suggest that Cpn10 is secreted by some species, but not by others.

An alternative, but not mutually exclusive, explanation for the presence of chaperonins outside the cell is that these proteins are released as the cell dies. This could be a kind of death message. In certain situations, it is possible that both secretion and release from dying cells might be responsible for extracellular chaperonins. That the broadcaster is dead does not diminish the power of the message, as readers of Shakespeare will no doubt agree.

6.4. Do all chaperonins transmit the same message?

The answer is a definite 'No'. The first indication that different chaperonins transmit different messages was the observation that *M. tuberculosis* Cpn10 induces lysis of bone, whereas Cpn60 from the same species is totally inactive [18]. Subsequently, it has been demonstrated *in vivo* [19] that *M. tuberculosis* Cpn60.1 actually blocks bone resorption. Even Cpn60s within the same species have different properties. For example, *Rhizobium leguminosarum* Cpn60.3 induces cytokine production by human monocytes, whereas Cpn60.1 is inactive [20]. In the case of *M. tuberculosis* Cpn60s, Cpn60.1 is a more powerful inducer of pro-inflammatory cytokines than Cpn60.2 [21]. These data indicate that different chaperonins, even though they may share very high sequence identity, carry different messages.

6.5. Who is in the audience?

This is one of the most intriguing aspects of this area of research. Effects on human peripheral blood monocytes have been observed by several different laboratories [20–25]. However, the audience also contains macrophages and endothelial cells [26, 27], epithelial cells [28], vascular smooth muscle cells [29], dendritic cells (DCs) [24, 30], bone cells such as osteoclasts [31, 32] and osteoblasts [18], and cells of the central nervous system (CNS) [8, 33]. The available evidence indicates that the audience is primarily from a myeloid background, although the inclusion of nerve cells suggests that this may be too narrow a definition. Chapter 15 describes in detail the interactions of chaperonins with nerve cells.

6.6. Is every chaperonin a molecular messenger?

Available evidence suggests that most species produce chaperonins which convey signalling messages to eukaryotic cells. For example, *M. tuberculosis* Cpn60.2 induces human monocytes to synthesise pro-inflammatory cytokines [20–25]. Many other Cpn60s from diverse bacterial species such as *E. coli* [23, 34], *Chlamydia spp.* [26, 34] and *Helicobacter pylori* [35] induce cells to synthesise pro-inflammatory cytokines. Even plant Cpn60s [20] can stimulate cytokine secretion. The significance of chaperonins became even more interesting when reports began to emerge that mammalian Cpn60s from rats, mice, hamsters and humans [24, 26, 34] also induce eukaryotic cells to synthesise pro-inflammatory cytokines. Such a ubiquitous property of chaperonins indicates that these molecules have a significant role to play in nature, one that we are only just beginning to understand.

6.7. What effect does the chaperonin signal have on the audience?

Chaperonins induce mammalian cells to produce pro-inflammatory cytokines [8, 18, 20–33]. For example, seven Cpn60s from different species all induce pro-inflammatory cytokine release from mouse macrophages [34]. However, data are available that suggest chaperonins have a greater significance than this in the world of non-folding biology: human Cpn10 has been shown to be early pregnancy factor [11, 36] and is present in red blood cells and in secretory granules [37]. Cpn60 from *Actinobacillus actinomycetemcomitans* [38], which is an oral pathogen in humans, induces proliferation of epithelial cells in 24 hours and increases the rate of epithelial cell death after prolonged incubation (144 hours). In cardiac muscle cells, Cpn10 and Cpn60 suppress ubiquitination of insulin-like growth factor-1 receptor and augment insulin-like growth factor-1 receptor signalling [39]. In the central nervous system, a chaperonin homologue is neuro-protective [33]. However, it is whole organ and animal studies that have revealed the truly startling biology of chaperonins.

In 1981 a major antigen of *M. tuberculosis* was identified by immunising mice with whole bacteria and generating the monoclonal antibody TB78 [40]. The antigen that bound to TB78 turned out to be Cpn60.2 (reviewed in [41]) which, curiously, can both induce and attenuate autoimmune arthritis and diabetes in animals [42, 43] (reviewed in [41]). Chaperonins are now being developed as vaccines for human diseases such as cancer [44] and diabetes [45]. Certain of the chapters in this volume deal with this in more detail (Chapters 16–18). It is thought that these properties are mediated by T lymphocytes. However, chaperonins have been called multiplex proteins [41], which means that they each have a number of parallel biological properties, of which immunogenicity and protein folding are but two.

The clearest example of a non-immunological, non-folding property of chaperonins is their ability to modulate bone formation. The first report of whole-organ bone modulation by chaperonins was that the Cpn60 proteins from *A. actinomycetemcomitans* and *E. coli* were potent stimulators of *in vitro* bone resorption [31]. Further work strengthened this observation, demonstrating that these proteins stimulated the proliferation and differentiation of myeloid precursor cells into mature bone-resorbing osteoclasts [32] and that the human Cpn60 protein was also a potent stimulator of bone resorption. In contrast, *M. tuberculosis* Cpn60s do not resorb bone [18], despite the fact that Cpn10 from this species is the primary bone resorbing molecule in the bacterium and may be responsible for the marked bone destruction that is seen in clinical cases of spinal tuberculosis, also known as Pott's disease.

However, in the rat adjuvant arthritis model, which is a T cell–driven disease with considerable osteoclastic bone remodeling [46], *M. tuberculosis* Cpn60.1 almost completely prevents bone destruction, whilst Cpn60.2 has no effect [19, 47]. The mechanism of action appears to be directly on bone formation (see the next section).

Another non-T cell–mediated effect is seen in the central nervous system. Rats that are given an intranasal dose of the Cpn60-like peptide activity-dependent neuroprotective protein (ADNP) are protected from neurodegeneration [33, 48]. An even stranger effect is a bacterial chaperonin in insect saliva which is used by the insect as a toxin to kill other insects [49].

This chapter will not review the T cell–mediated effects of chaperonins in autoimmune diabetes and arthritis because these are covered in the literature [41, 50, 51] and elsewhere in this volume. However, chaperonins seem to have effects in diseases that are not classically regarded as being T cell-mediated. For example, they are active in suppressing antigen-induced asthma in the mouse [52], despite the fact that asthma is an allergic condition. This is not to say that T cells play no part in asthma. The author and colleagues have found that *M. tuberculosis* Cpn60.1 blocks both the eosinophilia and the airway hyperresponsiveness found in this model and that it does so by a mechanism involving the priming of DCs. In contrast, Cpn60.2 was without effect ([52] and unpublished observations). These findings are supported by a recent publication which showed that *Mycobacterium leprae* Cpn60.2 is active in suppressing this asthma model, whereas *M. tuberculosis* Cpn60.2 is inactive [53].

Thus it seems that chaperonins remain full of biological surprises.

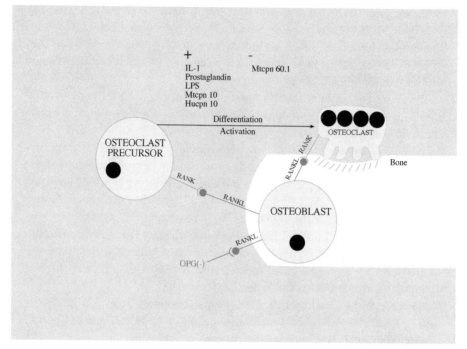

Figure 6.2. Osteoclast formation depends on the upregulation of the TNF family member RANKL (receptor activator of NF-κB ligand) on osteoblasts and the activation of osteoclast precursor cells through binding of RANKL to RANK. OPG binds RANKL and acts as a RANK antagonist, thereby inhibiting osteoclast formation (+ = molecules that enhance differentiation/activation; − = molecules that inhibit differentiation/activation).

6.8. Mechanism of action – cytokines and the cytokine ceiling

At the cellular level, there are two seemingly contradictory actions of chaperonins. Firstly, they stimulate the release of pro-inflammatory cytokines from eukaryotic cells. Secondly, they suppress cells that have been stimulated by either endogenous or exogenous agents. The effects of *M. tuberculosis* Cpn60.1 (MtCpn60.1) on bone are a case in point. This chaperonin stimulates cytokine release from cells [21], yet it suppresses bone resorption that has been triggered by exogenous bacterial lipopolysaccharide (LPS) or endogenous RANKL (receptor activator of NF-κB), a tumour necrosis factor (TNF) family member and key osteoclast-inducing cytokine. It is now recognised that the dynamic remodelling of bone, which is a consequence of the bone forming action of the osteoblast and the bone destroying activity of osteoclasts, is controlled by the regulation of RANKL expression on osteoblasts and its interaction with RANK on osteoclast precursor cells. Another TNF family member, osteoprotegerin (OPG), acts as a soluble RANK and antagonises RANKL/RANK interactions and inhibits osteoclast formation (Figure 6.2). In bone cell and explant culture studies it is now established that *M. tuberculosis* Cpn60.1 can block the activity of RANKL which interacts with RANK on osteoclast precursors [19]. Such interaction stimulates a complex signalling pathway involving TRAF-6, which is described in detail in Chapter 7. It has been established that inhibition is not due to (i) stimulation of synthesis of the negative regulator OPG, (ii) inhibition of the key MAP kinase c-JNK or (iii) binding of RANKL by Cpn60.1.

It is likely that DCs, the conductors of the immune system, are an important target for chaperonins and therein may hide secrets of their mechanism of action. Cpn60s can activate DCs to secrete pro-inflammatory cytokines and can induce these cells to mature [24]. Cpn60 induces the release of TNF-α, IL-12 and IL-1β from DCs [30], but only a small amount of IL-10, which suggests the presence of a bias towards Th1 (pro-inflammatory) responses. One interpretation of this is that Cpn60s may prime for destructive Th1 responses at sites of Cpn60 release [30]. An alternative hypothesis which has been suggested by the author (AC), called the 'cytokine ceiling hypothesis' suggests that whilst chaperonins do activate the innate immune system, they also suppress over-activation. This dampening down of, for example, a bacterial LPS-activated immune system, is an essential control mechanism which, if absent, would result in self-destruction by an innate immune system that is out of control. So, the 'cytokine ceiling' is an additional property of chaperonins. The innate immune system recognises bacterial pro-inflammatory signals (pattern-associated molecular patterns – PAMPs) as 'danger' molecules [54]; however, when the level of response reaches the 'cytokine ceiling', chaperonins induce suppression of an otherwise highly damaging immune response.

It is possible that some chaperonins are activators and others are suppressors (see Figure 6.3); however, it is also possible that one species of chaperonins can both activate and suppress the immune system. Furthermore, it likely that this paradigm applies to many other cell types such as bone and CNS cells, not just innate immune cells. For instance, *M. tuberculosis* chaperonins show three separate patterns of activity: (i) Cpn60.1 and Cpn60.2 both activate human monocytes [21], (ii) Cpn60.1 is a potent inhibitor of the differentiation of osteoclast precursor cells into mature bone resorbing osteoclasts [19] whilst Cpn60.2 is inactive in this regard and (iii) Cpn10 is a potent stimulator of osteoclast proliferation. Gram-negative bacterial and the human mitochondrial Cpn60 proteins also act as potent inducers of osteoclast formation. These data show that chaperonins can be very different from one another in terms of their interaction with selected target cells.

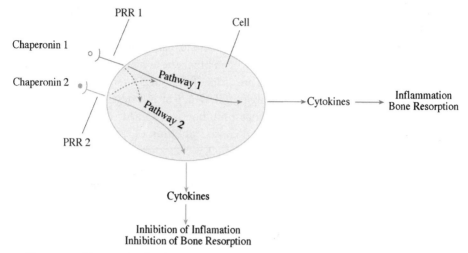

Figure 6.3. Chaperonins bind to PRRs. Shown here are hypothetical pathways 1 and 2 which lead to inflammation/bone resorption and inhibition, respectively.

6.9. Chaperonin receptors

Chaperonins activate cells by interacting with pattern-recognition receptors (PRRs) including CD14 and Toll-like receptors (TLRs) [26, 30, 55, 56] (Figure 6.3), and the interaction of Cpn60 proteins with the TLRs is discussed in detail in Chapter 7. However, chaperonins have such a diverse range of activities that it is likely that these receptors are probably the tip of the iceberg and it is anticipated that more receptors will be implicated soon. For example, although TLR4 is presumed by many workers to be the major (perhaps only) receptor for Cpn60 (see Chapter 7), early work on the Cpn60 protein of *A. actinomycetemcomitans* showed that it could activate myeloid cells from the TLR4 negative mutant, C3H/HeJ strain, of mouse [31]. Human and bacterial chaperonins seem to use the same receptors, and this suggests that bacteria may use this to modulate their host's immunity to their advantage.

Chaperonins can be considered PAMPs [57]. PAMPs are evolutionarily conserved components of pathogens recognised by non-clonal receptors of the host. A growing number of PRRs (recent examples being the TLRs) have been identified and are responsible for the ability of our innate immune system to rapidly recognise, and discriminate, infections by bacteria and fungi [58–60]. PAMPS include LPS, peptidoglycan, flagellae and CpG DNA [58–60] and are unique to pathogens. Interestingly, human chaperonins seem to use the same receptors as some of the bacterial PAMPs [55, 56]. This would be highly damaging to the host unless there was an upper cut-off point above which the chaperonins became suppressive, the so-called cytokine ceiling.

6.10. Is all the activity of chaperonins due to contaminating bacterial lipopolysaccharide?

Most recombinant chaperones are produced in *E. coli* and so are contaminated with LPS. In this and subsequent chapters the problems with such contamination will be discussed and evidence will be presented that the biological actions of molecular chaperones such as Cpn60 are not due to contaminating LPS. Many workers use commercially available columns with polymyxin B linked to agarose in order to remove LPS contaminating recombinant molecular chaperones. This has the disadvantage (as discussed in Chapter 10) of also resulting in the binding of the molecular chaperone to the column.

The author's (AC) group has solved this particular problem by washing affinity (e.g., Ni-NTA) columns containing recombinant chaperones with polymyxin B. This eliminates most of the contaminating LPS without removing the activity of the recombinant protein. Thus, powerful chaperonin activity can be detected when LPS is undetectable [21, 24]. If chaperonins are exposed to proteases or heat, their cell signalling activity is lost. Furthermore, when short peptides of chaperonins are made by chemical synthesis, chaperonin activity is seen [18]. In addition, *in vivo*, many activities of chaperonins cannot be reproduced by LPS. For example, *M. tuberculosis* Cpn60.1 blocks bone resorption [19]. This chaperonin also inhibits asthma in the mouse [52]. Single-amino-acid changes in another bacterial chaperonin render it toxic to insects [49] and a short mammalian chaperonin peptide protects rats from neurodegeneration [33, 48].

However, there are occasional reports of loss of activity of chaperonin which have been attributed to the absence of LPS [61]. In the author's experience, loss of activity of a chaperonin does occur from time to time; however, this is not due to absence of LPS, but rather from incorrect folding of the chaperonin or a technical problem associated with the biological assay. The involvement of LPS contamination in the observed biological activities of chaperonins has also been discussed elsewhere in this volume (Chapters 7, 8 and 10).

6.11. Conclusion

Chaperonins are multiplex molecules. Apart from helping cells to fold proteins, they have important non-folding activities, as well as being powerful antigens and cell-cell signalling molecules. Chaperonins are released from cells and can be detected in tissues and in the blood. They transmit messages from one cell to another, and messages transmitted by chaperonins from individual species can be different and distinct. The audience cells are primarily myeloid in origin; however, cells of the central nervous system are also involved.

The effects of the chaperonin signal are diverse. For example, chaperonins induce pro-inflammatory cytokine release, promote bone resorption, inhibit bone resorption, protect neural cells from degeneration, kill insects and protect mice from asthma. Chaperonins act through PRRs which include CD14 and Toll-like molecules. The mechanism of action has two main arms: firstly, stimulation of cytokine release, and secondly suppression. It is hypothesised that chaperonins suppress overactive cytokine responses at a certain high dangerous level, which is termed the cytokine ceiling. This protects the animal from damaging itself.

The extraordinary diversity of actions of chaperonins that are known today are almost certainly the tip of an iceberg.

REFERENCES

1. Coates A R M, Shinnick T M and Ellis J R. Chaperonin nomenclature. Mol Microb 1993, 8: 787.
2. Kong T H, Coates A R M, Butcher P D, Hickman C J and Shinnick T M. *Mycobacterium tuberculosis* expresses two chaperonin-60 homologs. Proc Natl Acad Sci USA 1993, 90: 2608–2612.
3. Cavanagh A C and Morton H. The purification of early-pregnancy factor to homogeneity from human platelets and identification as chaperonin 10. Eur J Biochem 1994, 222: 551–560.
4. Soltys B J and Gupta R S. Mitochondrial-matrix proteins at unexpected locations: are they exported? Trends Biochem Sci 1999, 24: 174–177.
5. Morton H. Early pregnancy factor: an extracellular chaperonin 10 homologue. Immunol Cell Biol 1998, 76: 483–496.
6. Goulhen F, Hafezi A, Uitto V J, Hinode D, Nakamura R, Grenier D and Mayrand D. Subcellular localization and cytotoxic activity of the GroEL-like protein isolated from *Actinobacillus actinomycetemcomitans*. Infect Immun 1998, 66: 5307–5313.
7. Kol A, Sukhova G K, Lichtman A H and Libby P. Chlamydial heat shock protein 60 localizes in human atheroma and regulates macrophage tumour necrosis factor-α and matrix metalloproteinase expression. Circulation 1998, 98: 300–307.
8. Bassan M, Zamostiano R, Giladi E, Davidson A, Wollman Y, Pitman J, Hauser J, Brenneman D E and Gozes I. The identification of secreted heat shock 60-like protein from rat glial cells and a human neuroblastoma cell line. Neurosci Lett 1998, 250: 37–40.
9. Coates A R M and Henderson B. Chaperonins in health and disease. Ann NY Acad Sci 1998, 851: 48–53.
10. Frisk A, Ison C A and Lagergard T. GroEL heat shock protein of *Haemophilus ducreyi*: association with cell surface and capacity to bind to eukaryotic cells. Infect Immun 1998, 66: 1252–1257.

11. Cavanagh A C. Identification of early pregnancy factor as chaperonin 10: implications for understanding its role. Rev Reprod 1996, 1: 28–32.

12. Esaguy N and Aguas A P. Subcellular localization of the 65-kDa heat shock protein in mycobacteria by immunoblotting and immunogold ultracytochemistry. J Submicrosc Cytol Pathol 1997, 29: 85–90.

13. Brenneman D E and Gozes I. A femtomolar-acting neuroprotective peptide. J Clin Invest 1996, 97: 2299–2307.

14. Sethna K B, Mistry N F, Dholakia Y, Antia N H and Harboe M. Longitudinal trends in serum levels of mycobacterial secretory (30 kD) and cytoplasmic (65 kD) antigens during chemotherapy of pulmonary tuberculosis patients. Scand J Infect Dis 1998, 30: 363–369.

15. Soltys B J and Gupta R S. Cell surface localization of the 60 kDa heat shock chaperonin protein (hsp60) in mammalian cells. Cell Biol Int 1997, 21: 315–320.

16. Lewthwaite J, Owen N, Coates A, Henderson B and Steptoe A. Circulating human heat shock protein 60 in the plasma of British civil servants. Circulation 2002, 106: 196–201.

17. Fossati G, Izzo G, Rizzi E, Gancia E, Niccolai N, Giannozzi E, Spiga O, Bono L, Marone P, Leone E, Mangili F, Harding S, Errington N, Walter C, Henderson B, Roberts M M, Coates A R M, Casetta B and Mascagni P. *Mycobacterium tuberculosis* chaperonin 10 is secreted in the macrophage phagolysosome: is secretion due to dissociation and the adoption of partially helical structure at the membrane? J Bact 2003, 185: 4256–4267.

18. Meghji S, White P, Nair S P, Reddi K, Heron K, Henderson B, Zaliani A, Fossati G, Mascagni P, Hunt J F, Roberts M M and Coates A R. *Mycobacterium tuberculosis* chaperonin 10 stimulates bone resorption: a potential contributory factor in Pott's disease. J Exp Med 1997, 186: 1241–1246.

19. Winrow V R, Meghji S, Mesher J, Coates A R M, Morris C J, Tormay P and Henderson B. *Mycobacterium tuberculosis* chaperonin 60.1, but not chaperonin 60.2, inhibits osteoclastic bone resorption by blocking RANKL activity.

20. Lewthwaite J, George R, Lund P A, Poole S, Tormay P, Sharp L, Coates A R M and Henderson B. *Rhizobium leguminosarum* chaperonin 60.3, but not chaperonin 60.1, induces cytokine production by human monocytes: activity is dependent on interaction with cell surface CD14. Cell Stress Chaperon 2002, 7: 130–136.

21. Lewthwaite J C, Coates A R M, Tormay P, Singh M, Mascagni P, Poole S, Roberts M, Sharp L and Henderson B. *Mycobacterium tuberculosis* chaperonin 60.1 is a more potent cytokine stimulator than chaperonin 60.2 (hsp 65) and contains a CD14-binding domain. Infect Immun 2001, 69: 7349–7355.

22. Friedland J S, Shattock R, Remick D G and Griffin G E. Mycobacterial 65-kD heat shock protein induces release of proinflammatory cytokines from human monocytic cells. Clin Exp Immunol 1993, 91: 58–62.

23. Tabona P, Reddi K, Khan S, Nair S P, Crean S J, Meghji S, Wilson M, Preuss M, Miller A D, Poole S, Carne S and Henderson B. Homogeneous *Escherichia coli* chaperonin 60 induces IL-1 and IL-6 gene expression in human monocytes by a mechanism independent of protein conformation. J Immunol 1998, 161: 1414–1421.

24. Bethke K, Staib F, Distler M, Schmitt U, Jonuleit H, Enk A H, Galle P R and Heike M. Different efficiency of heat shock proteins to activate human monocytes and dendritic cells: superiority of HSP60. J Immunol 2002, 169: 6141–6148.

25. Maguire M, Coates A R M and Henderson B. Cloning expression and purification of three chaperonin 60 homologues. J Chromatography 2003, 786: 117–125.
26. Kol A, Bourcier T, Lichtman A and Libby P. Chlamydial and human heat shock protein 60s activate human vascular endothelium, smooth muscle cells, and macrophages. J Clin Invest 1999, 103: 571–577.
27. Billack B, Heck D E, Mariano T M, Gardner C R, Sur R, Laskin D L and Laskin J D. Induction of cyclooxygenase-2 by heat shock protein 60 in macrophages and endothelial cells. Am J Physiol (Cell Physiol) 2002, 283: C1267–1277.
28. Zhang L, Pelech S L, Mayrand D, Grenier D, Heino J and Uitto V J. Bacterial heat shock protein-60 increases epithelial cell proliferation through the ERK1/2 MAP kinases. Exp Cell Res 2001, 266: 11–20.
29. Sasu S, LaVerda D, Qureshi N, Golenbock D T and Beasley D. *Chlamydia pneumoniae* and chlamydial heat shock protein 60 stimulate proliferation of human vascular smooth muscle cells via toll-like receptor 4 and p44/p42 mitogen-activated protein kinase activation. Circ Res 2001, 89: 244–250.
30. Flohé S B, Bruggemann J, Lendemans S, Nikulina M, Meierhoff G, Flohé S and Kolb H. Human heat shock protein 60 induces maturation of dendritic cells versus a Th1-promoting phenotype. J Immunol 2003, 170: 2340–2348.
31. Kirby A C, Meghji S, Nair S P, White P, Reddi K, Nishihara T, Nakashima K, Willis A C, Sim R, Wilson M and Henderson B. The potent bone-resorbing mediator of *Actinobacillus actinomycetemcomitans* is homologous to the molecular chaperone GroEL. J Clin Invest 1995, 96: 1185–1194.
32. Reddi K, Meghji S, Nair S P, Arnett T R, Miller A D, Preuss M, Wilson M, Henderson B and Hill P. The *Escherichia coli* chaperonin 60 (groEL) is a potent stimulator of osteoclast formation. J Bone Miner Res 1998, 13: 1260–1266.
33. Gozes I, Divinsky I, Pilzer I, Fridkin M, Brenneman D E and Spier A D. From vasoactive intestinal peptide (VIP) through activity-dependent neuroprotective protein (ADNP) to NAP: a view of neuroprotection and cell division. J Mol Neurosci 2003, 20: 315–322.
34. Habich C, Kempe K, van der Zee R, Burkart V and Kolb H. Different heat shock protein 60 species share pro-inflammatory activity but not binding sites on macrophages. FEBS Lett 2003, 533: 105–109.
35. Gobert A P, Bambou J C, Werts C, Balloy V, Chignard M, Moran A P and Ferrero R L. *Helicobacter pylori* heat shock protein 60 mediates interleukin-6 production by macrophages via a Toll-like receptor (TLR)-2-, TLR-4-, and myeloid differentiation factor 88–independent mechanism. J Biol Chem 2004, 279: 245–250.
36. Rolfe B, Cavanagh A, Forde C, Bastin F, Chen C and Morton H. Modified rosette inhibition test with mouse lymphocytes for detection of early pregnancy factor in human pregnancy serum. J Immunol Methods 1984, 70: 1–11.
37. Sadacharan S K, Cavanagh A C and Gupta R S. Immunoelectron microscopy provides evidence for the presence of mitochondrial heat shock 10-kDa protein (chaperonin 10) in red blood cells and a variety of secretory granules. Histochem Cell Biol 2001, 116: 507–517.
38. Zhang L, Pelech S and Uitto V J. Long-term effect of heat shock protein 60 from *Actinobacillus actinomycetemcomitans* on epithelial cell viability and mitogen-activated protein kinases. Infect Immun 2004, 72: 38–45.

39. Shan Y X, Yang T L, Mestril R and Wang P H. Hsp10 and Hsp60 suppress ubiquitination of insulin-like growth factor-1 receptor and augment insulin-like growth factor-1 receptor signaling in cardiac muscle: implications on decreased myocardial protection in diabetic cardiomyopathy. J Biol Chem 2003, 278: 45492–45498.

40. Coates A R M, Hewitt J, Allen B W, Ivanyi J and Mitchison D A. Antigenic diversity of *Mycobacterium tuberculosis* and *Mycobacterium bovis* detected by means of monoclonal antibodies. Lancet 1981, 2: 167–169.

41. Coates A R M. Immunological aspects of chaperonins. In Ellis, R. J. (Ed.) *The Chaperonins*. Academic Press, London 1996, pp 267–296.

42. van Eden W, Thole J E R, van der Zee R, Noordzij A, van Embden J D A, Hensen E J and Cohen I R. Cloning of the mycobacterial epitope recognized by T lymphocytes in adjuvant arthritis. Nature 1988, 331: 171–173.

43. Elias D, Markovits D, Reshef T, van der Zee R and Cohen I R. Induction and therapy of autoimmune diabetes in the non-obese diabetic mouse by a 65-kDa heat shock protein. Proc Natl Acad Sci USA 1990, 87: 1576–1580.

44. Srivastava P K. Immunotherapy of human cancer: lessons from mice. Nat Immunol 2000, 1: 363–366.

45. Raz I, Elias D, Avron A, Tamir M, Metzger M and Cohen I R. Beta-cell function in new-onset type 1 diabetes and immunomodulation with a heat-shock protein peptide (DiaPep277): a randomised, double-blind, phase II trial. Lancet 2001, 358: 1749–1753.

46. Billingham MEJ. Adjuvant arthritis: The first model. In Henderson, B., Edwards, J. C. W. and Pettipher, E. R. (Eds.) *Mechanisms and Models in Rheumatoid Arthritis*. Academic Press, London 1995, pp 389–409.

47. Winrow V R, Coates A R M, Tormay P, Henderson B, Singh M, Blake D R and Morris C J. Chaperonin 60.1 prevents bone destruction in Wistar rats with adjuvant-induced arthritis. Rheumatology 2002, 41 (abstr suppl 1): 47.

48. Gozes I and Brenneman D E. Activity-dependent neurotrophic factor (ADNF). An extracellular neuroprotective chaperonin? J Mol Neurosci 1996, 7: 235–244.

49. Yoshida N, Oeda K, Watanabe E, Mikami T, Fukita Y, Nishimura K, Komai K and Matsuda K. Protein function. Chaperonin turned insect toxin. Nature 2001, 411: 44.

50. Cohen I R. Peptide therapy for Type I diabetes: the immunological homunculus and the rationale for vaccination. Diabetologia 2002, 45: 1468–1474.

51. Cohen I R. The Th1/Th2 dichotomy, hsp60 autoimmunity, and type I diabetes. Clin Immunol Immunopathol 1997, 84: 103–106.

52. Riffo-Vasquez Y, Spina D, Page C, Tormay P, Singh M, Henderson B and Coates A R M. Effect of *Mycobacterium tuberculosis* chaperonins on bronchial eosinophilia and hyperresponsiveness in a murine model of allergic inflammation. Clin Exp Allergy 2004, 34: 712–719.

53. Rha Y-H, Taube C, Haczku A, Joeham A, Takeda K, Duez C, Siegel M, Ayditung M K, Born W K, Dakhama A and Gelfand E W. Effect of microbial heat shock proteins on airway inflammation and hyperresponsiveness. J Immunol 2002, 169: 5300–5307.

54. Matzinger P. An innate sense of danger. Semin Immunol 1998, 10: 399–415.

55. Asea A, Kraeft S-K, Kurt-Jones E A, Stevenson M A, Chen L B, Finberg R W, Koo G C and Calderwood S K. Hsp70 stimulates cytokine production through a

CD14-dependent pathway, demonstrating its dual role as a chaperone and cytokine. Nat Med 2000, 6: 435–442.

56. Vabulas R M, Ahmad-Nejad P, da Costa C, Miethke T, Kirschning C J, Hacker H and Wagner H. Endocytosed HSP60s use toll-like receptor 2 (TLR2) and TLR4 to activate the toll/interleukin-1 receptor signaling pathway in innate immune cells. J Biol Chem 2001, 276: 31332–31339.

57. Medzhitov R and Janeway C A J. Innate immunity: the virtues of a nonclonal system of recognition. Cell Stress Chaperon 1997, 91: 295–298.

58. Takeda K and Akira S. Toll receptors and pathogen resistance. Cell Microbiol 2003, 5: 143–153.

59. Girardin S E, Sansonetti P J and Philpott D J. Intracellular vs extracellular recognition of pathogens – common concepts in mammals and flies. Trends Microbiol 2002, 10: 193–199.

60. Colonna M. TREMS in the immune system and beyond. Nat Rev Immunol 2003, 3: 1–9.

61. Gao B and Tsan M F. Recombinant human heat shock protein 60 does not induce the release of tumor necrosis factor alpha from murine macrophages. J Biol Chem 2003, 278: 22523–22529.

7

Toll-Like Receptor-Dependent Activation of Antigen Presenting Cells by Hsp60, gp96 and Hsp70

Ramunas M. Vabulas and Hermann Wagner

7.1. Discovery of Toll-like receptors

The basic concept of the immune system postulates an ability to discriminate between self and non-self and to free the organism from the latter. Two major contributions advanced the comprehension of the cellular basis of self- versus non-self-discrimination. The first was the hypothesis regarding the expansion of antigen-recognising clones on encounter with a respective antigen, which allowed antigenic specificities of the resulting immune reactions to be explained. The co-stimulatory signal hypothesis represented another essential advancement. It postulated the necessity of a second, antigen-independent signal for lymphocyte activation. Its nature was put into an elegant metaphor of the 'immunologist's dirty little secret' [1], referring to substances of microbial origin that should be present concomitant with an antigen to prime an immune response to it.

Of a number of host receptors participating in detection of microbial constituents [2], Toll-like receptors (TLRs) currently represent the most interesting group. Their importance is assumed from the prominent cell activating capacity which they display after engagement with their cognate ligands. The name originates from the *Drosophila* homologue Toll, which was discovered as a part of the dorsoventral patterning cascade during the developmental larva stage of the fruit fly, and this seminal study established an additional, anti-microbial function for Toll in adult flies [3]. It demonstrated that mutants of the genes in the cassette between the Toll ligand Spätzle down to the IκB homologue Cactus showed a compromised inducibility of the anti-fungal peptide drosomycin upon fungal challenge and consequently succumbed to the infection. Moreover, there was an obvious specificity in discrimination of microbial classes, because the signalling cascade had no influence on the induction of anti-bacterial peptides.

Searches against the intracellular Toll domain were similarly performed and a human homologue hToll, now classified as the Toll-like receptor (TLR)4 was discovered [4]. Forced dimerisation of TLR4 induces a typical pro-inflammatory state, confirming not only structural but also functional similarity with the Toll of *Drosophila*. Ten human TLRs have now been described, and years of intense research have demonstrated the importance of this system for sensing a variety of microbial as well as endogenous products, including the main classes of human heat shock proteins.

7.2. Toll-like receptor structure

TLRs are type I membrane proteins. The extracellular (or lumenal) part of TLRs is composed of tandemly repeated modules enriched in leucines and, hence, called leucine-rich repeats (LRRs). LRRs are found in a variety of proteins with very different functions. The single LRR is usually 20–29 residues long and displays a characteristic leucine distribution pattern. By comparing sequences of LRRs from different proteins one can distinguish several classes which reflect the differences in the C-terminal part of the LRR module. On the other side, the leucine positions in the initial 10 amino acid stretches are well conserved and show an X-L-X-X-L-X-L-X-X-N pattern (X being any amino acid). The spatial visualisation of this conservation has been provided by crystal and nuclear magnetic resonance structures of several LRR proteins. They all indicate a horseshoe-like molecular shape in which parallel β-strands in perpendicular fashion line the concave horseshoe surface. β-strands fold out of the first, conserved LRR halves which explains the invariant appearance of the concave surface. The β-strand in each LRR module is followed by the less conserved stretch which forms more variable structures. The analysis of crystal structures of LRR proteins complexed with their ligands reveals that the concave surfaces and β–α loops provide the required interaction platform [5].

Unfortunately, no TLR structure has yet been solved, and the mechanistic aspects of activation therefore remain enigmatic. Because of the broad range of non-proteinaceous ligands implicated from functional studies, the structural determinants for TLR–ligand interactions might appear quite different to those already established. The recent bioinformatical analysis of the TLR family has implicated the insertions within the LRRs (typically at positions 10 and 15) to be important in controlling specificity of interaction [6]. Nevertheless, to test this and other hypotheses it will only be possible by solving X-ray structures of different TLR–ligand complexes.

The cytoplasmic domain of the TLR, the Toll/IL-1 receptor (TIR) domain, is responsible for all features of the receptor-mediated intracellular signalling of the TLR family. It was the TIR domain that provided the first evidence for the homology that spans such distant taxons as the fruit fly and the human and related proteins with such different functions as embryonic body polarisation and controlling the inflammatory response [7, 8]. Discovery of the TIR domain in proteins encoded by the plant disease-resistance genes underscore the conservation and, hence, the importance of this evolutionary 'invention' [9].

Together with the first insight into the heterogeneity of the TLR family, attempts were made to structurally analyse the TIR domain [10]. The $(\beta/\alpha)_5$ fold was predicted and the similarity to the bacterial chemotaxis regulator, CheY was implicated. The structural work proved the resemblance between human TIR domains and CheY and also revealed some specific properties [11]. Comparisons of TIR domain structures of different TLRs, including those that had been mutated or were non-functional, provided some insight into mechanistic aspects of the biology of these receptors. Two interaction interfaces on the domain surface appeared to be possible. The first of them, the R face, which varies in different receptors, appears to contribute to the specificity of the ligand recognition by specifying the oligomerisation of the receptors. Another one, the S face, which displays a conserved surface patch, contributes to the invariant aspect of the TLR family function by engaging the conserved intracellular machinery of cell activation.

7.3. Toll-like receptor signalling

After binding to their cognate extracellular ligands, many classes of membrane-bound receptors modulate or even re-programme cellular functions by recruiting and activating different sets of cytoplasmic mediators. The mediators, in turn, transmit inhibitory or activating signals to transcription factors, and these, consequently, alter the transcriptional profile and define the cellular response to the original external input.

The response to TLR engagement was first demonstrated by manipulating human TLR4 [4]. The extracellular part of the receptor was exchanged for the CD4 fragment, which drives spontaneous homodimerisation of the fusion protein. This forced ligand-independent signalling led to the paradigmatic state of immune alertness, namely, the activation of the transcription factor NF-κB and, consequently, the induction of pro-inflammatory cytokines and the upregulation of co-stimulatory molecules. Intense genetic and biochemical work ensued, and this has elucidated many aspects of the intracellular process. We will briefly discuss this work.

MyD88 is the first and essential adaptor molecule which becomes engaged after activation of every TLR, except for TLR3. MyD88-deficient mice show impaired responsiveness not only to most TLR ligands, but also to IL-1 and IL-18, thereby proving MyD88 to be a critical signalling component for any TIR domain-containing receptor [12, 13]. MyD88 possesses its own C-terminal TIR domain, which drives the heterodimerisation of the adaptor with the activated receptor. The N-terminal death domain of MyD88 then recruits IRAK1 and IRAK4 kinases. The IRAK4 phosphorylates and activates the IRAK1, which in turn initiates autophosphorylation and recruits TRAF6. Lack of IRAK4 reproduces the defects seen in MyD88 mutants [14]. The importance of IRAK1 is less clear, because its deficiency exhibits a less pronounced phenotype. In addition, there are two further IRAKs, neither of which have kinase activity – IRAK2 and IRAK-M. IRAK-M has been found to be a negative regulator of TLR signalling [15], whereas the exact role of IRAK2 remains unclear.

Of the family of six TRAF adaptor proteins, only TRAF6 is involved in TLR signalling. TRAF6, together with IRAK, dissociates from the activated receptor and binds to the preformed complex of TGF-β-activated kinase (TAK)1 and TAK1-binding proteins (TAB)1 and 2. TAK1 is a mitogen-activated protein kinase kinase kinase (MAP3K) involved in activation of IκB kinase (IKK). Activation of IKK appears to require atypical polyubiquitination [16, 17]. The TRAF6/TAK1/TAB1/TAB2 complex associates with the heterodimeric ubiquitin-conjugating enzyme Ubc13/Uev1A. This results in modification of TRAF6 with lysin63-linked polyubiquitin chain, which leads to IKK activation, IκB phosphorylation, ubiquitination and degradation. From the IκB released transcription factor NF-κB translocates to the nucleus and switches on the transcription of a large number of pro-inflammatory genes. TAK1 is also responsible for activation of MAP kinases (MAPKs) and c-Jun N-terminal kinases (JNKs), in this way broadening and diversifying transcriptional changes in response to receptor stimulation.

The specificity of the response to different ligands is assumed to arise from different adaptors associating with the respective receptors. In particular, and in contrast to other TLRs, TLR3 and TLR4 engagement is known to activate the transcription factor Interferon Regulatory Factor 3 (IRF3) and as a consequence to induce interferon-β. Furthermore, MyD88-deficient mice are not able to produce a number of cytokines upon challenge with lipopolysaccharide (LPS), whereas they still show NF-κB and JNK activation, albeit with slower kinetics [13]. This evidence prompted an intense search for additional adaptors.

The first candidate was a molecule named TIRAP/Mal; however, knock-out studies confirmed the function of the TIRAP/Mal in MyD88-dependent pathway [18, 19]. The specificity of another adaptor, the TRIF/Ticam1, was subsequently substantiated using classical and reverse genetic approaches [20, 21]. These studies proved the role of TRIF/Ticam in MyD88-independent effects upon TLR3 and TLR4 activation. The latest addition to the ever growing list of response specifiers is yet another TIR domain-containing molecule – TRAM. Analysis of TRAM-deficient mice located this adaptor in the exclusive position on the TLR4-triggered MyD88-independent pathway of cell activation [22].

There are a number of other receptor-proximal molecules implicated in one or another aspect of TLR function and further studies are awaited, for example, PI3-kinase, Tollip, Pellino or sterile alpml motif (SAM) and armadillo motif (ARM). The future will undoubtedly reveal additional complexities but, concomitantly, also a better understanding of TLR biology, thereby substantiating in molecular terms the hypothesis of the innate, and thus invariant system, of discrimination and the identification of 'foreign and dangerous'. The importance of this knowledge will become even more apparent during our further discussion showing that the TLR system has been accommodated to sense 'danger' by interacting with heat shock proteins, independently of their origin.

7.4. Heat shock protein signalling via Toll-like receptors

There are numerous circumstances in which the immune system becomes activated in the absence of infection. The most prominent examples could be aseptic necrosis or immunisation with syngeneic tumours. To explain these and other phenomena it has been postulated that the organism possesses its own (i.e., endogenous) danger signals that are normally concealed inside the cell. When released, these alert the organism that something is going wrong. Heat shock proteins, also called chaperones (see Chapter 1 for details on nomenclature) due to their assistance in protein folding and translocation, seem to be perfect candidates for endogenous danger signals. Firstly, most of them are essential and abundant. Secondly, some of them get strongly upregulated under stress conditions (hence, another name – stress proteins). Thirdly, they are normally located intracellularly. In addition, bacterial homologues have long been known to be immunodominant antigens in infections [23]. Pathogen heat shock protein–specific antibodies and T cell clones are often found in infected individuals, and their cross-reactivity with endogenous homologous molecules has been used to explain the pathogenesis of several autoimmune disorders. See Chapter 16 for more details of the immunology of chaperonins.

The essential role of heat shock proteins in a number of cellular processes [24] suggests that they should be evolutionarily conserved, and this is indeed the case. This sequence conservation also supports the cross-reactivity hypothesis and provides an additional basis for the proposition that heat shock proteins act as endogenous danger signals.

A strong thrust to the heat shock protein immunology field came with the discovery that some stress proteins are able to shuttle antigenic peptides into antigen presentation pathways and thereby prime adaptive immune responses [25–27]. Many details of this process have been elucidated including the characterisation of peptide association, the discovery of several receptors involved in the uptake of these proteins and some insights into the intracellular trafficking events. These aspects are discussed in Chapters 16–18. However, we will review here the stimulatory capacities of heat shock proteins and summarise what has been recently learned about their mechanisms.

7.4.1. Hsp60 signalling

In wishing to stress the point that heat shock proteins are more than an additional adjuvant for the immune system, it is salutary to start with Hsp60, the chaperone which does not have a documented ability to transport non-covalently bound antigens. In other chapters in this book the generic term chaperonin (Cpn)60 is used to describe this protein. The term Hsp60 tends to be used with eukaryotic (mitochondrial) proteins, but in this chapter it will be used to describe both human and bacterial proteins.

Hsp60 (also called GroEL – the *Escherichia coli* Hsp60) has the most impressive file on its role in immune responses among all chaperones. Immune reactions to bacterial Hsp60 become so prominent in some instances that they should be regarded as the characteristic sign of infection. Additionally, considerable interest in the inflammatory effects of Hsp60 arose from the field of atherosclerosis research. Chlamydial infection has been implicated in the pathogenesis of atherosclerosis, and chlamydial Hsp60 has been found in atherosclerotic lesions [28]. Interestingly, the human Hsp60 was co-localised with the chlamydial protein in the same study and it was shown to activate a number of different cell types. Subsequently, systematic analyses of human Hsp60 stimulatory features demonstrated a typical pro-inflammatory reaction pattern – specifically, the induction of TNF-α, IL-6, IL-12, IL-15 and nitric oxide (NO) production and synergy with IFN-γ – which was indistinguishable from stimulation with the classical bacterial inflammogen, LPS [29, 30]. These findings prompted the proposition that Hsp60 is an endogenous danger signal.

At the same time, evidence accumulated to establish the TLRs as the central axis of the infectious danger sensor. The question arose as to whether Hsp60 exploits the same receptor system. The first hint came with the analysis of CD14 involvement [31]. CD14 has been described as a high-affinity binding receptor for LPS [32] and for other bacterial products [33]. It was shown that cells, otherwise unresponsive to Hsp60, gained sensitivity to Hsp60 upon expression of CD14 [31]. Furthermore, anti-CD14 antibodies blocked Hsp60 activating potential on peripheral blood mononuclear cells. Attempts to analyse intracellular events following encounter with Hsp60 were undertaken and a transient activation of the transcription factor ATF2 was observed. ATF2 is a target of p38 MAPK, typically activated by a variety of bacterial products. CD14 is a glycosylphosphatidyl inositol (GPI)-anchored membrane protein without an intracellular domain and thus needs a partner to trigger signalling. CD14 can be found in complex with TLR4 on the cell surface and is thought to concentrate and pass ligands to it, thereby increasing TLR4 sensitivity [34]. Indeed, another study has provided evidence for the involvement of TLR4 in Hsp60-driven cell activation [35]. This study used C3H/HeJ mice, which are known to carry a mutation in the TLR4 TIR domain (P712H) rendering it unresponsive to TLR4 ligands [36]. C3H/HeJ mice failed to respond to Hsp60 stimulation in regard to each parameter tested, whereas C3H/HeN control mice showed normal reaction. Although there were no data on the direct intracellular events, the study gave clear indications for this kind of analysis, which followed soon afterwards.

The evidence provided by Vabulas et al. [37] remains the most direct data on Hsp60-triggered signalling. However, evidence for TLR4-independent signalling of Hsp60 proteins is presented in Chapter 6. Kinase activity measurements have shown the involvement of JNK and IKK in the macrophage activation process. Although the IKKβ subunit is typically responsible for the inducibility of IKK complex activity, IKKα activation was shown. Why this subunit is activated during Hsp60-induced signalling in macrophages and the consequences thereof require investigation.

As discussed earlier, IKK activation leads to IκB phosphorylation, ubiquitination and degradation allowing the NF-κB transcription factor to exit the cytoplasm and enter the nucleus. Analysis of IκB degradation in response to extracellular Hsp60 has shown good kinetic correlation to IKK activity. Together with NF-κB mobility shift assays [30], it has supported the involvement of the canonical pathway of NF-κB activation. Interestingly, analysis of different kinases measured either by kinase assays or by means of activated-state-specific antibodies have revealed delayed and sustained activation kinetics compared to most other bacterial stimuli ([37] and unpublished data). Only polyI:C, the surrogate double-stranded RNA which is the ligand of TLR3 [38], displays a similar kinetic pattern. Yet the involvement of TLR3 could be excluded by showing MyD88 dependence of Hsp60 signalling.

As discussed previously, the activity of all known TLRs, except TLR3, relies on the adaptor MyD88; hence, the demonstration of MyD88 dependence on JNK and IKK strongly supports the initial hypothesis that TLRs are involved in Hsp60 signalling. The definitive proof was supplied using the genetic reconstitution system. When Hsp60-unresponsive cells were transfected with cDNA encoding TLR2 or TLR4, the cells gained the sensitivity to Hsp60 as measured by an NF-κB–dependent luciferase reporter. Interestingly, TLR4-mediated signalling required co-transfection of MD2 co-receptor. MD2 is also known to be essential for LPS/TLR4 interaction [39]. The proof of specificity (i.e., the engagement of only selected receptors) provided the result from the reconstitution with TLR9. In this case Hsp60 failed to activate transfected cells. However, one should cautiously evaluate the latter data because of recent evidence on vesicular localisation of TLR9 [40] which possibly suggests that TLR9 is unavailable for interactions with extracellular ligands.

Another study, also aimed primarily at elucidation of heat shock protein signalling pathways, has supported and expanded previous findings [41]. The authors analysed the induction of cyclooxygenase-2 (COX2) and NO synthase-2 (NOS2) in macrophages and endothelial cells by Hsp60 by measuring the activity of respective luciferase reporters with intact and mutated transcription factor binding sites. They found that the inducibility of COX2 was dependent on the intact NF-κB binding site, cAMP-response element (CRE) and two NF-IL6 sites. One of the NF-IL6 elements was especially important for Hsp60 inducibility of the reporter. The mobility shift assays confirmed the earlier finding of NF-κB activation of others [30] and additionally validated the involvement of CRE binding protein. Furthermore, using phospho-specific antibodies and inhibitors of ERK and p38 MAPKs, the authors confirmed the involvement of ERK, JNK and p38 in cellular activation. Interestingly, IFN-γ was again shown to act synergistically with Hsp60 for the induction of NOS2, which is reminiscent of the synergy between IFN-γ and LPS.

The discussed studies and additional evidence [42–46] show that Hsp60 signalling, apart from the differences in the kinetics, is very similar to the signalling induced by other microbial ligands. This is in agreement with the involvement of TLRs. However, these similarities have prompted the proposition that the effects of heat shock proteins might result from contaminants in the heat shock protein preparations. This concern will be discussed later; however, at this point we would like to mention one aspect in advance, namely the requirement of Hsp60 uptake to initiate signalling which is, without exception, overlooked by protagonists of the contamination argument.

During the search for optimal conditions for Hsp60-driven macrophage activation, it was noted that serum exerts an inhibitory effect on the interaction [37]. Subsequent analysis demonstrated that the inhibition of Hsp60 signalling correlated with the blockade of Hsp60 uptake. Employing the clathrin-dependent endocytosis inhibitor monodansylcadaverin, the link between endocytosis of Hsp60 and its triggered signalling was established [37]. In contrast to the requirement for endocytosis in the signalling by Hsp60, LPS signalling is independent of uptake and can be initiated from the cell surface [40, 47].

7.4.2. Gp96 signalling

The connection between signalling and endocytosis makes sense in the case of heat shock proteins that are able to shuttle antigenic peptides. It is appealing to anticipate some economic rationale behind the evolutionary process having led to the situation in which endocytosis is a premise to evoke the signalling. Only those antigen-presenting cells that are able to internalise heat shock protein–peptide complexes become activated and thus acquire the immune response priming capacity. Gp96 and Hsp70 are chaperones that are capable of peptide shuttling. Analysis of their signalling capacity followed that of Hsp60 and soon allowed different aspects of the process, including the dependence on internalisation, to be compared.

Gp96 is an endoplasmic reticulum (ER)-resident chaperone, the function of which is less well understood than that of its cytoplasmic paralogue Hsp90. Nevertheless, gp96 has long been known to be able to confer tumour-specific immunity, a capacity that has been shown to be dependent on peptides with which it is associated [26, 27]. Further analysis revealed that gp96 can activate NF-κB and induce dendritic cell (DC) maturation [48, 49]. One study has claimed that this peptide-independent stimulatory capacity of gp96 is a major determinant of its anti-tumour activity [50].

Truncated forms of gp96 have been constructed, one lacking the ER-anchoring carboxy-terminal sequence Lys-Asp-Glu-Leu (KDEL) sequence and another without the complete C-terminal substrate binding domain. Both modifications render gp96 secretable, and the supernatants collected from the transfected cells were able to induce DC maturation in a similar manner to that induced by microbial products.

The interaction of human gp96 with TLRs has been investigated [51]. Resembling Hsp60-driven activation, it has been demonstrated that gp96 engages TLR2 and TLR4 to activate an NF-κB–dependent luciferase reporter. Again, MD2 appeared to be indispensable for TLR4 engagement. In regard to intracellular signalling, the activation of the typical inflammatory pattern of kinases was observed, namely, the activation of ERK, JNK and p38 MAP kinases. Because IκB degradation was detected, it was assumed that IKK also becomes activated. A clear dependency of the signalling on endocytosis was detected, as is the situation with Hsp60. However, given the capacity of gp96 to shuttle peptides, in this instance there was a more plausible basis to explain the requirement for cellular uptake in heat shock protein–mediated activation.

Comparison of Hsp60- and gp96-driven activation of immune cells suggested that the process is possibly more complicated than the simple ligand–receptor interaction. Bone marrow–derived DCs from TLR2-deficient and TLR4-mutant mice show partial defects in their response to Hsp60 as expected from genetic reconstitution data [37]. In contrast, similar experiments with gp96 unexpectedly showed that the response in DCs from mutant mice was completely controlled by TLR4 [51]. This finding could be explained by differences in the accessory proteins needed for response to ensue. Different uptake receptors could possibly act as critical specifiers of biological activity, and the requirement for these might differ between heat shock proteins and cell types. Thus, the cell type or even the developmental and maturational status of the same cell type seems to influence the type of reaction anticipated after stimulation with heat shock protein.

Additional evidence argues for interaction of gp96 and TLRs and comes from mutation studies of the 70Z/3 cell line [52]. The results from this study showed defects in the maturation of several cell surface receptors, notably, TLR1, TLR2 and TLR4, when gp96 was deleted or mutated. Interestingly, reconstitution of the missing gp96 or by the analogous parts or even the whole molecule of the cytoplasmic paralogue Hsp90 did not relieve the phenotype. However, distinct interactions between gp96 and TLRs in the ER as compared to the cell surface or endocytic compartments cannot be excluded and warrants further analysis.

7.4.3. Hsp70 signalling

Hsp70 is the second chaperone taking an exclusive place in immunology due to its role as a peptide shuttle [25]. The biology of this protein is also discussed in various other chapters in this volume. The analysis of Hsp70-triggered signalling events began with the proposal that CD14 is its receptor on the cell surface [53]. Besides demonstrating the inflammatory potential of human Hsp70 leading to NF-κB activation and the induction of pro-inflammatory cytokines IL-1β and IL-6, this study presented two additional, interesting and important findings. The first was the rapid induction of a transient cytoplasmic Ca^{2+} wave. This feature clearly distinguishes Hsp70 from LPS (described in more detail in Chapters 8 and 10). The second interesting observation was the unequal contribution of CD14 for the induction of different cytokines. Specifically, IL-1β and IL-6 were shown to depend on the presence of CD14, whereas TNF-α was produced independently of it. Unfortunately, the Ca^{2+} flux inhibitor turned out to be non-specific regarding CD14, in that CD14-dependent, IL-1β and IL-6 secretion, as well as CD14-independent TNF-α secretion upon Hsp70 stimulation, were compromised. Thus, the Ca^{2+} flux role in CD14-dependent effects remains an important, but still unresolved, issue. The study of human Hsp70 interactions with DCs followed and established this chaperone as a potent stimulus of innate immunity [54].

Finally, due to the established participation of CD14 and the insight gained with Hsp60 and gp96, the involvement of TLRs was investigated. Two studies simultaneously described TLR2 and TLR4 as receptors of Hsp70. Vabulas and colleagues [55] analysed the signalling pathway induced by Hsp70 in the macrophage cell line RAW264.7 by transfecting dominant negative forms of MyD88 and TRAF6 adaptors from the TLR signal pathway. Furthermore, a versatile tool for spatial dissection of signalling was used, namely, a cell line stably expressing functional MyD88-EGFP fusion protein. This allowed the visualisation of Hsp70-triggered signalling. The kinetics and pattern of MyD88 recruitment could be followed at the single-cell level. Once again, the results provided a strong argument against LPS contamination being responsible for the observed effects, since Hsp70 induced the intracellular MyD88 recruitment pattern, in contrast with LPS stimulation, which results in MyD88 recruitment to the macrophage surface [40, 47]. Furthermore, by employing transient transfections of TLR2 and TLR4 into an otherwise Hsp70-unresponsive fibroblast cell line, the authors demonstrated a specific involvement of those TLRs, because TLR9 transfection did not confer sensitivity to Hsp70.

Finally, to validate the results, MyD88-, TLR2- and TLR4-mutant mice were used. DCs from TLR4-mutant mice showed a complete defect in TNF-α and IL-12 production, which is reminiscent of the situation with gp96 and again implies the involvement of additional components in receptor complexes. A second study came to the same conclusion by using TLR stably transfected fibroblasts [56]. Apart from documenting the involvement of TLR2 and TLR4 in cellular activation by Hsp70, the authors confirmed their previous finding on CD14 involvement and showed TLR2/TLR4 synergism, which was established to be MyD88-independent.

Continuing work provides additional insights into Hsp70-induced responses and suggests broader implications. For example, an immediate release Hsp70 following heart surgery has been reported and might influence accompanying inflammatory effects [57]. Another study has demonstrated interactions of Hsp70 (and other heat shock proteins) with microglial cells, thereby suggesting an ability to evoke inflammation in the neural system [58]. The importance of understanding Hsp70-triggered cellular activation is therefore difficult to over-estimate.

7.4.4. Non-Toll-like receptor signalling

We have centred our discussion on heat shock protein–induced signalling around TLR-associated intracellular events because these have, to date, been more widely analysed and characterised. Nevertheless, TLR biology is still far from understood, especially with respect to the complexity of interactions with the diverse set of cell surface components [59]. In addition, given the documented versatility of heat shock protein uptake/scavenging mechanisms, it is tempting to speculate on additional non-TLR-driven signalling as a sequela to heat shock protein encounter. There has been some controversy regarding the route of gp96 uptake [60, 61]. Although gp96 uptake receptors under discussion (CD91, SR-A) are not primarily devoted to signalling, it is feasible to assume that they could introduce additional qualities into the primary reaction triggered by TLRs. The same holds true in the case of Hsp70, because CD91 has also been shown to participate in the uptake of Hsp70 [62] as has the scavenger receptor LOX1 [63]. Transient Ca^{2+} rises, induced by extracellular Hsp70 and discussed earlier, could represent one aspect of the TLR-independent process [53]. Interestingly, CD40 has been shown to interact with bacterial and human Hsp70 [64, 65]. The signalling potential of CD40 is well established and is highly relevant for the immune system. It would be interesting to explore this particular interaction in more detail. See Chapter 10 for a fuller discussion of CD40/Hsp70 interactions.

The peptides originating from microbial or endogenous heat shock proteins and their capacity to stimulate and expand peptide-specific T cell clones should also be highlighted [66]. Although such responses would involve different intracellular signalling cascades including the T cell receptor (TCR) signalling axis, the outcome could be as impressive as that following TLR engagement.

7.5. Contamination issue

Due to the extreme sensitivity of the immune system to some bacterial products, most notably LPS, concerns regarding reagent contamination inevitably accompany investigations of immune cell activation. This concern becomes of primary importance if the same receptor systems are implicated, as in the case of TLR2/TLR4 and heat shock proteins. From the beginning, much effort has been expended in order to exclude the possibility that contamination of heat shock protein preparations used for investigations accounts for the observed responses. Nevertheless, several reports proposing that some [67] or all of the stimulatory effects of heat shock proteins [68–70] originate from contaminating LPS have appeared. After examination of the available data, some major points of criticism of this 'contamination theory' should be mentioned:

1. Responsiveness to any ligand is dependent on the responsive cell type. Negative data obtained using a single cell line are not strong evidence on which to include or exclude an effect. The danger of ending up with a particular clone over short or long periods is well known and real. This unintended sub-cloning often results in loss of some and gain of additional properties, such as accessory receptors and defective signalling pathways to name but two. In some instances, the methods sections in publications acknowledge these problems. For example, in one study cell harvesting might be reported to require scraping, whereas in another the 'same' cell line is described as being only loosely attached or semi-adherent.

2. The possibility that different sub-units in receptor complexes are required for different heat shock proteins adds complexity to the analysis. If stimulation were to originate from LPS contamination, then the same DC would respond similarly. However, these cells respond to gp96 and Hsp70 irrespective of TLR2 expression and the response is fully dependent on TLR4. In contrast, Hsp60 uses both receptors in the same cells. These functional distinctions argue against the contamination theory.

3. Correct culture and, most importantly, activation conditions are of great concern. It is difficult to compare data collected from experiments in which stimulations are performed in serum-containing medium with those performed in the absence of serum because serum had been found to inhibit the process. At this point it is appropriate to mention the importance of generous upwards titrations in the event of negative results.

4. It is worth considering whether clarifying the source or even species of the protein would resolve some of these issues. Although homologous heat shock proteins are well conserved, one cannot exclude the possibility that small differences in primary sequence or in post-translational modification are critical for the activity.

5. Serum is known to enhance the sensitivity to LPS because of the presence of LPS binding protein (LBP), which interacts with LPS and passes it to CD14 on the myeloid cell surface [32]. If a considerable stimulatory capacity of heat shock protein preparations originates from contaminating LPS, how should one explain the strong inhibitory effect of serum on stimulation of macrophages with Hsp60 and gp96? Hsp70 has not been analysed in this respect by means of direct assays.

6. Last, but not least, is the endocytosis argument, an argument which is regrettably never discussed by proponents of the 'contamination theory'. LPS is known to signal from the cell surface and does not need endocytosis for its activity [40, 47]. In contrast, Hsp60 and gp96 clearly require uptake to initiate the signalling [37, 51]. Moreover, experiments with Hsp70 have shown that internalisation is required for TLR signalling, because the TLR-specific adaptor MyD88 is primarily recruited to the endocytic compartments, as visualised by means of MyD88-EGFP fusion protein [55]. This is in an obvious contrast to the cell surface recruitment of MyD88 after TLR4 engagement.

Even this incomplete list of concerns shows how important critical discussion and innovative experimental approaches are to advancing our understanding of the mechanisms of heat shock proteins and immune system interaction. Two recent publications demonstrate *in vivo* the adjuvant effects of Hsp70 and gp96, which simply can not be explained by contamination [71, 72]. Further evidence for the hypothesis that the biological effects of chaperones are not due to contaminants is provided in many of this book's chapters.

7.6. Conclusion

After presenting known facts on heat shock protein signalling we are now faced with two questions. The first question is – why TLRs? TLRs have evolved into a powerful system which is capable of sensing a broad range of invaders. TLRs not only mobilise cells for innate first line defence, they also condition the immune system for antigen-specific priming and thus provide a bridge between innate and adaptive immunity. Nothing else could be more suitable for alerting vertebrates to other kinds of danger. Endogenous heat shock proteins released as a consequence of tissue damage and signalling via TLR immediately attract all the resulting inflammatory power against the initiating insult. This scenario satisfies the original theoretical considerations [1, 73] and has received support from experimental data [74–77].

The second question, which asks how both pathogenic and host proteins can share the same receptor, is more difficult to answer. To gain some insight into this, one should remember that until recently it was believed that the only function of heat shock proteins was in the assistance in protein folding and prevention of aggregation. Binding and release of exposed or otherwise distorted regions on client polypeptides are features that are common to different classes of chaperones and constitute the mechanism of their biochemical action [78]. It is tempting to speculate that these common features account for the observed overlap in their interaction with TLRs. To test this and other assumptions at the molecular level is the challenge and the task for the future. Simultaneously, it is an opportunity to gain a better insight into the functions of TLRs and heat shock proteins, and this might reveal new avenues via which important physiological and pathological processes can be modulated.

REFERENCES

1. Janeway C A J. Approaching the asymptote? Evolution and revolution in immunology. Cold Spring Harb Symp Quant Biol 1989, 54 Pt 1: 1–13.
2. Gordon S. Pattern recognition receptors: doubling up for the innate immune response. Cell 2002, 111: 927–930.
3. Lemaitre B, Nicolas E, Michaut L, Reichhart J M and Hoffmann J A. The dorsoventral regulatory gene cassette spatzle/Toll/cactus controls the potent antifungal response in *Drosophila* adults. Cell 1996, 86: 973–983.
4. Medzhitov R, Preston-Hurlburt P and Janeway C A J. A human homologue of the *Drosophila* Toll protein signals activation of adaptive immunity. Nature 1997, 388: 394–397.
5. Kobe B and Deisenhofer J. A structural basis of the interactions between leucine-rich repeats and protein ligands. Nature 1995, 374: 183–186.
6. Bell J K, Mullen G E, Leifer C A, Mazzoni A, Davies D R and Segal D M. Leucine-rich repeats and pathogen recognition in Toll-like receptors. Trends Immunol 2003, 24: 528–533.
7. Gay N J and Keith F J. *Drosophila* Toll and IL-1 receptor. Nature 1991, 351: 355–356.
8. Schneider D S, Hudson K L, Lin T Y and Anderson K V. Dominant and recessive mutations define functional domains of Toll, a transmembrane protein required for dorsal-ventral polarity in the *Drosophila* embryo. Genes Dev 1991, 5: 797–807.
9. Whitham S, Dinesh-Kumar S P, Choi D, Hehl R, Corr C and Baker B. The product of the tobacco mosaic virus resistance gene N: similarity to toll and the interleukin-1 receptor. Cell 1994, 78: 1101–1115.
10. Rock F L, Hardiman G, Timans J C, Kastelein R A and Bazan J F. A family of human receptors structurally related to *Drosophila* Toll. Proc Natl Acad Sci USA 1998, 95: 588–593.
11. Xu Y, Tao X, Shen B, Horng T, Medzhitov R, Manley J L and Tong L. Structural basis for signal transduction by the Toll/interleukin-1 receptor domains. Nature 2000, 408: 111–115.
12. Adachi O, Kawai T, Takeda K, Matsumoto M, Tsutsui H, Sakagami M, Nakanishi K and Akira S. Targeted disruption of the MyD88 gene results in loss of IL-1- and IL-18-mediated function. Immunity 1998, 9: 143–150.
13. Kawai T, Adachi O, Ogawa T, Takeda K and Akira S. Unresponsiveness of MyD88-deficient mice to endotoxin. Immunity 1999, 11: 115–122.
14. Suzuki N, Suzuki S, Duncan G S, Millar D G, Wada T, Mirtsos C, Takada H, Wakeham A, Itie A, Li S, Penninger J M, Wesche H, Ohashi P S, Mak T W and Yeh W C. Severe impairment of interleukin-1 and Toll-like receptor signalling in mice lacking IRAK-4. Nature 2002, 416: 750–756.

15. Kobayashi K, Hernandez L D, Galan J E, Janeway C A J, Medzhitov R and Flavell R A. IRAK-M is a negative regulator of Toll-like receptor signaling. Cell 2002, 110: 191–202.

16. Deng L, Wang C, Spencer E, Yang L, Braun A, You J, Slaughter C, Pickart C and Chen Z J. Activation of the IkappaB kinase complex by TRAF6 requires a dimeric ubiquitin-conjugating enzyme complex and a unique polyubiquitin chain. Cell 2000, 103: 351–361.

17. Wang C, Deng L, Hong M, Akkaraju G R, Inoue J and Chen Z J. TAK1 is a ubiquitin-dependent kinase of MKK and IKK. Nature 2001, 412: 346–351.

18. Horng T, Barton G M, Flavell R A and Medzhitov R. The adaptor molecule TIRAP provides signalling specificity for Toll-like receptors. Nature 2002, 420: 329–333.

19. Yamamoto M, Sato S, Hemmi H, Sanjo H, Uematsu S, Kaisho T, Hoshino K, Takeuchi O, Kobayashi M, Fujita T, Takeda K and Akira S. Essential role for TIRAP in activation of the signalling cascade shared by TLR2 and TLR4. Nature 2002, 420: 324–329.

20. Hoebe K, Du X, Georgel P, Janssen E, Tabeta K, Kim S O, Goode J, Lin P, Mann N, Mudd S, Crozat K, Sovath S, Han J and Beutler B. Identification of Lps2 as a key transducer of MyD88-independent TIR signalling. Nature 2003, 424: 743–748.

21. Yamamoto M, Sato S, Hemmi H, Hoshino K, Kaisho T, Sanjo H, Takeuchi O, Sugiyama M, Okabe M, Takeda K and Akira S. Role of adaptor TRIF in the MyD88-independent toll-like receptor signaling pathway. Science 2003, 301: 640–643.

22. Yamamoto M, Sato S, Hemmi H, Uematsu S, Hoshino K, Kaisho T, Takeuchi O, Takeda K and Akira S. TRAM is specifically involved in the Toll-like receptor 4-mediated MyD88-independent signaling pathway. Nat Immunol 2003, 4: 1144–1150.

23. Kaufmann S H E. Heat shock proteins and the immune response. Immunol Today 1990, 11: 129–136.

24. Hartl F U and Hayer-Hartl M. Molecular chaperones in the cytosol: from nascent chain to folded protein. Science 2002, 295: 1852–1858.

25. Udono H and Srivastava P K. Heat shock protein 70-associated peptides elicit specific cancer immunity. J Exp Med 1993, 178: 1391–1396.

26. Arnold D, Faath S, Rammensee H-G and Schild H. Cross-priming of minor histocompatibility antigen-specific cytotoxic T cells upon immunization with the heat shock protein gp96. J Exp Med 1995, 182: 885–889.

27. Suto R and Srivastava P K. A mechanism for the specific immunogenicity of heat shock protein-chaperoned peptides. Science 1995, 269: 1585–1588.

28. Kol A, Sukhova G K, Lichtman A H and Libby P. Chlamydial heat shock protein 60 localizes in human atheroma and regulates macrophage tumour necrosis factor-a and matrix metalloproteinase expression. Circulation 1998, 98: 300–307.

29. Chen W, Syldath U, Bellmann K, Burkart V and Kold H. Human 60-kDa heat-shock protein: a danger signal to the innate immune system. J Immunol 1999, 162: 3212–3219.

30. Kol A, Bourcier T, Lichtman A and Libby P. Chlamydial and human heat shock protein 60s activate human vascular endothelium, smooth muscle cells, and macrophages. J Clin Invest 1999, 103: 571–577.

31. Kol A, Lichtman A H, Finberg R W, Libby P and Kurt-Jones E A. Heat shock protein (HSP) 60 activates the innate immune response: CD14 is an essential receptor for HSP60 activation of mononuclear cells. J Immunol 2000, 164: 13–17.

32. Wright S D, Ramos R A, Tobias P S, Ulevitch R J and Mathison J C. CD14, a receptor for complexes of lipopolysaccharide (LPS) and LPS binding protein. Science 1990, 249: 1431–1433.

33. Pugin J, Heumann I D, Tomasz A, Kravchenko VV, Akamatsu Y, Nishijima M, Glauser M P, Tobias P S and Ulevitch R J. CD14 is a pattern recognition receptor. Immunity 1994, 1: 509–516.

34. da Silva Correia J, Soldau K, Christen U, Tobias P S and Ulevitch R J. Lipopolysaccharide is in close proximity to each of the proteins in its membrane receptor complex. Transfer from CD14 to TLR4 and MD-2. J Biol Chem 2001, 276: 21129–21135.

35. Ohashi K, Burkart V, Flohé S and Kolb H. Heat shock protein 60 is a putative endogenous ligand of the Toll-like receptor-4 complex. J Immunol 2000, 164: 558–561.

36. Poltorak A, He X, Smirnova I, Liu M Y, van Huffel C, Du X, Birdwell D, Alejos E, Silva M, Galanos C, Freudenberg M, Ricciardi-Castagnoli P, Layton B and Beutler B. Defective LPS signaling in C3H/HeJ and C57BL/10ScCr mice: mutations in Tlr4 gene. Science 1998, 282: 2085–2088.

37. Vabulas R M, Ahmad-Nejad P, da Costa C, Miethke T, Kirschning C J, Hacker H and Wagner H. Endocytosed HSP60s use Toll-like receptor 2 (TLR2) and TLR4 to activate the toll/interleukin-1 receptor signaling pathway in innate immune cells. J Biol Chem 2001, 276: 31332–31339.

38. Alexopoulou L, Holt A C, Medzhitov R and Flavell R A. Recognition of double-stranded RNA and activation of NF-kappaB by Toll-like receptor 3. Nature 2001, 413: 732–738.

39. Shimazu R, Akashi S, Ogata H, Nagai Y, Fukudome K, Miyake K and Kimoto M. MD-2, a molecule that confers lipopolysaccharide responsiveness on Toll-like receptor 4. J Exp Med 1999, 189: 17777–17782.

40. Ahmad-Nejad P, Hacker H, Rutz M, Bauer S, Vabulas R M and Wagner H. Bacterial CpG-DNA and lipopolysaccharides activate Toll-like receptors at distinct cellular compartments. Eur J Immunol 2002, 32: 1958–1968.

41. Billack B, Heck D E, Mariano T M, Gardner C R, Sur R, Laskin D L and Laskin J D. Induction of cyclooxygenase-2 by heat shock protein 60 in macrophages and endothelial cells. Am J Physiol (Cell Physiol) 2002, 283: C1267–1277.

42. Sasu S, LaVerda D, Qureshi N, Golenbock D T and Beasley D. *Chlamydia pneumoniae* and chlamydial heat shock protein 60 stimulate proliferation of human vascular smooth muscle cells via toll-like receptor 4 and p44/p42 mitogen-activated protein kinase activation. Circ Res 2001, 89: 244–250.

43. Zhang L, Pelech S L, Mayrand D, Grenier D, Heino J and Uitto V J. Bacterial heat shock protein-60 increases epithelial cell proliferation through the ERK1/2 MAP kinases. Exp Cell Res 2001, 266: 11–20.

44. Bulut Y, Faure E, Thomas L, Karahashi H, Michelsen K S, Equils O, Morrison S G, Morrison R P and Arditi M. Chlamydial heat shock protein 60 activates macrophages and endothelial cells through Toll-like receptor 4 and MD2 in a MyD88-dependent pathway. J Immunol 2002, 168: 1435–1440.

45. Flohé S B, Bruggemann J, Lendemans S, Nikulina M, Meierhoff G, Flohé S and Kolb H. Human heat shock protein 60 induces maturation of dendritic cells versus a Th1-promoting phenotype. J Immunol 2003, 170: 2340–2348.

46. Zanin-Zhorov A, Nussbaum G, Franitza S, Cohen I R and Lider O. T cells respond to heat shock protein 60 via TLR2: activation of adhesion and inhibition of chemokine receptors. FASEB J 2003, 17: 1567–1569.
47. Latz E, Visintin A, Lien E, Fitzgerald K A, Monks B G, Kurt-Jones E A, Golenbock D T and Espevik T. Lipopolysaccharide rapidly traffics to and from the Golgi apparatus with the toll-like receptor 4-MD-2-CD14 complex in a process that is distinct from the initiation of signal transduction. J Biol Chem 2002, 277: 47834–47843.
48. Basu S, Binder R J, Suto R, Anderson K M and Srivastava P K. Necrotic but not apoptotic cell death releases heat shock proteins, which deliver a partial maturation signal to dendritic cells and activates the NF-kB pathway. Int Immunol 2000, 12: 1539–1546.
49. Singh-Jasuja H, Scherer H U, Hilf N, Arnold-Schild D, Rammensee H-G, Toes R E M and Schild H. The heat shock protein gp96 induces maturation of dendritic cells and down-regulation of its receptor. Eur J Immunol 2000, 30: 2211–2215.
50. Baker-LePain J C, Sarzotti M, Fields T A, Li C Y and Nicchitta C V. GRP94 (gp96) and GRP94 N-terminal geldanamycin binding domain elicit tissue nonrestricted tumor suppression. J Exp Med 2002, 196: 1447–1459.
51. Vabulas R M, Braedel S, Hilf N, Singh-Jasuja H, Herter S, Ahmad-Nejad P, Kirschning C J, Da Costa C, Rammensee H G, Wagner H and Schild H. The endoplasmic reticulum-resident heat shock protein Gp96 activates dendritic cells via the Toll-like receptor 2/4 pathway. J Biol Chem 2002, 277: 20847–20853.
52. Randow F and Seed B. Endoplasmic reticulum chaperone gp96 is required for innate immunity but not cell viability. Nat Cell Biol 2001, 3: 891–896.
53. Asea A, Kraeft S-K, Kurt-Jones E A, Stevenson M A, Chen L B, Finberg R W, Koo G C and Calderwood S K. Hsp70 stimulates cytokine production through a CD14-dependent pathway, demonstrating its dual role as a chaperone and cytokine. Nat Med 2000, 6: 435–442.
54. Kuppner M C, Gastpar R, Gelwer S, Nossner E, Ochmann O, Scharner A and Issels R D. The role of heat shock protein (hsp70) in dendritic cell maturation: hsp70 induces the maturation of immature dendritic cells but reduces DC differentiation from monocyte precursors. Eur J Immunol 2001, 31: 1602–1609.
55. Vabulas R M, Ahmad-Nejad P, Ghose S, Kirschning C J, Issels R D and Wagner H. HSP70 as endogenous stimulus of the Toll/interleukin-1 receptor signal pathway. J Biol Chem 2002, 277: 15107–15112.
56. Asea A, Rehli M, Kabingu E, Boch J A, Baré O, Auron P E, Stevenson M A and Calderwood S K. Novel signal transduction pathway utilized by extracellular HSP70. Role of Toll-like receptor (TLR) 2 and TLR4. J Biol Chem 2002, 277: 15028–15034.
57. Dybdahl B, Wahba A, Lien E, Flo T H, Waage A, Qureshi N, Sellevold O F, Espevik T and Sundan A. Inflammatory response after open heart surgery: release of heat-shock protein 70 and signaling through Toll-like receptor-4. Circulation 2002, 105: 685–690.
58. Kakimura J, Kitamura Y, Takata K, Umeki M, Suzuki S, Shibagaki K, Taniguchi T, Nomura Y, Gebicke-Haerter P J, Smith M A, Perry G and Shimohama S. Microglial activation and amyloid-beta clearance induced by exogenous heat-shock proteins. FASEB J 2002: 601–603.
59. Underhill D M. Toll-like receptors: networking for success. Eur J Immunol 2003, 33: 1767–1775.

60. Binder R J, Han D K and Srivastava P K. CD91: a receptor for heat shock protein gp96. Nat Immunol 2000, 1: 151–155.
61. Berwin B, Hart J P, Rice S, Gass C, Pizzo S V, Post S R and Nicchitta C V. Scavenger receptor-A mediates gp96/GRP94 and calreticulin internalization by antigen-presenting cells. EMBO J 2003, 22: 6127–6136.
62. Basu S, Binder R J, Ramalingam T and Srivastava P K. CD91 is a common receptor for heat shock proteins gp96, hsp90, hsp70 and calreticulin. Immunity 2001, 14: 303–313.
63. Delneste Y, Magistrelli G, Gauchat J, Haeuw J, Aubry J, Nakamura K, Kawakami-Honda N, Goetsch L, Sawamura T, Bonnefoy J and Jeannin P. Involvement of LOX-1 in dendritic cell-mediated antigen cross-presentation. Immunity 2002, 17: 353–362.
64. Wang Y, Kelly C G, Karttunen T, Whittall T, Lehner P J, Duncan L, MacAry P, Younson J S, Singh M, Oehlmann W, Cheng G, Bergmeier L and Lehner T. CD40 is a cellular receptor mediating mycobacterial heat shock protein 70 stimulation of CC-chemokines. Immunity 2001, 15: 971–983.
65. Becker T, Hartl F U and Wieland F. CD40, an extracellular receptor for binding and uptake of Hsp70-peptide complexes. J Cell Biol 2002, 158: 1277–1285.
66. van Eden W. Stress proteins as targets for anti-inflammatory therapies. Drug Discov Today 2000, 5: 115–120.
67. Reed R C, Berwin B, Baker J P and Nicchitta C V. GRP94/gp96 elicits ERK activation in murine macrophages. A role for endotoxin contamination in NF-kappa B activation and nitric oxide production. J Biol Chem 2003, 278: 31853–31860.
68. Bausinger H, Lipsker D, Ziylan U, Manie S, Briand J P, Cazenave J P, Muller S, Haeuw J F, Ravanat C, de la Salle H and Hanau D. Endotoxin-free heat-shock protein 70 fails to induce APC activation. Eur J Immunol 2002, 32: 3708–3713.
69. Gao B and Tsan M F. Endotoxin contamination in recombinant human Hsp70 preparation is responsible for the induction of TNFα release by murine macrophages. J Biol Chem 2003, 278: 174–179.
70. Gao B and Tsan M F. Recombinant human heat shock protein 60 does not induce the release of tumor necrosis factor alpha from murine macrophages. J Biol Chem 2003, 278: 22523–22529.
71. Millar D G, Garza K M, Odermatt B, Elford A R, Ono N, Li Z and Ohashi P S. Hsp70 promotes antigen-presenting cell function and converts T-cell tolerance to autoimmunity in vivo. Nat Med 2003, 9: 1469–1476.
72. Liu B, Dai J, Zheng H, Stoilova D, Sun S and Li Z. Cell surface expression of an endoplasmic reticulum resident heat shock protein gp96 triggers MyD88-dependent systemic autoimmune diseases. Proc Nat Acad Sci USA 2003, 100: 15824–15829.
73. Matzinger P. The Danger Model: A renewed sense of self. Science 2002, 296: 301–305.
74. Gallucci S, Lolkema M and Matzinger P. Natural adjuvants: endogenous activators of dendritic cells. Nat Med 1999, 11: 1249–1255.
75. Sauter B, Albert M L, Francisco L, Larsson M, Somersan S and Bhardwaj N. Consequences of cell death: exposure to necrotic tumour cells, but not primary tissue cells or apoptotic cells, induces the maturation of immunostimulatory dendritic cells. J Exp Med 2000, 191: 423–433.
76. Li M, Carpio D F, Zheng Y, Bruzzo P, Singh V, Ouaaz F, Medzhitov R M and Beg A A. An essential role of the NF-kappa B/Toll-like receptor pathway in induction of

inflammatory and tissue-repair gene expression by necrotic cells. J Immunol 2001, 166: 7128–7135.

77. Somersan S, Larsson M, Fonteneau J F, Basu S, Srivastava P and Bhardwaj N. Primary tumor tissue lysates are enriched in heat shock proteins and induce the maturation of human dendritic cells. J Immunol 2001, 167: 4844–4852.

78. Bukau B and Horwich A L. The Hsp70 and Hsp60 chaperone machines. Cell 1998, 92: 351–366.

8

Regulation of Signal Transduction by Intracellular and Extracellular Hsp70

Alexzander Asea and Stuart K. Calderwood

8.1. Introduction

There is a clear dichotomy between the effects of the 70-kDa heat shock protein (Hsp70) when expressed intracellularly and when released into the extracellular space. Intracellular Hsp70 is primarily implicated as a protein chaperone that transports and folds naïve, aberrantly folded, or mutated proteins, resulting in cytoprotection when cells are exposed to stressful stimuli – most notably heat shock itself. Intracellular Hsp70 also functions as a regulatory molecule and has a largely inhibitory function in cellular metabolism. In contrast, Hsp70 is implicated in immune activation, given that exposure of immunocompetent cells to exogenous Hsp70 triggers acute inflammatory responses, activates innate immunity and enhances anti-tumour surveillance. This review focuses on recent advances in understanding the contrasting roles of Hsp70 as an intracellular molecular chaperone and extracellular signalling ligand and highlights its relevance to host defence against pathogens and malignant transformation.

The formative studies of Hsp70, dating back over 30 years, suggested a strictly intracellular function for Hsp70 with the properties of a molecular chaperone – a protein that modulates the tertiary structures of other proteins and protects cells from stress [1, 2]. However, in 1989 Hightower and colleagues showed that cells in culture contain a pool of Hsp70 which is loosely associated with the cell and which could be released into the extracellular medium after heat shock or even after mild washing with tissue culture medium [3]. Subsequent *in vitro* studies indicated that such Hsp70 released from cells could be taken up by neuronal cells under which circumstances it enhances the molecular chaperoning power of these cells [4]. Later work identified the presence of extracellular Hsp70 in human subjects, as indicated by free Hsp70 and Hsp70 antibodies circulating in the bloodstream [5]; the biology of circulating molecular chaperones is discussed in Chapter 12. However, the studies that accelerated the interest in extracellular

Hsp70 and its functions were those of Srivastava and colleagues who showed that tumour-associated antigens bind stably to molecular chaperones such as Hsp70 and that these complexes can form the basis of an effective anti-cancer vaccine [6]. This aspect is discussed in Chapter 18. All aspects of the extracellular functions of Hsp70 are currently under active investigation, and we aim here to provide an overview of our present understanding.

8.2. Intracellular signalling functions of Hsp70   *Comparison Table*

As befits a molecular chaperone, intracellular Hsp70 plays a role in cellular signalling events that involves associations in high-molecular-weight complexes with other proteins [7, 8]. Hsp70 functions as a wide-spectrum negative regulator and plays a restraining role in a wide range of processes via its ability to inhibit the activities of protein kinases and transcription factors [7, 8]. Indeed, elevated Hsp70 inhibits a plethora of intracellular processes and, probably for this precise reason, Hsp70 levels are strictly regulated via negative control of its transcription factor heat shock factor 1 (HSF1) and its destabilisation at the mRNA level [9–12]. In addition, HSF1 itself can negatively regulate the promoters of cytokine genes and genes involved in cell proliferation [13, 14]. Particularly in the context of the inflammatory response and the innate immune response, therefore, intracellular Hsp70 and the heat shock system exhibit roles as negative regulators [13–16]. This property, as we shall discuss later, is in stark contrast to the pro-inflammatory and pro-immune functions of extracellular Hsp70.

8.3. Sources of extracellular Hsp70

It is evident that intracellular Hsp70 can escape into the interstitial fluid and is found at significant levels in the bloodstream [5, 17]. However, the cellular source of such Hsp70 and the mechanisms involved in the release of this intracellular protein are not well defined. See Chapter 2 for discussion of the cellular dispositions of mitochondrial Hsp70. As mentioned, cells in tissue culture can contain a pool of loosely associated Hsp70 which can be removed by gentle washing [3]. Such Hsp70 could be associated with the lipid rafts in the outer leaflet of the plasma membrane or could perhaps be loosely bound to other cell surface molecules [18–20]. This, of course, begs the question of how this intracellular protein is transported to the cell surface. One possible mechanism is via its release in association with proteins such as transferrin that are transported in exosomes bound to Hsp70 [21]. Details of this mechanism are provided in Chapter 3.

There is, in addition, considerable speculation that Hsp70 might be released into the extracellular space after destruction of the plasma membrane in cells

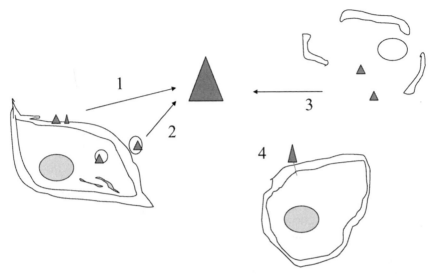

Figure 8.1. Origins of extracellular Hsp70. Extracellular Hsp70 (▲) may be derived from (1) a loosely bound pool on the plasma membrane, (2) through transport across the plasma membrane in exosomes, (3) from cells undergoing necrosis or (4) may be displayed in bound form on the cell surface.

undergoing necrotic death, and it might thereby act as a pro-inflammatory stimulus or a 'danger signal' for the immune system [22]. The extent to which this actually occurs *in vivo* is not clear, although, in the case of cancer cells, death is a continuous process and both apoptotic and necrotic cells are seen in the interior of tumours. It is also possible that Hsp70 is released after cancer therapy or as a consequence of ischaemic death following heart attacks or strokes [23].

A further source of extracellular Hsp70 appears to be brown fat from which Hsp70 is released in a quasi-endocrine manner, an effect that is exaggerated by exercise or behavioural stress [24, 25]. Such Hsp70 may be taken up by neuronal cells in the central nervous system that are chronically depleted of Hsp70 due to low endogenous HSF activity and the long distances involved in axonal transport of *de novo* synthesised proteins [4, 26, 27].

Yet another source of Hsp70 might be bound Hsp70 displayed on the surface of certain tumour cells [28]. Such bound Hsp70 is evidently able to interact with C-type lectin receptors on the surface of natural killer (NK) cells [29]. There is thus evidence for the existence of extracellular Hsp70 from a number of sources in the body which could act as a danger signal in the case of cancer or acute infection, as a target for tumour surveillance by NK cells, as a source of supplementary chaperones in the central nervous system or could play other as yet unknown functions [29–32] (Figure 8.1).

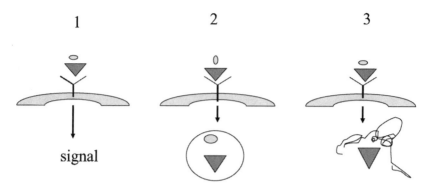

Figure 8.2. Consequences of Hsp70–cell surface binding. Hsp70 (▲) complexed with peptide cargo (●) (Hsp-PC) is shown binding to a cell surface receptor. Hsp70-PC is shown (1) activating intracellular signalling cascades, (2) entering the cell in endosomes through the process of receptor-mediated endocytosis and (3), after entering the cell as in (2), participating in molecular chaperone interactions with unfolded proteins.

8.4. Cell surface binding of Hsp70

Hsp70 and other chaperones released from cells appear able to bind with high affinity to cells such as B lymphocytes, peripheral blood monocytes, macrophages and dendritic cells (DCs), as well as to neuronal cells such as glial cells and astrocytes [29, 33–38]. The demonstration of high-affinity, saturable binding of Hsp70 to the cell surface has suggested the existence of specific 'Hsp70 receptors' (Figure 8.2). This hypothesis has been pursued, especially with regard to cells of the immune system, and a number of investigations have been carried out to determine specific binding structures. A number of candidate receptors have been proposed and their association with Hsp70 has been demonstrated. These include the α2-macroglobulin receptor CD91, the C-type lectin receptor and oxidised low-density lipoprotein (LDL) binding protein lectin-like oxidised low-density lipoprotein receptor-1 (LOX-1), and the tumour necrosis factor-α family protein CD40 [39–41]. Specific binding has been demonstrated in each case and Hsp70 binding is accompanied by endocytosis, which suggests a role for each ligand in both cell surface association and uptake of Hsp70 [39–41].

It is not as yet clear which receptor is the most important in the extracellular functions of Hsp70, although LOX-1 appears to be a promising candidate for an antigen-cross-presenting receptor [41]. At high concentrations, Hsp70 has been shown to associate with another member of the C-type lectin family, CD94 found on the surface of NK cells, and this interaction may underlie the enhanced ability of NK cells to kill tumour cells that express Hsp70 on the cell surface [29, 42]. In addition, Hsp70 has been found in lipid raft structures on the surface of cells concentrated together with important signalling receptors such as CD14, Toll-like receptors (TLRs) and chemokine receptors, and such interactions might be involved in the signalling events that are discussed in the next section [20]. At the present time, it is not clear whether cells possess a dedicated receptor for Hsp70, and Hsp70 thus remains an orphan ligand with, as yet, only hand-me-down receptors for its association with the cell surface. More details of Hsp70–receptor interactions are presented in Chapter 10.

8.5. Exogenous Hsp70-mediated signal transduction pathways

Much work on cell signalling has been carried out in macrophages, monocytes and DCs in which attempts have been made to determine signalling pathways upstream of the cytokine and antigen-presenting cell (APC) co-receptor gene expression which accompanies Hsp70 binding [34, 35, 43, 44]. Our group has evaluated the various steps involved in Hsp70-induced signal transduction and has demonstrated that Hsp70 binds with high affinity to the plasma membrane of APCs to elicit a rapid intracellular Ca^{2+} ($[Ca^{2+}]$) flux within 10 seconds [35]. This is an important signalling step which distinguishes Hsp70- from lipopolysaccharide (LPS)-induced signalling (discussed later), because the treatment of APCs with LPS does not result in an $[Ca^{2+}]_i$ flux [45].

The possibility that LPS contamination might confound our results was addressed by using Polymyxin B and Lipid IVa (LPS inhibitor) which abrogates LPS-induced, but not Hsp70-induced, cytokine expression. In addition, boiling for 1 hour abrogates Hsp70-induced, but not LPS-induced, cytokine expression. It is of note that Hsp70 also induces an increased Ca^{2+} flux in neuronal cells, suggesting that this is a general signalling response to Hsp70 binding [46].

We noted in studies on APCs that the rapid Hsp70-induced $[Ca^{2+}]_i$ flux is followed by the phosphorylation of I-κBα [35]. Activation of the transcription factor NF-κB is regulated by its cytoplasmic inhibitor I-κBα, via phosphorylation at Serine 32 (Ser-32) and 36 (Ser-36) which targets it for degradation by the proteosome and releases NF-κB to migrate to the nucleus and activate the promoter of target genes [47]. As early as 30 minutes after exposure to exogenous Hsp70, I-κBα is phosphorylated at Serine 32 (Ser-32) and 36 (Ser-36) in a calcium-dependent manner, and this results in the release and nuclear translocation of NF-κB [35]. Mechanistic studies using the HEK293 model system have revealed that Hsp70-induced NF-κB promoter activity is MyD88-dependent, CD14-dependent and is transduced via both TLR2 and TLR4 [43].

TLR2 and TLR4 are pattern-recognition receptors that recognise molecules associated with Gram-positive and Gram-negative bacteria, respectively [48, 49]. Hsp70 has the unique ability to interact with both of these signalling molecules

[43] and the presence of both TLR2 and TLR4 synergistically stimulates Hsp70-induced cytokine production [43]. Chapter 7 provides further discussion of these receptors. Interestingly, we have found that the synergistic activation of NF-κB promoter by co-expression of both TLR2 and TLR4 is MyD88-independent, which suggests an alterative pathway by which exogenous Hsp70 stimulates co-operation between TLR2 and TLR4 in cells of the immune system. As early as 2–4 hours after the exposure of APCs to exogenous Hsp70, there is significant release of TNF-α, IL-1β, IL-6 and IL-12 [35, 43]. By 3–5 days after exposure there is significant increase in proliferation of immature DCs and an augmentation in the expression of major histocompatibility complex (MHC) class II and the co-stimulatory molecule, CD86 [43]. These events might be highly significant to the pathways of antigen presentation, leading as they do to the maturation of DCs and antigen presentation to immune effector cells [31].

CD40 is a co-stimulatory molecule expressed on APCs that plays an important role in B lymphocyte function and autoimmunity [50] and CD40 binds Hsp70-peptide complexes via its exoplasmic domain [40]. The Hsp70–CD40 interaction is mediated by the NH_2-terminal ATPase domain of Hsp70 in its ADP-bound state and is augmented by the presence of substrate peptides in the COOH-terminal domain of Hsp70. The Hsp70–CD40 interaction is suppressed by Hip, a co-chaperone that is known to stabilise the Hsp70 ATPase domain in the ADP bound state [40].

Using the HEK293 cell model system, Hsp70–CD40 binding has been shown to stimulate signal transduction via the phosphorylation of p38 mitogen-activated protien kinase (previously shown to induce the release of TNF-α and secretion of IFN-γ [51]), and this results in the activation of NF-κB and uptake of peptide [40]. A detailed analysis of Hsp70–CD40 interactions is provided in Chapter 10. The oxidised LDL receptor LOX-1 on human DCs binds Hsp70, and incubation of cells with a neutralising anti-LOX-1 monoclonal antibody abrogates Hsp70 binding to DCs and suppresses Hsp70-induced antigen cross-presentation, although little is known of the signalling cascades that emanate from Hsp70–LOX-1 complexes [41]. However, LOX-1 is likely to exert a significant effect on Hsp70 uptake and antigen cross-presentation [41].

8.6. Conclusions

The data available in the literature therefore strongly support a role for extracellular Hsp70 as a danger signal for the immune response, as an emergency chaperone for the neuronal system and as playing some as yet undefined role in stress and exercise. We are, however, at an early stage in the exploration of this field and little is certain as yet. The uncertainties include the significance of the various sources of extracellular Hsp70, the receptor systems that permit cells to respond to this novel ligand, and the physiological role of free or bound extracellular Hsp70. What seems to be clear is that Hsp70, as a consequence of its capacity to carry processed peptides as cargo, permits the intracellular milieu of cells to be sampled by homeostatic cells such as APCs in a way similar to the MHC system [52]. This may represent a very primitive surveillance system which evolved many aeons prior to the development of the immune system [52]. In addition, cells may sense the structure of extracellular Hsp70 itself, regardless of peptide cargo, as a danger signal due to its capacity for massive induction by stress [35].

The relative physiological responses to heat shock proteins might depend on the amplitude and precise anatomical site of release. A gradual release of Hsp70 from a tissue in a regulated manner might be quite different from a massive Hsp70 release during toxic stress accompanied by other danger signals [25, 32]. Such signals might include other molecular chaperones such as Hsp60, Hsp90, gp96, Hsp110 and gp170, or molecules such as high mobility group box 1 protein (HMGB-1) and uric acid [36, 53–58]. Extracellular Hsp70 is potentially important in a number of diseases. In the case of cancer, Hsp70 offers an unique target for therapy with an agent that is already elevated in cancer and capable of capturing tumour-associated antigens, thus representing an Achilles heel that can be exploited by tumour immunotherapy approaches [6, 59, 60]. Extracellular Hsp70 might exert beneficial effects in neuronal and other cells in which extra chaperoning power can be transferred to vulnerable targets, permitting survival of acute stress [4, 27, 61].

Acknowledgements

We thank our colleagues Philip Auron, Rolph Issels, Hansjörg Schild, Chris Nichitta, Yue Xie, Betsy Repasky and John Subjeck for many helpful discussions and Dr. Auron in particular for invaluable materials. We also thank Olivia Bare, Maria Bausero and Edith Kabingu, for expert technical assistance. This work was supported in part by National Institutes of Health (NIH) Grants CA47407, CA31303, CA50642, CA77465 (to SKC) and the NIH grant RO1CA91889, Joint Center for Radiation Therapy Foundation Grant, Harvard Medical School and Institutional support from the Department of Medicine, Boston University School of Medicine (to AA).

REFERENCES

1. Li G C and Werb Z. Correlation between synthesis of heat shock proteins and development of thermotolerance in Chinese hamster fibroblasts. Proc Natl Acad Sci USA 1982, 79: 3218–3222.
2. Lindquist S and Craig E A. The heat-shock proteins. Ann Rev Genet 1988, 22: 631–677.
3. Hightower L E and Guidon P T. Selective release from cultured mammalian cells of heat-shock (stress) proteins that resemble glia-axon transfer proteins. J Cell Physiol 1989, 138: 257–266.
4. Tytell M, Greenberg S G and Lasek R J. Heat shock-like protein is transferred from glia to axon. Brain Res 1986, 363: 161–164.
5. Pockley A G, Shepherd J and Corton J. Detection of heat shock protein 70 (Hsp70) and anti-Hsp70 antibodies in the serum of normal individuals. Immunol Invest 1998, 27: 367–377.
6. Srivastava P K and Amato R J. Heat shock proteins: the 'Swiss Army Knife' vaccines against cancers and infectious agents. Vaccine 2001, 19: 2590–2597.
7. Nollen E A and Morimoto R I. Chaperoning signaling pathways: molecular chaperones as stress-sensing 'heat shock' proteins. J Cell Sci 2002, 115: 2809–2816.
8. Pratt W B and Toft D O. Regulation of signaling protein function and trafficking by the hsp90/hsp70-based chaperone machinery. Exp Biol Med 2003, 228: 111–133.
9. Chu B, Soncin F, Price B D, Stevenson M A and Calderwood S K. Sequential phosphorylation by mitogen-activated protein kinase and glycogen synthase kinase 3 represses transcriptional activation by heat shock factor-1. J Biol Chem 1996, 271: 30847–30857.
10. Feder J H, Rossi J M, Solomon J, Solomon N and Lindquist S. The consequences of expressing hsp70 in *Drosophila* cells at normal temperatures. Genes Dev 1992, 6: 1402–1413.
11. Wang X, Grammatikakis N, Siganou A and Calderwood S K. Regulation of molecular chaperone gene transcription involves the serine phosphorylation, 14-3-3 epsilon binding, and cytoplasmic sequestration of heat shock factor 1. Mol Cell Biol 2003, 23: 6013–6026.
12. Zhao M, Tang D, Lechpammer S, Hoffman A, Asea A, Stevenson M A and Calderwood S K. Double-stranded RNA-dependent protein kinase (pkr) is essential for thermotolerance, accumulation of HSP70, and stabilization of ARE-containing HSP70 mRNA during stress. J Biol Chem 2002, 277: 44539–44547.
13. Xie Y, Chen C, Stevenson M A, Hume D A, Auron P E and Calderwood S K. NF-IL6 and HSF1 have mutually antagonistic effects on transcription in monocytic cells. Biochem Biophys Res Commun 2002, 291: 1071–1080.
14. Xie Y, Zhong R, Chen C and Calderwood S K. Heat shock factor 1 contains two functional domains that mediate transcriptional repression of the c-fos and c-fms genes. J Biol Chem 2003, 278: 4687–4698.
15. Lau S S, Griffin T M and Mestril R. Protection against endotoxemia by HSP70 in rodent cardiomyocytes. Am J Physiol Heart Circ Physiol 2000, 278: H1439–1445.
16. McMillan D R, Xiao X, Shao L, Graves K and Benjamin I J. Targeted disruption of heat shock transcription factor 1 abolishes thermotolerance and protection against heat-inducible apoptosis. J Biol Chem 1998, 273: 7523–7528.

17. Wright B H, Corton J, El-Nahas A M, Wood R F M and Pockley A G. Elevated levels of circulating heat shock protein 70 (Hsp70) in peripheral and renal vascular disease. Heart Vessels 2000, 15: 18–22.
18. Broquet A H, Thomas G, Masliah J, Trugnan G and Bachelet M. Expression of the molecular chaperone Hsp70 in detergent-resistant microdomains correlates with its membrane delivery and release. J Biol Chem 2003, 278: 21601–21606.
19. Shin B K, Wang H, Yim A M, Le Naour F, Brichory F, Jang J H, Zhao R, Puravs E, Tra J, Michael C W, Misek D E and Hanash S M. Global profiling of the cell surface proteome of cancer cells uncovers an abundance of proteins with chaperone function. J Biol Chem 2003, 278: 7607–7616.
20. Triantafilou M, Miyake K, Golenbock D T and Triantafilou K. Mediators of innate immune recognition of bacteria concentrate in lipid rafts and facilitate lipopolysaccharide-induced cell activation. J Cell Sci 2002, 115: 2603–2611.
21. Mathew A, Bell A and Johnstone R M. Hsp-70 is closely associated with the transferrin receptor in exosomes from maturing reticulocytes. Biochem J 1995, 308: 823–830.
22. Shi Y and Rock K L. Cell death releases endogenous adjuvants that selectively enhance immune surveillance of particulate antigens. Eur J Immunol 2002, 32: 155–162.
23. Pockley A G. Heat shock proteins, inflammation and cardiovascular disease. Circulation 2002, 105: 1012–1017.
24. Campisi J and Fleshner M. Role of extracellular HSP72 in acute stress-induced potentiation of innate immunity in active rats. J Appl Physiol 2003, 94: 43–52.
25. Campisi J, Leem T H, Greenwood B N, Hansen M K, Moraska A, Higgins K, Smith T P and Fleshner M. Habitual physical activity facilitates stress-induced HSP72 induction in brain, peripheral, and immune tissues. Am J Physiol Regul Integr Comp Physiol 2003, 284: 1R520–R530.
26. Bechtold D A and Brown I R. Heat shock proteins Hsp27 and Hsp32 localize to synaptic sites in the rat cerebellum following hyperthermia. Brain Res Mol Brain Res 2000, 75: 309–320.
27. Guzhova I, Kislyakova K, Moskaliova O, Fridlanskaya I, Tytell M, Cheetham M and Margulis B. In vitro studies show that Hsp70 can be released by glia and that exogenous Hsp70 can enhance neuronal stress tolerance. Brain Res 2001, 914: 66–73.
28. Multhoff G and Hightower L E. Cell surface expression of heat shock proteins and the immune response. Cell Stress Chaperones 1996, 1: 167–176.
29. Multhoff G. Activation of natural killer cells by heat shock protein 70. Int J Hyperthermia 2002, 18: 576–585.
30. Bechtold D A, Rush S J and Brown I R. Localization of the heat-shock protein Hsp70 to the synapse following hyperthermic stress in the brain. J Neurochem 2000, 74: 641–646.
31. Noessner E, Gastpar R, Milani V, Brandl A, Hutzler P J, Kuppner M C, Roos M, Kremmer E, Asea A, Calderwood S K and Issels R D. Tumor-derived heat shock protein 70 peptide complexes are cross-presented by human dendritic cells. J Immunol 2002, 169: 5424–5432.
32. Todryk S M, Melcher A A, Dalgleish A G and Vile R G. Heat shock proteins refine the danger theory. Immunology 2000, 99: 334–337.

33. Arnold-Schild D, Hanau D, Spehner D, Schmid C, Rammensee H-G, de la Salle H and Schild H. Receptor-mediated endocytosis of heat shock proteins by professional antigen-presenting cells. J Immunol 1999, 162: 3757–3760.

34. Asea A, Kabingu E, Stevenson M A and Calderwood S K. Hsp70 peptide-bearing and peptide-negative preparations act as chaperokines. Cell Stress Chaperon 2000, 5: 425–431.

35. Asea A, Kraeft S-K, Kurt-Jones E A, Stevenson M A, Chen L B, Finberg R W, Koo G C and Calderwood S K. Hsp70 stimulates cytokine production through a CD14-dependent pathway, demonstrating its dual role as a chaperone and cytokine. Nat Med 2000, 6: 435–442.

36. Lipsker D, Ziylan U, Spehner D, Proamer F, Bausinger H, Jeannin P, Salamero J, Bohbot A, Cazenave J P, Drillien R, Delneste Y, Hanau D and de la Salle H. Heat shock proteins 70 and 60 share common receptors which are expressed on human monocyte-derived but not epidermal dendritic cells. Eur J Immunol 2002, 32: 322–332.

37. Reed R C and Nicchitta C V. Chaperone-mediated cross-priming: a hitchhiker's guide to vesicle transport. Int J Mol Med 2000, 6: 259–264.

38. Sondermann H, Becker T, Mayhew M, Wieland F and Hartl F U. Characterization of a receptor for heat shock protein 70 on macrophages and monocytes. Biol Chem 2000, 381: 1165–1174.

39. Basu S, Binder R J, Ramalingam T and Srivastava P K. CD91 is a common receptor for heat shock proteins gp96, Hsp90, Hsp70 and calreticulin. Immunity 2001, 14: 303–313.

40. Becker T, Hartl F U and Wieland F. CD40, an extracellular receptor for binding and uptake of Hsp70-peptide complexes. J Cell Biol 2002, 158: 1277–1285.

41. Delneste Y, Magistrelli G, Gauchat J, Haeuw J, Aubry J, Nakamura K, Kawakami-Honda N, Goetsch L, Sawamura T, Bonnefoy J and Jeannin P. Involvement of LOX-1 in dendritic cell-mediated antigen cross-presentation. Immunity 2002, 17: 353–362.

42. Gross C, Hansch D, Gastpar R and Multhoff G. Interaction of heat shock protein 70 peptide with NK cells involves the NK receptor CD94. Biol Chem 2003, 384: 267–279.

43. Asea A, Rehli M, Kabingu E, Boch J A, Baré O, Auron P E, Stevenson M A and Calderwood S K. Novel signal transduction pathway utilized by extracellular HSP70. Role of Toll-like receptor (TLR) 2 and TLR4. J Biol Chem 2002, 277: 15028–15034.

44. Vabulas R M, Braedel S, Hilf N, Singh-Jasuja H, Herter S, Ahmad-Nejad P, Kirschning C J, Da Costa C, Rammensee H G, Wagner H and Schild H. The endoplasmic reticulum-resident heat shock protein Gp96 activates dendritic cells via the Toll-like receptor 2/4 pathway. J Biol Chem 2002, 277: 20847–20853.

45. McLeish K R, Dean W L, Wellhausen S R and Stelzer G T. Role of intracellular calcium in priming of human peripheral blood monocytes by bacterial lipopolysaccharide. Inflammation 1989, 13: 681–692.

46. Smith P J, Hammar K and Tytell M. Effects of exogenous heat shock protein (Hsp70) on neuronal calcium flux. Biol Bull 1995, 189: 209–210.

47. Baeuerle P A and Baltimore D. I kB: a specific inhibitor of the NF-kB transcription factor. Science 1988, 242: 540–546.

48. Akira S and Sato S. Toll-like receptors and their signaling mechanisms. Scand J Infect Dis 2003, 35: 555–562.

49. Pulendran B, Palucka K and Banchereau J. Sensing pathogens and tuning immune responses. Science 2001, 293: 253–256.

50. Bodmer J L, Schneider P and Tschopp J. The molecular architecture of the TNF superfamily. Trends Biochem Sci 2002, 27: 19–26.

51. Pullen S S, Dang T T, Crute J J and Kehry M R. CD40 signaling through tumor necrosis factor receptor-associated factors (TRAFs). Binding site specificity and activation of downstream pathways by distinct TRAFs. J Biol Chem 1999, 274: 14246–14254.

52. Srivastava P. Interaction of heat shock proteins with peptides and antigen presenting cells: chaperoning of the innate and adaptive immune responses. Ann Rev Immunol 2002, 20: 395–425.

53. Manjili M H, Wang X Y, Chen X, Martin T, Repasky E A, Henderson R and Subjeck J R. HSP110-HER2/neu chaperone complex vaccine induces protective immunity against spontaneous mammary tumors in HER-2/neu transgenic mice. J Immunol 2003, 171: 4054–4061.

54. Park J S, Svetkauskaite D, He Q, Kim J Y, Strassheim D, Ishizaka A and Abraham E. Involvement of toll-like receptors 2 and 4 in cellular activation by high mobility group box 1 protein. J Biol Chem 2004, 279: 7370–7377.

55. Shi Y, Evans J E and Rock K L. Molecular identification of a danger signal that alerts the immune system to dying cells. Nature 2003, 425: 516–521.

56. Takata K, Kitamura Y, Tsuchiya D, Kawasaki T, Taniguchi T and Shimohama S. Heat shock protein-90-induced microglial clearance of exogenous amyloid-β1-42 in rat hippocampus *in vivo*. Neurosci Lett 2003, 344: 87–90.

57. Wang X Y, Kazim L, Repasky E A and Subjeck J R. Immunization with tumor-derived ER chaperone grp170 elicits tumor-specific CD8$^+$ T-cell responses and reduces pulmonary metastatic disease. Int J Cancer 2003, 105: 226–231.

58. Singh-Jasuja H, Toes R E M, Spee P, Münz C, Hilf N, Schoenberger S P, Ricciardi-Castagnoli P, Neefjes J, Rammensee H-G, Arnold-Schild D and Schild H. Cross-presentation of glycoprotein 96-associated antigens on major histocompatibility complex molecules requires receptor-mediated endocytosis. J Exp Med 2000, 191: 1965–1974.

59. Blagosklonny M V. Re: Role of the heat shock response and molecular chaperones in oncogenesis and cell death. J Natl Cancer Inst 2001, 93: 239–240.

60. Ciocca D R, Clark G M, Tandon A K, Fuqua S A, Welch W J and McGuire W L. Heat shock protein hsp70 in patients with axillary lymph node-negative breast cancer: prognostic implications. J Natl Cancer Inst 1993, 85: 570–574.

61. Yenari M A. Heat shock proteins and neuroprotection. Adv Exp Med Biol 2002, 513: 281–299.

9

Hsp72 and Cell Signalling

Vladimir L. Gabai and Michael Y. Sherman

9.1. Introduction

Many signalling molecules such as steroid hormone receptors and other receptors, protein kinases and phosphatases are found associated with various types of heat shock proteins, including Hsp90, Hsp70, Hsp40 and other co-chaperones. The functional role of these associations appears to be multi-faceted and the association of signalling proteins with these chaperone cohorts plays a pivotal role in initial folding and maturation of steroid hormone receptors and many kinases (e.g., Src). In addition, association with Hsp90 and its co-chaperones is critical for the stability of signalling proteins, because inhibition of Hsp90 by geldanamycin and other specific inhibitors leads to rapid ubiquitin-dependent degradation of Raf-1, Akt and other kinases that normally associate with Hsp90 [1, 2]. In fact, the anti-cancer activities of Hsp90 inhibitors could be related to the degradation and downregulation of signalling pathways that are controlled by these kinases [1, 3]. In contrast to Hsp90, which protects from degradation, an association with Hsp70 might target these proteins for rapid ubiquitination (usually via a ubiquitin ligase CHIP) followed by proteolysis [4].

In addition to their critical role in folding, maturation and stability of various signalling components, chaperones may be directly involved in regulation of their activities. In fact, it appears that Hsp70 and other chaperones play a regulatory role in the activation of many signalling pathways that are elicited by heat shock and other stresses.

There are multiple members of the Hsp70 protein family. Many of them, such as Hsc73, are expressed constitutively and serve a number of housekeeping functions, such as the folding of newly synthesised proteins and the transport of proteins to various organelles, and are involved in the ubiquitin-proteasome-dependent degradation pathway. Other members of this protein family, such as

Hsp72, are typically not expressed in normal, unstressed cells; however, their expression is rapidly induced to high levels upon exposure of the cells to heat shock and other stresses that cause protein damage. These proteins are believed to prevent aggregation of stress-damaged polypeptides and to promote their rapid refolding [5, 6].

The stress-induced Hsp70 has also been shown to play an important role in controlling multiple signalling pathways, and it appears that these interactions are directly related to a well-known cell protective function of Hsp70. Furthermore, these interactions can contribute to a phenomenon of acquired stress tolerance, in which cells that have been exposed to mild heat stress followed by recovery become tolerant to a wide range of stressful treatments.

This chapter will focus on the regulatory role of Hsp70 family members in activation of several kinase pathways and on the implications of Hsp70-mediated cell signalling in tumour development.

9.2. Hsp70 and MAP kinase signalling pathways

A role for Hsp70 in the regulation of mitogen-activated protein (MAP) kinase pathways has been studied for several years. Early work demonstrated that artificial expression of recombinant Hsp70 at high levels in human lymphoid cells dramatically diminished stress-induced activation of two MAP kinases, p38 and c-Jun N-terminal kinase (JNK) [7, 8]. The initial hypothesis arising from these findings was that abnormal proteins accumulating in cells under stressful treatments activate stress signalling cascades, and that Hsp70 can indirectly suppress the activity of these kinases by inhibiting the accumulation of abnormal proteins and facilitating protein refolding. Accordingly, it was shown that the accumulation of abnormal proteins induced by amino-acid analogues or proteasome inhibitors are powerful activators of p38 and JNK cascades [7, 9–11]. On the other hand, in addition to those stresses that can elicit protein damage (such as heat shock, ethanol, oxidative stress), a number of other stimuli, including tumour necrosis factor (TNF), IL-1, UV irradiation, or osmotic stress, can activate p38 and JNK without causing apparent proteotoxicity. Hsp70 expressed at high levels can also reduce the activation of the JNK and p38 kinase cascades induced by these stimuli [7, 12, 13]. Therefore, it appears that the effects of Hsp70 on these pathways may not be related to the handling of damaged proteins that accumulate after stressful treatment and may not even be related to the chaperone activities of Hsp70.

An understanding of the inhibitory effect of Hsp70 on JNK activation is complicated by the fact that there are at least two major pathways of activation of JNK. Of note, there are two similarly regulated major isoforms of JNK in non-neuronal cells, JNK1 and JNK2, and a neuron-specific isoform, JNK3 [14]. Activation of JNK by UV, osmotic stress and cytokines has been shown to proceed through a signal transduction pathway which involves a cascade of protein kinases, starting from mitogen activated protein kinase kinase kinase (MEKKs), followed by dual-specificity kinases, SAPK/ERK kinase/stress activated protein kinase (SEK1) mitogen activated protein kinase kinase (MKK4) and MKK7, which in turn phosphorylate JNK at tyrosine and threonine, thus activating it [15]. Several kinases other than MEKKs, including apoptosis signal regulating kinase 1 (ASK1) and mixed lineage kinase (MLKs), can also activate MKK4 and MKK7, and therefore there is a network of signalling pathways that contribute to JNK activation. p38 and extracellular signal-regulated kinases (ERKs), the third group of MAP kinase pathways, are activated via homologous kinase cascades [15, 16].

Interestingly, heat shock, oxidative stress, and other protein-damaging stresses activate MAP kinases via a novel pathway that involves an inhibition of their de-phosphorylation [17]. Multiple phosphatases with varying specificities, including dual-specificity phosphatases, PP2C, PP2A and others, are involved in the de-phosphorylation of MAP kinases [18, 19], the major contributors to which are dual-specificity phosphatases. For example, a stress-inducible phosphatase map kinase phosphatase (MKP-1) can de-phosphorylate all types of MAP kinases [20, 21]. VHI-recated phosphatase (VHR) and M3/6 are more specific to JNK [22–24], whereas MKP-3 and MKP-2 are specific to ERKs [22, 23, 25].

Many of these phosphatases are very sensitive to heat shock and other protein-damaging stresses. For example, upon exposure of cells to even mild heat shock, M3/6 becomes inactive and rapidly aggregates, leading to an inhibition of the de-phosphorylation of JNK and therefore an increase in the activity of this kinase [26]. Other, as yet unknown phosphatases that de-phosphorylate JNK in cells are probably also very heat-sensitive, because heat shock inactivates JNK de-phosphorylation almost entirely [17].

Interestingly, dual-specificity phosphatases have an essential cysteine in the active site, which potentially can be easily oxidised. The presence of this cysteine may explain the high sensitivity of these phosphatases to oxidative stress [27, 28]. Unexpectedly, JNK phosphatase M3/6 has also been found to aggregate, not only as a result of heat denaturation but also upon accumulation in cells of an abnormal polypeptide, a fragment of huntingtin with expanded polyglutamine domain (a cause of neurodegenerative Huntington's disease) [29]. The mechanism of M3/6 aggregation under these conditions is unclear. It might be that M3/6 phosphatase is so unstable that it requires molecular chaperones to maintain

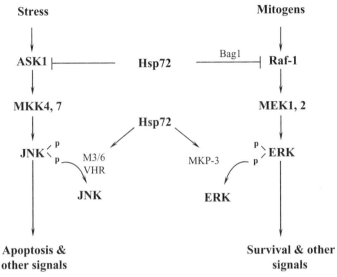

Figure 9.1. Effects of Hsp72 on MAP kinase pathways.

its normal conformation even at normal temperature. Then, accumulation of the mutant huntingtin fragment may bind chaperones and titrate them from a complex with the M3/6 phosphatase, leading to its aggregation and inactivation. Interestingly, endogenous levels of Hsp70 are strongly reduced in cells expressing the mutant huntingtin [29], which has been proposed to contribute to the M3/6 phosphatase aggregation and JNK activation.

In suppressing JNK activation by heat shock and other protein-damaging stresses, Hsp70 alleviates the inhibition of phosphatases (Figure 9.1). Similarly, over-production of Hsp70 protects the M3/6 phosphatase in cells that express the mutant huntingtin fragment with extended polyglutamine [29]. This function of Hsp70 is clearly dependent on its chaperone activity, because an Hsp70 mutant with a short C-terminal deletion, which abrogates the refolding activity, fails to preserve JNK dephosphorylation in heat-shocked cells [30]. Interestingly, protection of JNK phosphatases is not the only site of Hsp70 action in the JNK signalling pathway. In fact, in suppressing JNK activation induced by TNF and other non-protein-damaging stimuli, Hsp70 inhibits the upstream kinase cascade [13]. In contrast to protection of JNK phosphatases, for Hsp70-mediated inhibition of the kinase cascade, the refolding activity is dispensable, because the C-terminal deletion mutant of Hsp70 is as efficient in inhibiting the cascade as normal Hsp70 [13].

In line with these observations, it has been demonstrated that Hsp70 can directly interact and suppress activity of ASK1, an upstream component of p38 and JNK signalling cascades, which is activated by stimuli such as TNF or UV irradiation [31]. Hsp70 associated with ASK1 via an ATP-binding domain and deletion of this domain of Hsp70 abrogates interactions with ASK1 [31]. It is not clear whether interactions between Hsp70 and ASK1 are direct or are mediated by a distinct factor. It is possible that a co-chaperone such as Bag-1 may participate in these interactions, as it does with interactions between Hsp70 and a kinase Raf-1 (see discussion following).

The effect of proteotoxic stresses on MAP kinase de-phosphorylation seems to be quite general. Indeed, a dual-specificity phosphatase, MKP-3, which is involved in inactivation of ERK1/2, is also highly sensitive to heat shock and rapidly aggregates under these conditions [32]. Furthermore, Hsp70, but not the C-terminal mutant, was able to prevent heat shock–induced MKP-3 aggregation, thus suppressing ERK1/2 activation. As with the JNK pathway, Hsp70 can inhibit the ERK signalling pathway, also at a distinct site in the kinase cascade, and this activity of Hsp70 does not require its chaperone function [32]. Song and colleagues have reported that a co-chaperone, Bag-1, which can form a complex with a component of ERK-activating cascade, Raf-1, enhances Raf-1 activity [33]. Hsp70, when expressed at high levels, binds to Bag-1 and titrates it from the complex with Raf-1, leading to inactivation of the latter [33]. Interestingly, Bag-1 is known to interact with Hsp70 via its ATPase domain [34], which is intact in the C-terminal deletion mutant. This interaction may explain why the chaperone activity of Hsp70 is dispensable for the regulation of ERK kinase cascade. Further studies are needed to establish whether a similar Bag-1-dependent mechanism operates in Hsp70-mediated inhibition of the JNK signalling cascade.

Data from several laboratories indicate that this novel capacity of Hsp70 to control JNK signalling is important for the protection of cells from apoptosis induced by certain stimuli. For example, triggering apoptosis via specific activation of JNK via the expression of an active form of an upstream kinase MEKK1 can be efficiently blocked by Hsp70 [35]. In these cells, tolerance to UV-induced apoptosis caused by mild heat shock pre-treatment was clearly associated with Hsp70-mediated JNK inhibition [35]. In suppression of H_2O_2-induced apoptosis, or apoptosis induced by a constitutively active form of ASK1, JNK inhibition was probably associated with blocking of ASK1 by Hsp70 [31]. Furthermore, the ATPase domain of Hsp70, which is critical for the control of ASK1, is also essential for the inhibition of apoptosis under these conditions [31]. Although inhibition of JNK, which could be achieved by Hsp70 mutants lacking the chaperone function, is sufficient for suppressing apoptosis caused by UV irradiation [35], it does not block heat-induced apoptosis [13, 30]. Indeed, Hsp70 mutants

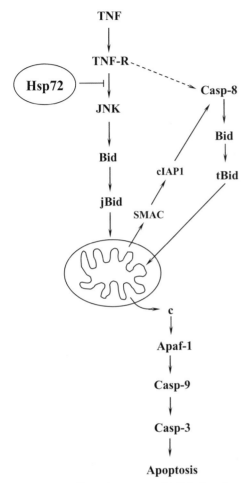

Figure 9.2. Effect of Hsp72 on TNF-induced apoptotic pathway.

that lack chaperone function efficiently suppressed activation of JNK by heat shock but were unable to protect cells from the consequences [13, 30].

The best-studied example of Hsp70-mediated inhibition of apoptosis is TNF-induced apoptosis of primary human fibroblasts, in which the suppression of JNK plays the major protective role [13]. The activation of JNK by TNF leads to caspase-8 independent cleavage of a BH3-domain protein Bid, which regulates mitochondrial integrity, and the release of a caspase regulator Smac/Diablo from mitochondria, with the subsequent activation of caspase-8 [36]. Hsp70-mediated inhibition of JNK in TNF-treated cells led to a suppression of Bid cleavage and an inhibition of subsequent apoptotic events [13] (Figure 9.2). Interestingly, neither JNK or Hsp70 could regulate the Bid-independent apoptotic pathway, which takes over at later time points [13].

Of note is that the suppression of JNK may not be the only target of Hsp70 in its inhibition of apoptotic signalling. For example, Hsp70 inhibits formation of the apoptosome [37, 38]. However, these effects of Hsp70 were studied in *in vitro* experiments, and the relevance of these data to the effects of Hsp70 *in vivo* have been challenged [74, 75].

An important development in recent years has been the finding that Hsp70-mediated JNK suppression is involved not only in apoptosis, but also in necrosis. In contrast to apoptosis, necrosis has always been considered to be a spontaneous, unregulated process. However, it now appears that some forms of necrosis may be tightly regulated and programmed [39]. For example, ischaemia-induced necrosis of cardiac cells is controlled by several signalling pathways, including JNK and p38 [40–42]. In the myogenic cell H9c2, simulated ischaemia (transient ATP depletion) leads to necrosis involving rapid JNK-dependent mitochondrial de-energisation. Expression of Hsp70 (or its C-terminal mutant) prevents JNK activation, mitochondrial de-energisation, and cell necrosis [32, 42]. This Hsp70-mediated suppression of stress kinases in ischaemic cells may be the basis for the protective effects of Hsp70 over-expression that have been demonstrated in a number of organs, including heart, brain, liver or kidney.

9.3. Control of the heme-activated kinase by Hsc73

The first example of a kinase that is directly regulated by an Hsp70 family member was HRI, a heme-regulated kinase of the α-subunit of eukaryotic translation initiation factor 2 (eIF-2α). HRI is present in reticulocytes (and possibly other cell types; R. Matts, personal communication). As with many other kinases, the initial folding and maturation of HRI requires an ensemble of chaperones, including Hsp90, Hsc73, Cdc37 and others [43, 44]. Upon maturation, HRI has low activity under normal conditions. However, upon heme limitation, HRI is phosphorylated and activated, and phosphorylates eIF-2α, leading to an inhibition of translation [45, 46]. Alternatively, HRI can be activated by heat shock, oxidative stress or other stresses that cause protein damage [45, 47] (Figure 9.3). Therefore, HRI shuts down synthesis of globin, the main protein produced in reticulocytes, under conditions that may cause improper production of haemoglobin (i.e., heme limitation or chaperone overload).

A breakthrough in understanding the mechanisms of HRI regulation was the finding that the activation of HRI by protein-damaging stresses is likely to be mediated by a build-up of damaged polypeptides. In fact, in reticulo-cyte lysates, the influence of heat shock on HRI activation can be mimicked by the addition of denatured, reduced carboxymethylated bovine serum albu-min (BSA), but not of normal BSA [48]. Elegant *in vitro* experiments have demonstrated that normally mature HRI associates with a constitutively ex-pressed member of the Hsp70 family, Hsc73. This association maintains HRI in a latent form in haemin-supplied lysates. Denatured proteins accumulated in cells after heat shock or denatured proteins added to lysates interact with Hsc73 and competitively inhibit association of Hsc73 with HRI. This interaction shifts the equilibrium, leading to dissociation of Hsc73 from HRI and activation of the latter [44]. Hsc73-mediated regulation of HRI is parallel with, and is inde-pendent of regulation of HRI by haemin, since neither association of Hsc73 with HRI nor its dissociation after heat shock are influenced by haemin [43] (Figure 9.3).

It is not clear whether Hsc73 interacts with HRI directly or whether there is an adaptor protein that mediates these interactions. Interestingly, there are two distinct protein kinases, PKR and PERK, that also regulate translation via phos-phorylation of eIF-2α. Although the main regulator of PKR is double-stranded RNA, this kinase, like HRI, can also be activated by heat shock and other protein-damaging stresses [49, 50] (Figure 9.3). The mechanism of this regulation is cur-rently unknown, although the involvement of Hsp70 has been suggested. In fact, an important regulator of PKR is the inhibitor p58[ipk], a protein with J-domain and multiple tetratricopeptide repeat (TPR) domains, which can bind Hsp70 pro-teins [51, 52]. Interactions between p58[ipk] and Hsp70 can be mediated by Hsp40 [53]. p58[ipk] can function as a co-chaperone of DnaJ type, because it stimulates ATPase activity of Hsp70 *in vitro* [52]. Interestingly, interactions between p58[ipk] and Hsp70 play an important role in regulation of PKR, because the J-domain of p58[ipk] is indispensable for this regulation [52]. It has been hypothesised that p58[ipk] recruits Hsp70 into the complex with PKR, which leads to inactivation of this kinase [54] (Figure 9.3).

In addition, p58[ipk] can also inhibit PERK, another eIF-2α kinase [55, 56] (Fig-ure 9.3). Therefore, it appears that all major kinases that regulate translation via the phosphorylation of eIF-2α could be controlled by Hsp70 family members.

9.4. Hsp70, cell signalling and cancer

Normal tissues usually express a constitutive member of the Hsp70 family, Hsc73, but not the inducible form Hsp70 (Hsp72). In contrast, tumours often express both Hsp70 and Hsc73, and a number of reports indicate that high expression of Hsp70 in human tumours, especially of epithelial origin, correlates

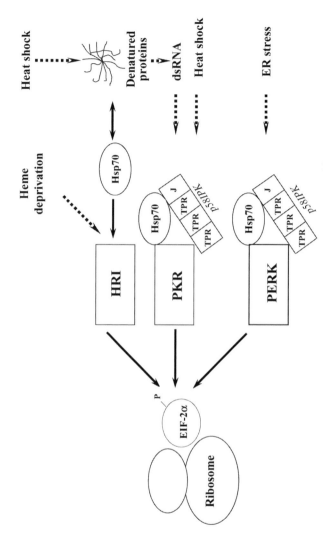

Figure 9.3. Regulation of HRI, PKR ad PERK kinases by Hsp70 and p58[ipk].

with high invasiveness, metastatic disease, poor prognosis of the disease and resistance to chemotherapy [57–60]. For example, in colorectal and lung cancers, expression of Hsp70 highly correlates with advanced clinical stages and positive lymph node involvement [61, 62]. These findings suggest that Hsp70 provides a selective advantage to tumour cells during cancer progression. One possible involvement of Hsp70 in tumour development could be its capacity to suppress apoptosis (see preceding discussion). Indeed, many oncogenes (e.g., myc) when activated can induce apoptosis in normal cells or enhance sensitivity of normal cells to various apoptotic stimuli [63, 64]. This response is considered to be an intrinsic cellular anti-cancer mechanism. Therefore, to avoid oncogene-activated apoptosis, upon cancer transformation cells develop special mechanisms for anti-apoptotic protection such as an over-expression of bcl-2, mutations in p53 or mutations in PTEN phosphatase.

In line with this observation, the expression of anti-apoptotic factors (e.g., bcl-2) significantly promotes cell transformation *in vitro* and tumour development *in vivo* following the expression of myc oncogene [64–66]. Hsp70 may serve as an alternative potent anti-apoptotic factor in the promotion of cancer transformation. In fact, the expression of Hsp70 in Rat-1 fibroblasts leads to the formation of foci, anchorage-independent growth, and formation of tumours in nude mice [67], and an over-production of Hsp70 in fibrosarcoma cells significantly enhances their tumourigenic potential in athymic mice [68]. Furthermore, transgenic mice that express human Hsp70 at high levels developed multiple lymphomas [69].

An important advance of recent years has been that, besides its possible role in cell transformation, Hsp70 may be critical for viability and proliferation of established cancer cell lines. Indeed, it has been demonstrated that the depletion of Hsp70 by adenovirus-encoded Hsp70 anti-sense RNA causes rapid death of various types of tumour cells, whereas non-transformed cells are resistant to such treatment [70]. Treatment with adenovirus expressing Hsp70 anti-sense RNA is also effective in human glioblastoma, breast and colon carcinoma grown as xenografts [71].

However, a caveat to these studies is that an adenoviral vector for anti-sense RNA delivery was employed, because adenovirus by itself causes subtle toxicity to tumour cells. This toxicity, combined with the effects of Hsp70 depletion, may lead to cancer cell killing. In fact, when we and others down-regulated Hsp70 in a milder way by retroviral expression of Hsp70 anti-sense RNA or siRNA, no significant tumour cell killing was observed; rather there was a specific sensitisation of cells to some drugs and stresses (Gabai et al, in press). Using retrovirus vector expressing Hsp70 in an anti-sense orientation, clones of PC-3 human prostate adenocarcinoma cells (PC-3/AS) which express 4-fold lower levels of Hsp70 than parental cells have been generated. These clones concomitantly expressed normal levels of Hsc73. As expected, PC-3/AS cells were significantly more sensitive to heat-induced apoptosis, as well as to apoptosis induced by TNF, cis-platinum, vinblastine or taxol. Surprisingly, these cells had a decreased etoposide-induced apoptosis, indicating that the effects of Hsp70 depletion are specific to certain types of apoptosis. Importantly, reduced levels of Hsp70 not only influence apoptotic cell death, but also overall cell survival. Indeed, short-term treatments with certain anti-cancer drugs or radiation did not induce apoptosis; rather it led to mitotic catastrophe and loss of colony-forming ability of cells. Down-regulation of Hsp70 markedly enhanced such a mitotic catastrophe.

In a variety of cells, activation of JNK and p38 has been shown to stimulate apoptosis, whereas stimulation of ERK and Akt kinases promotes survival. As discussed earlier, over-expression of Hsp70 suppresses stress-induced activation of JNK, p38, ASK1 and ERK kinases [7, 31, 33, 72] and hence we anticipated that Hsp70 downregulation in cancer cells should increase the activities of at least some of these kinases. However, no upregulation of p38 and JNK activities has been found. Furthermore, depletion of Hsp72 in PC-3 cell clones leads to a dramatic downregulation of endogenous and stress-induced ERK1/2 activity. This effect of Hsp72 depletion is associated with strong downregulation of the ERK-activation kinase cascade, probably at the level of Raf-1. It was previously demonstrated that treatment of cells with geldanamycin, an inhibitor of Hsp90, leads to ERK1/2 inactivation because of rapid ubiquitin-proteasome-dependent degradation of Raf-1 [1]. Although Hsp72 acts as a co-factor of Hsp90 in protein refolding, the levels of Raf-1 expression increased rather than decreased in PC-3/AS cells, indicating that Hsp72 downregulation did not reduce Raf-1 stability. At the same time, de-phosphorylation of the inhibitory site Ser-259 (which is required for Raf-1 activation [73]) was significantly reduced in PC-3/AS cells.

Therefore, it seems that a certain level of Hsp72 expression in tumour cells is necessary for maintaining activity of the major cell survival pathway, ERK, and downregulation of Hsp72 leads to a loss of ERK activity. These findings can potentially explain the increased sensitivity of PC-3/AS cells to stresses and drugs, because suppression of ERK can significantly affect viability of cells exposed to harsh conditions. Indeed, the specific inhibitor of ERK (U0126) made parental PC-3 cells as sensitive to heat shock as PC-3/AS cells. Taken together it would appear that the effects of Hsp72 on various signalling systems play a major role in tumour development and survival.

REFERENCES

1. Hostein I, Robertson D, DiStefano F, Workman P and Andrew Clarke P. Inhibition of signal transduction by the Hsp90 inhibitor 17-allylamino-17-demethoxygeldanamycin results in cytostasis and apoptosis. Cancer Res 2001, 61: 4003–4009.

2. Fujita N, Sato S, Ishida A and Tsuruo T. Involvement of Hsp90 in signaling and stability of 3-phosphoinositide-dependent kinase-1. J Biol Chem 2002, 277: 10346–10353.

3. Neckers L. Hsp90 inhibitors as novel cancer chemotherapeutic agents. Trends Mol Med 2002, 8: S55–61.

4. Wiederkehr T, Bukau B and Buchberger A. Protein turnover: a CHIP programmed for proteolysis. Cur Biol 2002, 12: R26–28.

5. Michels A A, Kanon B, Konings A W, Ohtsuka K, Bensaude O and Kampinga H H. Hsp70 and Hsp40 chaperone activities in the cytoplasm and the nucleus of mammalian cells. J Biol Chem 1997, 272: 33283–33289.

6. Nollen E A, Brunsting J F, Roelofsen H, Weber L A and Kampinga H H. *In vivo* chaperone activity of heat shock protein 70 and thermotolerance. Mol Cell Biol 1999, 19: 2069–2079.

7. Gabai V L, Meriin A B, Mosser D D, Caron A W, Rits S, Shifrin VI and Sherman M Y. HSP70 prevent activation of stress kinases: a novel pathway of cellular thermotolerance. J Biol Chem 1997, 272: 18033–18037.

8. Mosser D D, Caron A W, Bourget L, Denis-Larose C and Massie B. Role of the human heat shock protein hsp70 in protection against stress-induced apoptosis. Mol Cell Biol 1997, 17: 5317–5327.

9. Meriin A, Gabai V, Yaglom J, Shifrin V and Sherman M. Proteasome inhibitors activate stress-kinases and induce Hsp72: diverse effects on apoptosis. J Biol Chem 1998, 273: 6373–6379.

10. Hideshima T, Mitsiades C, Akiyama M, Hayashi T, Chauhan D, Richardson P, Schlossman R, Podar K, Munshi N C, Mitsiades N and Anderson K C. Molecular mechanisms mediating antimyeloma activity of proteasome inhibitor PS-341. Blood 2003, 101: 1530–1534.

11. Almond J B and Cohen G M. The proteasome: a novel target for cancer chemotherapy. Leukemia 2002, 16: 433–443.

12. Yaglom J A, Gabai V L, Meriin A B, Mosser D D and Sherman M Y. The function of HSP72 in suppression of c-Jun N-terminal kinase activation can be dissociated from its role in prevention of protein damage. J Biol Chem 1999, 274: 20223–20228.

13. Gabai V L, Mabuchi K, Mosser D D and Sherman M Y. Hsp72 and stress kinase c-jun N-terminal kinase regulate the bid-dependent pathway in tumor necrosis factor-induced apoptosis. Mol Cell Biol 2002, 22: 3415–3424.

14. Davis R J. Signal transduction by the JNK group of MAP kinases. Cell 2000, 103: 239–252.

15. Kyriakis J M a J A. Mammalian mitogen-activated protein kinase signal transduction pathways activated by stress and inflammation. Physiol Rev 2001, 81: 807–869.

16. Su B and Karin M. Mitogen-activated protein kinase cascades and regulation of gene expression. Cur Opin Immunol 1996, 8: 402–411.

17. Meriin A B, Yaglom J A, Gabai V L, Mosser D D, Zon L and Sherman M Y. Protein damaging stresses activate JNK via inhibition of its phosphatase: a novel pathway controlled by Hsp72. Mol Cell Biol 1999, 19: 2547–2555.

18. Millward T A, Zolnierowicz S and Hemmings B A. Regulation of protein kinase cascades by protein phosphatase 2A. Trends Biochem Sci 1999, 24: 186–191.

19. Saxena M and Mustelin T. Extracellular signals and scores of phosphatases: all roads lead to MAP kinase. Semin Immunol 2000, 12: 387–396.

20. Keyse S M. Protein phosphatases and the regulation of MAP kinase activity. Semin Cell Develop Biol 1998, 9: 143–152.

21. Keyse S M. An emerging family of dual specificity MAP kinase phosphatases. Biochim Biophys Acta 1995, 1265: 152–160.

22. Muda M, Theodosiou A, Gillieron C, Smith A, Chabert C, Camps M, Boschert U, Rodrigues N, Davies K, Ashworth A and Arkinstall S. The mitogen-activated protein kinase phosphatase-3 N-terminal noncatalytic region is responsible for tight substrate binding and enzymatic specificity. J Biol Chem 1998, 273: 9323–9329.

23. Muda M, Theodosiou A, Rodrigues N, Boschert U, Camps M, Gillieron C, Davies K, Ashworth A and Arkinstall S. The dual specificity phosphatases M3/6 and MKP-3 are highly selective for inactivation of distinct mitogen-activated protein kinases. J Biol Chem 1996, 271: 27205–27208.

24. Todd J L, Rigas J D, Rafty L A and Denu J M. Dual-specificity protein tyrosine phosphatase VHR down-regulates c-Jun N-terminal kinase (JNK). Oncogene 2002, 21: 2573–2583.

25. Chu Y, Solski P A, Khosravi-Far R, Der C J and Kelly K. The mitogen-activated protein kinase phosphatases PAC1, MKP-1, and MKP-2 have unique substrate specificities and reduced activity *in vivo* toward the ERK2 sevenmaker mutation. J Biol Chem 1996, 271: 6497–6501.

26. Palacios C, Collins M K and Perkins G R. The JNK phosphatase M3/6 is inhibited by protein-damaging stress. Cur Biol 2001, 11: 1439–1443.

27. Chen Y R, Shrivastava A and Tan T H. Down-regulation of the c-Jun N-terminal kinase (JNK) phosphatase M3/6 and activation of JNK by hydrogen peroxide and pyrrolidine dithiocarbamate. Oncogene 2001, 20: 367–374.

28. Theodosiou A and Ashworth A. Differential effects of stress stimuli on a JNK-inactivating phosphatase. Oncogene 2002, 21: 2387–2397.

29. Merienne K, Helmlinger D, Perkin G R, Devys D and Trottier Y. Polyglutamine expansion induces a protein-damaging stress connecting heat shock protein 70 to the JNK pathway. J Biol Chem 2003, 278: 16957–16967.

30. Mosser D D, Caron A W, Bourget L, Meriin A B, Sherman M Y, Morimoto R I and Massie B. The chaperone function of hsp70 is required for protection against stress-induced apoptosis. Mol Cell Biol 2000, 20: 7146–7159.

31. Park H-S, Cho S-G, Kim C K, Hwang H S, Noh K T, Kim M-S, Huh S-H, Kim M J, Ryoo K, Kim E K, Kang W J, Lee J-S, Seo J-S, Ko Y-G, Kim S and Choi E-J. Heat shock protein Hsp72 is a negative regulator of apoptosis signal-regulating kinase 1. Mol Cell Biol 2002, 22: 7721–7730.

32. Yaglom J A, Ekhterae D, Gabai V L and Sherman M Y. Regulation of necrosis of H9C2 myogenic cells upon transient energy deprivation: rapid de-energization of mitochondria precedes necrosis and is controlled by reactive oxygen species, stress-kinase JNK, HSP72 and ARC. J Biol Chem 2003, 278: 50483–50496.

33. Song J, Takeda M and Morimoto R I. Bag1-Hsp70 mediates a physiological stress signalling pathway that regulates Raf-1/ERK and cell growth. Nat Cell Biol 2001, 3: 276–282.

34. Bimston D, Song J, Winchester D, Takayama S, Reed J C and Morimoto R I. BAG-1, a negative regulator of Hsp70 chaperone activity, uncouples nucleotide hydrolysis from substrate release. EMBO J 1998, 17: 6871–6878.

35. Park H S, Lee J S, Huh S H, Seo J S and Choi E J. Hsp72 functions as a natural inhibitory protein of c-Jun N-terminal kinase. EMBO J 2001, 20: 446–456.

36. Deng Y, Ren X, Yang L, Lin Y and Wu X. A JNK-dependent pathway is required for TNFα-induced apoptosis. Cell 2003, 115: 61–70.

37. Beere H M, Wolf B B, Cain K, Mosser D D, Mahboubi A, Kuwana T, Tailor P, Morimoto R I, Cohen G M and Green D R. Heat-shock protein 70 inhibits apoptosis by preventing recruitment of procaspase-9 to the Apaf-1 apoptosome. Nat Cell Biol 2000, 2: 469–475.

38. Saleh A, Srinivasula S, Balkir L, Robbins P and Alnemri E. Negative regulation of the Apaf-1 apoptosome by Hsp70. Nat Cell Biol 2000, 2: 476–483.

39. Proskuryakov S Y, Konoplyannikov A G and Gabai V L. Necrosis: a specific form of programmed cell death? Exp Cell Res 2003, 283: 1–16.

40. Ma X L, Kumar S, Gao F, Louden C S, Lopez B L, Christopher T A, Wang C, Lee J C, Feuerstein G Z and Yue T L. Inhibition of p38 mitogen-activated protein kinase decreases cardiomyocyte apoptosis and improves cardiac function after myocardial ischemia and reperfusion. Circulation 1999, 99: 1685–1691.

41. Mackay K and Mochly-Rosen D. An inhibitor of p38 mitogen-activated protein kinase protects neonatal cardiac myocytes from ischemia. J Biol Chem 1999, 274: 6272–6279.

42. Gabai V L, Meriin A B, Yaglom J A, Wei J Y, Mosser D D and Sherman M Y. Suppression of stress kinase JNK is involved in Hsp72-mediated protection of myogenic cells from transient energy deprivation. Hsp72 alleviates the stress-induced inhibition of JNK dephosphorylation. J Biol Chem 2000, 275: 38088–38094.

43. Uma S, Thulasiraman V and Matts R L. Dual role for Hsc70 in the biogenesis and regulation of the heme-regulated kinase of the alpha subunit of eukaryotic translation initiation factor 2. Mol Cell Biol 1999, 19: 5861–5871.

44. Thulasiraman V, Xu Z, Uma S, Gu Y, Chen J J and Matts R L. Evidence that Hsc70 negatively modulates the activation of the heme-regulated eIF-2 alpha kinase in rabbit reticulocyte lysate. Eur J Biochem 1998, 255: 552–562.

45. Chen J J and London I M. Regulation of protein synthesis by heme-regulated eIF-2 alpha kinase. Trends Biochem Sci 1995, 20: 105–108.

46. Hinnebusch A G. The eIF-2 alpha kinases: regulators of protein synthesis in starvation and stress. Semin Cell Biol 1994, 5: 417–426.

47. Matts R L, Xu Z, Pal J K and Chen J J. Interactions of the heme-regulated eIF-2 alpha kinase with heat shock proteins in rabbit reticulocyte lysates. J Biol Chem 1992, 267: 18160–18167.

48. Matts R L, Hurst R and Xu Z. Denatured proteins inhibit translation in hemin-supplemented rabbit reticulocyte lysate by inducing the activation of the heme-regulated eIF-2 alpha kinase. Biochemistry 1993, 32: 7323–7328.

49. Williams B R. PKR; a sentinel kinase for cellular stress. Oncogene 1999, 18: 6112–6120.

50. Brostrom C O, Prostko C R, Kaufman R J and Brostrom M A. Inhibition of translational initiation by activators of the glucose-regulated stress protein and heat shock protein stress response systems. Role of the interferon-inducible double-stranded RNA-activated eukaryotic initiation factor 2 alpha kinase. J Biol Chem 1996, 271: 24995–25002.

51. Tang N M, Ho C Y and Katze M G. The 58-kDa cellular inhibitor of the double stranded RNA-dependent protein kinase requires the tetratricopeptide repeat 6 and DnaJ motifs to stimulate protein synthesis *in vivo*. J Biol Chem 1996, 271: 28660–28666.

52. Melville M W, Tan S L, Wambach M, Song J, Morimoto R I and Katze M G. The cellular inhibitor of the PKR protein kinase, P58(IPK), is an influenza virus-activated co-chaperone that modulates heat shock protein 70 activity. J Biol Chem 1999, 274: 3797–3803.

53. Melville M W, Hansen W J, Freeman B C, Welch W J and Katze M G. The molecular chaperone hsp40 regulates the activity of P58IPK, the cellular inhibitor of PKR. Proc Nat Acad Sci USA 1997, 94: 97–102.

54. Melville M W, Katze M G and Tan S L. P58IPK, a novel co-chaperone containing tetratricopeptide repeats and a J-domain with oncogenic potential. Cell Mol Life Sci 2000, 57: 311–322.

55. van Huizen R, Martindale J L, Gorospe M and Holbrook N J. P58IPK, a novel endoplasmic reticulum stress-inducible protein and potential negative regulator of eIF2 alpha signaling. J Biol Chem 2003, 278: 15558–15564.

56. Yan W, Frank C L, Korth M J, Sopher B L, Novoa I, Ron D and Katze M G. Control of PERK eIF2 alpha kinase activity by the endoplasmic reticulum stress-induced molecular chaperone P58IPK. Proc Nat Acad Sci USA 2002, 99: 15920–15925.

57. Ciocca D R, Clark G M, Tandon A K, Fuqua S A, Welch W J and McGuire W L. Heat shock protein hsp70 in patients with axillary lymph node-negative breast cancer: prognostic implications. J Natl Cancer Inst 1993, 85: 570–574.

58. Nanbu K, Konishi I, Mandai M, Kuroda H, Hamid A A, Komatsu T and Mori T. Prognostic significance of heat shock proteins Hsp70 and Hsp90 In endometrial carcinomas. Cancer Detect Prevent 1998, 22: 549–555.

59. Costa M J M, Rosas S L B, Chindano A, Lima P D S, Madi K and Carvalho M D D. Expression of heat shock protein 70 and P53 in human lung cancer. Oncol Rep 1997, 4: 1113–1116.

60. Vargasroig L M, Fanelli M A, Lopez L A, Gago F E, Tello O, Aznar J Z and Ciocca D R. Heat shock proteins and cell proliferation in human breast cancer biopsy samples. Cancer Detect Prevent 1997, 21: 441–451.

61. Hwang T S, Han H S, Choi H K, Lee Y J, Kim Y-J, Han M-Y and Park Y-M. Differential, stage-dependent expression of Hsp70, Hsp110 and Bcl-2 in colorectal cancer. J Gastroenterol Hepatol 2003, 18: 690–700.

62. Volm M, Koomagi R, Mattern J and Efferth T. Protein expression profile of primary human squamous cell lung carcinomas indicative of the incidence of metastases. Clin Exp Metastasis 2002, 19: 385–390.

63. Prendergast G C. Mechanisms of apoptosis by c-Myc. Oncogene 1999, 18: 2967–2987.

64. Nilsson J A and Cleveland J L. Myc pathways provoking cell suicide and cancer. Oncogene 2003, 22: 9007–9021.

65. Pelengaris S, Khan M and Evan G I. Suppression of Myc-induced apoptosis in beta cells exposes multiple oncogenic properties of Myc and triggers carcinogenic progression. Cell 2002, 109: 321–334.

66. Pelengaris S, Khan M and Evan G. c-MYC: more than just a matter of life and death. Nat Rev Cancer 2002, 269: 764–776.

67. Volloch V Z and Sherman M Y. Oncogenic potential of Hsp72. Oncogene 1999, 18: 3648–3651.

68. Jaattela M. Over-expression of hsp70 confers tumorigenicity to mouse fibrosarcoma cells. Int J Cancer 1995, 60: 689–693.

69. Seo J S, Park Y M, Kim J I, Shim E H, Kim C W, Jang J J, Kim S H and Lee W H. T cell lymphoma in transgenic mice expressing the human Hsp70 gene. Biochem Biophys Res Commun 1996, 218: 582–587.

70. Nylandsted J, Rohde M, Brand K, Bastholm L, Elling F and Jaattela M. Selective depletion of heat shock protein 70 (Hsp70) activates a tumor-specific death program that is independent of caspases and bypasses Bcl-2. Proc Natl Acad Sci USA 2000, 97: 7871–7876.

71. Nylandsted J, Wick W, Hirt U A, Brand K, Rohde M, Leist M, Weller M and Jaattela M. Eradication of glioblastoma, and breast and colon carcinoma xenografts by Hsp70 depletion. Cancer Res 2002, 62: 7139–7142.

72. Yaglom J, O'Callaghan-Sunol C, Gabai V and Sherman M Y. Inactivation of dual-specificity phosphatases is involved in the regulation of extracellular signal-regulated kinases by heat shock and Hsp72. Mol Cell Biol 2003, 23: 3813–3824.

73. Dhillon A S, Meikle S, Yazici Z, Eulitz M and Kolch W. Regulation of Raf-1 activation and signalling by dephosphorylation. EMBO J 2002, 21: 64–71.

74. Nylandsted J, Gyrd-Hansen M, Danielewicz A, Fehrenbacher N, Lademann U, Hoyer-Hansen M, Weber E, Multhoff G, Rohde M and Jaattela M. Heat shock protein 70 promotes cell survival by inhibiting lysosomal membrane permeabilization. J Exp Med 2004, 200:425-435.

75. Steel R, Doherty J P, Buzzard K, Clemons N, Hawkins C J and Anderson R L. Hsp72 inhibits apoptosis upstream of the mitochondria and not through interactions with Apaf-1. J Biol Chem 2004, 279:51490–51490.

10

Heat Shock Proteins, Their Cell Surface Receptors and Effects on the Immune System

Thomas Lehner, Yufei Wang, Trevor Whittall
and Lesley A. Bergmeier

10.1. Introduction

Heat shock or stress proteins are important intracellular protein chaperones that control their trafficking. The function of heat shock proteins in the immunopathology of infections, tumours and autoimmune diseases has been the subject of numerous experimental and clinical investigations over the past few decades and some of their properties are summarised in Table 10.1. Because there is extensive homology between mammalian and microbial heat shock proteins, immunological cross-reactions were considered to account for a number of autoimmune diseases. However, although the biological significance of lipopolysaccharide (LPS) found in Gram-negative bacteria has been well appreciated, heat shock proteins, which are found more widely in Gram-negative and -positive bacteria, especially those in the gut, have received more limited attention. A relatively new phase in this area of biology was initiated by the discovery of specific heat shock protein receptors and by rapid advances in our understanding of the signalling pathways. This chapter will deal with the receptors used by heat shock proteins, with particular reference to Hsp70, involvement of these proteins in the stimulation of chemokine production, maturation of dendritic cells (DCs), their intrinsic adjuvanticity and capacity to enhance immunogenicity.

10.2. Structural features of Hsp70

Although the overall three-dimensional structure of Hsp70 is not known, the structures of the two domains from various members of the family have been solved separately. The crystal structures of the ATPase domains of bovine Hsc70 (heat shock constitutive protein) and human Hsp70 have been determined [1, 2]. The domain consists of two approximately equal-sized lobes with a deep

Table 10.1. Properties of Heat Shock Proteins

---

1. Heat shock proteins are intracellular chaperones binding unfolded polypeptides to prevent misfolding and aggregation.
2. They bind peptides with a hydrophobic motif by non-covalent linkage.
3. Hsp70 and Hsp90 deliver exogenous antigen into the MHC class I as well as class II pathways, playing an important role in cross-priming free or antigen released from apoptotic cells.
4. Heat shock proteins stimulate production of the CC-chemokines, CCL3, CCL4 and CCL5 and cytokines, especially TNF-α, IL-12 and NO.
5. Heat shock proteins stimulate maturation of DCs in a similar way to that of CD40L.
6. They exert robust adjuvant function when linked to antigens and they are effective when administered systemically or by the mucosal route.
7. Generation of IL-12 and TNF-α by Hsp70 induces a Th1-α polarised adjuvant function.
8. Peptide epitopes within Hsp70 exert diverse immunomodulating functions.
9. Hsp70 can function as an alternative ligand to CD4$^+$ T cells, in activating the CD40–CD40L (CD154) co-stimulatory pathway, thereby enhancing immunity.
10. Tumour- or virus-specific peptides, non-covalently bound to Hsp70 or gp96, elicit CTL responses and exert protection against the specific tumours or viruses.
11. Heat shock proteins elicit innate immunity which may drive adaptive immune responses.

---

cleft between them. ATP binds at the base of the cleft. Two crystal forms of the human ATPase fragment which differ by a shift of 1–2 Å in one of the sub-domains have been obtained. This shift might be important in ATP binding and ADP release, and its presence indicates some degree of flexibility in this domain.

The crystal structure of the substrate-binding domain of DnaK from *Escherichia coli* with bound substrate (a seven-residue peptide with the sequence NRLLLTG) has been determined [3], and this consists of a β-sandwich sub-domain followed by an α-helical sub-domain. The β-sandwich sub-domain is formed by two stacked anti-parallel four-stranded β-sheets. The upper sheet forms the substrate binding site with loops L1,2 and L3,4 (between β-strands 1 and 2 and between β-strands 3 and 4, respectively), forming the sides of a channel that is the primary site of interaction with substrate. In the outer loop, L4,5 stabilises L1,2 by hydrogen bonds and hydrophobic interaction while L5,6 forms hydrogen bonds that stabilise L3,4. The α-helical sub-domain comprises five helices with the first and second helices (αA and αB) forming hydrophobic side-chain contacts with the β-sandwich. Helix αB extends over the entire substrate binding site and may stabilise substrate binding by interacting with all four loops that form this site; however, it does not interact directly with substrate.

The peptide substrate (NRLLLTG) complexed to DnaK adopts an extended conformation in which main-chain atoms form hydrogen bonds with DnaK, whereas side-chain contacts are predominantly hydrophobic [3]. The central

Table 10.2. Major receptors interacting with heat shock proteins and stimulating cytokine and chemokine production, and DC maturation

Receptor	Heat shock protein	Function	Reference
CD14	Human and chlamydial Hsp60 Human Hsp70	Stimulation of TNF-α, IL-12 and IL-6	[4, 5]
CD40	Microbial and human Hsp70	Stimulation of chemokines, TNF-α, IL-12 DC Maturation	[6–11]
CD91	Human Hsp70, Hsp90, gp96 and calreticulin (α2M)	Stimulation of TNF α, IL-12 and IL-1β	[12, 13]
TLRs	Human Hsp60, Hsp70 and Hsp90	Stimulation of TNF-α and IL-12 DC maturation	[14–17]
LOX-1	Human Hsp70(LDL)	Scavenger receptor	[18, 19]

residue (Leu 4) is buried in a relatively large hydrophobic pocket of DnaK and, together with Leu 3, contributes most of the contacts with the protein. Binding is almost completely determined by a five-residue core (RLLLT) which is centred on Leu 4. The specificity of binding is principally determined by an interaction of the residue at the centre of the core sequence with the hydrophobic pocket of DnaK and it has been suggested that Ile, Met, Thr, Ser, and possibly Phe could be accommodated as alternatives to Leu. Hydrophobic residues are preferred at the positions adjacent to the central residues and, although there are fewer constraints on residues at the ends of the motif, negative charges are generally excluded. However, some differences in substrate specificity between members of the Hsp70 family have been described and these might reflect functional differences.

10.3. Heat shock protein receptors and co-receptors

A number of receptors that bind different heat shock protein family members have been identified and these are shown in Table 10.2. Despite their high degree of conservation, there is a diversity of receptor usage, in that different members of the heat shock protein family may use the same receptor whereas others may use different receptors. Interactions between heat shock proteins and their receptors might elicit two different, but related, functions: (a) non-specific stimulation of antigen-presenting cells generates production of chemokines [20] and cytokines [7, 21] and (b) internalisation of heat shock protein–peptide complexes by endocytosis and the translocation of heat shock proteins into the human leukocyte antigen (HLA) class I or II pathway [22, 23].

10.3.1. CD14

CD14 is a glycosylphosphatidylinositol-anchored protein expressed on the cell surface of monocytes and macrophages and, to a lesser extent, on other myeloid cells. It is a receptor for LPS and serum LPS-binding protein [24], in association with TLR molecules [25]. Because CD14 lacks a transmembrane domain, signalling is dependent on the associated molecules in multi-receptor complexes [26, 27].

CD14 was characterised as a receptor for human (hu) and chlamydial Hsp60 by Kol and colleagues in 2000 [4]. They demonstrated that Hsp60 treatment of peripheral blood mononuclear cells resulted in the production of IL-6 and activation of p38 mitogen-activated protein kinase (MAPK). This could be inhibited by antibody to CD14. However, transfection of CD14 into CHO cells resulted in these cells acquiring the ability to respond to LPS, but not to Hsp60. They suggested that CD14 is necessary, but not sufficient, for cellular responsiveness to Hsp60, and that other molecules are required, possibly for transduction of the activation signal. The possibility has been raised that inducible forms of human Hsp70 may also stimulate human monocytes to produce TNF-α, IL-1β and IL-6 [5]. This involves TLR2 and TLR4 and calcium-dependent activation of MyD88, IRAK and NF-κB [28]. In apparent contradiction, it has been reported that CD14$^-$ DCs are capable of internalising Hsp70 [29], although it is not clear that internalisation and stimulation are necessarily mediated by the same receptors. The reader is referred to Chapter 7 for more details on CD14–TLR–Hsp interactions.

10.3.2. CD40

CD40 is a 40–50-kDa glycoprotein, a member of the tumour necrosis factor (TNF) receptor superfamily, and is primarily expressed on B lymphocytes, monocytes and DCs [30]. CD40 can also be found on epithelial cells, some cancer cells and activated CD8$^+$ T cells [31, 32]. CD40 plays an important role in T cell–mediated immune responses. It is crucial for T cell–dependent B cell activation, differentiation and immunoglobulin class switching and germinal centre formation [30]. CD40 is also involved in the activation of antigen-presenting cells and mediates DC maturation. It induces CD8$^+$ cytotoxic T lymphocytes and the generation of memory CD8$^+$ T cells [32–34]. The natural ligand for CD40 is CD40 ligand (CD40L, CD154), and this is expressed by activated T cells.

We have reported that CD40 is a receptor for microbial Hsp70 (mHsp70) [6] and this was later confirmed and extended to include human Hsp70 [9].

We have put forward the hypothesis that the adjuvanticity of mHsp70 and mHsp65 is accounted for by their capacity to stimulate production of the CC-chemokines CCL3, CCL4 and CCL5, all of which attract the entire repertoire of immune cells [20]. Because both of the major co-stimulatory pathways, CD80/86–CD28 and CD40–CD40L, stimulate these CC-chemokines [35–37], we explored the possibility that heat shock proteins might interact with one of the co-stimulatory molecules [6]. Whereas antibodies to CD80 or CD86 had no effect, those to CD40 blocked Hsp70 stimulation of CC-chemokine production. Further in-depth investigations using HEK293 cells (human embryonic kidney cell lines) revealed that Hsp70 stimulated the production of CC-chemokines only if cells were transfected with human CD40, but not with control molecules. Immunoprecipitation studies revealed that Hsp70 physically associates with cell membrane CD40 when incubated with CD40-expressing cells, and surface plasmon resonance showed that Hsp70 can directly bind to CD40 molecules [6]. Hsp70–peptide complexes binding to CD40 deliver the peptide into the MHC class I pathway and this process is dependent on the ADP-loaded state of Hsp70 [9].

CD40 also mediates Hsp70 stimulation of monocytes and bone marrow-derived DCs, which are the principle antigen-presenting cells in priming $CD4^+$ and $CD8^+$ T cells responses. Treatment of human monocyte-derived immature DCs for two days with mHsp70 induces dramatic changes in the expression of phenotypes, including an increase in the expression of major histocompatibility complex (MHC) class II molecules, the co-stimulatory molecules CD80, CD86 and the CD83, CCR7 maturation markers [7]. The CC-chemokines and Th1-polarising cytokines – IL-12 and TNF-α – are also produced. The C-terminal portion of the molecule (Hsp70 359–610) is a more potent inducer of cytokine expression and DC maturation than the full-length Hsp70 molecule, whereas the N-terminal ATPase domain of Hsp70 exhibits no such biological effects. Indeed, there is evidence that the preceding functions can be induced by the peptide binding domain of mHsp70 (aa 359–494), and a stimulatory epitope (aa 407–426) has been identified [8]. Human and microbial Hsp70 bind the CD40 receptor; however, surprisingly, the human ATPase domain of Hsp70 binds to one site, whereas a microbial C-terminal domain binds another site of the CD40 molecule [6, 9]. It is not clear whether CD40L shares one of the two receptor sites on CD40 or whether it binds yet another site. Indeed, two other functional domains have been identified in the cytoplasmic tail of CD40; one is involved in the induction of extra-follicular B cells and another is required for germinal centre formation [38].

10.3.3. Toll-like receptors (TLRs)

Toll-like receptors (TLRs) are examples of pattern-recognition receptors, are expressed by the innate immune system and recognise specific pathogen-associated molecular patterns (PAMPs) expressed on microbial components [39]. To date, about 10 TLRs have been described within the TLR family and each receptor appears to recognise different microbial pathogenic elements. TLRs are primarily expressed in those cell types that are involved in the first line of defence, such as DCs, monocytes, neutrophils, epithelial cells and endothelial cells. Activation of TLRs leads to production of inflammatory cytokines, chemokines, nitric oxide (NO), complement proteins, enzymes (such as cyclooxygenase-2), adhesion molecules and immune receptors [40]. These innate immune responses are essential for the elimination of pathogens and the regulation of adaptive immunity.

TLRs are primary candidates as receptors for heat shock proteins, because these proteins are highly conserved among microbial organisms and may serve as PAMPs to activate the innate immune system by interacting with pattern recognition receptors. However, only human Hsp60 and Hsp70, but not microbial Hsp, have so far been found to stimulate TLRs [14, 15]. Bone marrow-derived macrophages from the mouse strain C3H/HeJ, which carry a mutant TLR4, do not respond to Hsp60 [14].

It is, however, not clear whether TLRs act as receptors for heat shock proteins or are involved in signalling cellular activation. Some studies suggest that cell surface TLR4 is essential for human Hsp60 activation of monocytes [14], whereas others indicate that endocytosis of Hsp60 or Hsp70 is a prerequisite for activation of the intracellular TLR2 and TLR4 signalling pathways and that the endocytosis process is independent of these receptors [16, 17]. There is no evidence to indicate that there is a direct interaction between Hsp and TLRs. However, the possibility that human heat shock proteins can activate TLRs suggests that these not only serve as receptors for PAMPs derived from pathogenic microbes but that they also recognise endogenous ligands. Endogenous Hsp are present predominantly in the cell cytoplasm and can be induced and released in pathological conditions, such as injury, necrosis and stress. The observation that intracellular TLR2 and TLR4 may interact with internalised bacteria [41] is consistent with interaction of internalised heat shock proteins with TLR or other receptors in phagosomes. Indeed, small Hsp70 fragments or peptides have potent functional activity [7, 8]. Induction of local inflammatory responses by TLRs and endogenous heat shock protein interactions in the milieu around damaged tissues is probably helpful for tissue repair and wound healing, but it can also play an important part in chronic inflammatory diseases. Chapter 7 should be read in the context of this paragraph because Vabulas and Wagner take a different line with regard to the importance of TLRs in the response of cells to chaperones.

10.3.4. CD91

CD91 is a receptor for α2 macroglobulin (α2M) which is a protease inhibitor that binds to microbial pathogens and mediates phagocytosis by monocytes [42] but is poorly expressed on DCs. Several members of the heat shock protein families, including gp96, Hsp90, Hsp70 and calreticulin, despite being structurally distinct, interact with CD91 and are internalised by the receptor and the heat shock protein–bound peptides. This interaction is critical to the induction of CD8[+] cytotoxic T lymphocytes by heat shock protein–mediated cross-priming mechanisms [13, 43]. The heat shock protein gp96 binds directly to CD91, and this can be inhibited either by antibodies to CD91 or α2M [13]. It is not clear whether CD91 engagement by heat shock proteins activates antigen-presenting cells to produce cytokines, express co-stimulatory molecules or induces DC maturation [44].

10.3.5. LOX-1

LOX-1 is a scavenger receptor which may bind human heat shock proteins, especially Hsp70 [19]. It is found on endothelial cells, monocytes, immature DCs and smooth muscle cells [18, 19]. LOX-1 is a cell surface glycoprotein which binds modified lipoproteins and modified LDL (Ox-LDL), apoptotic cells and bacterially derived cell wall components [45, 46]. LOX-1 induces Hsp70-mediated cross-priming of CD8[+] cytotoxic T lymphocytes [19]. However, it is not clear whether LOX-1 engagement by heat shock proteins activates antigen-presenting cells to produce cytokines, express co-stimulatory molecules or induce DC maturation.

10.4. LPS contamination of heat shock protein preparations

Heat shock proteins stimulate innate immune cells to produce inflammatory cytokines, including TNF-α, IL-1β, IL-12, GM-CSF, NO and chemokines, such as MIP-1α, MIP-1β, RANTES, MCP-1 and MCP-2. These activities of heat shock proteins are similar to those of LPS which can contaminate heat shock protein preparations, and it is essential to exclude the possibility that the observed effects are elicited by contaminating LPS. Currently it is difficult to prepare LPS-free heat shock protein preparations, especially those expressed in *E. coli*. LPS activity is abrogated or greatly reduced by polymixin B treatment [5, 6], whereas heat shock protein stimulation is reduced by heat denaturation [5, 47]. Other reagents, such as RSLP derived from *Rhodopseudomonas spheroids* and lipid IVa, also inhibit LPS activity and have no effect on heat shock protein stimulation [5, 47]. As the stimulating activity of Hsp70 is calcium-dependent, the intracellular calcium chelator 1,2-bis(2-aminopheroxy)ethane-N,N,N, 'N'-tetraacetic acid-acetoxymethylester (BAPTA-AM) has been used to differentiate between the functions of LPS and Hsp70 [5, 6, 48]. Furthermore, treatment with proteinase K also inhibits Hsp70, but not LPS-stimulating activity [48]. The C3H/HeJ and C57BL/10ScCr inbred mouse strains are homozygous for a mutant *Lps* allele (*Lps$^{d/d}$*) which confers hyporesponsiveness to LPS challenge [49] and provides a model to study immunological functions of heat shock proteins. Genetic analysis revealed that C3H/HeJ mice have a point mutation within the coding region of the TLR4 gene, whereas C57BL/10ScCr mice exhibit a deletion of TLR4 [50]. In studies with monocytes and DCs using microbial Hsp70, inhibition with antibodies to CD40, but not to CD14, should discriminate between heat shock protein and LPS [6, 7].

Most of the LPS contamination (95–99%) in heat shock protein preparations can be removed using polymixin B immobilised on agarose affinity column. Polymixin B is a cationic cyclopeptide that can neutralise the biological activity of LPS by binding to its lipid A portion. It is noteworthy that heat shock proteins contain a hydrophobic domain that can interact with the affinity column and result in protein loss. This may be enhanced by heat shock protein binding to LPS on the column [27, 51]. Our experience with this method indicates that 60–70% of proteins can be recovered after one treatment with polymixin B and there is further protein loss if a second treatment is required. The residual LPS concentration in recovered heat shock protein preparations is typically less than 5 U per mg protein, as determined by the Limulus amoebocyte lysate assay [7]. Although the significance of this low level of LPS in Hsp70 preparations depends on the function under investigation, such low levels of LPS do not stimulate production of cytokines, CC-chemokines or maturation of DCs [6, 7]. The removal of LPS by washing columns of recombinant proteins with polymyxin B is described in Chapter 6. This overcomes the losses of protein described.

10.5. Signalling pathways

Engagement of heat shock proteins with receptors, such as CD40 and TLRs, might induce intracellular signalling which varies with the level of cell surface expression of the receptors on DCs and monocytes [52]. CD40 engagement with its natural ligand CD154 forms trimeric clusters and recruits adaptor proteins known as TNF receptor-associated factors (TRAFs) to the cytoplasmic tail [53]. CD40 has two cytoplasmic domains for binding TRAF. The TRAF6 binding site is within the membrane proximal cytoplasmic (Cmp) region, whereas TRAF2/3/5 is in the membrane distal cytoplasmic region (Cmd). See Chapter 7 for a more detailed discussion of the role of TRAFs in Hsp70-induced cell signalling cascades.

Binding of TRAF2, 3 and 5 results in formation of a signalling complex which includes multiple kinases, such as the MAPK family, NF-κB inducing kinase or receptor interacting protein (reviewed in [54]). Human Hsp70 uses the CD40 receptor and activates p38 MAPK [9]. However, another report has suggested that huHsp70 activates CD14 [28], as does huHsp60, with TLR2 and TLR4 receptors signalling via myeloid differentiation protein 88 and IL-1 receptor associated kinase (IRAK) [14, 15, 17, 55]. The activated IRAK associates with TRAF6 and activates p38 MAPK or NF-κB [28].

10.6. Interface between immunity and tolerance

The finding that CD40 is a receptor for mycobacterial Hsp70 (mHsp70) [6] and huHsp70 [9] might have profound implications for the development of acquired immune responses at the interface between adaptive immunity and tolerance. The interaction of CD40 with its ligand CD154 (or Hsp70) plays an important role in the development of the quality and magnitude of humoral and cellular immunity. Interference in the interaction between CD40 and CD154 can induce allogeneic and xenogeneic graft tolerance (reviewed in [54]). Indeed, there is evidence that CD154 blockade of immature DCs with specific antibodies might induce a state of antigen-specific tolerance [56].

Exposure to necrotic cells can induce maturation of DCs and immune stimulation via the release of antigens and heat shock proteins which are preferentially taken up by DCs [12, 57–60]. However, a state of tolerance might be induced by DCs that have captured apoptotic cells [61]. These findings and the cytokine stimulating activity of heat shock proteins suggest that heat shock proteins might play a role in the interface between immunity and tolerance. The state of maturation of DCs might be an important determining factor in the induction of immunity or tolerance. However, DCs within lymphoid tissue are able to form MHC–peptide complexes in the absence of maturation signals. On recognition of their ligands, naïve T cells divide and might undergo clonal deletion, resulting in a state of tolerance [62, 63]. However, if maturation signals are co-administered with antigen, an immune response develops. DCs in the absence of maturation signals are also able to down-modulate established immune responses [64–67], probably by inducing regulatory T cells [62, 63].

The capacity of CD40 signalling to induce DC maturation has been demonstrated in a transgenic CD40$^{-/-}$ DC model in which TNF-α fails to rescue the immune deficiency, despite its ability to induce maturation of DCs [68]. CD40 ligation of lymphoid DCs might abrogate their tolerogenic activity to a normally tolerogenic peptide [69]. Indeed, Hsp70 has been shown to stimulate CD40 and convert T cell tolerance, in the presence of lymphocytic choriomeningitis virus (LCMV) peptide, into autoimmune diabetes [10]. Another study has also implicated an inducible form of Hsp70 in the presentation of a major autoantigen in multiple sclerosis [70]. Steinmann and Nussenzweig [71] suggested that chronic infection might result from tolerance being exploited by a pathogen, as when persistent micro-organisms are taken up by DCs that fail to mature. The DC might then induce tolerance either by deleting T cells or by inducing T regulatory cells. They suggested that this might occur in HIV infection, when a large amount of virus is produced and steady-state DCs express receptors for the virus, such as dendritic cell-specific ICAM-3 grabbing non-integrin (DC-SIGN).

10.7. Interface between innate and adaptive immunity

Utilisation of the co-stimulatory CD40 molecule as a receptor for Hsp70 raises the paradigm that Hsp70 might function at the interface between innate and adaptive immunity. The B7 (CD80/CD86) and CD28 co-stimulatory interaction plays a central role in providing the second signal necessary for the adaptive immune function between HLA peptide and TCR [72]. The HLA-bound peptide and co-stimulatory molecule CD40 on an antigen-presenting cell interact with the corresponding TCR and CD40L on T cells, respectively, to elicit an effective immune response. It is of interest to note that ligation of CD40 by CD40L or Hsp70 and of CD80/86 by CD28 elicits CC-chemokines [35–37] and this suggests the presence of a non-cognate immune response which is responsible for attracting the immunological repertoire of cells (monocytes, immature DC, T and B cells). The interaction between CD40 and CD40L or Hsp70 also elicits production of some cytokines (such as IL-12 and TNF-α) and may induce Th1 polarisation of the immune response [7]. Hsp70 may thus function as an alternative ligand to CD40L, stimulating the major co-stimulatory pathway CD40–CD40L. Hsp70 functions as a multi-purpose molecule by acting as a carrier of antigens, inducing immune cells to the Hsp70 antigen site and by eliciting the maturation of DCs. There is also the possibility that Hsp70-bound antigen is processed by antigen-presenting cells and chaperoned by Hsp70 into the MHC presentation pathway for recognition by T cells [73].

10.8. The role of heat shock proteins in innate and adaptive immunity and vaccination

An essential component of vaccines against infections and tumours is an adjuvant activity which can elicit innate immune responses that can drive the development of specific adaptive immunity. Microbial Hsp70 and Hsp65 have been used as carrier molecules or adjuvants to enhance systemic immune responses when covalently linked to synthetic peptides [74–76]. Indeed, these and gp96 can be fused, covalently linked or loaded with peptides to elicit specific immunity to tumours or viruses [77–80]. The adjuvanticity of microbial Hsp70 and Hsp65 has been demonstrated not only by systemic but also by mucosal immunisation in non-human primates [20]. Both systemic and mucosal adjuvanticity are dependent on stimulating the production of three CC-chemokines – CCL3, 4 and 5 (or MIP-1α and MIP-1β and RANTES). CCL5 is a potent chemoattractant for monocytes, CD4$^+$ T cells and activated CD8$^+$ T cells [81–84]. CCL3 and CCL4 attract CD4$^+$ T and B cells [85] and all three chemokines attract immature DCs [86]. Monocytes and DCs internalise antigens that are processed and presented on the cell surface. DCs then undergo maturation and migrate to the regional lymph nodes, in which they present the processed antigen to T and B cells and elicit cell-mediated and humoral immune responses.

In addition to the CC-chemokines, IL-12, TNF-α and NO are also elicited by Hsp70 or, more efficiently, by its C-terminal portion [7]. Because IL-12 is one of the most potent cytokines for inducing Th1 polarisation [87] this might be responsible for the Th1-polarised adjuvanticity. The C-terminal portion of Hsp70-linked peptide elicits higher serum IgG$_{2a}$ and IgG$_3$ subclasses of antibodies than the native Hsp70-bound peptide, which is consistent with a Th1-polarising activity [7]. Furthermore, the Th2-type cytokine (IL-4) was not produced in immunised macaques. Thus, the C-terminal portion might be used as a microbial adjuvant that attracts the entire immunological repertoire of cells by virtue of stimulating the production of CC-chemokines and eliciting a Th1 response by generating IL-12.

The presence of Hsp70 and Hsp65 in most microorganisms [88, 89] and their capacity to generate CC-chemokines raises the possibility that the well-recognised immunogenicity of whole organisms, as compared with a subunit antigen, is mediated by CC-chemokines generated by heat shock proteins [20]. This is consistent with the principle that the innate immune system might drive adaptive immunity [90, 91]. Heat shock proteins in micro-organisms might function as a natural adjuvant generating CC-chemokines and cytokines. This concept is also consistent with the 'danger hypothesis' of infection [92], with the heat shock protein inducing the innate system to secrete CC-chemokines and mobilise the repertoire of cells required to generate specific immune responses against the invading organism.

The significance of Hsp70 as an alternative ligand to CD40L [6] stimulating the major co-stimulatory pathway CD40–CD40L has been highlighted. In mice lacking CD40 (CD40 knockout mice, $CD40^{-/-}$) the production of IL-12 by bone marrow–derived DCs was substantially reduced following Hsp70 stimulation [9, 10]. In these $CD40^{-/-}$ mice, Hsp70 failed to enhance DC function or to prime $CD4^+$ and $CD8^+$ T cell antigen-specific responses, and protection from *Mycobacterium tuberculosis* infection depended on the alternative Hsp70–CD40 co-stimulatory pathway [11]. An intriguing report that over-expression of Hsp70 in *M. tuberculosis* reduces the level of infection observed during the chronic phase [93] might also be interpreted as reflecting an enhanced immunity to the organism resulting from the interaction between increased amounts of Hsp70 with CD40 expressed on macrophages and DCs. In another report, co-administration of the tolerogenic LCMV peptide with human Hsp70 might reverse tolerance and promote the induction of autoimmune diabetes by DCs [10].

The application of Hsp70 as a carrier of HIV gp120 and peptides derived from CCR5 in mucosal vaccination has been recently demonstrated in rhesus macaques [94]. Significant protection against SHIV 89.6P has been associated with the induction of specific serum and secretory antibodies, IL-2 and IFN-γ stimulated by the vaccine components, and a raised concentration of CC-chemokines which was inversely correlated with the proportion of $CCR5^+$ cells [94].

$CD8^+$ cytotoxic T lymphocytes (CTLs) can be generated by loading LCMV peptides onto human Hsp70, and this has been shown to elicit protective anti-viral immunity in mice [80]. Human anti-influenza CTLs have been generated by pulsing DCs with mHsp70 loaded with peptides from influenza virus; the resulting CTL response is significantly greater than that induced by pulsing DCs with peptides alone [48]. There is also evidence that natural killer (NK) cells can be stimulated to proliferate by human Hsp70 and that this function resides in the C-terminal portion of Hsp70 [95]. Cell-surface-bound Hsp70 found on some tumour cells may induce migration of, and cytolysis by, $CD56^+CD94^+$ NK cells [96].

These investigators identified a peptide (aa 450–463) within the sequence of huHsp70 which enhances NK cell activity. A signal peptide derived from Hsp60 which binds HLA-E and interferes with CD94/NKG2A recognition, and enables NK cells to detect stressed cells, has also been identified [97]. Hsp70 and Hsp65 upregulate $\gamma\delta^+$ T cells, both *in vitro* and *in vivo* in non-human primates, and induce CD8-suppressor factors and CC-chemokines [98]. Indeed, a significant increase in $\gamma\delta^+$ T cells was found in rectal mucosal tissue and the draining lymph nodes in macaques immunised with SIVgp120 and p27 and protected from rectal mucosal challenge by SIV [98].

Acknowledgments

We wish to acknowledge the support received from the European Union (Grant No: LSHP-CT-2003-503240)

REFERENCES

1. Flaherty K M, Deluca-Flaherty C and McKay D B. Three-dimensional structure of the ATPse fragment of a 70K heat-shock cognate protein. Nature 1990, 346: 623–628.
2. Osipiuk J, Walsh M A, Freeman B C, Morimoto R I and Joachimiak A. Structure of a new crystal form of human hsp70 ATPse domain. D. Biol. Crystallogr. 1999, D55: 1105–1107.
3. Zhu X, Zhao X, Burkholder W F, Gragerov A, Ogata C M, Gottesman M E and Hendrickson W A. Structural analysis of substrate binding by the molecular chaperone DnaK. Science 1996, 272: 1606–1614.
4. Kol A, Lichtman A H, Finberg R W, Libby P and Kurt-Jones E A. Heat shock protein (HSP) 60 activates the innate immune response: CD14 is an essential receptor for HSP60 activation of mononuclear cells. J Immunol 2000, 164: 13–17.
5. Asea A, Kraeft S-K, Kurt-Jones E A, Stevenson M A, Chen L B, Finberg R W, Koo G C and Calderwood S K. Hsp70 stimulates cytokine production through a CD14-dependent pathway, demonstrating its dual role as a chaperone and cytokine. Nat Med 2000, 6: 435–442.
6. Wang Y, Kelly C G, Karttunen J T, Whittall T, Lehner P J, Duncan L, MacAry P, Younson J S, Singh M, Oehlmann W, Cheng G, Bergmeier L and Lehner T. CD40 is a cellular receptor mediating mycobacterial heat shock protein 70 stimulation of CC-chemokines. Immunity 2001, 15: 971–983.
7. Wang Y, Kelly C G, Singh M, McGowan E G, Carrara A S, Bergmeier L A and Lehner T. Stimulation of Th1-polarizing cytokines, C-C chemokines, maturation of dendritic cells, and adjuvant function by the peptide binding fragment of heat shock protein 70. J Immunol 2002, 169: 2422–2429.
8. Wang Y, Whittal T, McGowan E, Younson J, Kelly C, Bergmeier L A, Singh M, Lehner T. Identification of stimulating and inhibitory epitopes within the Hsp70 molecule which modulate cytokine production and maturation of dendritic cells. J Immunology 2005, 174: 3306–3316.
9. Becker T, Hartl F U and Wieland F. CD40, an extracellular receptor for binding and uptake of Hsp70-peptide complexes. J Cell Biol 2002, 158: 1277–1285.
10. Millar D G, Garza K M, Odermatt B, Elford A R, Ono N, Li Z and Ohashi P S. Hsp70 promotes antigen-presenting cell function and converts T-cell tolerance to autoimmunity in vivo. Nat Med 2003, 9: 1469–1476.

11. Lazarevic V, Myers A J, Scanga C A and Flynn J L. CD40, but not CD40L, is required for the optimal priming of T cells and control of aerosol M. tuberculosis infection. Immunity 2003, 19: 823–835.

12. Basu S, Binder R J, Suto R, Anderson K M and Srivastava P K. Necrotic but not apoptotic cell death releases heat shock proteins, which deliver a partial maturation signal to dendritic cells and activates the NF-κB pathway. Int Immunol 2000, 12: 1539–1546.

13. Binder R J, Han D K and Srivastava P K. CD91: a receptor for heat shock protein gp96. Nat Immunol 2000, 1: 151–155.

14. Ohashi K, Burkart V, Flohé S and Kolb H. Heat shock protein 60 is a putative endogenous ligand of the Toll-like receptor-4 complex. J Immunol 2000, 164: 558–561.

15. Vabulas R M, Ahmad-Nejad P, da Costa C, Miethke T, Kirschning C J, Hacker H and Wagner H. Endocytosed HSP60s use toll-like receptor 2 (TLR2) and TLR4 to activate the toll/interleukin-1 receptor signaling pathway in innate immune cells. J Biol Chem 2001, 276: 31332–31339.

16. Vabulas R M, Ahmad-Nejad P, Ghose S, Kirschning C J, Issels R D and Wagner H. HSP70 as endogenous stimulus of the Toll/interleukin-1 receptor signal pathway. J Biol Chem 2002, 277: 15107–15112.

17. Vabulas R M, Braedel S, Hilf N, Singh-Jasuja H, Herter S, Ahmad-Nejad P, Kirschning C J, Da Costa C, Rammensee H G, Wagner H and Schild H. The endoplasmic reticulum-resident heat shock protein Gp96 activates dendritic cells via the Toll-like receptor 2/4 pathway. J Biol Chem 2002, 277: 20847–20853.

18. Draude G, Hrboticky N and Lorenz R L. The expression of the lectin-like oxidized low-density lipoprotein receptor (LOX-1) on human vascular smooth muscle cells and monocytes and its down-regulation by lovastatin. Biochem Pharmacol 1999, 57: 383–386.

19. Delneste Y, Magistrelli G, Gauchat J, Haeuw J, Aubry J, Nakamura K, Kawakami-Honda N, Goetsch L, Sawamura T, Bonnefoy J and Jeannin P. Involvement of LOX-1 in dendritic cell-mediated antigen cross-presentation. Immunity 2002, 17: 353–362.

20. Lehner T, Bergmeier L A, Wang Y, Tao L, Singh M, Spallek R and van der Zee R. Heat shock proteins generate β-chemokines which function as innate adjuvants enhancing adaptive immunity. Eur J Immunol 2000, 30: 594–603.

21. Retzlaff C, Yamamoto Y, Hoffman P S, Friedman H and Klein T W. Bacterial heat shock proteins directly induce cytokine mRNA and interleukin-1 secretion in macrophage cultures. Infect Immun 1994, 62: 5689–5693.

22. Castellino F, Boucher P E, Eichelberg K, Mayhew M, Rothman J E, Houghton A N and Germain R N. Receptor-mediated uptake of antigen/heat shock protein complexes results in major histocompatibility complex class I antigen presentation via two distinct pathways. J Exp Med 2000, 191: 1957–1964.

23. Fujihara S M and Nadler S G. Intranuclear targeted delivery of functional NF-κB by 70kDa heat shock protein. EMBO J 1999, 18: 411–419.

24. Wright S D, Ramos R A, Tobias P S, Ulevitch R J and Mathison J C. CD14, a receptor for complexes of lipopolysaccharide (LPS) and LPS binding protein. Science 1990, 249: 1431–1433.

25. Yang R B, Mark M R, Gurney A L and Godwski P J. Signalling events induced by lipopolysaccharide-activated toll-like receptor 2. J Immunol 1999, 163: 639–643.
26. da Silva Correia J, Soldau K, Christen U, Tobias P S and Ulevitch R J. Lipopolysaccharide is in close proximity to each of the proteins in its membrane receptor complex. Transfer from CD14 to TLR4 and MD-2. J Biol Chem 2001, 276: 21129–21135.
27. Triantafilou K, Triantafilou M and Dedrick R L. A CD14-independent LPS receptor cluster. Nat Immunol 2001, 2: 338–344.
28. Asea A, Rehli M, Kabingu E, Boch J A, Baré O, Auron P E, Stevenson M A and Calderwood S K. Novel signal transduction pathway utilized by extracellular HSP70. Role of Toll-like receptor (TLR) 2 and TLR4. J Biol Chem 2002, 277: 15028–15034.
29. Lipsker D, Ziylan U, Spehner D, Proamer F, Bausinger H, Jeannin P, Salamero J, Bohbot A, Cazenave J P, Drillien R, Delneste Y, Hanau D and de la Salle H. Heat shock proteins 70 and 60 share common receptors which are expressed on human monocyte-derived but not epidermal dendritic cells. Eur J Immunol 2002, 32: 322–332.
30. van Kooten C and Banchereau J. Functions of CD40 on B cells, dendritic cells and other cells. Cur Opin Immunol 1997, 9: 330–337.
31. Young L S, Eliopoulos A G, Gallagher N J and Dawson C W. CD40 and epithelial cells: across the great divide. Immunol Today 1998, 19: 502–506.
32. Bourgeois C, Rocha B and Tanchot C. A role for CD40 expression on CD8+ T cells in the generation of CD8+ T cell memory. Science 2002, 297: 2060–2063.
33. Bennett S R, Carbone F R, Karamalis F, Flavell R A, Miller J F and Heath W R. Help for cytotoxic-T cell responses is mediated by CD40 signalling. Nature 1998, 393: 478–480.
34. Schoenberger S P, Toes R E, van der Voort E I, Offringa R and Melief C J. T cell help for cytotoxic T lymphocytes is mediated by CD40-CD40L interactions. Nature 1998, 393: 480–483.
35. Herold K C, Lu J, Rulifson I, Vezys V, Taub D, Grusby M J and Bluestone J A. Regulation of C-C chemokine production by murine T cells by CD28/B7 costimulation. J Immunol 1997, 159: 4150–4153.
36. Kornbluth R S, Kee K and Richman D D. CD40 ligand (CD154) stimulation of macrophages to produce HIV-1 suppressive chemokines. Proc Natl Acad Sci USA 1998, 99: 5205–5210.
37. McDyer J F, Dybul M, Goletz T J, Kinter A L, Thomas E K, Berzofsky J A, Fauci A S and Seder R A. Differential effects of CD40 ligand/trimer stimulation on the ability of dendritic cells to replicate and transmit HIV infection: evidence for CC-chemokine-dependent and -independent mechanisms. J Immunol 1999, 162: 3711–3717.
38. Yasui T, Muraoka M, Takaoka-Shichijo Y, Ishida I, Takegahara N, Uchida J, Kumanogoh A, Suematsu S, Suzuki M and Kikutani H. Dissection of B cell differentiation during primary immune responses in mice with altered CD40 signals. Int Immunol 2002, 14: 319–329.
39. Janeway C A J and Medzhitov R. Innate immune recognition. Ann Rev Immunol 2002, 20: 197–216.
40. Medzhitov R. Toll-like receptors and innate immunity. Nat Rev Immunol 2001, 1: 135–145.

41. Uronen-Hansson H, Allen J, Osman M, Squires G, Klein N and Callard R E. Toll-like receptor 2 (TLR2) and TLR4 are present inside human dendritic cells, associated with microtubules and the Golgi apparatus but are not detectable on the cell surface: integrity of microtubules is required for interleukin-12 production in response to internalized bacteria. Immunology 2004, 111: 173–178.

42. Armstrong P B and Quigley J P. α2 Macroglobulin: an evolutionary conserved arm of the innate immune system. Dev Comp Immunol 1999, 23: 375–390.

43. Basu S, Binder R J, Ramalingam T and Srivastava P K. CD91 is a common receptor for heat shock proteins gp96, hsp90, hsp70 and calreticulin. Immunity 2001, 14: 303–313.

44. Srivastava P. Roles of heat-shock proteins in innate and adaptive immunity. Nat Rev Immunol 2002, 2: 185–194.

45. Krieger M. The other side of scavenger receptors: pattern recognition for host defense. Cur Opin Lipidol 1997, 8: 275–280.

46. Gough P J and Gordon S. The role of scavenger receptors in the innate immune system. Microbes Infect 2000, 2: 305–311.

47. Panjwani N N, Popova L and Srivastava P K. Heat shock proteins gp96 and hsp70 activate the release of nitric oxide by APCs. J Immunol 2002, 168: 2997–3003.

48. MacAry P A, Javid B, Floto R A, Smith K G C, Singh M and Lehner P J. HSP70 peptide binding mutants separate antigen delivery from dendritic cell stimulation. Immunity 2004, 20: 95–106.

49. Coutinho A and Meo T. Genetic basis for unresponsiveness to lipopolysaccharide in C57BL/10Cr mice. Immunogenetics 1978, 7: 17–24.

50. Quershi S T, Larivière L, Leveque G, Clermont S, Moore K J, Gros P and Malo D. Endotoxin-tolerant mice have mutations in Toll-like receptor 4 (TLR4). J Exp Med 1999, 189: 615–625.

51. Randow F and Seed B. Endoplasmic reticulum chaperone gp96 is required for innate immunity but not cell viability. Nat Cell Biol 2001, 3: 891–896.

52. Kadowaki N, Ho S, Antonenko S, Malefyt R W, Kastelein R A and Bazan F. Subsets of human dendritic cell precursors express different toll-like receptor and respond to different microbial antigens. J Exp Med 2001, 163: 5786–5795.

53. Lee H H, Dempsey P W, Parks T P, Zhu X, Baltimore D and Cheng G. Specificities of CD40 signaling: involvement of TRAF2 in CD40-induced NF-κB activation and intercellular adhesion molecule-1 up-regulation. Proc Natl Acad Sci USA 1999, 96: 1421–1426.

54. Quezada S A, Jarvinen L Z, Lind E F and Noelle R J. CD40/CD154 interactions at the interface of tolerance and immunity. Ann Rev Immunol 2004, 22: 307–328.

55. Vabulas R M, Wagner H and Schild H. Heat shock proteins as ligands of Toll-like receptors. Cur Topics Microbiol Immunol 2002, 270: 169–184.

56. Markees T G, Phillips N E, Noelle R J, Shultz L D, Mordes J P, Greiner D L and Rossini A A. Prolonged survival of mouse skin allografts in recipients treated with donor splenocytes and antibody to CD40 ligand. Transplantation 1997, 64: 329–335.

57. Gallucci S, Lolkema M and Matzinger P. Natural adjuvants: endogenous activators of dendritic cells. Nat Med 1999, 11: 1249–1255.

58. Sauter B, Albert M L, Francisco L, Larsson M, Somersan S and Bhardwaj N. Consequences of cell death: exposure to necrotic tumour cells, but not primary tissue cells

or apoptotic cells, induces the maturation of immunostimulatory dendritic cells. J Exp Med 2000, 191: 423–433.

59. Shi Y and Rock K L. Cell death releases endogenous adjuvants that selectively enhance immune surveillance of particulate antigens. Eur J Immunol 2002, 32: 155–162.

60. Shi Y, Zhang W and Rock K L. Cell injury releases endogenous adjuvants that stimulate cytotoxic T cell responses. Proc Natl Acad Sci USA 2000, 97: 14590–14595.

61. Steinman R M, Turley S, Mellman I and Inaba K. The induction of tolerance by dendritic cells that have captured apoptotic cells. J Exp Med 2000, 191: 411–416.

62. Steinman R M, Hawiger D and Nussenzweig M C. Tolerogenic dendritic cells. Ann Rev Immunol 2003, 21: 685–711.

63. Lutz M B and Schuler G. Immature, semi-mature and fully mature dendritic cells: which signals induce tolerance or immunity. Trends Immunol 2002, 23: 445–449.

64. Menges M, Rößner S, Voigtländer C, Schindler H, Kukutsch N A, Bogdan C, Erb K, Schuler G and Lutz M B. Repetitive injections of dendritic cells matured with tumor necrosis factor alpha induce antigen-specific protection of mice from autoimmunity. J Exp Med 2002, 195: 15–21.

65. Dhodapkar M V, Steinman R M, Krasovsky J, Munz C and Bhardwaj N. Antigen-specific inhibition of effector T cell function in humans after injection of immature dendritic cells. J Exp Med 2001, 193: 233–238.

66. Legge K L, Gregg R K, Maldonado-Lopez R, Li L, Caprio J C, Moser M and Zaghouani H. On the role of dendritic cells in peripheral T cell tolerance and modulation of autoimmunity. J Exp Med 2002, 196: 217–227.

67. Ferguson T A, Herndon J, Elzey B, Griffith T S, Schoenberger S and Green D R. Uptake of apoptotic cells by lymphoid dendritic cells and cross-priming of CD8$^+$ T cells produce active immune unresponsiveness. J Immunol 2002, 168: 5589–5595.

68. Miga A J, Masters S R, Durell B G, Gonzalez M, Jenkins M K, Maliszewski C, Kikutani H, Wade W F and Noelle R J. Dendritic cell longevity and T cell persistence is controlled by CD154-CD40 interactions. Eur J Immunol 2001, 31: 959–965.

69. Grohmann U, Fallarino F, Silla S, Biachi R, Belladonna M L, Vacca C, Micheletti A, Fioretti M C and Puccetti P. CD40 ligation abrogates the tolerogenic potential of lymphoid dendritic cells. J Immunol 2001, 166: 277–283.

70. Mycko M P, Cwiklinska H, Szymanski J, Szymanska B, Kudla G, Kilianek L, Odyniec A, Brosnan C F and Selmaj K W. Inducible heat shock protein 70 promotes myelin autoantigen presentation by the HLA class II. J Immunol 2004, 172: 202–213.

71. Steinman R M and Nussenzweig M C. Avoiding horror autotoxicus: the importance of dendritic cells in peripheral T cell tolerance. Proc Natl Acad Sci USA 2002, 99: 351–358.

72. Bretscher P A. A two-step, two-signal model for the primary activation of precursor helper T cells. Proc Natl Acad Sci USA 1999, 96: 185–190.

73. Roth S, Willcox N, Rzepka R, Mayer M P and Melchers I. Major differences in antigen-processing correlate with a single Arg71↔Lys substitution in HLA-DR molecules predisposing to rheumatoid arthritis and with their selective interactions with 70-kDa heat shock protein chaperones. J Immunol 2002, 169: 3015–3020.

74. Lussow A R, Barrios C, van Embden J, van der Zee R, Verdini A S, Pessi A, Louis J A, Lambert P-H and Del Giudice G. Mycobacterial heat-shock proteins as carrier molecules. Eur J Immunol 1991, 21: 2297–2302.

75. Barrios C, Lussow J A, van Embden J, van der Zee R, Rappouli R, Costantino P, Louis J A, Lambert P-H and Del Giudice G. Mycobacterial heat-shock proteins as carrier molecules. II. The use of the 70kDa mycobacterial carrier for conjugated vaccines can circumvent the need for adjuvants and Bacillus Calmette Guerin priming. Eur J Immunol 1992, 22: 1365–1372.
76. Perraut R, Lussow A R, Gavoille S, Garraud O, Matile H, Tougne C, van Embden J, van der Zee R, Lambert P-H, Gysin J and Del Giudice G. Successful primate immunization with peptide conjugated to purified protein derrivative or mycobacterial heat shock proteins in the absence of adjuvants. Clin Exp Immunol 1993, 93: 382–386.
77. Suzue K and Young R A. Adjuvant-free hsp70 fusion protein system elicits humoral and cellular immune responses to HIV-1 p24. J Immunol 1996, 156: 873–879.
78. Udono H and Srivastava P K. Heat shock protein 70-associated peptides elicit specific cancer immunity. J Exp Med 1993, 178: 1391–1396.
79. Nieland T J F, Tan M C A A, Monee-van Muijen M, Koning F, Kruisbeek A M and van Bleek G M. Isolation of an immunodominant viral peptide that is endogenously bound to the stress protein GP96/GRP94. Proc Natl Acad Sci USA 1996, 93: 6135–6139.
80. Ciuputu A T, Petersson M, O'Donnell C L, Williams K, Jindal S, Kiessling R and Welsh R M. Immunization with a lymphocytic choriomeningitis virus peptide mixed with heat shock protein 70 results in protective antiviral immunity and specific cytotoxic T lymphocytes. J Exp Med 1998, 187: 685–691.
81. Schall T J, Bacon K, Toy K J and Goedell D V. Selective attraction of monocytes and T lymphocytes of the memory phenotype by cytokine RANTES. Nature 1990, 347: 669–671.
82. Murphy W J, Taub D D, Anver M, Conlon K, Oppenheim J J, Kelvin D J and Longo D L. Human RANTES induces the migration of human T lymphocytes into the peripheral tissues of mice with severe combined immune deficiency. Eur J Immunol 1994, 24: 1823–1827.
83. Meurer R, van Riper G, Feeney W, Cunningham P, Hora D J, Springer M S, MacIntyre D E and Rosen H. Formation of eosinophilic and monocytic intradermal inflammatory sites in the dog by injection of human RANTES but not human monocyte chemoattractant protein 1, human macrophage inflammatory protein 1 α, or human interleukin 8. J Exp Med 1993, 178: 1913–1921.
84. Kim J J, Nottingham L K, Sin J I, Tsai A, Morrison L, Oh J, Dang K, Hu Y, Kazahaya K, Bennett M, Dentchev T, Wilson D M, Chalian A A, Boyer J D, Agadjanyan M G and Weiner D B. CD8 positive influence antigen-specific immune responses through the expression of chemokines. J Clin Invest 1998, 102: 1112–1124.
85. Schall T J, Bacon K, Camp R D, Kaspari J W and Goeddel D V. Human macrophage inflammatory protein alpha (MIP-1α) and MIP-1β chemokines attract distinct populations of lymphocytes. J Exp Med 1993, 177: 1821–1826.
86. Dieu M C, Vanbervliet B, Vicari A, Bridon J M, Oldham E, Ait-Yahia S, Briere F, Zlotnik A, Lebecque S and Caux C. Selective recruitment of immature and mature dendritic cells by distinct chemokines expressed in different anatomic sites. J Exp Med 1998, 188: 373–386.
87. Trinchieri G. Interleukin-12: a cytokine produced by antigen presenting cells with immunoregulatory functions in the generation of T-helper cells type 1 and cytotoxic lymphocytes. Blood 1994, 84: 4008–4027.

88. Thole J E, van Schooten W C, Keulen W J, Hermans P W, Janson A A, de Vries R R, Kolk A H and van Embden J D. Use of recombinant antigens expressed in *Escherichia coli* K-12 to map B-cell and T-cell epitopes on the immunodominant 65-kilodalton protein of *Mycobacterium bovis* BCG. Infect Immun 1988, 56: 1633–1640.

89. Ivanyi J, Sharp K, Jackett P and Bothamley G. Immunological study of defined constituents of mycobacteria. Springer Semin Immunopathol 1988, 10: 279–300.

90. Medzhitov R M and Janeway C A J. Innate immunity: impact on the adaptive immune response. Cur Opin Immunol 1997, 9: 4–9.

91. Fearon D T and Locksley R M. The instructive role of innate immunity in the acquired immune response. Science 1996, 272: 50–53.

92. Matzinger P. Tolerance, danger, and the extended family. Ann Rev Immunol 1994, 12: 991–1045.

93. Stewart G R, Snewin V A, Walzl G, Hussell T, Tormay P, O'Gaora P, Goyal M, Betts J, Brown I N and Young D B. Overexpression of heat-shock proteins reduces survival of *Mycobacterium tuberculosis* in the chronic phase of infection. Nat Med 2001, 7: 732–737.

94. Bogers W M, Bergmeier L A, Ma J, Oostermeijer H, Wang Y, Kelly C G, Ten Haaft P, Singh M, Heeney J L and Lehner T. A novel HIV-CCR5 receptor vaccine strategy in the control of mucosal SIV/HIV infection. AIDS 2004, 18: 25–36.

95. Multhoff G, Mizzen L, Winchester C C, Milner C M, Wenk S, Eissner G, Kampinga H H, Laumbacher B and Johnson J. Heat shock protein 70 (Hsp70) stimulates proliferation and cytolytic activity of natural killer cells. Exp Hematol 1999, 27: 1627–1636.

96. Gastpar R, Gross C, Rossbacher L, Ellwart J, Riegger J and Multhoff G. The cell surface-localized heat shock protein 70 epitope TKD induces migration and cytolytic activity selectively in human NK cells. J Immunol 2004, 172: 972–980.

97. Michaëlsson J, Teixeira de Matos C T, Achour A, Lanier L L, Kärre K and Söderström K. A signal peptide derived from hsp60 binds HLA-E and interferes with CD94/NKG2A recognition. J Exp Med 2002, 196: 1403–1414.

98. Lehner T, Mitchell E, Bergmeier L, Singh M, Spallek R, Cranage M, Hall G, Dennis M, Villinger F and Wang Y. The role of $\gamma\delta$ T cells in generating antiviral factors and β-chemokines in protection against mucosal simian immunodeficiency virus infection. Eur J Immunol 2000, 30: 2245–2256.

11

Molecular Chaperone–Cytokine Interactions at the Transcriptional Level

Anastasis Stephanou and David S. Latchman

11.1. Introduction

The heat shock proteins (Hsps) are a group of highly conserved proteins that have major physiological roles in protein homeostasis [1, 2]. In most cell types, 1–2% of total proteins consist of heat shock proteins even prior to stress, which suggests important roles for these proteins in the biology and physiology of the unstressed cell. These roles particularly concern regulating the folding and unfolding of other proteins. The term 'heat shock proteins' was coined because these proteins were first identified on the basis of their increased synthesis following exposure to elevated temperatures [3]. Subsequently, it has been clearly shown that they can be induced following a variety of stressful stimuli. Some heat shock proteins, such as Hsp90 (each heat shock protein is named according to its mass in kilodaltons – see Chapter 1 for more details), are detectable at significant levels in unstressed cells and increase in abundance following a suitable stimulus, whereas others such as Hsp70 exist in both constitutively expressed and inducible forms [4, 5].

The dual role of heat shock proteins in both normal and stressed cells evidently requires the existence of complex regulatory processes which ensure that the correct expression pattern is produced. Indeed, such processes must be operative at the very earliest stages of embryonic development, since the genes encoding Hsp70 and Hsp90 are amongst the first embryonic genes to be transcribed [6, 7].

The induction of heat shock proteins in response to various stresses is dependent on the activation of specific members of a family of transcription factors, the heat shock factors that bind to the heat shock element in the promoters of the genes encoding heat shock proteins [8]. Four heat shock factors (HSF-1 to -4) have been cloned from a number of organisms, and their roles have now been characterised (Table 11.1). HSF-1 and HSF-3 have been shown to be involved

Table 11.1. Functional role of different heat shock factors in regulating heat shock proteins

	Intracellular function	Knock-out phenotype
HSF-1	Stress-induced heat shock protein gene expression	Defective heat shock response
HSF-2	Non-stress-induced heat shock protein gene expression	NA
HSF-2α	HSF-2α expressed predominantly in adult tissue	
HSF-2β	HSF-2β expressed during early development	
HSF-3	HSF-3 also involved in stress-induced heat shock protein gene expression and has a higher threshold in response to heat shock than HSF-1	NA
HSF-4α	HSF-4α acts as a repressor of heat shock protein gene expression	NA
HSF-4β	HSF-4β is a transactivator of heat shock protein gene expression	

in regulating heat shock proteins in response to thermal stress, whereas HSF-2 and HSF-4 are involved in heat shock protein regulation in unstressed cells, and their levels are regulated in response to a wide variety of biological processes such as immune activation and cellular differentiation [8] (Figure 11.1). In general however, the stimuli that induce such alterations in heat shock protein gene expression under non-stress conditions are poorly characterised, and the mechanisms by which they act are also unclear. Heat shock proteins are not only

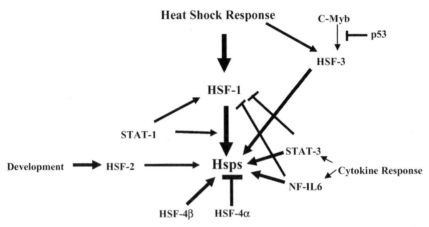

Figure 11.1. Heat shock factor pathways and their interaction or cooperation in modulating heat shock proteins.

regulated by heat shock factors, and this chapter will discuss heat shock factors and transcription factors that are able to interact or co-operate with HSF-1 to modulate the transcriptional regulation of heat shock proteins in response to non-stressful stimuli.

11.2. Transcriptional regulation of heat shock proteins by the HSF family

11.2.1. HSF-1

As mentioned earlier, HSF-1 has been identified as the heat shock factor that mediates stress-induced heat shock protein gene expression in response to environmental stressors. Such stresses induce HSF-1 oligomerisation and nuclear translocalisation followed by enhanced DNA binding on the heat shock protein DNA promoters. HSF-1 is negatively regulated by Hsp70 and Hsp90 which is suggestive of a negative-feedback loop for the regulation of Hsp70 and Hsp90 genes following a heat shock response [9, 10].

The phosphorylation of HSF-1 also modulates its activity, and constitutive phosphorylation is important for negatively regulating the activity of HSF-1 under normal growth conditions [11]. The kinases responsible for phosphorylating HSF-1 on several serine sites include glycogen synthase kinase 3β and c-jun N-terminal kinase [12, 13]. Although a positive role of HSF-1 phosphorylation in stress-induced activation of heat shock protein gene expression is also known to occur, the kinases involved and the phosphorylation sites on HSF-1 have not yet been characterised.

Cells from HSF-1 knock-out mice exhibit defects in heat shock protein induction following exposure to heat shock [14]. Moreover, cells lacking HSF-1 are susceptible to apoptotic cell death following exposure to heat stress [14]. Mice lacking HSF-1 also exhibit elevated levels of tumour necrosis factor-α, which results in an increased mortality after endotoxin and inflammatory challenge [14]. Interestingly, HSF-1 has also been shown to modulate other genes such as IL-1β and c-fos [15, 16], suggesting a role for HSF-1 in regulating other stress-responsive genes.

11.2.2. HSF-2

As mentioned earlier, heat shock protein gene expression is crucial not only for the survival of cells exposed to extracellular stress stimuli, but also for normal cellular physiological events such as embryonic development and cellular differentiation. HSF-2 has now been described as the factor involved in regulating heat shock proteins under non-stressful conditions. For example, Hsp70 expression is induced by haemin in K562 cells, and this causes them to differentiate; this process requires activation of HSF-2 [17]. HSF-2 exists as two isoforms, HSF-2α and HSF-2β, due to alternative splicing. The HSF-2α isoform is predominantly expressed in adult tissue, whereas the HSF-2β isoform is predominantly expressed in embryonic tissue [18]. HSF-2 DNA binding activity is high during early embryogenesis in tissues such as the heart, central nervous system and testis [18]. The importance of HSF-2 in development will become more apparent when the HSF-2 knock-out animals become available.

11.2.3. HSF-3 and HSF-4

HSF-3 was originally identified in avian cells and, like HSF-1, is also heat stress responsive [19]. However, the threshold temperatures required for the activation of HSF-3 and HSF-1 are different in that HSF-1 is activated by less severe heat shock than HSF-3 [19]. No reports have yet described HSF-3 in other organisms. Previously, HSF-3 was reported to bind to c-Myb, a transcription factor involved in cellular proliferation and required for the G1/S transition of the cell cycle, which also paralleled the expression of Hsp70 [20]. These studies suggest that HSF-3/cMyb interaction might be involved in cell-cycle-dependent expression of heat shock proteins. Furthermore, it has also been shown that HSF-3/c-Myb association is disrupted by direct binding of p53 tumour suppressor transcription factor, resulting in inhibition of Hsp70 expression [21].

In contrast to other heat shock factors, HSF-4 has been reported to function as a repressor of heat shock protein gene expression [22]. HSF-4 also exists as two isoforms, HSF-4α and HSF-4β [22, 23], and it was the HSF-4α isoform that was cloned and used in the original study reporting it to be a repressor. However, HSF-4β has subsequently been shown to activate heat shock protein gene expression, which suggests that the HSF-4 gene is able to generate both an activator and a repressor of heat shock genes [23].

11.3. The role of non-HSF transcription factors in modulating heat shock protein gene expression

The phenotype of mice lacking HSF-1 is normal in the absence of stress, and the expression of Hsp70 and Hsp90 in cells lacking HSF-1 is similar to that in wild-type cells, despite the fact that they exhibit a defect in the heat shock response following heat stress [14]. These studies suggest that other heat shock factors might compensate for the lack of HSF-1 and/or that other factors are also responsible for the expression of heat shock proteins under normal growth

Table 11.2. Functional role of STATs and their phenotype observed in the STAT knockout animal

	Induce	Knockout phenotype
STAT-1	Interferons, IL-6	Viable – defects in immune responses to microbes
STAT-2	Interferons	Viable, defects in INF responses
STAT-3	IL-6 family	Embryonic lethal
STAT-4	IL-12	Viable – defects in immune responses
STAT-5α	Numerous	Viable – defects in mammary gland development due to loss of responses to growth hormone
STAT-5β	Numerous	Viable – defects in responses to growth hormone and prolactin as well as defects in T cell responses
STAT-6	IL-4	Viable – IL-4 responses abolished resulting in defects in immune responses

conditions. Studies from our laboratory have unravelled a separate group of transcription factors that are activated by distinct cytokines and are able to modulate Hsp70 and Hsp90 gene expression. These factors include STAT-1, STAT-3 and NF-IL6, and their functional roles are described in the next section.

11.3.1. STATs

The signal transducers and activators of transcription (STATs) are a family of cytoplasmic transcription factors which mediate intracellular signalling initiated at cytokine cell-surface receptors and transmitted to the nucleus (Table 11.2). STATs are activated by phosphorylation on conserved tyrosine and serine residues on their C-terminal domains by the Janus kinases (JAKs) and mitogen-activated protein kinase families, respectively. These allow the STATs to dimerise and translocate to the nucleus and thereby regulate gene expression (for review see [24]). Interferon-γ is a potent activator of STAT-1, whilst the interleukin-6 (IL-6) family members including IL-6, leukaemia inhibitory factor and CT-1 primarily activate STAT-3 [24].

Studies from our laboratory have shown STAT-1 and STAT-3 to have opposing action on apoptotic cell death in various cell types [25]. We have reported that over-expression of STAT-1 is able to enhance apoptotic cell death in cardiac myocytes exposed to ischaemia-reperfusion, whereas over-expression of STAT-3 plus STAT-1 reduces the level of STAT-1–induced cell death following ischaemia-reperfusion by modulating the expression of pro- and anti-apoptotic genes [26]. Furthermore, these effects on apoptosis require serine-[727] (but not tyrosine-[701])

phosphorylation on the C-terminal transactivation domain of STAT-1 [27, 28]. We have subsequently shown that STAT-1 is able to modulate the activity of p53 and its effects on apoptosis [29]. These effects involve STAT-1/p53 protein–protein interaction with STAT-1 acting as a co-activator for p53 [29].

11.3.2. NF-IL6

The cytokine IL-6 is known to stimulate two distinct signalling pathways which results in the activation of two different classes of cellular transcription factors [30] (Figure 11.2). Thus, initial studies showed that a variety of IL-6–inducible genes contained binding sites for a transcription factor named NF-IL6 (nuclear factor IL-6), which showed high homology with the rat-liver nuclear factor C/EBP (CCAAT-enhancer-binding protein), and is therefore also known as C/EBPβ [31]. Subsequently, a second member of the C/EBP family, known as NF-ILβ or C/EBPδ, was identified and shown to form heterodimers with NF-IL6, resulting in a synergistic transcriptional effect [32]. After exposure of cells to IL-6, NF-IL6 is phosphorylated, resulting in its enhanced ability to stimulate transcription [32], whereas NF-IL6β is synthesised *de novo* [32]. As mentioned earlier, the second pathway that is stimulated by IL-6 is the JAK/STAT-3 signalling pathway.

It is generally accepted that the NF-IL6/NF-IL6β and STAT-3 signalling pathways allow IL-6 to activate two distinct sets of genes, each of which is responsive to one of these pathways. Thus, class 1 acute-phase proteins (such as α_1-acid glycoprotein, haptoglobin, C-reactive protein and serum amyloid) contain response elements for NF-IL6 and NF-IL6β, and these factors have been shown to be involved in the activation of these genes following IL-6 treatment [33]. In agreement with this idea, these genes are stimulated by exposure of cells to IL-1, which also stimulates NF-IL6/NF-IL6β activity without affecting STAT-3 [33]. In contrast, type 2 acute-phase genes such as fibrinogen, thiostatin and α_2-microglobulin, are not inducible by IL-1 and lack binding sites for NF-IL6/NF-IL6β. Instead, these genes contain binding sites allowing binding of STAT-3, which is responsible for activation of these genes in response to IL-6 [33].

11.4. Role of STAT-1, STAT-3 and NF-IL6 factors in modulating heat shock proteins

We have reported [34] that IL-6 can induce increased expression of Hsp90 in a variety of different cell types. The Hsp90β gene promoter is responsive to IL-6

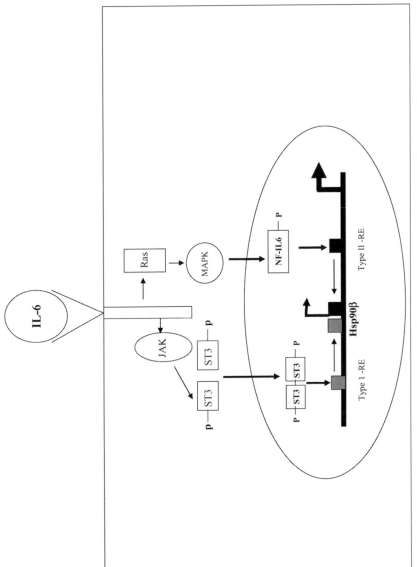

Figure 11.2. Distinct pathways activated by IL-6 that stimulate either type I or type II genes and heat shock proteins which are modulated by both pathways.

and can also be activated by NF-IL6 or NF-IL6β [34]. Moreover, a short region of the promoter containing an NF-IL6 binding site is essential for activation of the promoter by both IL-6 and NF-IL6 and can confer responsiveness both to IL-6 and to over-expression of NF-IL6 on a heterologous promoter. These findings suggest that Hsp90 is a member of the class of IL-6–responsive genes that are activated by NF-IL6/NF-IL6β.

Interestingly, this short region of the promoter also contains binding sites for STAT-3, and the Hsp90 promoter can also be activated by this factor. Moreover, over-expression of NF-IL6 and STAT-3 has a synergistic effect on the Hsp90 promoter and both these signalling pathways appear to be required for activation of the Hsp90 promoter by IL-6 [35]. However, despite their synergistic action in IL-6 signalling, these two pathways have opposite effects on the heat shock–mediated regulation of the Hsp90 promoter. Thus, STAT-3 reduces the stimulatory effect of heat shock, whereas NF-IL6 enhances it. When applied together, heat shock and IL-6 produce only weak activation of the Hsp90 promoter compared with either stimulus alone, indicating that the inhibitory effect of STAT-3 on heat shock factor predominates under these conditions [35]. In contrast, IL-1, which activates only the NF-IL6 pathway, synergises with heat shock to produce strong activation of Hsp90 [35]. These results therefore open up a new aspect of Hsp90 gene regulation which is additional to, and interacts with, the heat shock–activated pathway.

We have also examined whether STAT-1 is able to modulate heat shock protein expression. We have shown that IFN-γ treatment increases the levels of Hsp70 and Hsp90 and also enhances the activity of the Hsp70 and Hsp90β promoters, with these effects being dependent on activation of the STAT-1 transcription factor by IFN-γ [36]. These effects were not seen in a STAT-1–deficient cell line, indicating that IFN-γ modulates heat shock protein induction via a STAT-1–dependent pathway. The effect of IFN-γ/STAT-1 is mediated via a short region of the Hsp70/Hsp90 promoters, which also mediates the effects of NF-IL6 and STAT-3 and can bind STAT-1 [36].

This region also contains a binding site for the stress-activated transcription factor HSF-1. We have shown that STAT-1 and HSF-1 interact with one another via a protein–protein interaction to produce a strong activation of transcription [36]. This is in contrast to our previous finding that STAT-3 and HSF-1 antagonise each other, and we showed that STAT-3 and HSF-1 do not interact directly. To our knowledge, this was the first report of HSF-1 interacting directly via a protein–protein interaction with another transcription factor. Such protein–protein interactions and the binding of a number of different stress and cytokine-activated transcription factors to a short region of the Hsp90 and Hsp70 gene promoters are likely to play a very important role in heat shock protein gene activation by non-stressful stimuli and the integration of these responses with the stress response of these genes.

11.5. Linking STAT-1, STAT-3 and NF-IL6 elevation to pathological states

A number of disease states have been shown to exhibit elevated levels of heat shock proteins [37]. This includes patients with systemic lupus erythematosus (SLE), in whom elevated levels of Hsp90 are present. Interestingly, elevated levels of circulating IL-6 have also been reported in patients with SLE [38], and levels have been shown to be correlated with disease activity, being highest in patients with active disease. Moreover, spontaneous production of IgG by normal and SLE-derived B lymphocytes in culture can be enhanced by the addition of exogenous IL-6 and inhibited by antibody to IL-6 [39]. These findings therefore suggest that IL-6 might play a role in the pathogenesis of autoimmune diseases and the infusion of an antibody to IL-6 can relieve disease symptoms in lupus-prone NZB/NZW F1 mice [40].

In order to directly test the role of IL-6 in regulating Hsp90 expression *in vivo* we have used mice that have been artificially engineered to express elevated levels of IL-6 either by being made transgenic for extra copies of the IL-6 gene [41] or by inactivation of the gene encoding the transcription factor C/EBPβ, which also results in elevation of IL-6 levels in these mice [42]. In these experiments, elevated levels of Hsp90 were observed in both the IL-6 transgenic and the C/EBPβ knock-out mice [43]. Hence, the elevated IL-6 levels induced in these animals are indeed paralleled by increased levels of Hsp90 compared to normal control mice. In addition, elevated Hsp90 was associated with the specific production of autoantibodies to Hsp90 in both IL-6 transgenic and C/EBPβ knock-out animals. It is also of interest that inactivation of the IL-6 gene in the C/EBPβ knock-out mice results in the suppression of Castleman-like disease normally observed in these animals and a reduction in the production of autoantibodies.

Furthermore, elevated levels of Hsp90 in SLE correlates with levels of IL-6 and autoantibodies to Hsp90 [44]. These results support a model in which elevated levels of IL-6 in SLE patients induce elevated levels of Hsp90 protein, which in turn results in the production of autoantibodies to this protein. Additionally, IL-10 is also elevated in SLE, and IL-10 was demonstrated to enhance Hsp90 gene expression [45]. Therefore, these studies strongly suggest that IL-6 and IL-10 are likely to play a critical role in the regulation of Hsp90 levels and autoantibody production, both in autoimmune disease states and potentially in normal cells *in vivo*.

11.6. Conclusion

In this chapter, studies demonstrating the modulation of heat shock proteins by a group of transcription factors other than the traditional heat shock factor family under normal non-stressful conditions and also in several disease states have been reviewed. The finding that the responses to these factors occur around the HSF DNA binding site suggests that HSF-1, as well as other heat shock factors, are able to interact or cooperate with STATs or NF-IL6 family members. Further studies to identify novel protein interacting partners for heat shock factors will also provide insight into the regulation of heat shock proteins. Unravelling the mechanistic basis to this cooperation will undoubtedly enhance our understanding of the interdependent relationship between distinct heat shock factors and their interaction with other factors in these complex regulatory processes, which ensure that the correct heat shock protein expression pattern is produced under different physiological states.

REFERENCES

1. Ellis R J and van der Vies S M. Molecular chaperones. Ann Rev Biochem 1991, 60: 321–347.
2. Mathew A and Morimoto R I. Role of the heat-shock response in the life and death of proteins. Ann NY Acad Sci 1998, 851: 99–111.
3. Ritossa F A. A new puffing pattern induced by temperature shock and DNP in *Drosophila*. Experientia 1962, 18: 571–573.
4. Morimoto R I, Kline M P, Bimston D N and Cotto J J. The heat-shock response: regulation and function of heat-shock proteins and molecular chaperones. Essays Biochem 1997, 32: 17–29.
5. Jolly C and Morimoto R I. Role of the heat shock response and molecular chaperones in oncogenesis and cell death. J Natl Cancer Inst 2000, 92: 1564–1572.
6. Neidhardt F C, VanBogelen R, A, and Vaughn V. The genetics and regulation of heat-shock proteins. Ann Rev Genet 1984, 18: 295–329.
7. Walsh D, Li Z, Wu Y and Nagata K. Heat shock and the role of the HSPs during neural plate induction in early mammalian CNS and brain development. Cell Mol Life Sci 1997, 53: 198–211.
8. Pirkkala L, Nykanen P and Sistonen L. Roles of the heat shock transcription factors in regulation of the heat shock response and beyond. FASEB J 2001, 15: 1118–1131.
9. Shi Y, Mosser D D and Morimoto R I. Molecular chaperones as HSF1-specific transcriptional repressors. Genes Develop 1998, 12: 654–666.
10. Ali A, Bharadwaj S, O'Carrol R and Ovsenek N. HSP90 interacts with and regulates the activity of heat shock factor 1 in Xenopus oocytes. Mol Cell Biol 1998, 18: 4949–4960.
11. Xia W and Voellmy R. Hyperphosphorylation of heat shock transcription factor 1 is correlated with transcriptional competence and slow dissociation of active factor trimers. J Biol Chem 1997, 272: 4094–4102.

12. Xavier I J, Mercier P A, McLoughlin C M, Ali A, Woodgett J R and Ovsenek N. Glycogen synthase kinase 3β negatively regulates both DNA-binding and transcriptional activities of heat shock factor 1. J Biol Chem 2000, 275: 29147–29152.

13. Park J and Liu A Y. JNK phosphorylates the HSF1 transcriptional activation domain: role of JNK in the regulation of the heat shock response. J Cell Biochem 2001, 82: 326–338.

14. Xiao X, Zuo X, Davis A A, McMillan D R, Curry B B, Richardson J A and Benjamin I J. HSF1 is required for extra-embryonic development, postnatal growth and protection during inflammatory responses in mice. EMBO J 1999, 18: 5943–5952.

15. Xie Y, Chen C, Stevenson M A, Auron P E and Calderwood S K. Heat shock factor 1 represses transcription of the IL-1β gene through physical interaction with the nuclear factor of interleukin 6. J Biol Chem 2002, 277: 11802–11810.

16. Chen C, Xie Y, Stevenson M A, Auron P E and Calderwood S K. Heat shock factor 1 represses Ras-induced transcriptional activation of the c-fos gene. J Biol Chem 1997, 272: 26803–26806.

17. Sistonen L, Sarge K D, Phillips B, Abravaya K and Morimoto R I. Activation of heat shock factor 2 during hemin-induced differentiation of human erythroleukemia cells. Mol Cell Biol 1992, 12: 4104–4111.

18. Goodson M L, Park-Sarge O K and Sarge K D. Tissue-dependent expression of heat shock factor 2 isoforms with distinct transcriptional activities. Mol Cell Biol 1995, 15: 5288–5293.

19. Nakai A and Morimoto R I. Characterization of a novel chicken heat shock transcription factor, heat shock factor 3, suggests a new regulatory pathway. Mol Cell Biol 1993, 13: 1983–1997.

20. Kanei-Ishii C, Tanikawa J, Nakai A, Morimoto R I and Ishii S. Activation of heat shock transcription factor 3 by c-Myb in the absence of cellular stress. Science 1997, 277: 246–248.

21. Tanikawa J, Ichikawa-Iwata E, Kanei-Ishii C, Nakai A, Matsuzawa S, Reed J C and Ishii S. p53 suppresses the c-Myb-induced activation of heat shock transcription factor 3. J Biol Chem 2000, 275: 15578–15585.

22. Nakai A, Tanabe M, Kawazoe Y, Inazawa J, Morimoto R I and Nagata K. HSF4, a new member of the human heat shock factor family which lacks properties of a transcriptional activator. Mol Cell Biol 1997, 17: 469–481.

23. Tanabe M, Sasai N, Nagata K, Liu X D, Liu P C, Thiele D J and Nakai A. The mammalian HSF4 gene generates both an activator and a repressor of heat shock genes by alternative splicing. J Biol Chem 1999, 274: 27845–27856.

24. Ihle J N. The Stat family in cytokine signaling. Cur Opin Cell Biol 2001, 13: 211–217.

25. Battle T E and Frank D A. The role of STATs in apoptosis. Cur Mol Med 2002, 2: 381–392.

26. Stephanou A, Brar B K, Scarabelli T, Jonassen A K, Yellon D M, Marber M S, Knight R A and Latchman D S. Ischaemia-induced STAT-1 expression and activation plays a critical role in cardiac myocyte apoptosis. J Biol Chem 2000, 275: 10002–10008.

27. Stephanou A, Scarabelli T, Brar B K, Nakanishi Y, Matsumura M, Knight R A and Latchman D S. Induction of apoptosis and Fas/FasL expression by ischaemia/reperfusion in cardiac myocytes requires serine 727 of the STAT1 but not tyrosine 701. J Biol Chem 2001, 276: 28340–28347.

28. Stephanou A, Scarabelli T, Townsend P A, Bell R, Yellon D M, Knight R A and Latchman D S. The carboxyl-terminal activation domain of the STAT-1 transcription factor enhances ischaemia/reperfusion-induced apoptosis in cardiac myocytes. FASEB J 2002, 16: 1841–1843.

29. Townsend P, Scarabelli T M, Davidson S M, Knight R A, Latchman D S and Stephanou A. STAT-1 interacts with p53 to enhance DNA damage-induced apoptosis. J Biol Chem 2004, 279: 5811–5822.

30. Kishimoto T, Akira S, Narazaki M and Taga T. Interleukin-6 family of cytokines and gp130. Blood 1995, 86: 1243–1254.

31. Nakajima T, Kinoshita S, Sasagawa T, Sasaki K, Naruto M, Kishimoto T and Akira S. Phosphorylation at threonine-235 by ras-dependent mitogen-activated protein kinase cascade is essential for transcription factor NF-IL6. Proc Natl Acad Sci USA 1993, 90: 2207–2211.

32. Kinoshita S, Akira S and Kishimoto T. A member of the C/EBP family, NF-IL6 beta, forms a heterodimer and transcriptionally synergizes with NF-IL6. Proc Natl Acad Sci USA 1992, 89: 1473–1476.

33. Ganter U, Arcone R, Toniatti C, Morrone G and Ciliberto G. Dual control of C-reactive protein gene expression by interleukin-1 and interleukin-6. EMBO J 1989, 8: 3773–3779.

34. Stephanou A, Amin V, Isenberg D A, Akira S, Kishimoto T and Latchman D S. IL-6 activates heat shock protein 90 gene expression. Biochem J 1997, 321: 103–106.

35. Stephanou A, Isenberg D A, Akira S, Kishimoto T and Latchman D S. NF-IL6 and STAT-3 signalling pathways co-operate to mediate the activation of the Hsp90β gene by IL-6 but have opposite effects on its inducibility by heat shock. Biochem J 1998, 330: 189–195.

36. Stephanou A, Isenberg D A, Nakajima K and Latchman D S. Signal transducer and activator of transcription-1 and heat shock factor-1 interact and activate the transcription of the Hsp-70 and Hsp-90β gene promoters. J Biol Chem 1999, 274: 1723–1728.

37. Twomey B M, Dhillon V B, McCallum S, Isenberg D A and Latchman D S. Elevated levels of the 90kD heat shock protein in patients with systemic lupus erythematosus are dependent upon enhanced transcription of the Hsp90 gene. J Autoimmunity 1993, 6: 495–506.

38. De Benedetti F, Massa M, Robbion R, Ravelli A, Burgio G R and Martini A. Correlation of serum IL-6 levels with joint involvement and thrombocytosis in systemic juvenile rheumatoid arthritis. Arthritis Rheum 1991, 34: 1158–1163.

39. Linker-Israeli M, Deans R J, Wallace D J, Prehn J, Ozeri-Chen T and Kinenberg J R. Elevated levels of endogenous IL-6 in SLE. A putative role in pathogenesis. J Immunol 1991, 147: 117–123.

40. Finck B K, Chan B and Wofsy D. Interleukin 6 promotes murine lupus in NZB/NZW F1 mice. J Clin Invest 1994, 945: 585–591.

41. Suematsu S, Matsuda T, Aozasa K, Akira S, Nakano N, Ohno S, Miyazaki J-I, Yamamura K-I, Hirano T and Kishimoto T. IgG1 plasmacytosis in IL-6 transgenic mice. Proc Natl Acad Sci USA 1989, 86: 7547–7551.

42. Screpanti I, Romani L, Musiani P, Modesti A, Fattoro E, Lazzaro D, Sellitto C, Scarpa S, Bellavia D, Lattanzio G, Bistoni F, Frati L, Cortese R, Gulino A, Ciliberto G,

Costani F and Poli V. Lymphoproliferative disorder and imbalanced T-helper response in C/EBP-deficient mice. EMBO J 1995, 14: 1932–1941.

43. Stephanou A, Conroy S, Isenberg D A, Poli V, Ciliberto G and Latchman D S. Elevation of IL-6 in transgenic mice results in increased levels of the 90KD heat shock protein and production of the anti-Hsp90 antibodies. J Autoimmunity 1998, 11: 249–253.

44. Ripley B J, Isenberg D A and Latchman D. S. Elevated levels of the 90 kDa heat shock protein [hsp90] in SLE correlate with levels of IL-6 and autoantibodies to hsp90. J Autoimmunity 2001, 17: 341–346.

45. Ripley B J, Stephanou A, Isenberg D A and Latchman D S. Interleukin-10 activates heat-shock protein 90β gene expression. Immunology 1999, 97: 226–231.

Extracellular Biology of Molecular Chaperones: Physiological and Pathophysiological Signals

12

Heat Shock Protein Release and Naturally Occurring Exogenous Heat Shock Proteins

Johan Frostegård and A. Graham Pockley

12.1. Introduction

Although for many years the perception has been that mammalian heat shock proteins are intracellular molecules that are only released into the extracellular environment in pathological situations such as necrotic cell death, it is now known that these molecules can be released from a variety of viable (non-necrotic) cell types [1–4]. Moreover, we and a number of others have reported Hsp60 and/or Hsp70 to be present in the peripheral circulation of normal individuals [5–12]. These observations have profound implications for the perceived role of these proteins as pro-inflammatory intercellular 'danger' signalling molecules and have prompted a re-evaluation of the functional significance and role(s) of these ubiquitously expressed and highly conserved families of molecules. The reader should refer to Chapter 2, which discusses the intracellular dispositions of molecular chaperones and also touches on the release of heat shock proteins, and Chapter 3, in which novel pathways of protein release are described.

The mechanism(s) leading to the release of heat shock proteins are unknown, as is the source of circulating heat shock proteins in the peripheral circulation and their physiological and pathophysiological role(s). The inverse relationship between levels of circulating Hsp70 and the progression of carotid atherosclerosis [13], or the presence of coronary artery disease (CAD) [14], appears to be inconsistent with the concept that this molecule is a danger signal and an *in vitro* activator of innate and pro-inflammatory immunity [15]. Although a great deal of attention has focussed on the capacity of exogenous heat shock proteins to act as inflammatory activators of innate and adaptive immunity (discussed in detail in many of the chapters in this volume), exogenous heat shock proteins have also been shown to have a number of anti-inflammatory and non-immunological, cytoprotective effects on a variety of cells types (see, for example, Chapters 13, 14 and 16).

This chapter reviews the evolving evidence that heat shock proteins are present in, and can be released into, the extracellular compartment under physiological conditions and summarises the functional versatility of such exogenous proteins. Further insight into the functionality and significance of actively released and circulating heat shock proteins might reveal hitherto unknown physiological and pathophysiological roles for these ubiquitously expressed families of proteins.

12.2. Heat shock protein release – *in vitro* studies

12.2.1. Historical perspective

During the course of experiments between 1996 and 1997 in which the Pockley laboratory was investigating the influence of different physicochemical stressors on stress protein induction and expression by human peripheral blood mononuclear cells [16], it became apparent that the heat shock proteins Hsp60 and Hsp70 were present in the plasma of normal individuals. These findings were counter-intuitive to the proposition that heat shock proteins were only present in the extracellular milieu in the event of pathological processes that involved cellular necrosis and were viewed with scepticism. However, the literature revealed that these and other heat shock proteins can be released from a variety of intact cells and that this release appears to be via active, rather than passive, processes.

One of the earliest papers documenting heat shock protein release came from Tytell and colleagues, who reported the transfer of glia-axon transfer proteins, which include Hsp70, Hsc70 and Hsp100, from adjacent glial cells into the squid giant axon [17]. This finding prompted the suggestion that the release of such proteins might be a mechanism by which glial cells, which are capable of generating effective stress protein-mediated resistance to physical and metabolic insults, can protect adjacent neuronal cells, which exhibit a deficient response to stress. The capacity of glial cells to export Hsp70 has subsequently been demonstrated in a human system using T98G human glioma cells and stress-sensitive, differentiated LA-N-5 human neuroblastoma cells [18].

One of the seminal studies in the area of heat shock protein release came from Hightower and Guidon (see Chapter 19) when they reported that heat shock proteins could be released from cultured rat embryo cells [1]. Heat treatment increased the number of proteins released from a small set, which included the constitutively expressed Hsc70, to include the inducible Hsp70 and Hsp110 molecules. Although uncertain, the proposition was that the release of heat shock proteins might have resulted from changes in pH and gas tension, a disruption of the diffusion layer at the cell surface or by mechanical stresses that were associated with the manipulations that are an inevitable consequence of prolonged *in vitro* cell culture techniques [1]. The release of heat shock proteins did not appear to be mediated via the common secretory pathway, because it was not blocked by the inhibitors colchicine and monensin [1]. Nor was heat shock protein release due to cell lysis, because Hsp70 was not readily released from cells exposed to low concentrations of non-ionic detergents [1]. Rather, a selective release mechanism was suggested, and this was supported by the observation that Hsp70 synthesised in the presence of the lysine amino acid analogue aminoethyl cysteine was not released from cells. This was probably due to an alteration in the structure and/or function of the molecule which prevents its correct interaction with the specific release mechanism [1]. A number of studies have since reported the release of heat shock proteins from a range of cell types. See Chapter 3 for a detailed discussion of protein secretion mechanisms.

12.2.2. Heat shock proteins are released from a number of different cell types

Heat shock induces a four-fold increase in the levels of Hsp60 in the medium from cultured human islet cells [2], and an Hsp60-like protein is released from insulin-secreting β-cells [19].

An Hsp60-like protein has also been detected in conditioned media derived from cultured rat cortical astrocytes and a human neuroblastoma cell line [3]. In the case of neuroblastoma cells, extracellular Hsp60-like immunoreactivity is increased three-fold in the presence of the neuropeptide vasoactive intestinal peptide (VIP), and this increase occurs concomitantly with a two- to three-fold reduction in intracellular levels. Levels of exogenous Hsp60 are also increased two-fold after temperature elevation, and the effects are additive when VIP and thermal stress are combined [3]. As with most studies, no lactate dehydrogenase activity, an exclusively intracellular enzyme, was observed in the extracellular compartment, thus excluding the presence of cellular necrosis/damage [3]. The ability of cells to secrete heat shock proteins appears to be dependent on the cell type, because the levels of extracellular Hsp60-like protein generated by a human keratinocyte-derived cell line are at least 10-fold less than those generated by human neuroblastoma cells [3]. Readers should refer to Chapter 15 for further discussion of the interactions among neuronal cells, VIP and Hsp60.

Heat shock proteins are also released from cultured vascular smooth muscle cells subjected to oxidative stress by treatment with the naphthoquinolinedione LY83583. Sequential chromatography and tandem mass spectrometry has identified several proteins that are specifically secreted in response to such a stress, one of which is Hsp90α [4]. The release of heat shock proteins appears to be selective, because Hsp90β is not secreted under such conditions [4].

Murine and human prostate cancer cell lines secrete Hsp70, and secretion can be increased by forcing the expression of the protein by transfecting cells with a vector coding for murine Hsp70 [20]. Data from a number of sources would suggest that the release of such proteins from tumour cells might have implications for the development of tumour immunity (see Chapter 18), and indeed the forced over-expression of Hsp70 delays tumour growth and extends the survival of mice administered such Hsp70 transformed cells [20]. Another study has shown that IFN-γ can induce the active release of Hsc70 from K562 erythroleukaemic cells and that this was mirrored by a concomitant reduction in the amount of Hsc70 present on the surface of these cells [21]. The impact that such a release might have on the development of immunity is unclear given the evidence that Hsc70 appears to be unable to mediate the induction of tumour-specific immunity [22].

As detailed in Chapter 14, the presence of cell-free BiP has been reported in the synovial fluid and serum of patients with rheumatoid arthritis, and also in the synovial fluid of patients with other joint diseases. Hsp70 levels are dramatically increased in the synovial fluid of patients with rheumatoid arthritis and, to a much lesser extent, in the synovial fluid of patients with osteoarthritis and gout [23]. Given that BiP has anti-inflammatory actions, the release of this protein in rheumatoid arthritis might be part of a natural anti-inflammatory mechanism involving molecular chaperones.

12.3. Mechanisms of heat shock protein release

Although cellular necrosis inevitably leads to the non-specific release of intra-cellular proteins, the mechanism(s) via which heat shock proteins are actively released, either constitutively or in response to various factors from viable (non-necrotic) cells, has yet to be fully elucidated. As described in Chapter 3, 'non-classical' secretion of proteins that lack the typical N-terminal signal peptide sequences has been observed for a number of proteins such as fibroblast growth factors 1 and 2, IL-1 as well as viral proteins [24] and the mechanisms involved in such a process are reviewed in that chapter. A number of mechanisms might be involved in the release of heat shock proteins.

12.3.1. Release of heat shock proteins via classical or non-classical secretory pathways?

The original report of heat shock protein release from rat embryo cells suggested that it was not influenced by the inhibitors of the common secretory pathway, colchicine and monensin [1]. Inhibiting the common secretory pathway using brefeldin A also appears to have no effect on the release of Hsp70 from neuroblastoma cells treated with VIP [3], nor does it influence the release of Hsp70 from prostate cancer cell lines [20].

One study showing that a pharmacological inhibitor of phospholipase C activity (U731222) induces the release of Hsp70 from the A431 human carcinoma cell line suggests that phospholipase C inhibition might be one such mechanism [25]. The release of Hsp70 induced by inhibiting phospholipase C activity might occur via vesicular transport, because in the same publication the authors refer to unpublished data that indicate that the inhibition of vesicular transport with brefeldin A prevents Hsp70 release [25].

Another observation has been that a large proportion of the Hsp70 released by A431 human carcinoma cells is ubiquitinylated [25]. In addition to signalling for proteosome-dependent degradation, ubiquitination has been shown to serve as a trigger for different transport events [26, 27].

12.3.2. Release of heat shock proteins via exosomes?

Exosomes are small membrane vesicles that form within late endocytic compartments called multi-vesicular bodies (MVBs) and are distinct to apoptotic vesicles in that they differ in their mode of production and protein composition [28]. Further details on exosomes can be found in Chapter 3. The fusion of MVBs with the plasma membrane leads to the release of exosomes into the extracellular space. Various haematopoietic and non-haematopoietic cell types secrete exosomes, including reticulocytes, B and T lymphocytes, mast cells, platelets, macrophages, alveolar lung cells, tumour cells, intestinal epithelial cells and professional antigen-presenting cells (APCs) such as dendritic cells (DCs), and their function in different physiological processes depends on their origin [29]. DC- and tumour-derived exosomes are enriched in Hsp70, Hsc70 and Hsp90 [30, 31] and exosomes released from reticulocytes contain Hsp70 [32]. It might be that the release of heat shock proteins from cells is achieved via such a route.

12.3.3. Release of heat shock proteins via lipid rafts?

Lipid rafts might also be involved in the localisation of Hsp70 to the cell surface and its secretion into the extracellular environment [33]. Lipid rafts are specialised membrane domains enriched in sphingolipids, cholesterol and proteins that have been primarily characterised in polarised epithelial cells. Many functions have been attributed to lipid rafts, including cholesterol transport, membrane sorting, endocytosis and signal transduction [34, 35], and they can be isolated as detergent-resistant microdomains (DRMs) [36]. Hsp70 and Hsp90 and other molecules that are implicated in lipopolysaccharide (LPS)-mediated cellular activation are present in DRMs following LPS stimulation [37].

In unstressed Caco-2 human colonic adenocarcinoma epithelial cells, heat shock proteins (especially Hsp70) are present in a major Triton X-100 soluble form and a minor detergent insoluble form, which is associated with DRMs. Levels of Hsc70 and chaperones that are typically resident in the endoplasmic reticulum are low or undetectable in DRMs [33]. The translocation of Hsp70 into DRMs can be enhanced by heat shock or by increasing intracellular Ca^{2+} levels [33]. Although the incorporation of Hsp70 into the DRMs and the release of Hsp70 from Caco-2 cannot be inhibited by blockade of the common secretory pathway using brefeldin A or monensin, Hsp70 release can be blocked by disrupting lipid rafts using methyl-β-cyclodextrin [33].

Hypotheses about the extracellular biology of molecular chaperones find lack of support from referees and granting bodies because of a perception that because there are no defined mechanisms for the secretion of these proteins they, therefore, cannot be secreted. This position has been clearly defined by John Ellis, one of the pioneers of chaperone biology, in Chapter 1. However, it must be recognised that we know surprisingly little about protein secretions. It is only within the past decade or so that a number of extremely important and novel pathways of protein secretion in bacteria have been identified. The secretion pathways by which key mediators like IL-1 are released from eukaryotic cells are still not fully identified. The development of the biology of extracellular molecular chaperones will depend on the elucidation of the pathways by which these proteins are released from cells, and the authors hope that readers of this volume will take up this challenge.

12.4. Circulating heat shock proteins in health and disease

12.4.1. Circulating heat shock proteins in normal individuals

Hsp60 and Hsp70 are present in the serum of clinically normal individuals, in some instances at levels that are likely to elicit biological effects (>1000 ng/mL; [5, 8, 10, 12]. Circulating Hsp60 levels are not associated with cardiovascular risk factors such as body mass index, blood pressure and smoking status [5]; however, higher levels of circulating Hsp60 have been noted in individuals exhibiting an unfavourable lipid profile, as indicated by a low HDL cholesterol and a high total/HDL cholesterol ratio [10]. Levels are also associated with levels of the inflammatory cytokine tumour necrosis factor (TNF)-α [10]. In another study, serum Hsp60, but not Hsp70, levels have been shown to be associated with VLDL and triglyceride levels [5].

Exercise has been shown to induce the release of Hsp70 into the peripheral circulation of normal individuals [38, 39]. In one study, the observed increase in Hsp70 levels preceded any increases in Hsp70 protein and gene expression in contracting muscle, thereby arguing against contracting muscle being the source of the circulating Hsp70 [38]. Subsequent studies have demonstrated that the release of Hsp70 from splanchnic tissues during exercise is responsible, in part at least, for the elevated systemic concentrations of this protein [39].

12.4.2. Circulating heat shock proteins and ageing

Increasing age is associated with a reduced capacity to maintain homeostasis in all physiological systems, and it might be that this results, in part at least, from a parallel and progressive decline in the ability to produce heat shock proteins. If this is so, an attenuated heat shock protein response could contribute to the increased susceptibility to environmental challenges and the more prevalent morbidity and mortality which is seen in aged individuals [40, 41].

In vitro studies have shown that Hsp70 expression in heat-stressed lung cells [42], hepatocytes and liver [43, 44], splenocytes [45], myocardium [46] and mononuclear cells is reduced with increasing age [40], as is the induction of Hsp70 expression in response to ischaemia [47] and mitogenic stimulation [48]. Hsp70 gene expression declines during normal aging in human retina [49], and heat shock–induced Hsp70 expression is decreased in senescent and late-passage cells, both of which suggest that the process of aging itself might be associated with reduced Hsp70 production [50–52].

In keeping with the reduced capacity of cells and organisms to generate stress responses with aging, Hsp70 levels in peripheral blood lymphocytes decline with age, as do serum levels of Hsp60 and Hsp70 [9, 53]. The biological and physiological relevance of declining levels of circulating heat shock proteins with increasing age are unclear; however, intuitively, one consequence might be a reduced resistance to stress and the accumulation of damage.

12.4.3. Circulating heat shock proteins in cardiovascular disease

Evidence suggests that the inflammatory component to atherosclerosis might, at least in part, involve immune reactivity to heat shock proteins [54, 55], and a number of investigators have measured circulating levels of heat shock proteins in a variety of cardiovascular disease states. Hsp60 and Hsp70 are present in the serum of clinically normal individuals [5, 8, 10], and we and others have shown that serum Hsp60 levels are associated with early atherosclerosis in such individuals [5, 8]. Hsp60 has also been detected in the circulation of patients with acute coronary syndromes and chronic stable angina [56], and Hsp70 levels are elevated in patients with peripheral and renal vascular disease [57].

12.4.3.1. Circulating heat shock proteins as attenuators of cardiovascular disease?

A study of 218 subjects with established hypertension has shown that increases in carotid intima-media (IM) thicknesses (a measure of atherosclerosis) at a four-year follow-up are significantly less prevalent (odds ratio 0.42; p < 0.008) in those individuals having high serum Hsp70 levels (75th percentile) at enrollment [13]. A similar, albeit non-significant, trend for Hsp60 levels (odds ratio 0.6; p = 0.10) has also been observed. The relationship between Hsp70 levels and changes in IM thickness is independent of age, smoking habits and blood lipids.

A cross-sectional study that measured serum Hsp70 levels in 421 individuals evaluated for CAD by coronary angiography found that serum Hsp70 levels are significantly higher in patients without evidence of CAD, which supports our findings [14]. Again, the association of high Hsp70 levels with lack of CAD was independent of any relationship with traditional risk factors [14]. These findings indicate that circulating Hsp70 levels predict the development of atherosclerosis, at least in subjects with established hypertension and, arguably more importantly, suggest that Hsp70 influences its progression. The mechanism(s) by which such effects are manifested are currently unclear. However, atherosclerosis is an inflammatory condition, and Hsp70 is known to be capable of attenuating inflammatory responses by inducing self-heat shock protein–specific Th2-type CD4$^+$ T cells producing the regulatory cytokines IL-4 and IL-10 [58–60], and mycobacterial Hsp70 induces the secretion of IL-10 from peripheral blood monocytes [61].

Although we have previously reported elevated levels of Hsp70 in the peripheral circulation of patients with peripheral and renal vascular disease [57], we have not observed any relationship between Hsp70 levels and IM thickness in subjects with established hypertension [11], nor have we observed any relationship between Hsp70 levels and IM thickness in subjects with borderline hypertension [5]. Elevated levels of Hsp70 in peripheral and renal vascular disease [57] might result from the inflammatory response that is associated with established atherosclerotic disease. This proposition is supported by the observation that, although higher than controls, Hsp70 levels in patients with localised renal vascular disease are significantly lower than those in patients with more disseminated peripheral vascular disease [57]. It appears to be difficult to draw parallels between the events leading to elevated Hsp70 levels in overt and clinically established symptomatic vascular disease with those involved in the more subtle changes associated with increases in IM thickness.

12.4.3.2. Circulating heat shock proteins as promoters of cardiovascular disease?

A subset of CD4$^+$ T cells which lacks expression of the CD28 co-stimulation antigen (CD4$^+$CD28^{null}) is expanded in the circulation of patients with unstable angina and can comprise up to 50% of this cell population [62]. That these cells exhibit characteristics of natural killer cells, can produce high levels of IFN-γ and are present in ruptured atherosclerotic plaques suggests that they might have a role in the events that lead to plaque destabilisation and acute coronary syndrome (ACS) [63]. Human Hsp60 (which is present in the peripheral circulation of these individuals [56]) induces CD4$^+$CD28^{null} T cells from patients with ACS to express mRNA for IFN-γ and the cytolytic molecule perforin, whereas CD4$^+$CD28^{null} cells obtained from normal individuals or patients with chronic stable angina do not respond to Hsp60 [56]. The influence of Hsp60-reactive CD4$^+$CD28^{null} cells on the events leading to ACS remains to be more fully determined.

12.4.4. Circulating heat shock proteins in diabetes

12.4.4.1. Type 1 diabetes

Although they did not detect them in plasma from normal individuals, Finotti and colleagues have reported that Hsp70 and grp94 (gp96) are present at high concentrations in the plasma of patients with type 1 diabetes and that they are complexed with IgG and albumin [64, 65]. Vascular complications in patients with type 1 diabetes are reflected, often independently, by (i) glycaemic control, (ii) alterations in proteolytic enzyme action and inhibition and (iii) a higher than normal proteolytic activity of plasma [66]. Grp94 entirely accounts for the proteolytic activity of plasma from diabetic patients and the proteolytic form of circulating grp94 appears to lack the glycosylation exhibited by its ER-derived counterpart [64, 65]. Alpha$_1$-anti-trypsin (α_1AT) is the most important circulating inhibitor of serine protease activity and is complexed with grp94 in the plasma of patients with type 1 diabetes [64, 65].

12.4.4.2. Type 2 diabetes

The presence of Hsp70 in the plasma of patients with type 2 diabetes has been reported by Williams and colleagues [67]. Patients with type 2 diabetes are subject to oxidative stress as a consequence of their hyperglycaemic state, and this might contribute to the vascular complications that are experienced by these individuals [68, 69]. Oxidative stress also results from the elevated homocysteine levels that are often found in patients with type 2 diabetes, high levels of which are a significant risk factor for diabetes [70]. Hsp70 levels in patients that are not taking insulin are higher than those that are present in patients taking insulin, and the reduction of serum homocysteine levels by the administration of the antioxidant folic acid, which reduces oxidative stress *in vivo*, significantly lowers circulating Hsp70 levels [67]. Hsp70 might therefore be a suitable marker of the severity of this clinical condition and be useful for monitoring patients with type 2 diabetes.

12.4.5. Circulating heat shock proteins and stress

During the course of a study investigating levels of circulating Hsp60 in the plasma of 229 healthy British civil servants taking part in the Whitehall II study, a prospective study aimed at identifying risk factors for coronary heart disease [71], Henderson and colleagues identified a significant association between elevated levels of Hsp60, low socioeconomic status and social isolation in males and females, as well as psychological distress in women [10]. Some insight into the mechanism by which this occurs has been provided by experimental animal studies.

Psychological stress induced by exposing male rats to a cat without physical contact increases serum levels of Hsp70, concomitant with an induction of intracellular expression of Hsp70 in the hypothalamus and dorsal vagal complex [72]. This effect appears to be mediated by adrenal hormones, as the induction on intracellular expression and circulating levels of Hsp70 elicited by cat exposure does not occur, or is attenuated in adrenalectomised animals [72].

12.4.6. Circulating heat shock proteins and infection

Elevated serum levels of Hsp70 have been found in patients with acute infections, and Hsp70 levels correlate with levels of the inflammatory markers IL-6 and TNF-α, as well as with levels of the anti-inflammatory cytokine IL-10 [12].

12.4.7. Circulating heat shock proteins after surgery and trauma

Surgical procedures increase circulating levels of heat shock proteins. Plasma concentrations of Hsp70 and IL-6 markedly increase in patients undergoing liver resection and are significantly associated with post-operative infection [73]. Hsp70 is also associated with hepatic ischaemic time and with the degree of post-operative organ dysfunction [73]. Although the observed relationship between Hsp70 and organ dysfunction would suggest that, rather than being cytoprotective, circulating Hsp70 is involved in the development of organ dysfunction, another study has shown there to be no relationship between Hsp70 levels and organ dysfunction, nor between Hsp70 levels and the severity of the post-injury inflammatory response following severe trauma [74]. Indeed, in the latter study, high levels of Hsp70 appear to be associated with an improved survival under such circumstances [74].

Hsp70 is released into the circulation following coronary artery bypass grafting (CABG) [75]. The observation that levels of Hsp70 peak immediately after surgery and before those of IL-6 (which peak at 5 hours) prompted the suggestion that Hsp70 release might contribute to the inflammatory response which is reflected by elevated IL-6 levels [75]. The release of Hsp70 following CABG appears to be related to the use of a heart-lung machine, because on-pump procedures result in plasma levels of Hsp70 that are approximately four times greater than those present after off-pump procedures [76]. Nevertheless, Hsp70 levels after off-pump procedures are significantly higher than pre-operative levels [76]. Interestingly, on-pump procedures also induce high levels of IL-10, and the highest levels of IL-10 are present in those individuals with the highest levels of Hsp70 [76]. Although this would appear to be counterintuitive given the perceived role of Hsp70 as an inflammatory agent, it is consistent with the observations that Hsp70 can induce regulatory Th2-type CD4$^+$ T cells producing the cytokines IL-4 and IL-10 ([58–60] and see Section 12.7) and that mycobacterial Hsp70 can induce the production of IL-10 from monocytes [61].

12.5. Sequence versus functional conservation in the heat shock protein families

The ability of heat shock proteins to influence the activities of the innate and adaptive immune systems independently of chaperoned peptides has been demonstrated using both microbial- and endogenously derived (self) heat shock proteins (micHsp and enHsp, respectively). One of the dogmas of heat shock protein biology is that the high degree of sequence homology between equivalent heat shock protein family members derived from prokaryotes and eukaryotes (~50%) is reflected in a high degree of functional conservation. However, the rigidity of this concept is questioned by a number of studies because, despite their high degree of phylogenetic conservation, the biological activities of highly homologous heat shock proteins can differ considerably (see Section 6.4 in Chapter 6). Immune responses to micHsp and enHsp are tightly controlled, differentially controlled, and quantitatively and qualitatively different. A small number of bacteria, one of which is *Mycobacterium tuberculosis*, contain multiple genes encoding Hsp60 (chaperonin 60, cpn) genes, and despite having greater than 73% amino acid similarity, mycobacterial Cpn60.1 is between 10- and 100-fold more active in inducing cytokine secretion than Cpn60.2

(otherwise known as Hsp65) [77]. In addition, whereas Cpn60.3 from *Rhizobium leguminosarum* induces the production of a range of cytokines from human monocytes, Cpn60.1, which has a 74% amino acid sequence homology with Cpn60.3, exhibits no cytokine-inducing activity [78]. Thirdly, Hsp60 from the oral bacterium *Actinobacillus actinomycetemcomitans* and *Escherichia coli* are potent stimulators of bone resorption [79, 80], whereas equivalent molecules from mycobacteria are not [79, 81].

Despite having similar potency for stimulating TNF-α production in mouse macrophages, phylogenetically separate Hsp60 species interact with murine macrophages via different recognition systems [82]. The same might be true for members of other heat shock protein families, and it has been shown that human Hsp70 binds to murine macrophages via the CD40 molecule, but at a binding site which is distinct to that used by the bacterial Hsp70 homologue DnaK [83].

T cells appear to be capable of distinguishing enHsp60 and micHsp60, because the phenotype of T cells responding to eukaryotic and prokaryotic Hsp60 and their cytokine secretion profile differ. Whereas human Hsp60 activates CD45RA$^+$RO$^-$ (naïve) human peripheral blood T cells, bacterial-specific peptides activate CD45RA$^-$RO$^+$ (memory) T cells, and bacterial Hsp60 (which contains both conserved (human) and non-conserved (bacterial) sequences) activates CD45RA$^+$RO$^-$ and CD45RA$^-$RO$^+$ T cells [84].

The phenotype of the immune response to enHsp and micHsp60 also differs because T cells isolated from the synovial fluid of patients with rheumatoid arthritis respond to enHsp60 by predominantly producing regulatory Th2-type cytokine responses, whereas the response to micHsp60 produces higher levels of IFN-γ, which is consistent with a pro-inflammatory Th1-type response [85]. In addition, T cell lines generated from the synovial fluid of patients with rheumatoid arthritis in response to enHsp60 suppress the production of the pro-inflammatory cytokine TNF-α by peripheral blood mononuclear cells, whereas cells generated using mycobacterial Hsp65 have no such regulatory effect [85].

12.6. Functional consequences of exogenous heat shock proteins: immunological

Much attention has focussed on the capacity of heat shock proteins to interact with and influence the activities of innate and adaptive immune cells via a number of different receptors, including those of the Toll-like receptor family, and these activities are reviewed in great detail elsewhere in this volume (see Chapters 5 to 11).

Heat shock protein expression and immune reactivity towards heat shock proteins have been implicated in autoimmune diseases such as arthritis [86–88], multiple sclerosis [89–91], diabetes [92–94] and cardiovascular disease [55], and the administration of mammalian Hsp70 has been shown to prevent the induction of tolerance and promote the development of autoimmune disease *in vivo* [95].

The general perception is that exogenous heat shock proteins act as inflammatory mediators. However, from an evolutionary perspective it would not seem reasonable that mammalian responses to bacterial heat shock proteins (which presumably evolved as a defence) should also occur against ubiquitously expressed mammalian heat shock proteins, especially given that these proteins are present in the extracellular compartment under non-pathological conditions. A number of observations question the proposition that immune reactivity to self-heat shock proteins necessarily has a direct pro-inflammatory role in inflammatory disease.

The induction of T cell reactivity to enHsp60 and enHsp70 down-regulates disease in a number of experimental arthritis models, by a mechanism that appears to involve the induction of Th2-type CD4$^+$ T cells producing the regulatory cytokines IL-4 and IL-10 [58–60, 96–100]. The clinical relevance of these findings has been confirmed by studies that have reported an inverse association between the severity of disease and the production of regulatory cytokines such as IL-4 and IL-10 by T cells stimulated with Hsp60 in patients with rheumatoid arthritis [101–103]. The anti-inflammatory capacity of heat shock proteins also appears to be effective at the level of the APC, because mycobacterial Hsp70 induces the production of IL-10 by synoviocytes from patients with arthritis, and from monocytes from both patients and healthy controls [61]. IL-10 production by synoviocytes is accompanied by a decrease in the production of TNF-α [61].

The anti-inflammatory capacity of enHsp60 reactivity appears to dominate, because the administration of whole mycobacterial Hsp65, which contains the epitope that induces T cell activation and can induce arthritis in rats when administered alone, does not induce the disease. It therefore appears that the concomitant presence of conserved (self) epitopes can dominantly down-regulate the arthritogenic capacity of the non-conserved (non-self) epitopes [98]. The capacity of heat shock proteins to regulate inflammatory disease and the potential mechanisms by which this is achieved are reviewed in Chapter 16.

Much less studied has been the regulatory role of heat shock proteins in transplant rejection [104]; however, immunising recipient mice with enHsp60, or Hsp60 peptides that have the capacity to shift Hsp60 reactivity from a pro-inflammatory Th1 phenotype towards a regulatory Th2 phenotype, can delay murine skin allograft rejection [105]. In the clinical situation, it also appears that the development of immune responses to enHsp60 can regulate the allograft rejection response, in that in the late post-transplantation period (longer than one year) IL-10 production in response to enHsp60 peptides is increased [106]. At this time, the recognition of peptides from the intermediate and C-terminal regions of the protein appears to dominate [106].

A number of other heat shock proteins have been shown to exhibit anti-inflammatory activity. Human Hsp10 is now known to be early pregnancy factor [107, 108] and has been shown to be capable of inhibiting inflammation in the animal model of multiple sclerosis, experimental allergic encephalomyelitis [109, 110] (see Chapter 5). Extracellular Hsp27 stimulates IL-10 secretion by monocytes [111], and the anti-inflammatory properties of this protein are reviewed in Chapter 13. The protein BiP (grp78), a member of the Hsp70 family of molecules, is an autoantigen in rheumatoid arthritis [112], and its anti-inflammatory effects are reviewed in Chapter 14.

Taken together, these findings suggest that, rather than being pro-inflammatory, reactivity to endogenous (self) heat shock proteins is part of a normal immunoregulatory response that has the potential to dominantly control pro-inflammatory responses and inflammatory disease. The report that the treatment of human monocytes with enHsp60 suppresses their production of TNF-α following re-stimulation with enHsp60 or treatment with LPS, yet enhances their production of IL-1β, and that it down-regulates the expression of HLA-DR, CD86 and Toll-like receptor 4 [113], highlights the complexity of heat shock protein–mediated immunoregulation and the difficulties that will be encountered in attempting to understand the balance between the ability of these proteins to control inflammatory and regulatory responses.

12.7. Functional consequences of exogenous heat shock proteins: non-immunological

Although much attention has focussed on the capacity of exogenous heat shock proteins to interact with and influence the activities of innate and adaptive immune cells, a number of studies have also considered non-immunological consequences of heat shock protein interactions with a range of cell types. For example, exogenous Hsp/Hsc70 can change the differentiation patterns of the U937 promonocytic cell line [114] and has been shown to have a number of cytoprotective and other activities. Exogenous Hsp27, which is a member of the smaller heat shock protein family, inhibits *in vitro* culture-induced neutrophil apoptosis [115].

12.7.1. Cytoprotection

The cytoprotective effects of intracellular heat shock proteins have been appreciated for some considerable time [116]; however, extracellular heat shock proteins also appear to have cytoprotective effects (see Chapter 9). Some of the earliest evidence that heat shock proteins might have a therapeutic potential arose from the observations that exogenous members of the Hsp70 family protect spinal sensory neurons from axotomy-induced death and cultured aortic cells from heat stress [117, 118]. Subsequent work using a neonatal mouse model, in which spinal sensory (dorsal root ganglion) and motor neurons are induced to die by transection of their peripheral projections, demonstrated that exogenous Hsc70 can prevent axotomy-induced death of spinal sensory neurons [117]. The protective effect was selective, because treatment had no effect on the survival of motor neurons [117].

The capacity of glial cells to export Hsp70, and of Hsp70 to protect stressed neural cells, has also been demonstrated in a human system using T98G human glioma cells and stress-sensitive, differentiated LA-N-5 human neuroblastoma cells [18]. It might therefore be the case that heat shock protein release is an altruistic response on the part of one cell which is aimed at the protection of its more vulnerable neighbours [1].

Evidence that exogenous Hsp70 has cytoprotective effects on vascular-derived cells arose from the observations that exogenous Hsp70 protects heat-stressed cynomolgus macaque aortic cells [118] and serum-deprived rabbit arterial smooth muscle cells [119], the latter by a mechanism which involved cell association, but not internalisation. The mechanism by which such protection is induced is unknown, and the cell surface receptors involved have not been identified. However, some insight has been provided by studies that have shown exogenous Hsp70 to increase intracellular Hsp70 levels, which in turn delays the decline in viability of stressed cells [120], and that the accumulation of Hsp70 protects a range of cell types from apoptotic cell death induced by a number of apoptotic stimuli [121–125]. Hsp70 can inhibit apoptosis downstream of cytochrome c release, but upstream of caspase-3 cleavage, and the carboxyl-terminal region containing the peptide-binding domain is sufficient to inhibit caspase-3 activation [126].

Exogenous Hsp/Hsc70 renders neuroblastoma cells more resistant to staurosporine-induced apoptosis [18] and U937 pre-monocytes more resistant to cell death and apoptosis induced by TNF-α [114]. In human carcinoma cells, the release of Hsp70 induced by an inhibitor of phospholipase C activity leads to a concomitant reduction in intracellular levels of the protein, which in turn renders cells more sensitive to the apoptogenic effects of hydrogen peroxide

[25]. However, other studies in which Hsp70 expression has been enhanced by transfection have reported that elevated release of Hsp70 does not result in reduced intracellular levels [20].

12.7.2. Other activities

It must be emphasised that the biology of extracellular molecular chaperones is still in its infancy and the authors suggest that many and varied effects of these proteins will be reported. As examples from the current literature, *M. tuberculosis* Cpn10 is a potent inducer of bone resorption and the major osteolytic component of this organism [81] (see Chapter 5). As highlighted earlier, Hsp60 from the oral bacteria *A. actinomycetemcomitans* and *E. coli* are potent stimulators of bone resorption [79, 80], although equivalent molecules from mycobacteria are not [79, 81]. Human Hsp70, bacterial Hsp60 and their mycobacterial homologues induce ion-conducting pores across planar lipid bilayers at low or neutral pH [127]. A final example of the evolutionary plasticity of Hsp60 is the report that the endosymbiotic bacterium, *Enterobacter aerogenes*, present in the saliva of the antlion (a hunting insect), is the source of the neurotoxin produced by this insect. This insect toxin is none other than our old friend Hsp60 [128].

12.8. Conclusions

Heat shock proteins are extremely versatile and potent molecules, the importance of which to biological processes is highlighted by the high degree to which their structure and function are phylogenetically conserved. Our knowledge of the physiological role of heat shock proteins is currently limited; however, a better understanding of their function and thereby the acquisition of the capacity to harness their power might lead to their use as therapeutic agents and revolutionise clinical practice in a number of areas.

The observations that heat shock proteins can be released, and that they can directly or indirectly elicit potent immunoregulatory activities, requires that a new perspective on the roles of heat shock proteins and anti–heat shock protein reactivity in autoimmunity, transplantation, vascular disease and other conditions must be considered. It is the qualitative nature of the response to, or induced by, heat shock proteins rather than its presence *per se* that is important, and future experimental and clinical studies attempting to associate heat shock proteins in disease pathogenesis need to be structured and designed to address these issues. It is also important to definitively define the specificity of any responses, so that its outcome can be attributed to self- or non-self-reactivity. By doing this, the contribution of infective agents to pathogenic processes such as autoimmunity and vascular disease can be truly evaluated.

Acknowledgements

Work in the authors' laboratories has been funded by Boehringer-Ingelheim, Sweden, the Swedish Heart Lung Foundation, the King Gustav V 80th Birthday Fund, the Swedish Society of Medicine, the Swedish Rheumatism Association, the Söderberg Foundation, the Swedish Science Fund (JF), and the National Heart, Lung and Blood Institute, the Association for International Cancer Research and the British Heart Foundation (AGP).

REFERENCES

1. Hightower L E and Guidon P T. Selective release from cultured mammalian cells of heat-shock (stress) proteins that resemble glia-axon transfer proteins. J Cell Physiol 1989, 138: 257–266.
2. Child D F, Williams C P, Jones R P, Hudson P R, Jones M and Smith C J. Heat shock protein studies in type 1 and type 2 diabetes and human islet cell culture. Diabetic Med 1995, 12: 595–599.
3. Bassan M, Zamostiano R, Giladi E, Davidson A, Wollman Y, Pitman J, Hauser J, Brenneman D E and Gozes I. The identification of secreted heat shock 60-like protein from rat glial cells and a human neuroblastoma cell line. Neurosci Letters 1998, 250: 37–40.
4. Liao D-F, Jin Z-G, Baas A S, Daum G, Gygi S P, Aebersold R and Berk B C. Purification and identification of secreted oxidative stress-induced factors from vascular smooth muscle cells. J Biol Chem 2000, 275: 189–196.
5. Pockley A G, Wu R, Lemne C, Kiessling R, de Faire U and Frostegård J. Circulating heat shock protein 60 is associated with early cardiovascular disease. Hypertension 2000, 36: 303–307.
6. Pockley A G, Bulmer J, Hanks B M and Wright B H. Identification of human heat shock protein 60 (Hsp60) and anti-Hsp60 antibodies in the peripheral circulation of normal individuals. Cell Stress Chaperon 1999, 4: 29–35.
7. Pockley A G, Shepherd J and Corton J. Detection of heat shock protein 70 (Hsp70) and anti-Hsp70 antibodies in the serum of normal individuals. Immunol Invest 1998, 27: 367–377.
8. Xu Q, Schett G, Perschinka H, Mayr M, Egger G, Oberhollenzer F, Willeit J, Kiechl S and Wick G. Serum soluble heat shock protein 60 is elevated in subjects with atherosclerosis in a general population. Circulation 2000, 102: 14–20.
9. Rea I M, McNerlan S and Pockley A G. Serum heat shock protein and anti-heat shock protein antibody levels in aging. Exp Gerontol 2001, 36: 341–352.
10. Lewthwaite J, Owen N, Coates A, Henderson B and Steptoe A. Circulating human heat shock protein 60 in the plasma of British civil servants. Circulation 2002, 106: 196–201.

11. Pockley A G, de Faire U, Kiessling R, Lemne C, Thulin T and Frostegård J. Circulating heat shock protein and heat shock protein antibody levels in established hypertension. J Hypertension 2002, 20: 1815–1820.
12. Njemini R, Lambert M, Demanet C and Mets T. Elevated serum heat-shock protein 70 levels in patients with acute infection: use of an optimized enzyme-linked immunosorbent assay. Scand J Immunol 2003, 58: 664–669.
13. Pockley A G, Georgiades A, Thulin T, de Faire U and Frostegård J. Serum heat shock protein 70 levels predict the development of atherosclerosis in subjects with established hypertension. Hypertension 2003, 42: 235–238.
14. Zhu J, Quyyumi A A, Wu H, Csako G, Rott D, Zalles-Ganley A, Ogunmakinwa J, Halcox J and Epstein S E. Increased serum levels of heat shock protein 70 are associated with low risk of coronary artery disease. Arterioscler Thromb Vasc Biol 2003, 23: 1055–1059.
15. Todryk S M, Gough M J and Pockley A G. Facets of heat shock protein 70 show immunotherapeutic potential. Immunology 2003, 110: 1–9.
16. Bulmer J, Bolton A E and Pockley A G. Effect of combined heat, ozonation and ultraviolet light (VasoCare™) on heat shock protein expression by peripheral blood leukocyte populations. J Biol Reg Homeostatic Agents 1997, 11: 104–110.
17. Tytell M, Greenberg S G and Lasek R J. Heat shock-like protein is transferred from glia to axon. Brain Res 1986, 363: 161–164.
18. Guzhova I, Kislyakova K, Moskaliova O, Fridlanskaya I, Tytell M, Cheetham M and Margulis B. In vitro studies show that Hsp70 can be released by glia and that exogenous Hsp70 can enhance neuronal stress tolerance. Brain Res 2001, 914: 66–73.
19. Brudzynski K and Martinez V. Synaptophysin-containing microvesicles transport heat-shock protein Hsp60 in insulin-secreting β cells. Cytotechnology 1993, 11: 23–33.
20. Wang M H, Grossmann M E and Young C Y. Forced expression of heat-shock protein 70 increases the secretion of Hsp70 and provides protection against tumour growth. Brit J Cancer 2004, 90: 926–931.
21. Barreto A, Gonzalez J M, Kabingu E, Asea A and Fiorentino S. Stress-induced release of HSC70 from human tumors. Cell Immunol 2003, 222: 97–104.
22. Ménoret A, Patry Y, Burg C and Le Pendu J. Co-segregation of tumor immunogenicity with expression of inducible but not constitutive Hsp70 in rat colon carcinomas. J Immunol 1995, 155: 740–747.
23. Martin C A, Carsons S E, Kowalewski R, Bernstein D, Valentino M and Santiago-Schwarz F. Aberrant extracellular and dendritic cell (DC) surface expression of heat shock protein (Hsp)70 in the rheumatoid joint: possible mechanisms of Hsp/DC-mediated cross-priming. J Immunol 2003, 171: 5736–5742.
24. Cleves A E. Protein transports: the nonclassical ins and outs. Curr Biol 1997, 7: R318–R320.
25. Evdonin A L, Guzhova I V, Margulis B A and Medvedeva N D. Phospholipase C inhibitor, U73122, stimulates release of hsp-70 stress protein from A431 human carcinoma cells. Cancer Cell Int 2004, 4: 2.
26. Strous G J and Gent J. Dimerization, ubiquitylation and endocytosis go together in growth hormone receptor function. FEBS Letters 2002, 529: 102–109.
27. Katzmann D J, Odorizzi G and Emr S D. Receptor downregulation and multivesicular-body sorting. Nat Rev Mol Cell Biol 2002, 3: 893–905.

28. Théry C, Boussac M, Véron P, Ricciardi-Castagnoli P, Raposo G, Garin G and Amigorena S. Proteomic analysis of dendritic cell-derived exosomes: a secreted subcellular compartment distinct from apoptotic vesicles. J Immunol 2001, 166: 7309–7318.

29. Denzer K, Kleijmeer M J, Heijnen H F, Stoorvogel W and Geuze H J. Exosome: from internal vesicle of the multivesicular body to intercellular signalling device. J Cell Sci 2000, 113 Pt 19: 3365–3374.

30. Théry C, Zitvogel L and Amigorena S. Exosomes: composition, biogenesis and function. Nat Rev Immunol 2002, 2: 569–579.

31. Chaput N, Taïeb J, Schartz N E, Andre F, Angevin E and Zitvogel L. Exosome-based immunotherapy. Cancer Immunol Immunother 2004, 53: 234–239.

32. Mathew A, Bell A and Johnstone R M. Hsp-70 is closely associated with the transferrin receptor in exosomes from maturing reticulocytes. Biochem J 1995, 308: 823–830.

33. Broquet A H, Thomas G, Masliah J, Trugnan G and Bachelet M. Expression of the molecular chaperone Hsp70 in detergent-resistant microdomains correlates with its membrane delivery and release. J Biol Chem 2003, 278: 21601–21606.

34. Pralle A, Keller P, Florin E L, Simons K and Horber J K. Sphingolipid-cholesterol rafts diffuse as small entities in the plasma membrane of mammalian cells. J Cell Biol 2000, 148: 997–1008.

35. Vereb G, Matko J, Vamosi G, Ibrahim S M, Magyar E, Varga S, Szollosi J, Jenei A, Gaspar R J, Waldmann T A and Damjanovich S. Cholesterol-dependent clustering of IL-2Rα and its colocalization with HLA and CD48 on T lymphoma cells suggest their functional association with lipid rafts. Proc Natl Acad Sci USA 2000, 97: 6013–3018.

36. Horejsi V, Cebecauer M, Cerny J, Brdicka T, Angelisova P and Drbal K. Signal transduction in leucocytes via GPI-anchored proteins: an experimental artefact or an aspect of immunoreceptor function? Immunol Lett 1998, 63: 63–73.

37. Triantafilou M, Miyake K, Golenbock D T and Triantafilou K. Mediators of innate immune recognition of bacteria concentrate in lipid rafts and facilitate lipopolysaccharide-induced cell activation. J Cell Sci 2002, 115: 2603–2611.

38. Walsh R C, Koukoulas I, Garnham A, Moseley P L, Hargreaves M and Febbraio M A. Exercise increases serum Hsp72 in humans. Cell Stress Chaperon 2001, 6: 386–393.

39. Febbraio M A, Ott P, Nielsen H B, Steensberg A, Keller C, Krustrup P, Secher N H and Pedersen B K. Exercise induces hepatosplanchnic release of heat shock protein 72 in humans. J Physiol 2002, 544: 957–962.

40. Richardson A and Holbrook N J. Aging and the cellular response to stress: reduction in the heat shock response. In Holbrook, N. J., Martin, G. R. and Lockshin, R. A. (Eds.) Cellular Aging and Cell Death. Wiley-Liss, New York 1996, pp 67–79.

41. Shelton D N, Chang E, Whittier P S, Choi D and Funk W D. Microarray analysis of replicative senescence. Curr Biol 1999, 9: 939–945.

42. Fargnoli J, Kunisada T, Fornace A J J, Schneider E L and Holbrook N J. Decreased expression of heat shock protein 70 mRNA and protein after heat treatment in cells of aged rats. Proc Natl Acad Sci USA 1990, 87: 846–850.

43. Heydari A R, Conrad C C and Richardson A. Expression of heat shock genes in hepatocytes is affected by age and food restriction in rats. J Nutrition 1995, 125: 410–418.

44. Hall D, Xu L, Drake V J, Oberley L W, Oberley T D, Museley P L and Kregel K C. Aging reduces adaptive capacity and stress protein expression in the liver after heat stress. J Appl Physiol 2000, 2: 749–759.

45. Pahlavani M A, Denny M, Moore S A, Weindruch R and Richardson A. The expression of heat shock protein 70 decreases with age in lymphocytes from rats and rhesus monkeys. Expl Cell Res 1995, 218: 310–318.

46. Gray C C, Amrani M, Smolenski R T, Taylor G I and Yacoub M H. Age dependence of heat stress mediated cardioprotection. Ann Thoracic Surg 2000, 2: 621–626.

47. Nitta Y, Abe K, Aoki M, Ohno I and Isoyama S. Diminished heat shock protein 70 mRNA induction in aged rats after ischemia. Am J Physiol 1994, 267: H1795–1803.

48. Faassen A E, O'Leary J J, Rodysill K J, Bergh N and Hallgren H M. Diminished heat-shock protein synthesis following mitogen stimulation of lymphocytes from aged donors. Exp Cell Res 1989, 183: 326–334.

49. Bernstein S L, Liu A M, Hansen B C and Somiari R I. Heat shock cognate-70 gene expression declines during normal aging of the primate retina. Invest Ophthalmol Vis Sci 2000, 10: 2857–2862.

50. Liu A Y, Lin Z, Choi H, Sorhage F and Li B. Attenuated induction of heat shock gene expression in aging diploid fibroblasts. J Biol Chem 1989, 264: 12037–12045.

51. Luce M C and Cristofalo V J. Reduction in heat shock gene expression correlates with increased thermosensitivity in senescent human fibroblasts. Exp Cell Res 1992, 202: 9–16.

52. Effros R B, Zhu X and Walford R L. Stress response of senescent T lymphocytes: reduced hsp70 is independent of the proliferative block. J Gerontol 1994, 49: B65–70.

53. Jin X, Wang R, Xiao C, Cheng L, Wang F, Yang L, Feng T, Chen M, Chen S, Fu X, Deng J, Wang R, Tang F, Wei Q, Tanguay R M and Wu T. Serum and lymphocyte levels of heat shock protein 70 in aging: a study in the normal Chinese population. Cell Stress Chaperon 2004, 9: 69–75.

54. Wick G, Kleindienst R, Schett G, Amberger A and Xu Q. Role of heat shock protein 65/60 in the pathogenesis of atherosclerosis. Int Arch Allergy Immunol 1995, 107: 130–131.

55. Pockley A G. Heat shock proteins, inflammation and cardiovascular disease. Circulation 2002, 105: 1012–1017.

56. Zal B, Kaski J C, Arno G, Akiyu J P, Xu Q, Cole D, Whelan M, Russell N, Madrigal J A, Dodi I A and Baboonian C. Heat-shock protein 60-reactive CD4+CD28null T cells in patients with acute coronary syndromes. Circulation 2004, 109: 1230–1235.

57. Wright B H, Corton J, El-Nahas A M, Wood R F M and Pockley A G. Elevated levels of circulating heat shock protein 70 (Hsp70) in peripheral and renal vascular disease. Heart Vessels 2000, 15: 18–22.

58. Kingston A E, Hicks C A, Colston M J and Billingham M E J. A 71-kD heat shock protein (hsp) from *Mycobacterium tuberculosis* has modulatory effects on experimental rat arthritis. Clin Exp Immunol 1996, 103: 77–82.

59. Tanaka S, Kimura Y, Mitani A, Yamamoto G, Nishimura H, Spallek R, Singh M, Noguchi T and Yoshikai Y. Activation of T cells recognizing an epitope of heat-shock protein 70 can protect against rat adjuvant arthritis. J Immunol 1999, 163: 5560–5565.

60. Wendling U, Paul L, van der Zee R, Prakken B, Singh M and van Eden W. A conserved mycobacterial heat shock protein (hsp) 70 sequence prevents adjuvant arthritis upon

nasal administration and induces IL-10-producing T cells that cross-react with the mammalian self-hsp70 homologue. J Immunol 2000, 164: 2711–2717.

61. Detanico T, Rodrigues L, Sabritto A C, Keisermann M, Bauer M E, Zwickey H and Bonorino C. Mycobacterial heat shock protein 70 induced interleukin-10 production: immunomodulation of synovial cell cytokine profile and dendritic cell maturation. Clin Exp Immunol 2004, 135: 336–342.

62. Liuzzo G, Kopecky S L, Frye R L, O' Fallon W M, Maseri A, Goronzy J J and Weyand C M. Perturbation of the T-cell repertoire in patients with unstable angina. Circulation 1999, 100: 2135–2139.

63. Nakajima T, Schulte S, Warrington K J, Kopecky S L, Frye R L, Goronzy J J and Weyand C M. T-cell-mediated lysis of endothelial cells in acute coronary syndromes. Circulation 2002, 105: 570–575.

64. Pagetta A, Folda A, Brunati A M and Finotti P. Identification and purification from the plasma of type 1 diabetic subjects of a proteolytically active Grp94. Evidence that Grp94 is entirely responsible for plasma proteolytic activity. Diabetologia 2003, 46: 996–1006.

65. Finotti P and Pagetta A. A heat shock protein 70 fusion protein with α_1-antitrypsin in plasma of type 1 diabetic subjects. Biochem Biophys Res Comm 2004, 315: 297–305.

66. Finotti P, Carraro P and Calderan A. Purification of proteinase-like and Na^+/K^+-ATPase stimulating substance from plasma of insulin-dependent diabetics and its identification as α_1-antitrypsin. Biochim Biophys Acta 1992, 1139: 122–132.

67. Hunter-Lavin C, Hudson P R, Mukherjee S, Davies G K, Williams C P, Harvey J N, Child D F and Williams J H H. Folate supplementation reduces serum Hsp70 levels in patients with type 2 diabetes. Cell Stress Chaperone, 2004, 9: 344–349.

68. Sampson M J, Gopaul N, Davies I R, Hughes D A and Carrier M J. Plasma F2 isoprostanes: direct evidence of increased free radical damage during acute hyper-glycemia in type 2 diabetes. Diabetes Care 2002, 25: 537–541.

69. Spanheimer G R. Reducing cardiovascular risk in diabetes. Which factors to modify first? Postgrad Med 2001, 109: 33–36.

70. Stehouwer C D, Gall M A, Hougaard P, Jakobs C and Parving H H. Plasma homo-cysteine concentration predicts mortality in non-insulin-dependent diabetic patients with and without microalbuminuria. Kidney Int 1999, 55: 308–314.

71. Marmot M G, Smith G D, Stansfeld S, Patel C, North F, Head J, White I, Brunner E and Feeney A. Health inequalities among British civil servants: the Whitehall II study. Lancet 1991, 337: 1387–1393.

72. Fleshner M, Campisi J, Amiri L and Diamond D M. Cat exposure induces both intra- and extracellular Hsp72: the role of adrenal hormones. Psychoneuroendocrinology 2004, 29: 1142–1152.

73. Kimura F, Itoh H, Ambiru S, Shimizu H, Togawa A, Yoshidome H, Ohtsuka M, Shimamura F, Kato A, Nukui Y and Miyazaki M. Circulating heat-shock protein 70 is associated with postoperative infection and organ dysfunction after liver resection. Am J Surg 2004, 187: 777–784.

74. Pittet J F, Lee H, Morabito D, Howard M B, Welch W J and Mackersie R C. Serum levels of Hsp 72 measured early after trauma correlate with survival. J Trauma 2002, 52: 611–617.

75. Dybdahl B, Wahba A, Lien E, Flo T H, Waage A, Qureshi N, Sellevold O F, Espevik T and Sundan A. Inflammatory response after open heart surgery: release of

heat-shock protein 70 and signaling through Toll-like receptor-4. Circulation 2002, 105: 685–690.

76. Dybdahl B, Wahba A, Haaverstad R, Kirkeby-Garstad I, Kierulf P, Espevik T and Sundan A. On-pump versus off-pump coronary artery bypass grafting: more heat-shock protein 70 is released after on-pump surgery. Eur J Cardiothorac Surg 2004, 25: 985–992.

77. Lewthwaite J C, Coates A R M, Tormay P, Singh M, Mascagni P, Poole S, Roberts M, Sharp L and Henderson B. *Mycobacterium tuberculosis* chaperonin 60.1 is a more potent cytokine stimulator than chaperonin 60.2 (Hsp 65) and contains a CD14-binding domain. Infect Immun 2001, 69: 7349–7355.

78. Lewthwaite J, George R, Lund P A, Poole S, Tormay P, Sharp L, Coates A R M and Henderson B. *Rhizobium leguminosarum* chaperonin 60.3, but not chaperonin 60.1, induces cytokine production by human monocytes: activity is dependent on interaction with cell surface CD14. Cell Stress Chaperon 2002, 7: 130–136.

79. Kirby A C, Meghji S, Nair S P, White P, Reddi K, Nishihara T, Nakashima K, Willis A C, Sim R, Wilson M and Henderson B. The potent bone-resorbing mediator of *Actinobacillus actinomycetemcomitans* is homologous to the molecular chaperone GroEL. J Clin Invest 1995, 96: 1185–1194.

80. Reddi K, Meghji S, Nair S P, Arnett T R, Miller A D, Preuss M, Wilson M, Henderson B and Hill P. The *Escherichia coli* chaperonin 60 (groEL) is a potent stimulator of osteoclast formation. J Bone Miner Res 1998, 13: 1260–1266.

81. Meghji S, White P, Nair S P, Reddi K, Heron K, Henderson B, Zaliani A, Fossati G, Mascagni P, Hunt J F, Roberts M M and Coates A R. *Mycobacterium tuberculosis* chaperonin 10 stimulates bone resorption: a potential contributory factor in Pott's disease. J Exp Med 1997, 186: 1241–1246.

82. Habich C, Kempe K, van der Zee R, Burkart V and Kolb H. Different heat shock protein 60 species share pro-inflammatory activity but not binding sites on macrophages. FEBS Letters 2003, 533: 105–109.

83. Becker T, Hartl F U and Wieland F. CD40, an extracellular receptor for binding and uptake of Hsp70-peptide complexes. J Cell Biol 2002, 158: 1277–1285.

84. Ramage J M, Young J L, Goodall J C and Hill Gaston J S. T cell responses to heat shock protein 60: differential responses by CD4[+] T cell subsets according to their expression of CD45 isotypes. J Immunol 1999, 162: 704–710.

85. van Roon J A G, van Eden W, van Roy J L A M, Lafeber F J P G and Bijlsma J W J. Stimulation of suppressive T cell responses by human but not bacterial 60-kD heat shock protein in synovial fluid of patients with rheumatoid arthritis. J Clin Invest 1997, 100: 459–463.

86. Res P C, Schaar C G, Breedveld F C, van Eden W, van Embden J D S, Cohen I R and De Vries R R P. Synovial fluid T cell reactivity against 65 kDa heat shock protein of mycobacteria in early chronic arthritis. Lancet 1988, ii: 478–480.

87. Gaston J S H, Life P F, Jenner P J, Colston M J and Bacon P A. Recognition of a mycobacteria-specific epitope in the 65kD heat shock protein by synovial fluid derived T cell clones. J Exp Med 1990, 171: 831–841.

88. de Graeff-Meeder E R, van der Zee R, Rijkers G T, Schuurman H J, Kuis W, Bijlsma J W J, Zegers B J M and van Eden W. Recognition of human 60 kD heat shock protein by mononuclear cells from patients with juvenile chronic arthritis. Lancet 1991, 337: 1368–1372.

89. Wucherpfennig K, Newcombe J, Li H, Keddy C and Cuzner M L. γδ T cell receptor repertoire in acute multiple sclerosis lesions. Proc Natl Acad Sci USA 1992, 89: 4588–4592.

90. Georgopoulos C and McFarland H. Heat shock proteins in multiple sclerosis and other autoimmune diseases. Immunology Today 1993, 14: 373–375.

91. Stinissen P, Vandevyver C, Medaer R, Vandegaar L, Nies J, Tuyls L, Hafler D A, Raus J and Zhang J. Increased frequency of γδ T cells in cerebrospinal fluid and peripheral blood of patients with multiple sclerosis: reactivity, cytotoxicity, and T cell receptor V gene rearrangements. J Immunol 1995, 154: 4883–4894.

92. Elias D, Markovits D, Reshef T, van der Zee R and Cohen I R. Induction and therapy of autoimmune diabetes in the non-obese diabetic mouse by a 65-kDa heat shock protein. Proc Natl Acad Sci USA 1990, 87: 1576–1580.

93. Child D, Smith C and Williams C. Heat shock protein and the double insult theory for the development of insulin-dependent diabetes. J Royal Soc Med (Eng) 1993, 86: 217–219.

94. Tun R Y M, Smith M D, Lo S S M, Rook G A W, Lydyard P and Leslie R D G. Antibodies to heat shock protein 65 kD in type 1 diabetes mellitus. Diabetic Medicine 1994, 11: 66–70.

95. Millar D G, Garza K M, Odermatt B, Elford A R, Ono N, Li Z and Ohashi P S. Hsp70 promotes antigen-presenting cell function and converts T-cell tolerance to autoimmunity *in vivo*. Nat Med 2003, 9: 1469–1476.

96. van den Broek M F, Hogervorst E J M, van Bruggen M C J, van Eden W, van der Zee R and van den Berg W. Protection against streptococcal cell wall induced arthritis by pretreatment with the 65kD heat shock protein. J Exp Med 1989, 170: 449–466.

97. Thompson S J, Rook G A W, Brealey R J, van der Zee R and Elson C J. Autoimmune reactions to heat shock proteins in pristane induced arthritis. Eur J Immunol 1990, 20: 2479–2484.

98. Anderton S M, van der Zee R, Prakken B, Noordzij A and van Eden W. Activation of T cells recognizing self 60-kD heat shock protein can protect against experimental arthritis. J Exp Med 1995, 181: 943–952.

99. Anderton S M and van Eden W. T lymphocyte recognition of Hsp60 in experimental arthritis. In van Eden, W. and Young, D. (Eds.) *Stress Proteins in Medicine*. Marcel Dekker, New York 1996, pp 73–91.

100. Paul A G A, van Kooten P J S, van Eden W and van der Zee R. Highly autoproliferative T cells specific for 60-kDa heat shock protein produce IL-4/IL-10 and IFN-γ and are protective in adjuvant arthritis. J Immunol 2000, 165: 7270–7277.

101. de Graeff-Meeder E R, van Eden W, Rijkers G T, Prakken B J, Kuis W, Voorhorst Ogink M M, van der Zee R, Schuurman H J, Helders P J and Zegers B J. Juvenile chronic arthritis: T cell reactivity to human HSP60 in patients with a favorable course of arthritis. J Clin Invest 1995, 95: 934–940.

102. van Roon J, van Eden W, Gmelig-Meylig E, Lafeber F and Bijlsma J. Reactivity of T cells from patients with rheumatoid arthritis towards human and mycobacterial hsp60. FASEB 1996, 10: A1312.

103. Macht L M, Elson C J, Kirwan J R, Gaston J S H, Lamont A G, Thompson J M and Thompson S J. Relationship between disease severity and responses by blood mononuclear cells from patients with rheumatoid arthritis to human heat-shock protein 60. Immunology 2000, 99: 208–214.

104. Pockley A G. Heat shock proteins, heat shock protein reactivity and allograft rejection. Transplantation 2001, 71: 1503–1507.
105. Birk O S, Gur S L, Elias D, Margalit R, Mor F, Carmi P, Bockova J, Altmann D M and Cohen I R. The 60-kDa heat shock protein modulates allograft rejection. Proc Nat Acad Sci USA 1999, 96: 5159–5163.
106. Caldas C, Spadafora-Ferreira M, Fonseca J A, Luna E, Iwai L K, Kalil J and Coelho V. T-cell response to self HSP60 peptides in renal transplant recipients: a regulatory role? Transplant Proc 2004, 36: 833–835.
107. Cavanagh A C. Identification of early pregnancy factor as chaperonin 10: implications for understanding its role. Rev Reprod 1996, 1: 28–32.
108. Rolfe B, Cavanagh A, Forde C, Bastin F, Chen C and Morton H. Modified rosette inhibition test with mouse lymphocytes for detection of early pregnancy factor in human pregnancy serum. J Immunol Methods 1984, 70: 1–11.
109. Zhang B, Walsh M D, Nguyen K B, Hillyard N C, Cavanagh A C, McCombe P A and Morton H. Early pregnancy factor treatment suppresses the inflammatory response and adhesion molecule expression in the spinal cord of SJL/J mice with experimental autoimmune encephalomyelitis and the delayed-type hypersensitivity reaction to trinitrochlorobenzene in normal BALB/c mice. J Neurol Sci 2003, 212: 37–46.
110. Athanasas-Platsis S, Zhang B, Hillyard N C, Cavanagh A C, Csurhes P A, Morton H and McCombe P A. Early pregnancy factor suppresses the infiltration of lymphocytes and macrophages in the spinal cord of rats during experimental autoimmune encephalomyelitis but has no effect on apoptosis. J Neurol Sci 2003, 214: 27–36.
111. De A K, Kodys K M, Yeh B S and Miller-Graziano C. Exaggerated human monocyte IL-10 concomitant to minimal TNF-α induction by heat-shock protein 27 (Hsp27) suggests Hsp27 is primarily an anti-inflammatory stimulus. J Immunol 2000, 165: 3951–3958.
112. Corrigall V M, Bodman-Smith M D, Fife M S, Canas B, Myers L K, Wooley P, Soh C, Staines N A, Pappin D J, Berlo S E, van Eden W, van der Zee R, Lanchbury J S and Panayi G S. The human endoplasmic reticulum molecular chaperone BiP is an autoantigen for rheumatoid arthritis and prevents the induction of experimental arthritis. J Immunol 2001, 166: 1492–1498.
113. Kilmartin B and Reen D J. HSP60 induces self-tolerance to repeated HSP60 stimulation and cross-tolerance to other pro-inflammatory stimuli. Eur J Immunol 2004, 34: 2041–2051.
114. Guzhova I V, Arnholdt A C, Darieva Z A, Kinev A V, Lasunskaia E B, Nilsson K, Bozhkov V M, Voronin A P and Margulis B A. Effects of exogenous stress protein 70 on the functional properties of human promonocytes through binding to cell surface and internalization. Cell Stress Chaperon 1998, 3: 67–77.
115. Sheth K, De A, Nolan B, Friel J, Duffy A, Ricciardi R, Miller-Graziano C and Bankey P. Heat shock protein 27 inhibits apoptosis in human neutrophils. J Surg Res 2001, 99: 129–133.
116. Hightower L E. Heat shock, stress proteins, chaperones and proteotoxicity. Cell 1991, 66: 191–197.
117. Houenou L J, Li L, Lei M, Kent C R and Tytell M. Exogenous heat shock cognate protein Hsc70 prevents axonomy-induced death of spinal sensory neurons. Cell Stress Chaperon 1996, 1: 161–166.

118. Johnson A D, Berberian P A and Bond M G. Effect of heat shock proteins on survival of isolated aortic cells from normal and atherosclerotic cynomolgus macaques. Atherosclerosis 1990, 84: 111–119.

119. Johnson A D and Tytell M. Exogenous Hsp70 becomes cell associated, but not internalised by stressed arterial smooth muscle cells. In Vitro Cell Dev Biol 1993, 29A: 807–812.

120. Berberian P, Johnson A and Bond M. Exogenous 70kD heat shock protein increases survival of normal and atheromatous arterial cells. FASEB J 1990, 4: A1031.

121. Jäättelä M, Wissing D, Bauer P A and Li G C. Major heat shock protein Hsp70 protects tumor cells from tumor necrosis factor cytotoxicity. EMBO J 1992, 11: 3507–3512.

122. Simon M M, Reikerstorfer A, Schwarz A, Krone C, Luger T A, Jäättelä M and Schwarz T. Heat shock protein 70 overexpression affects the response to ultraviolet light in murine fibroblasts. Evidence for increased cell viability and suppression of cytokine release. J Clin Invest 1995, 95: 926–933.

123. Samali A and Cotter T G. Heat shock proteins increase resistance to apoptosis. Exp Cell Res 1996, 223: 163–170.

124. Lasunskaia E B, Fridlianskaia I I, Guzhova I V, Bozhkov V M and Margulis B A. Accumulation of major stress protein 70kDa protects myeloid and lymphoid cells from death by apoptosis. Apoptosis 1997, 2: 156–163.

125. Mosser D D, Caron A W, Bourget L, Denis-Larose C and Massie B. Role of the human heat shock protein Hsp70 in protection against stress-induced apoptosis. Mol Cell Biol 1997, 17: 5317–5327.

126. Li C-Y, Lee J-S, Ko Y-G, Kim K-I and Seo J-S. Heat shock protein 70 inhibits apoptosis downstream of cytochrome c release and upstream of caspase-3 activation. J Biol Chem 2000, 275: 25665–25671.

127. Alder G M, Austen B M, Bashford C L, Mehlert A and Pasternak C A. Heat shock proteins induce pores in membranes. Biosci Rep 1990, 10: 509–518.

128. Yoshida N, Oeda K, Watanabe E, Mikami T, Fukita Y, Nishimura K, Komai K and Matsuda K. Protein function. Chaperonin turned insect toxin. Nature 2001, 411: 44.

13

Hsp27 as an Anti-inflammatory Protein

Krzysztof Laudanski, Asit K. De and Carol L. Miller-Graziano

13.1. Introduction

As discussed in other chapters in this volume, heat shock proteins are traditionally viewed as protein chaperones rather than immunomodulators [1–3]. However, recent data suggest that heat shock proteins might also be ancestral danger signals which activate adaptive and innate immune responses [3]. The majority of studies examining the immunomodulatory activities of heat shock proteins have focused on the large heat shock proteins, Hsp60, Hsp70 and gp96, and these proteins have been shown to stimulate the innate immune system via binding to a variety of cellular receptors, particularly on monocytes, and to play an important role in health and disease [1–6].

This chapter focuses on the small heat shock protein, Hsp27, which, although shown to have some role in resistance to chemotherapeutic drugs, cytokine-induced cytotoxicity and to have been described as a prognostic marker in serum of breast cancer patients, has not been well characterised as an immunomodulator [7–13]. Hsp27 has been reported as present in increased amounts in the serum of patients with several human diseases, as well as being necessary for activation of the signal transduction pathway leading to monocyte production of the anti-inflammatory and immunoinhibitory cytokine IL-10 [1–3, 14]. These data led to our interest in investigating the possible immunomodulatory activity of Hsp27 on different human monocyte functions, which are pivotal in both the development of inflammatory responses as well as the triggering of lymphocyte-specific immunity.

In contrast to the immunomodulatory effects of Hsp60 or Hsp70, the influence of which on monocytes is primarily pro-inflammatory, we found Hsp27 to have potent monocyte anti-inflammatory and immune inhibitory effects [15]. Since Hsp27 has been identified as circulating in the serum of breast cancer patients as well as burn victims, it has the potential to act as an exogenous mediator

[2–4, 7]. Exogenous Hsp27 could function as an anti-inflammatory and/or immunoinhibitory monocyte mediator through a number of mechanisms, which are not mutually exclusive. Hsp27 binding to leukocytes could activate other cytokines/mediators which then influence innate and specific immunity, in a similar manner to large heat shock proteins which induce monocyte production of pro-inflammatory cytokines such as tumour necrosis factor (TNF)-α [2, 3, 16, 17]. Hsp27 binding could also alter monocyte surface receptor expression, thereby changing their activation potential or differentiation capacity. Our data support the concept that Hsp27 has all of these activities. In the next chapter (Chapter 14) a similar set of actions are ascribed to the large chaperone, BiP.

Cytokine production by monocytes is key to their inflammatory and lymphocyte and antigen-presenting cell (APC) immune-activating activities, as well as their differentiation to either the most potent APC, the dendritic cell (DC), or to the end-stage inflammatory macrophage. Consequently, Hsp27 modulation of cytokine production by monocytes could dramatically impact their inflammatory and immune functions.

13.2. Hsp27 as an inducer of anti-inflammatory/immunomodulatory cytokines

We first assessed Hsp27 for its stimulation of monocyte mediators, because large heat shock proteins induce highly elevated pro-inflammatory cytokine production by monocytes [2, 3]. All assays were performed in polymyxin B–containing media to prevent any lipopolysaccharide (LPS) present in the recombinant Hsp27 from binding to lipopolysaccharide binding protein (LBP) and/or to the LPS receptor complex of CD14 and Toll-like receptor 4 (TLR4), thereby stimulating LPS rather than Hsp27-mediated responses [2]. Hsp27 induced exaggerated production of IL-10 by monocytes, but equivalent low levels of TNF-α as compared to other non-LPS bacterial stimuli such as *Staphylococcus aureus* enterotoxin B (SEB) or muramyl dipeptide (MDP), both of which are also unaffected by polymyxin B (Figure 13.1). Hsp27 induction of IL-10 is dose-dependent and inhibited by addition of a neutralising antibody to Hsp27, thereby further confirming that Hsp27 induces monocyte cytokine production (Figure 13.1). We have also shown that the activation of monocyte p38 mitogen-activated protein (MAP) kinase signalling pathway is prolonged when compared to LPS-induced activation [9]. This prolongation is crucial for the stimulation of large quantities of IL-10 by its alteration of the IL-10 induction kinetics [15, 18].

Hsp27-induced production of IL-10 by monocytes can inhibit the production of other pro-inflammatory cytokines such as IL-1β, TNF-α, IL-6 and

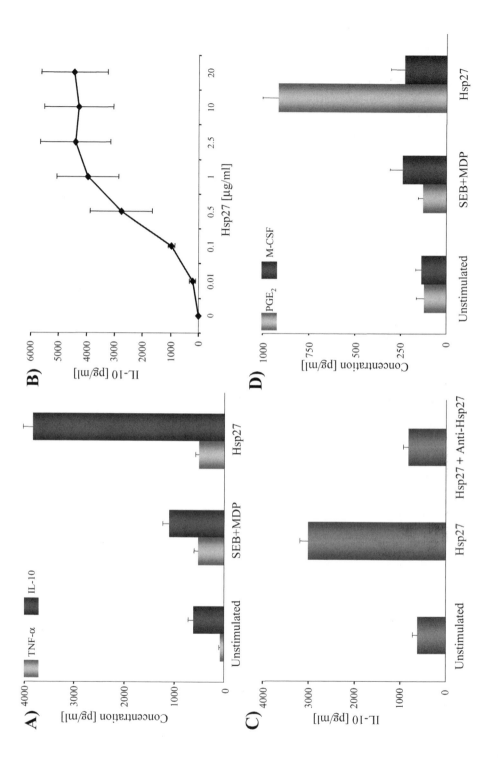

IL-12 and can decrease the expression of co-stimulatory molecules on DCs, diminish immunoglobulin production by B cells and prevent development of APCs [18]. These effects are all crucially important activities for the development and persistence of inflammatory responses. Therefore, Hsp27-mediated induction of IL-10 can result in a profound and widespread inhibition of inflammation and immunity. Hsp27 has been shown to induce high levels of monocyte prostaglandin E_2 (PGE_2), which is another mediator having well-described anti-inflammatory properties [19] (Figure 13.1). PGE_2 suppresses antigen- and mitogen-induced proliferation of T and B cells, as well as antibody production by B cells [19]. PGE_2 also decreases oxidative and phagocytic responses of monocytes, and their cytokine production. These activities suggest that the negative modulation of immune response after bacterial stimulation is one of its primary roles in host defence [19].

Finally, Hsp27 induces monocyte production of M-CSF. M-CSF has both pro- and anti-inflammatory actions on monocytes. M-CSF boosts phagocytosis, superoxide production, cytotoxicity and secondary cytokine-secreting macrophages, and suppresses immune response in pregnancy, facilitating HIV infection and depleting APC precursors from the peripheral blood pool of monocytes [20, 21]. M-CSF drives the emergence of terminally differentiated, activated, tissue macrophages from the monocyte population, thereby decreasing the DC precursor population by depleting available monocyte precursors. This can result in immunoparalysis and contribute to the monocyte-related pathology which has been described in trauma and in other patient groups [20–22].

13.3. Hsp27 can modulate the expression of monocyte receptors

Both monocyte activation and differentiation are dependent on the level and combination of surface receptors expressed [23–26]. Among the crucial receptors influencing monocyte activation and differentiation are the pattern recognition receptors such as the TLRs [23, 27].

Many authors have demonstrated the binding of large heat shock proteins to TLR4 [5, 28], and the interested reader should refer to Chapters 7, 8 and

---

Figure 13.1. Anti-inflammatory properties of Hsp27 in comparison to bacterial stimuli (SEB – staphylococcus enterotoxin B; MDP – muramyl dipeptide): (A) Hsp27 induced higher level of IL-10 in comparison to SEB + MDP, with only a minimal induction of TNF. (B) Induction of IL-10 is dose-dependent, reaching maximum levels at 5 μg/ml. (C) Neutralising antibody to Hsp27 significantly reduces Hsp27-mediated IL-10 induction. (D) Hsp27 is also a potent inducer of PGE and M-CSF in human monocytes.

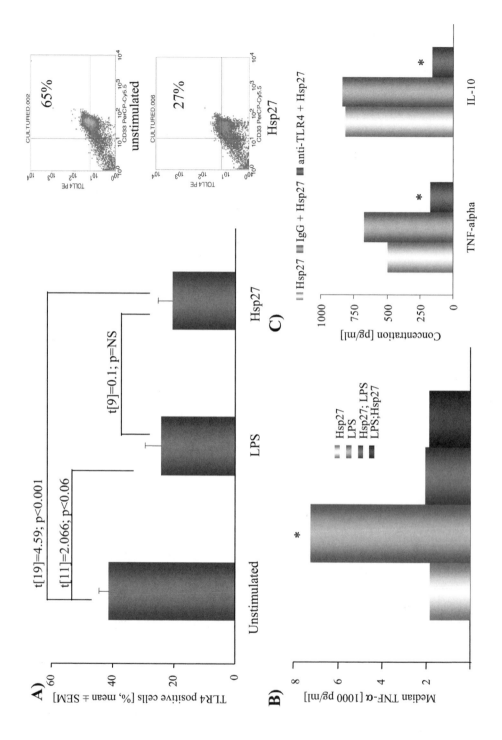

224

10 for more details. TLR4 is the receptor for not only LPS but also for other circulating mediators [24]. The variable detection and transduction of different ligand-specific signals by monocytes depends on the formation of co-receptor clusters around TLR4 [23–25, 29]. The specificity of this TLR4 receptor combination determines how the surface signal from LPS, Hsp60 or other TLR4 ligands is transduced and results in differential activation of kinases followed by signal-specific cytokine release [24, 29]. Based on the ability of large heat shock proteins to signal through TLR4 or other innate immune receptors, we hypothesise that Hsp27 might also interact with a TLR4 receptor cluster, thereby resulting in the production of its unique cytokine pattern, which is distinctive for this small chaperone. Binding of LPS to TLR4 on monocytes down-regulates this receptor and contributes to the development of LPS tolerance [24, 29], in which a subsequent dose of LPS elicits diminished cytokine production due to receptor changes, activation of intracellular inhibitory pathways, and the release of inhibitory cytokines (IL-10, TGF-β). LPS tolerance plays an important role in sepsis, autoimmune disease and allergy [1, 24–26]. Although it is a beneficial mechanism, under certain circumstances it can impair the ability to respond to pathogenic challenge [26, 30].

We have shown that exogenous Hsp27 is at least as potent as LPS in decreasing surface expression of the TLR4 receptor (Figure 13.2). In contrast, surface expression of TLR1 remains unchanged after monocyte stimulation with exogenous Hsp27, and TLR2 receptor expression is slightly up-regulated. The decrease of TLR4 levels on monocytes after Hsp27 stimulation *in vitro* might reflect classical down-regulation of surface expression, since intracellular staining of Hsp27-treated monocytes reveals no significant differences in total intracellular TLR4 concentrations. The precise mechanism by which Hsp27 influences TLR4 expression is undefined, although there are several possibilities.

Hsp27 actions on TLR4 could be mediated by direct binding to TLR4 followed by internalisation of the TLR4–Hsp27 complex into the cytoplasmic compartment. Although several authors have demonstrated that TLR4 is the binding receptor for large heat shock proteins, there is no conclusive proof to show that the same is true for Hsp27 [2, 3, 5, 24, 31]. Alternatively, Hsp27 could

←——

Figure 13.2. Hsp27 may interact with TLR4 expressed on monocytes. Hsp27 induces a profound down-regulation of TLR4 receptors similar to that produced by LPS stimulation. Pre-treatment with Hsp27 modulates the magnitude of monocyte LPS TNF responses. Hsp27 added 2 hours prior to LPS stimulation reduces TNF production to 27% of LPS alone, whereas LPS pre-treatment did not affect Hsp27-induced TNF levels. Blocking anti-TLR4 antibody reduced Hsp27-induced pro- and anti-inflammatory cytokine production (* $p < 0.05$).

bind to other monocyte receptors such as RAGE, CD91, CD36 and chemokine receptors, and this might then indirectly stimulate down-regulation of TLR4 expression on the surface of monocytes.

Since TLR4 surface expression is significantly down-regulated after stimulation of monocytes with exogenous Hsp27, the monocyte response on subsequent exposure to LPS might also be diminished in a manner similar to LPS-induced tolerance, should TLR4 be a major Hsp27 receptor on monocytes [3, 26, 30]. When human monocytes are treated with Hsp27 followed by LPS (2-hour interval), or pre-treated with LPS followed by Hsp27 *in vitro*, the TNF-α response pattern is the same as that of Hsp27 alone, thereby demonstrating a reduced LPS response, which is similar to that observed for LPS tolerance [26] (Figure 13.2). Hsp27 pre-treatment or post-treatment reduces LPS-induced monocyte TNF-α levels, whereas Hsp27-induced TNF-α levels remain constant. One explanation for this finding is that Hsp27 induced early increases (8 hours after stimulation) of IL-10, which down-regulate LPS-induced TNF-α levels. LPS-induced IL-10 normally peaks 18 hours after stimulation, which is considerably later than TNF-α induction, thereby allowing an enhanced TNF-α induction [18]. However, treating monocytes with a neutralising antibody to TLR4 diminishes their Hsp27-induced IL-10 and TNF-α production, which indicates some type of Hsp27–TLR4 interactions (Figure 13.2).

The failure of monocyte LPS pre-treatment to reduce Hsp27-induced TNF-α implies that additional or other elements of the LPS binding receptor cluster might be affected. Antibody to TLR4 could inhibit Hsp27 by steric hindrance of the LPS receptor complex, rather than by directly interfering with Hsp27–TLR4 binding. The LPS receptor cluster involves TLR4, CD14, MD-2, CD11c, CD18 and CD36 arranged in a specific spatial pattern [24, 29]. Stimulation with other TLR4 ligands besides LPS results not only in different spatial composition of co-receptors for LPS receptor cluster, but also different composition of co-stimulatory receptors [22]. Whatever the mechanism, the ability of Hsp27 stimulation to decrease monocyte TLR4 levels will attenuate monocyte stimulation by other TLR4-dependent mediators and further influence the anti-inflammatory/inhibitory effects of Hsp27.

13.4. Hsp27 stimulation alters monocyte differentiation

Monocytes are pluripotent cells that can differentiate into a range of myeloid cell populations including immature DCs (iDCs) and macrophages, depending on the local microenvironmental signals [22]. IL-4 plus granulocyte macrophage-colony stimulating factor (GM-CSF), or similar cytokine combinations, drive monocytes to differentiate into iDCs, whereas stimulation by M-CSF, GM-CSF

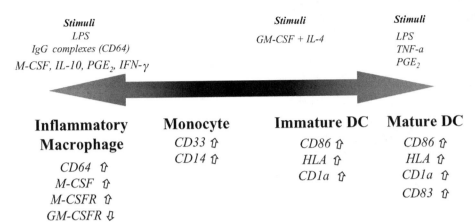

Figure 13.3. Mediator effects on differentiation of monocytes into terminally differentiated macrophage or DCs.

alone, or LPS induces monocyte-to-macrophage differentiation (Figure 13.3) [6, 12, 27]. The emergence of mature DCs (mDCs) is a two-phase process which requires distinct signals that induce differentiation to iDCs and their subsequent maturation. IL-4 and GM-CSF stimulate the differentiation of the iDCs, and these undergo terminal specialisation following exposure to LPS, TNF-α, PGE$_2$, or other TLR ligands. A relative lack of IL-4 or GM-CSF, or the presence of LPS, PGE$_2$, IL-10 or TNF-α at the initial differentiation stage of iDC development, results in the induction of apoptosis or a failure of the differentiation of monocytes into iDCs [22]. The absence of maturation stimuli for iDC development into fully competent mDCs also results in an aborted differentiation and iDC apoptosis. Surprisingly, mediators that inhibit the initial differentiation of monocytes into iDCs often subsequently induce the maturation of iDCs, and the timing of stimulation must therefore be well orchestrated. Monocyte-derived DCs play a pivotal role in T lymphocyte activation [22]. DCs are the only APC which is capable of activating naïve T cells, and they are crucial for the transition of activated T lymphocytes to memory T cells. DC defects have been described as major contributors to immunoaberrations in the regulation of T cells and the attenuation of specific immunity [21, 22, 32].

We have demonstrated that Hsp27 is a potent inducer of M-CSF, PGE$_2$ and IL-10. All of these mediators inhibit monocyte differentiation into DCs while promoting the differentiation of monocytes into macrophages. Hsp27 might therefore exert significant immunosuppressive effects on monocytes by inhibiting their differentiation into DCs. In a series of experiments, we have shown that the stimulation of monocytes with Hsp27 inhibits their differentiation into

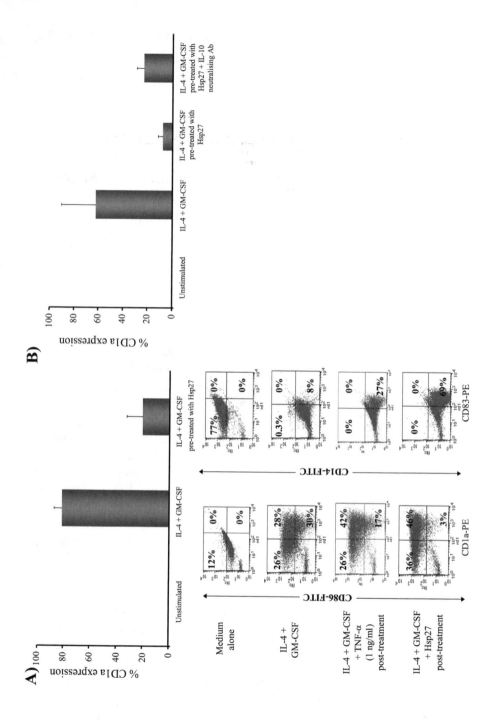

iDCs (CD1a$^+$ cells) after IL-4/GM-CSF co-stimulation, but promotes iDC maturation into mDCs (CD83$^+$ cells) (Figure 13.4). Hsp27-induced monocyte production of M-CSF, PGE$_2$ and IL-10 could be responsible for the inhibition of iDC differentiation after IL-4 plus GM-CSF stimulation [21, 22, 32]. IL-10 and M-CSF promote the differentiation of monocytes to macrophages and are secreted by monocytes after exposure to Hsp27 (Figures 13.2, 13.4). The action of M-CSF is especially critical since this cytokine induces autocrine secretion of both itself (positive feedback loop) and IL-10 [20, 21]. In healthy volunteers, neutralising antibody to IL-10 partially reverses the inhibitory effect of Hsp27 on iDC differentiation (Figure 13.4). This suggests that an increase in IL-10 levels is at least partially responsible for the diminished differentiation of monocytes to iDC, which is mediated by Hsp27. However, the inhibitory effects of Hsp27 on monocyte differentiation to iDCs naturally involve other inhibitory effects, perhaps elevated PGE$_2$ and M-CSF levels.

In contrast to its inhibitory effect on the differentiation of monocytes into iDCs, Hsp27 is a potent inducer of iDC maturation (Figure 13.4). Neither M-CSF nor IL-10 are particularly potent maturation factors for iDC, and PGE$_2$ is only a co-stimulator of maturation [19, 22, 32]. LPS stimulation is suggested to act as a DC maturation signal via TLR4 induction of TNF-α, IL-1β and, to a lesser degree, TGF-β. Since Hsp27 is a relatively poor inducer of TNF-α compared to LPS, the mechanism by which it induces DC maturation awaits further dissection. Nevertheless, these data illustrate that Hsp27 has both immunosuppressive and immunostimulatory capacities *in vitro*, and these will be dependent on when and where it interacts with differentiating monocytes.

13.5. Hsp27 in disease

We have shown exogenous Hsp27 to have potent inhibitory effects on monocyte inflammatory cytokine production and immunostimulatory activities *in vitro*. However, it is unclear whether Hsp27 has similar activities *in vivo*. Hsp27 is predominantly an intracellular molecule [3] and, to date, the secretion of Hsp27 under physiological conditions has not been demonstrated. However, it is released from cells undergoing necrosis [3, 7, 14, 33, 34]. This implies that Hsp27 released into tissue or serum after cell injury or death could then interact with leukocytes.

Figure 13.4. Hsp27 treatment affects differentiation and maturation of DCs: (A) Exogenous Hsp27 significantly inhibits the differentiation of iDCs from monocytes but augments the maturation of mature DCs from iDCs. (B) Addition of antibodies to monocytes during iDC differentiation partially reduces the inhibitory effects of Hsp27.

Elevated serum levels of Hsp27 have been demonstrated in breast cancer patients as well as in sufferers of other malignancies [7, 11]. Apoptosis and necrosis are increased in malignant cells and this possibly contributes to the release of Hsp27 and elevated circulating levels. Circulating Hsp27 also appears to trigger an immune response, as anti-Hsp27 antibodies are also detected in the serum of some cancer patients [9, 13]. Interestingly, increased serum titres of anti-Hsp27 antibodies correlate with an improved survival, which suggests that exogenous circulatory Hsp27 might itself be exerting activities that suppress anti-cancer immune functions. The capacity of Hsp27 to induce the immunosuppressive mediators IL-10, M-CSF and PGE_2, and to inhibit the differentiation of monocytes into DCs, might impair anti-tumour immunity. Hsp27 has also been implicated as a regulator of oestrogen receptor expression in breast cancer patients [35]. Further characterisation of the role of Hsp27 in cancer host defence is clearly required.

Since elevated circulatory Hsp27 levels are apparent in subjects with extensive tissue damage, we hypothesise that large quantities of Hsp27 can be released into the bloodstream in trauma. In a small pilot study, we have shown that Hsp27 can be detected in the serum of trauma patients at levels approximately three times higher than those found in healthy volunteers (24 ± 5.38 vs. 7.32 ± 5.1 ng/ml, respectively; $p < 0.05$). This finding suggests a new investigative area in which the capacity of circulating Hsp27 to modulate inflammatory events and immune function in the severely injured patient should be defined. Trauma patients experience monocyte paralysis, which is typified by LPS tolerance and a significant defect in the differentiation of monocytes to iDCs [21, 26]. Down-regulation of TLR4 as well as other monocyte receptors is also characteristic of these patients. Our laboratory is investigating the contribution of elevated exogenous Hsp27 in the development and maintenance of these monocyte aberrations in trauma patients.

13.6. Conclusions

Hsp27 is unusual among heat shock proteins in that it has the potential to suppress aspects of the immune system by multiple mechanisms. Induction of monocyte production of predominantly inhibitory cytokines (IL-10, M-CSF, PGE_2) with a relatively small release of pro-inflammatory molecules (TNF-α) clearly distinguishes Hsp27 from Hsp60, Hsp70 and other members of the large shock protein family. Hsp27 might also interfere with the response to bacterial endotoxins by down-regulating TLRs. However, the clinical importance of these activities remains to be established. Hsp27 is a potent inhibitor of monocyte differentiation into iDCs, and this activity could profoundly impair the induction of specific adaptive immunity. Further studies are necessary to determine the precise mechanisms by which exogenous Hsp27 elicits immunoinhibitory and anti-inflammatory activities and their impact. Nevertheless, like many members of the large heat shock protein family, Hsp27 is emerging as a major modulator of host defence as well as a molecular chaperone and substrate in the p38 MAP kinase signalling pathway.

Acknowledgements

This work was supported by grant GM036214 from the National Institutes of Health, U.S.A.

REFERENCES

1. Wick G, Knoflach M and Xu Q. Autoimmune and inflammatory mechanisms in atherosclerosis. Ann Rev Immunol 2004, 22: 361–403.
2. Tsan M F and Gao B. Cytokine function of heat shock proteins. Am J Physiol Cell Physiol 2004, 286: C739–C744.
3. Feder M E and Hofmann G E. Heat-shock proteins, molecular chaperones, and the stress response: evolutionary and ecological physiology. Ann Rev Physiol 1999, 61: 243–282.
4. Jäättelä M and Wissing D. Emerging role of heat shock proteins in biology and medicine. Ann Med 1992, 24: 249–258.
5. Ohashi K, Burkart V, Flohé S and Kolb H. Heat shock protein 60 is a putative endogenous ligand of the Toll-like receptor-4 complex. J Immunol 2000, 164: 558–561.
6. Zanin-Zhorov A, Nussbaum G, Franitza S, Cohen I R and Lider O. T cells respond to heat shock protein 60 via TLR2: activation of adhesion and inhibition of chemokine receptors. FASEB J 2003, 17: 1567–1569.
7. Fanelli M A, Cuello Carrion F D, Dekker J, Schoemaker J and Ciocca D R. Serological detection of heat shock protein hsp27 in normal and breast cancer patients. Canc Epidemiol Biomarkers Prev 1998, 7: 791–795.
8. Wissing D and Jäättelä M. HSP27 and HSP70 increase the survival of WEHI-S cells exposed to hyperthermia. Int J Hyperthermia 1996, 12: 125–138.
9. Korneeva I, Bongiovanni A M, Girotra M, Caputo T A and Witkin S S. IgA antibodies to the 27-kDa heat-shock protein in the genital tracts of women with gynecologic cancers. Int J Cancer 2000, 87: 824–828.
10. Garrido C, Ottavi P, Fromentin A, Hammann A, Arrigo A P, Chauffert B and Mehlen P. Hsp27 as a mediator of confluence-dependent resistance to cell death induced by anticancer drugs. Cancer Res 1997, 57: 2661–2667.
11. Korneeva I, Caputo T A and Witkin S S. Cell-free 27 kDa heat shock protein (hsp27) and hsp27-cytochrome c complexes in the cervix of women with ovarian or endometrial cancer. Int J Cancer 2002, 102: 483–486.

12. Jäättelä M and Wissing D. Heat-shock proteins protect cells from monocyte cytotoxicity: possible mechanism of self-protection. J Exp Med 1993, 177: 231–236.
13. Conroy S E, Sasieni P D, Amin V, Wang D Y, Smith P, Fentiman I S and Latchman D S. Antibodies to heat-shock protein 27 are associated with improved survival in patients with breast cancer. Br J Cancer 1998, 77: 1875–1879.
14. Carter Y, Liu G, Stephens W B, Carter G, Yang J and Mendez C. Heat shock protein (HSP72) and p38 MAPK involvement in sublethal hemorrhage (SLH)-induced tolerance. J Surg Res 2003, 111: 70–77.
15. De A K, Kodys K M, Yeh B S and Miller-Graziano C. Exaggerated human monocyte IL-10 concomitant to minimal TNF-α induction by heat-shock protein 27 (Hsp27) suggests Hsp27 is primarily an anti-inflammatory stimulus. J Immunol 2000, 165: 3951–3958.
16. Asea A, Rehli M, Kabingu E, Boch J A, Baré O, Auron P E, Stevenson M A and Calderwood S K. Novel signal transduction pathway utilized by extracellular HSP70. Role of Toll-like receptor (TLR) 2 and TLR4. J Biol Chem 2002, 277: 15028–15034.
17. Srivastava P. Interaction of heat shock proteins with peptides and antigen presenting cells: chaperoning of the innate and adaptive immune responses. Ann Rev Immunol 2002, 20: 395–425.
18. Moore K W, de Waal Malefyt R, Coffman R L and O'Garra A. Interleukin-10 and the interleukin-10 receptor. Ann Rev Immunol 2001, 19: 683–765.
19. Hwang D. Fatty acids and immune responses – a new perspective in searching for clues to mechanism. Ann Rev Nutr 2000, 20: 431–456.
20. Hashimoto S, Yamada M, Motoyoshi K and Akagawa K S. Enhancement of macrophage colony-stimulating factor-induced growth and differentiation of human monocytes by interleukin-10. Blood 1997, 89: 315–321.
21. De A K, Laudanski K and Miller-Graziano C L. Failure of monocytes of trauma patients to convert to immature dendritic cells is related to preferential macrophage-colony-stimulating factor-driven macrophage differentiation. J Immunol 2003, 170: 6355–6362.
22. Banchereau J, Briere F, Caux C, Davoust J, Lebecque S, Liu Y J, Pulendran B and Palucka K. Immunobiology of dendritic cells. Ann Rev Immunol 2000, 18: 767–811.
23. Dobrovolskaia M A and Vogel S N. Toll receptors, CD14, and macrophage activation and deactivation by LPS. Microbes Infect 2002, 4: 903–914.
24. Triantafilou M and Triantafilou K. Lipopolysaccharide recognition: CD14, TLRs and the LPS-activation cluster. Trends Immunol 2002, 23: 301–304.
25. van Amersfoort E S, van Berkel T J and Kuiper J. Receptors, mediators, and mechanisms involved in bacterial sepsis and septic shock. Clin Microbiol Rev 2003, 16: 379–414.
26. West M A and Heagy W. Endotoxin tolerance: a review. Crit Care Med 2002, 30(1 Supp): S64–S73.
27. Kirschning C J and Schumann R R. TLR2: cellular sensor for microbial and endogenous molecular patterns. Curr Top Microbiol Immunol 2002, 270: 121–144.
28. Habich C, Baumgart K, Kolb H and Burkart V. The receptor for heat shock protein 60 on macrophages is saturable, specific, and distinct from receptors for other heat shock proteins. J Immunol 2002, 168: 569–576.

29. Heine H, El-Samalouti V T, Notzel C, Pfeiffer A, Lentschat A, Kusumoto S, Schmitz G, Hamann L and Ulmer A J. CD55/decay accelerating factor is part of the lipopolysaccharide-induced receptor complex. Eur J Immunol 2003, 33: 1399–1408.

30. Volk H D, Reinke P and Docke W D. Clinical aspects: from systemic inflammation to 'immunoparalysis'. Chem Immunol 2000, 74: 162–177.

31. Vabulas R M, Ahmad-Nejad P, da Costa C, Miethke T, Kirschning C J, Hacker H and Wagner H. Endocytosed HSP60s use toll-like receptor 2 (TLR2) and TLR4 to activate the toll/interleukin-1 receptor signaling pathway in innate immune cells. J Biol Chem 2001, 276: 31332–31339.

32. Guermonprez P, Valladeau J, Zitvogel L, Théry C and Amigorena S. Antigen presentation and T cell stimulation by dendritic cells. Ann Rev Immunol 2002, 20: 621–667.

33. Stevens T R, Winrow V R, Blake D R and Rampton D S. Circulating antibodies to heat-shock protein 60 in Crohn's disease and ulcerative colitis. Clin Exp Immunol 1992, 90: 271–274.

34. Gao Y L, Raine C S and Brosnan C F. Humoral response to hsp 65 in multiple sclerosis and other neurological conditions. Neurology 1994, 44: 941–946.

35. O'Neill P A, Shaaban A M, West C R, Dodson A, Jarvis C, Moore P, Davies M P, Sibson D R and Foster C S. Increased risk of malignant progression in benign proliferating breast lesions defined by expression of heat shock protein 27. Br J Cancer 2004, 90: 182–188.

14

Bip, a Negative Regulator Involved in Rheumatoid Arthritis

Valerie M. Corrigall and Gabriel S. Panayi

14.1. Introduction

The heat shock protein (Hsp) 70 family is a collection of evolutionarily conserved, ubiquitous proteins that are either constitutively expressed and/or stress induced and which are nominally defined by their molecular weight (Hsp70, Hsc73, BiP (binding immunoglobulin protein, or glucose regulated protein (grp) 78)). Historically, these proteins have been perceived to function as intracellular molecular chaperones that ensure the correct folding of nascent proteins and are involved in the translocation of proteins and assist in protein degradation through the proteasome [1]. At times of physical or chemical stress, such chaperones are upregulated by the unfolded protein response and provide protection against the accumulation and aggregation of denatured proteins.

In contrast to this long-standing perception, there is now increasing interest in an intercellular signalling role for these proteins and, as a consequence, they have been termed 'chaperokines' in light of their cytokine-like qualities [2, 3]. The interaction between heat shock proteins and specific cell surface receptors that signal the release of inflammatory mediators has revealed a link between the innate and adaptive immune response. A wide range of extracellular receptors for human Hsp70 has been identified. These include CD14 [2, 4, 5], Toll-like receptor (TLR) 4, TLR2 [4, 6], CD91 [7, 8] and CD40 [9, 10] on monocytes, and scavenger receptors such as LOX-1 on dendritic cells (DCs) [11]. The role of these receptors is detailed in Chapters 7 and 10. Hsp70 and Hsp60 stimulation of monocytes via these receptors induces predominantly pro-inflammatory cytokine/chemokine production, including TNF-α, IL-1β and IL-12 [6, 12].

The present assessment of heat shock proteins is that they act as 'danger signals' alerting the adaptive immune system to raise a Th1 immune response [13]. However, immunological homeostasis needs to be maintained. For this, other proteins, including heat shock proteins, must induce a counter-regulatory

Th2 response or the production of cytokines and other factors with anti-inflammatory properties. Our preliminary studies have shown that BiP does not bind to any of the previously identified receptors for Hsp70 despite the high degree of homology between the two molecules. There is now accumulating evidence that BiP, like Hsp27 (reviewed in Chapter 13) [14], might serve such an immunomodulatory role partially through the secretion of IL-10.

This chapter proposes that BiP is an immunoregulator of the innate and adaptive immune systems which may prevent inappropriate damaging responses to antigenic challenge. These immune functions of BiP have only recently been identified by us, and information is limited. The evidence reviewed in the following indicates that BiP is a natural stimulator of anti-inflammatory cytokines from mononuclear cells. BiP may also prevent DC maturation and stimulate regulatory T cells, all of which may collectively play a key role in regulating the immune system and thus maintaining homeostasis.

14.2. Glucose regulated protein 78 or binding immunoglobulin protein (BiP)

Glucose regulated protein 78 (grp78) or binding immunoglobulin protein (BiP) is classified as a member of the Hsp70 family [1, 15, 16]. As evident from its name, BiP was first identified as the chaperone protein that was involved in the folding of the H and L chains of the immunoglobulin molecule to generate the complete immunoglobulin molecule prior to its exit from the endoplasmic reticulum (ER) [1, 16, 17]. It is now known that BiP is involved in the correct folding of all nascent proteins [1, 17]. BiP binds to the sequences of seven amino acids with exposed hydrophobic residues that are only seen in denatured proteins to prevent intracellular damage [1, 17]. It is from this function, perhaps, that the misconception arose that BiP binds to all proteins. In fact, BiP binds only to the hydrophobic regions of the polypeptide chains that are obscured as the protein is folded [1, 17].

BiP is regulated at two levels: firstly, to maintain a basal constitutive level which is sufficient for intracellular protein folding functions, and secondly, at an induced level which is upregulated when the cell is stressed. Constitutive BiP is tethered in the lumen of the ER by the 3′ carboxyl terminal amino acid sequence, KDEL, which binds to the ERD2 receptor [18]. Upregulation occurs when the cell is stressed, particularly under conditions of reduced glucose or oxygen levels, calcium flux or increased concentrations of reactive oxygen species [19]. These conditions are often associated with inflammation, especially with the late stages of acute inflammation or chronic inflammation, and are also prevalent within the joints of patients with rheumatoid arthritis (RA) [20–22].

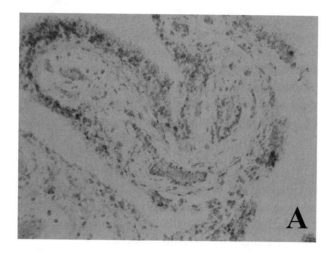

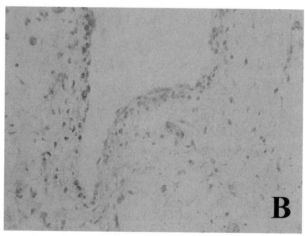

Figure 14.1. BiP is over-expressed in RA synovial membrane. Synovial membrane from patients with (A) rheumatoid arthritis or (B) osteoarthritis was stained with 1/100 dilution of anti-BiP antibody with vector red alkaline phosphatase substrate kit (Dako). Magnification 20×.

Indeed, we and Blass and colleagues [23] have shown that BiP is over-expressed in the synovial membrane of patients with rheumatoid arthritis when compared with membranes from osteoarthritis patients (Figure 14.1).

Upregulation of BiP by stress *in vitro* causes its translocation from the ER and its expression on the cell surface. Delpino and colleagues [24] have reported cell surface BiP expression in response to thapsigargin treatment, which is an inhibitor of ER calcium ATPase and thus a powerful stress inducer. In addition, Gagnon and colleagues [25] have shown that, during endocytosis, ER proteins may become exposed on the cell surface as the pinocytotic cup becomes inverted at the point of internalisation (see Chapter 3). There is *in vivo* evidence for cell surface expression of BiP and heat shock proteins by tumour cells [26–28], and the secretion of heat shock proteins at levels that can be detected by enzyme immunoassays in sera has also been reported [29]. These findings reinforce the view that these molecules have important functions as intercellular messengers outwith their established intracellular functions.

In vitro studies into the extracellular functions of the Hsp70 family have used either recombinant human/bacterial heat shock proteins [6, 10, 12, 30] produced in a bacterial system or mammalian heat shock proteins isolated from cell lysates [23, 31–33]. Protein glycosylation and tertiary structure-dependent differences in their functional ability have not been directly compared. Furthermore, the role of contaminating endotoxin in the heat shock protein preparations used has not always been adequately addressed (a point debated in many chapters in this volume). Much work on heat shock proteins has been performed in animal models [30, 32] or has used mouse and/or human cell lines [6, 10, 12, 30, 33]. Although this approach has considerably simplified the analysis and reduced the variability of the results obtained, extrapolation of these findings into an entirely human system, especially *in vivo*, must be made with caution. At present, data from experiments assessing the interactions of human heat shock proteins with primary human cells are limited and much of the work cited in this chapter involves human peripheral blood mononuclear cells (PBMCs) in conjunction with a recombinant human BiP (rhuBiP) from an *Escherichia coli* source [34]. This same preparation of rhuBiP has been used throughout the animal studies.

14.3. Studies in animals

14.3.1. Induction of arthritis

On identification of BiP as an autoantigen in RA, we determined whether BiP was arthritogenic. A wide range of concentrations of BiP in complete Freund's adjuvant was administered to several strains of rats and mice. However, all failed to induce arthritis [34]. There are two possible explanations for this. The first is that the correct genetic strain of mouse or rat had not been chosen, because it is known that the induction of arthritis exhibits a high genetic dependency. The second possibility is that BiP is unable to stimulate the induction of arthritogenic Th1 T cells.

Human *in vitro* studies provided the clue that the second possibility was the most likely. We showed that, although BiP stimulated the proliferation of T cells from the RA joint, the proliferation was low and was not accompanied by the secretion of IFN-γ [34]. It is therefore possible that BiP is not a 'pathogenic' autoantigen but rather plays some other, possibly regulatory, role in RA synovitis. This hypothesis was supported by the finding that intravenous BiP prevented collagen-induced arthritis (CIA) in the DBA/1 or HLA-DR1$^{+/+}$ transgenic mouse [34].

14.3.2. Prevention of arthritis

Work with CIA and adjuvant arthritis (AA) animal models confirmed that BiP may have anti-inflammatory properties. A single intravenous injection of BiP totally protects mice from the onset of CIA and reduces the incidence and severity of AA in rats when compared with animals similarly treated with a control recombinant protein (β-galactosidase) or the PBS vehicle [34].

14.3.3. Therapy of collagen-induced arthritis

Although protection from disease is of scientific interest, it is of limited clinical value. Therapeutic studies were therefore initiated in which a single dose of BiP was administered via three different routes, namely intravenous (IV), subcutaneous (SC) and intranasal (IN). In these studies, BiP was administered to DBA/1 or HLA-DR$^{+/+}$ transgenic mice in a single dose at the time of onset of CIA, as judged by paw swelling. IV- and SC-administered BiP caused significant therapeutic activity with the minimal effective IV dose being 1 μg/mouse, and that for SC dose being 50 μg/mouse [35]. IN administration was without therapeutic effect. The reason for the failure is presently unknown.

14.3.4. Animal studies – *in vitro* experiments

14.3.4.1. Cellular studies

In vitro experiments on spleen and lymph node single-cell preparations from CIA mice were carried out in parallel with the disease studies. Incubation of cells from the DBA/1 mice (not previously injected with BiP) with rhuBiP stimulated a strong proliferative response when compared to that induced by a control protein (bovine serum albumin). Although interferon (IFN)-γ was present in the supernatants, the level of IL-4 was significantly increased. It should be noted that it is difficult to induce IL-4 secretion from T cells of this strain of mouse. Analysis of the splenocyte and lymph node cell responses to either rhuBiP or collagen type II (CII) showed that those animals that had been treated with BiP produced high levels of the Th2 cytokines IL-10, IL-4 and IL-5, spontaneously and when re-stimulated with BiP or CII [35, 36]. In contrast, the response to CII was characterised by a significantly greater secretion of IFN-γ.

Ongoing work in our laboratory is aimed at determining whether BiP-treated cells can transfer protection and evaluating whether this is dependent on the production of anti-inflammatory cytokines or by cell-to-cell contact.

14.3.4.2. Antibody studies

The analysis of serum from the mice used in the CIA prevention study not only demonstrated that anti-CII antibody levels were reduced in BiP-treated mice but also that these were predominantly of the IgG₁ isotype, an isotype that is typically associated with Th2-driven immune responses [34].

14.4. Human studies

If BiP has intercellular signalling functions, then two requirements must be satisfied. Firstly, cell-free BiP must be found *in vivo*; secondly, there must be the presence of a cell surface receptor capable of binding to BiP and transducing intracellular signals.

14.4.1. Cell-free BiP in synovial fluid

Cell-free BiP, in either synovial fluid (SF) or serum, could originate from cells by several different routes. In the SF, it is known that cell death occurs and this would certainly lead to the release of BiP. In addition, BiP can be expressed on the cell surface [26] and has been detected on the cell surface of single-cell suspensions from RA synovial membranes (unpublished results). Proteolytic enzymes, which are plentiful in the SF, may cleave BiP from the cell surface. Alternatively, upregulation of BiP expression under stress may induce alternative splicing of the BiP gene which, by omitting the retaining KDEL sequence, allows extracellular release. BiP has been detected in the SF of 74% of patients with RA and 38% of individuals with other inflammatory joint diseases. Interestingly, no KDEL was detected on the BiP, despite the fact that other KDEL-containing proteins were detected. Thus, it appears that the mechanism leading to the presence of BiP in the SF is distinct from that used by other ER proteins. The possibilities for such differences include cleavage, and alternative splicing of BiP was discussed earlier.

14.4.2. BiP receptor-like molecule (BiPRL)

We have shown that a variety of cells express a BiP-binding, receptor-like molecule (BiPRL) on their surface such that >95% monocytes, <29% B cells, <10% T cells and 85% fibroblast-like synoviocytes bind fluorescein isothiocyanate-conjugated BiP (BiP.FITC). Specificity of binding is suggested by the difference in percentage of BiPRL$^+$ cells of the different cell types. No competitive binding for BiPRL has been seen with α_2-macroglobulin or anti-bodies to TLR2, TLR4, CD40, CD91 or CD14 (manuscript submitted). The latter have already been identified as receptors for human Hsp70, grp94 and Hsp60 (see preceding discussion). Following binding to the BiPRL on monocytes, BiP.FITC is internalised and co-localises with the vesicle membrane protein LAMP-1 (CD107a) (manuscript submitted). This internalisation is similar to the process that has been seen with Hsp70 [33] and grp94 [31].

The localisation of BiP within the endosome suggests that BiP could be degraded and processed and peptides derived therefrom presented to T cells by major histocompatibility complex (MHC) class II molecules. The pathway used for the degradation of BiP, either exogenous or endogenous, has not yet been elucidated; however, it is known that there is leakage in this process between peptides that are eventually presented either by MHC class I molecules to CD8$^+$ T cells or by MHC class II molecules to CD4$^+$ T cells. This important observation would explain the development of both CD4$^+$ and CD8$^+$ clones from BiP-stimulated human (PBMCs [37]; (see following text). Chapter 6 also describes the role of internalisation of chaperones in myeloid cell modulation.

14.4.3. T cells

The stimulation of T cells *in vitro* is used as a measure of T cell responsiveness to nominal antigen. However, the length of time taken for the response to become apparent must be carefully considered when reviewing such results. Proliferation that is maximal after seven days is most likely to be due to the induction of a primary response during the culture. In contrast, proliferation that occurs in less than seven days may be considered to be due to the induction of a secondary response by previously primed memory T cells. These considerations are important when reviewing work with BiP. Thus Blass and colleagues [23] have demonstrated borderline *primary* proliferative responses by peripheral blood CD45RA$^+$ T cells to BiP that were HLA-DR restricted. However, the low stimulation index of these cells renders the interpretation of the results difficult. We have shown that RA SF T cells proliferate to exogenous rhuBiP with the kinetics of a *secondary* response. In contrast to Blass et al.'s study we found little, or no, proliferative response by normal PBMCs and a limited proliferative response in PBMCs, even those from RA patients whose SF T cells showed a response [34]. The response to BiP appeared to be specifically in RA SF T cells, because those from other inflammatory diseases, including ankylosing spondylitis and psoriatic arthritis, showed no proliferative response to BiP. One peculiarity of this response was the lack of either IFN-γ or IL-2 secretion; the predominant cytokines produced were IL-10 and IL-4 [34].

There are at least four possible interpretations for these findings: firstly, the depressed proliferation was due to the IL-10 produced; secondly, BiP was activating Th2 T cells; thirdly, BiP stimulated the development of regulatory T cells; or fourthly, BiP drives alternative activation of monocytes to produce IL-10 and anti-inflammatory mediators and promote Th2 cell differentiation. With respect to the first possibility, there are many reports in the literature that describe depressed T cell proliferation caused by excess IL-10 [38]. We have shown that there is a correlation between PBMC proliferation to tuberculin purified protein derivative (PPD) and the ratio of IL-2 to IL-10 produced. This was particularly striking in patients with RA, the PBMC cultures from whom exhibited a reduced IL-2 production [39]. Because the cultures stimulated by BiP produced no IL-2 but high IL-10 levels, it is not surprising that proliferation was low. However, the addition of neutralising anti–IL-10 antibody to these cultures did not totally reverse the poor proliferation, thereby indicating that BiP had anti-proliferative properties that are independent of IL-10 [40]. Interestingly, in cultures in which IL-10 was produced by alternatively activated monocytes, the reduction of PBMC proliferation remained unchanged following the addition of neutralising anti–IL-10 antibody [41].

With respect to the second possibility – that BiP was generating a Th2 response – this is at present being investigated. Evidence from the animal work and the fact that CD4+ and CD8+ BiP specific T cell clones are secreting Th2 cytokines [37] would support this contention. As to the third possibility, we are currently investigating the regulatory capacity of BiP activated T cells. The fourth option, the stimulation of alternatively activated monocytes, is discussed next.

14.4.4. CD4 and CD8 human T cell clones

Additional evidence for the priming of T cells by BiP peptides is provided by the isolation of CD4+ and CD8+ BiP-specific T cell clones from normal PBMCs [37]. The generation of these clones suggests that BiP is processed via both endogenous and exogenous pathways, as has been found for Hsp70 [42]. The CD8+ clones produce IL-10, IL-4 and IL-5, strongly suggesting that they are Th2 cells [37].

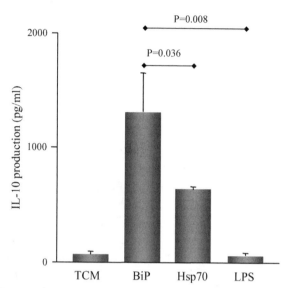

Figure 14.2. IL-10 production stimulated by BiP and Hsp70. IL-10 production by peripheral blood mononuclear cells following 24 hours of stimulation with BiP, Hsp70 (both at 20 μg/ml) or LPS (20 ng/ml). Polymyxin B (10 μg/ml) was added to all cultures.

14.4.5. Monocytes

An important function of BiP is its ability to stimulate human monocytes. As already noted, >95% of monocytes express BiPRL. In PBMC cultures, BiP-stimulated monocytes secrete large amounts of IL-10, significantly greater than that induced by an equivalent amount of Hsp70 (Figure 14.2). In addition, preliminary cytokine gene array analysis of BiP-stimulated, purified peripheral blood monocytes has indicated that BiP up- and/or downregulates many cytokine/chemokine/cytokine receptor genes. The pattern of genes activated indicates that BiP may be alternatively activating the monocytes [43] to produce an anti-inflammatory macrophage producing IL-10. This is similar to the process described with Hsp27 in Chapter 13. The natural inhibitors of IL-1β and TNF-α, IL-1 receptor antagonist and soluble TNF receptor II, respectively, are both upregulated by BiP stimulation (Table 14.1), as is macrophage migration inhibition factor and alternative monocyte activated chemokine (AMAC). In addition, IL-8, granulocyte macrophage-colony stimulating factor (GM-CSF) and epithelial neutrophic activating peptide (ENA)-78, all of which have the potential to increase the inflammatory influx of cells into the synovium, are downregulated following BiP stimulation. In addition, BiP induced a prolonged downregulation of CD86 and HLA-DR expression, in contrast to the intense but transient

Table 14.1. Cytokine gene array data analysed by densitometry

Upregulation			Downregulation		
Cytokine mRNA	BiP	PMA + IONO	Cytokine mRNA	BiP	PMA + IONO
TIMP	65.5 ± 19	30.7	GROα	29.8 ±18.6	57
IL-6	17 ± 18.2	Not detected	GROβ	35.4 ± 26	59
MIF	20.2 ± 10.6	3.4	GROγ	35.7 ± 25	64
TNF RII	4.1 ± 2.2	Not detected	CCL1	Not detected	8.8
			IL-8	57 ± 20.4	85.8
			GMCSF	3.9 ± 3.2	17.2
			Osteopontin	Not detected	14.7
			Urokinase R	50.8 ± 11	80.4
			LIGHT	6.6 ± 1.1	19.2
	BiP	Unstimulated		BiP	Unstimulated
IL-1Ra	26.7 ± 19.7	12.2	ENA-78	4.2 ± 16.2	55
Chem23	15.3 ± 15.9	7.2	GROγ	35.6 ± 25	47
IL6	17 ± 18.2	6.7	IL-8	57.8 ± 20.4	80
AMAC	18.3 ± 5.3	not detected	LDGF	5.4 ± 0.7	52
TNF RII	4.1 ± 2.2	not detected			

Note: Peripheral blood monocytes were isolated and stimulated with either BiP (20 μg/ml) or phorbol myristic acid (PMA; 20 ng/ml) + inomycin (IONO; 500 ng/ml) or left unstimulated for 24 hours. The cytokine gene array autoradiographs were analysed by densitometry and normalised to give a percentage expression of maximum (100%). The results shown are from two subjects for BiP-stimulated monocytes and one subject each for unstimulated and maximally stimulated cells (PMA + IONO). The results are shown as either upregulation or downregulation of BiP-stimulated cytokine mRNA, in both samples, when compared with the control cells.

downregulation observed in the presence of rhuIL-10 [40]. Thus, BiP may have direct anti-inflammatory effects outwith its ability to secrete anti-inflammatory mediators such as IL-10.

14.4.6. Dendritic cells

Maturation of monocytes into DCs is of paramount importance to the development of an efficient adaptive immune system. *In vitro* studies, which use GM-CSF plus IL-4 to drive the differentiation of immature DCs (iDCs), have been performed in the presence and absence of BiP. In concurrence with the hypothesised immunoregulatory functions of BiP, BiP inhibited the differentiation of monocytes to iDCs [44]. Failure of this development was accompanied by a depressed ability to induce the proliferation of allogeneic T cells. This corresponded with the production of high levels of IL-10 and could be reversed either by neutralising IL-10 or by blocking the IL-10 receptor by neutralising monoclonal antibodies. In contrast to BiP suppression of antigen stimulation of PBMCs, the reduced allogeneic proliferative response was reversed by the neutralisation of IL-10 [44]. Interestingly, this property of BiP is in complete contrast to Hsp60. Flohé and colleagues [45] have shown that a mouse macrophage cell line cultured with Hsp60 upregulates CD86 and CD40 and enhances iDC maturation. The antigen-presenting quality of the Hsp60-stimulated cells was increased [45].

14.5. Summary of BiP functions

In this chapter we have attempted to provide an overview of our studies with BiP. In summary, we have found that BiP, which is found cell-free in synovial fluid and can therefore exert intercellular activity, has immunoregulatory functions mediated via a receptor-like molecule that is different from those described for other members of the human Hsp70 family, which include

- anti-inflammatory properties, in distinction from other members of the Hsp70 family;
- the stimulation of PBMCs to produce IL-10 and T cell clones to produce IL-10, IL-4 and IL-5;
- the suppression of differentiation of iDCs from monocytes; and
- the prevention and treatment of CIA in DBA/1 and HLA-DR1$^{+/+}$ mice.

14.6. Conclusion

In conclusion, all the data gathered from the animal *in vivo* and *in vitro* studies and the human *in vitro* studies indicate that BiP is an immunomodulatory protein. This is in contrast to the data obtained with Hsp70, a molecule with which BiP has regions of high homology. Competitive binding studies show that BiP does not bind to any of the confirmed Hsp70 receptors. This supports the fact that BiP is also different functionally. We hypothesise, therefore, that in normal circumstances cell surface expressed and/or secreted BiP will regulate the development and cytokine profile of immune cells maintaining homeostasis and prevent an inappropriate inflammatory immune reaction (Figure 14.3). However, in a chronic inflammatory focus, such as the RA synovium, the overwhelming presence of pro-inflammatory agents (cytokines, chemokines, reactive oxygen species) will prevent upregulated BiP from being an effective means of reducing inflammation. In these circumstances, addition of exogenous BiP might prove to be an effective immunotherapy.

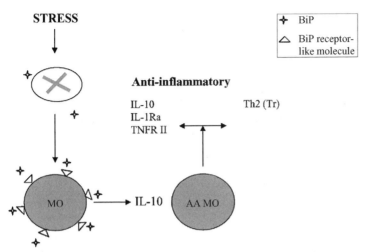

Figure 14.3. Stress, such as hypoxia, induces BiP release from the stressed cell. Cell-free BiP, after binding to a proposed BiP receptor-like molecule stimulates monocytes (MOs) to secrete IL-10. This deviates monocyte differentiation to alternatively activated monocytes (aaMOs). AaMOs induce an anti-inflammatory response via the secretion of anti-inflammatory molecules (such as IL-10, IL-1Ra, TNFR II) and regulatory T cells, thus controlling the inflammatory response.

REFERENCES

1. Gething M J. Role and regulation of the ER chaperone BiP. Semin Cell Dev Biol 1999, 10: 465–472.
2. Asea A, Kraeft S-K, Kurt-Jones E A, Stevenson M A, Chen L B, Finberg R W, Koo G C and Calderwood S K. Hsp70 stimulates cytokine production through a CD14-dependent pathway, demonstrating its dual role as a chaperone and cytokine. Nat Med 2000, 6: 435–442.
3. Asea A, Kabingu E, Stevenson M A and Calderwood S K. HSP70 peptide-bearing and peptide-negative preparations act as chaperokines. Cell Stress Chaperon 2000, 5: 425–431.
4. Kol A, Lichtman A H, Finberg R W, Libby P and Kurt-Jones E A. Heat shock protein (HSP) 60 activates the innate immune response: CD14 is an essential receptor for HSP60 activation of mononuclear cells. J Immunol 2000, 164: 13–17.
5. Lewthwaite J C, Coates A R M, Tormay P, Singh M, Mascagni P, Poole S, Roberts M, Sharp L and Henderson B. *Mycobacterium tuberculosis* chaperonin 60.1 is a more potent cytokine stimulator than chaperonin 60.2 (hsp 65) and contains a CD14-binding domain. Infect Immun 2001, 69: 7349–7355.
6. Asea A, Rehli M, Kabingu E, Boch J A, Baré O, Auron P E, Stevenson M A and Calderwood S K. Novel signal transduction pathway utilized by extracellular HSP70. Role of Toll-like receptor (TLR) 2 and TLR4. J Biol Chem 2002, 277: 15028–15034.

7. Basu S, Binder R J, Ramalingam T and Srivastava P K. CD91 is a common receptor for heat shock proteins gp96, hsp90, hsp70 and calreticulin. Immunity 2001, 14: 303–313.

8. Binder R J, Han D K and Srivastava P K. CD91: a receptor for heat shock protein gp96. Nat Immunol 2000, 1: 151–155.

9. Becker T, Hartl F U and Wieland F. CD40, an extracellular receptor for binding and uptake of Hsp70-peptide complexes. J Cell Biol 2002, 158: 1277–1285.

10. Wang Y, Kelly C G, Karttunen T, Whittall T, Lehner P J, Duncan L, MacAry P, Younson J S, Singh M, Oehlmann W, Cheng G, Bergmeier L and Lehner T. CD40 is a cellular receptor mediating mycobacterial heat shock protein 70 stimulation of CC-chemokines. Immunity 2001, 15: 971–983.

11. Delneste Y, Magistrelli G, Gauchat J, Haeuw J, Aubry J, Nakamura K, Kawakami-Honda N, Goetsch L, Sawamura T, Bonnefoy J and Jeannin P. Involvement of LOX-1 in dendritic cell-mediated antigen cross-presentation. Immunity 2002, 17: 353–362.

12. Habich C, Baumgart K, Kolb H and Burkart V. The receptor for heat shock protein 60 on macrophages is saturable, specific, and distinct from receptors for other heat shock proteins. J Immunol 2002, 168: 569–576.

13. Matzinger P. The Danger Model: A renewed sense of self. Science 2002, 296: 301–305.

14. De A K, Kodys K M, Yeh B S and Miller-Graziano C. Exaggerated human monocyte IL-10 concomitant to minimal TNF-α induction by heat-shock protein 27 (Hsp27) suggests Hsp27 is primarily an anti-inflammatory stimulus. J Immunol 2000, 165: 3951–3958.

15. Haas I G. BiP (GRP78), an essential hsp70 resident protein in the endoplasmic reticulum. Experientia 1994, 50: 1012–1020.

16. Kozutsumi Y, Normington K, Press E, Slaughter C, Sambrook J and Gething M J. Identification of immunoglobulin heavy chain binding protein as glucose-regulated protein 78 on the basis of amino acid sequence, immunological cross-reactivity, and functional activity. J Cell Sci Suppl 1989, 11: 115–137.

17. Knarr G, Gething M J, Modrow S and Buchner J. BiP binding sequences in antibodies. J Biol Chem 1995, 270: 27589–27594.

18. Janson I M, Toomik R, O'Farrell F and Ek P. KDEL motif interacts with a specific sequence in mammalian erd2 receptor. Biochem Biophys Res Commun 1998, 247: 447–451.

19. Kozutsumi Y, Segal M, Normington K, Gething M J and Sambrook J. The presence of malfolded proteins in the endoplasmic reticulum signals the induction of glucose-regulated proteins. Nature 1988, 332: 462–464.

20. Mapp P I, Grootveld M, C, and Blake D R. Hypoxia, oxidative stress and rheumatoid arthritis. Br Med Bull 1995, 51: 419–436.

21. Tak P P, Zvaifler N J, Green D R and Firestein G S. Rheumatoid arthritis and p53: how oxidative stress might alter the course of inflammatory diseases. Immunol Today 2000, 21: 78–82.

22. Maurice M M, Nakamura H, van der Voort E A, van Vliet A I, Staal F J, Tak P P, Breedveld F C and Verweij C L. Evidence for the role of an altered redox state in hyporesponsiveness of synovial T cells in rheumatoid arthritis. J Immunol 1997, 158: 1458–1465.

23. Blass S, Union A, Raymackers J, Schumann F, Ungethum U, Muller-Steinbach S, De Keyser F, Engel J M and Burmester G R. The stress protein BiP is overexpressed and is a major B and T cell target in rheumatoid arthritis. Arthritis Rheum 2001, 44: 761–771.

24. Delpino A and Castelli M. The 78 kDa glucose-regulated protein (GRP78/BIP) is expressed on the cell membrane, is released into cell culture medium and is also present in human peripheral circulation. Biosci Reports 2002, 22: 407–420.

25. Gagnon E, Duclos S, Rondeau C, Chevet E, Cameron P H, Steele-Mortimer O, Paiement J, Bergeron J J and Desjardins M. Endoplasmic reticulum-mediated phago-cytosis is a mechanism of entry into macrophages. Cell Biol Int 2002, 110: 119–131.

26. Shin B K, Wang H, Yim A M, Le Naour F, Brichory F, Jang J H, Zhao R, Puravs E, Tra J, Michael C W, Misek D E and Hanash S M. Global profiling of the cell surface proteome of cancer cells uncovers an abundance of proteins with chaperone function. J Biol Chem 2003, 278: 7607–7616.

27. Barreto A, Gonzalez J M, Kabingu E, Asea A and Fiorentino S. Stress-induced release of HSC70 from human tumors. Cell Immunol 2003, 222: 97–104.

28. Dai J, Liu B, Caudill M, Zheng H, Qiao Y, Podack E R and Li Z. Cell surface expression of heat shock protein gp96 enhances cross-presentation of cellular anti-gens and the generation of tumor-specific T cell memory. Cancer Immun 2003, 3: 1–5.

29. Lewthwaite J, Owen N, Coates A, Henderson B and Steptoe A. Circulating human heat shock protein 60 in the plasma of British civil servants. Circulation 2002, 106: 196–201.

30. Habich C, Kempe K, van der Zee R, Burkart V and Kolb H. Different heat shock pro-tein 60 species share pro-inflammatory activity but not binding sites on macrophages. FEBS Lett 2003, 533: 105–109.

31. Wassenberg J J, Dezfulian C and Nicchitta C V. Receptor mediated and fluid phase pathways for internalization of the ER Hsp90 chaperone grp94 in murine macrophages. J Cell Sci 1999, 112: 2167–2175.

32. Binder R J, Harris M L, Ménoret A and Srivastava P K. Saturation, competition, and specificity in interaction of heat shock proteins (hsp) gp96, hsp90, and hsp70 with CD11b+ cells. J Immunol 2000, 165: 2582–2587.

33. Arnold-Schild D, Hanau D, Spehner D, Schmid C, Rammensee H-G, de la Salle H and Schild H. Receptor-mediated endocytosis of heat shock proteins by professional antigen-presenting cells. J Immunol 1999, 162: 3757–3760.

34. Corrigall V M, Bodman-Smith M D, Fife M S, Canas B, Myers L K, Wooley P, Soh C, Staines N A, Pappin D J, Berlo S E, van Eden W, van der Zee R, Lanchbury J S and Panayi G S. The human endoplasmic reticulum molecular chaperone BiP is an autoantigen for rheumatoid arthritis and prevents the induction of experimental arthritis. J Immunol 2001, 166: 1492–1498.

35. Brownlie R, Sattar Z, Corrigall V M, Bodman-Smith M D, Panayi G S and Thompson S. Immunotherapy of collagen induced arthritis with BiP. Rheumatology (Oxford) 2003, 42 suppl: 13.

36. Sattar Z, Brownlie R, Corrigall V M, Bodman-Smith M D, Staines N A, Panayi G S et al. CD4+ T cells specific for the stress protein BiP modulate the development of collagen induced arthritis. Rheumatology (Oxford) 2003, 42 suppl: 124.

37. Bodman-Smith M D, Corrigall V M, Kemeny D M and Panayi G S. BiP, a putative autoantigen in rheumatoid arthritis, stimulates IL-10-producing CD8$^+$ T cells from normal individuals. Rheumatology (Oxford) 2003, 42: 637–644.

38. Krakauer T. Differential inhibitory effects of interleukin-10, interleukin-4, and dexamethasone on staphylococcal enterotoxin-induced cytokine production and T cell activation. J Leuk Biol 1995, 57: 450–454.

39. Corrigall V M, Garyfallos A and Panayi G S. The relative proportions of secreted interleukin-2 and interleukin-10 determine the magnitude of rheumatoid arthritis T-cell proliferation to the recall antigen tuberculin purified protein derivative. Rheumatology (Oxford) 1999, 38: 1203–1207.

40. Corrigall V M, Bodman-Smith M D, Brunst M, Cornell H and Panayi G S. Inhibition of antigen-presenting cell function and stimulation of human peripheral blood mononuclear cells to express an antiinflammatory cytokine profile by the stress protein BiP: relevance to the treatment of inflammatory arthritis. Arthritis Rheum 2004, 50: 1164–1171.

41. Schebesch C, Kodelja V, Muller C, Hakij N, Bisson S, Orfanos C E and Goerdt S. Alternatively activated macrophages actively inhibit proliferation of peripheral blood lymphocytes and CD4$^+$ T cells in vitro. Immunology 1997, 92: 478–486.

42. Castellino F, Boucher P E, Eichelberg K, Mayhew M, Rothman J E, Houghton A N and Germain R N. Receptor-mediated uptake of antigen/heat shock protein complexes results in major histocompatibility complex class I antigen presentation via two distinct pathways. J Exp Med 2000, 191: 1957–1964.

43. Gordon S. Alternative activation of macrophages. Nat Rev Immunol 2003, 3: 23–35.

44. Vittecoq O, Corrigall V M, Bodman-Smith M D and Panayi G S. The molecular chaperone BiP (GRP78) inhibits the differentiation of normal human monocytes into immature dendritic cells. Rheumatology (Oxford) 2003, 42 suppl: 43.

45. Flohé S B, Bruggemann J, Lendemans S, Nikulina M, Meierhoff G, Flohé S and Kolb H. Human heat shock protein 60 induces maturation of dendritic cells versus a Th1-promoting phenotype. J Immunol 2003, 170: 2340–2348.

Extracellular Biology of Molecular Chaperones: Molecular Chaperones as Therapeutics

15

Neuroendocrine Aspects of the Molecular Chaperones ADNF and ADNP

Illana Gozes, Inna Vulih, Irit Spivak-Pohis and Sharon Furman

15.1. Introduction

Vasoactive intestinal peptide (VIP), which was originally discovered in the intestine as a 28–amino acid peptide and shown to induce vasodilation, was later found to be a major brain peptide with neuroprotective activities *in vivo* [1–5]. To exert neuroprotective activity in the brain, VIP requires glial cells that secrete protective proteins such as activity-dependent neurotrophic factor (ADNF [6]). ADNF, isolated by sequential chromatographic methods, was named activity-dependent neurotrophic factor because it protects neurons from death associated with the blockade of electrical activity.

ADNF is a 14-kDa protein, and structure-activity studies have identified femtomolar-active neuroprotective peptides, ADNF-14 (VLGGGSALLRSIPA) [6] and ADNF-9 (SALLRSIPA) [7]. ADNF-9 exhibits protective activity in Alzheimer's disease–related systems (β-amyloid toxicity [7], presenilin 1 mutation [8], apolipoprotein E deficiencies [9] – genes that have been associated with the onset and progression of Alzheimer's disease (AD)). Other studies have indicated protection against oxidative stress via the maintenance of mitochondrial function and a reduction in the accumulation of intracellular reactive oxygen species [10]. In the target neurons, ADNF-9 regulates transcriptional activation associated with neuroprotection (nuclear factor-κB [11]), promotes axonal elongation through transcriptionally regulated cAMP-dependent mechanisms [12] and increases chaperonin 60 (Cpn60/Hsp60) expression, thereby providing cellular protection against the β-amyloid peptide [13].

Longer peptides that include the ADNF-9 sequence (e.g., ADNF-14) activate protein kinase C and mitogen-associated protein kinase kinase and protect developing mouse brain against excitotoxicity [14]. In neocortical synaptosomes, ADNF-9 enhances basal glucose and glutamate transport and attenuates oxidative impairment of glucose and glutamate transport induced by the β-amyloid peptide and Fe^{2+} [15]. In hippocampal neurons, ADNF-9 stimulates synapse formation as demonstrated by glutamate responses of excitatory neurons and morphological development [16]. In this hippocampal culture system, ADNF-9 induces the secretion of neurotrophin 3 (NT-3). Because both NT-3 and ADNF-9 regulate the NMDA receptor subunits 2A (NR2A) and NR2B, these results suggest *in vivo* effects of ADNF-9 on learning and behaviour in the adult nervous system. Indeed, in a rat model of cholinodeficiency, intranasal ADNF-9 enhances performance in a water maze, which is indicative of spatial learning and memory [17].

Antibody studies suggest that ADNF-like molecules mediate VIP neuroprotective and neurotrophic activities [12, 16, 18]. Preparation of ADNF-9–like analogues have resulted in the discovery of neuroprotective activity in an all D-amino acids ADNF-9 (D-ADNF-9, D-Ser-D-Ala-D-Leu-D-Leu-D-Arg-D-Ser-D-Ile-D-Pro-D-Ala) which suggests a non-chiral mode of action [17, 19]. Studies on ADNF-9 originated in our laboratory and Dr. D.E. Brenneman's laboratory [6, 7, 20, 21]. ADNF neuroprotective activity, at very low concentrations in models relevant to AD in particular and neurodegeneration in general, were independently corroborated in laboratories all over the world, for example, by Mattson and colleagues [8, 11, 15, 21], Gressens and colleagues [14], Hashimoto and colleagues [22] and Ramirez and colleagues [23].

Activity-dependent neuroprotective protein (ADNP) is another glial mediator of VIP-associated neuroprotection [9]. Antibodies that recognise ADNF-9 also recognise ADNP, and this has allowed the isolation of ADNP cDNA by expression cloning. An active eight–amino acid peptide (NAP, NAPVSIPQ) derived from ADNP, which shares structural and functional similarities with ADNF-9 in cell culture, has thus been identified [9]. ADNP was implicated in the maintenance of cell survival via a modulation of p53 expression [24]. A 100-fold more potent VIP analogue which provides neuroprotection is stearyl-Nle17-VIP (SNV [2–5]) and recent studies have now identified ADNP as a molecule which may mediate protection offered by SNV against ischaemic cell death [25].

As for NAP, *in vitro* experiments have shown that NAP protects neurons against numerous toxins and cellular stresses [9, 18, 26–30] including the AD neurotoxin (the β-amyloid peptide), excitotoxicity, the toxic envelope protein of the human immunodeficiency virus [9], electrical blockade [9], oxidative stress [18], dopamine toxicity [26], decreased glutathione [26], glucose deprivation [27] and tumour necrosis factor–associated toxicity [30].

NAP also has neuroprotective activity in a variety of animal models including the learning-deficient apolipoprotein E knockout mice (a model related to atherosclerosis [4, 9]), mouse paradigms of traumatic head injury (a risk factor for AD which exhibits some similar stroke-like secondary outcomes [30, 31]) and fetal alcohol syndrome (a model of oxidative damage [32]). In two rat paradigms, a model of cholinotoxicity and normal middle-aged animals treated daily by intranasal NAP administration, significant improvements in short-term spatial memory have been observed [17, 29]. NAP has a short structure, is active at exceptionally low concentrations (femtomolar), is water soluble, is bioavailable, is easily delivered via intranasal inhalation and is unusually stable. No NAP toxicity has been observed to date [33].

NAP has been studied in several independent laboratories. Busciglio and colleagues [34] have shown that the *in vitro* degeneration of Down syndrome neurons is prevented by ADNF-derived peptides; Shohami and colleagues have shown protection in head trauma [30], Brenneman and colleagues have shown protection in fetal alcohol syndrome [32], Leker and colleagues have shown protection in a model of mid cerebral artery occlusion [35], Offen and colleagues have shown protection against glutathione depletion [26] and Smith-Swintosky and colleagues have shown that NAP promotes neurite outgrowth in rat hippocampal and cortical cultures [36]. All of these findings corroborate the original description of NAP's neuroprotective properties [9]. The efficacy of NAP administration (μg to mg/kg, depending on the indication) has been demonstrated in animals using a variety of administration routes including intranasal [17], intraperitoneal [32], intravenous [35] and subcutaneous [31], and intact NAP has been detected in the brain 30 minutes and even 1 hour after administration [17, 32, 35].

Significant steps have been made recently towards understanding the mode of action of NAP. These include the initial identification of specific binding molecules and cells [37, 38], the identification of potential signal transduction pathways such as cGMP production [39], an interference with inflammatory mechanisms [30, 31] and a protective effect against apoptotic processes [35]. Protection against oxidative stress [18, 26, 32] and glucose deprivation [27] also suggests interference with fundamental processes.

The primary interest of the current chapter is the relationship between ADNF-9 and NAP with the chaperone family of proteins. Results have shown that ADNF-9, while having a structure similar to a short sequence in Hsp60, is directly associated with Hsp60 metabolism in the cell. These results are further discussed with regards to NAP which is highly homologous to ADNF-9 and provides an extracellular chaperone function. Because of the complexity of the systems used a brief description of the experimental methods used has been provided.

15.2. NAP and Chaperonin 60: the practical base

15.2.1. Cell culture and anti-sense oligodeoxynucleotide treatment

Cerebral cortical astrocytes were derived from newborn rat brains by trypsinisation of the cortices and growing the cells in conventional culture medium [9]. To determine the role of Hsp60, the anti-sense oligodeoxynucleotide (5'-TGT GGG TAG TCG AAG CAT-3') which has a sequence that is complementary to the 6–24 position on the rat liver Hsp60 cDNA and has previously been used to inhibit the neosynthesis of endogenous Hsp60 [40], was used. Anti-sense oligodeoxynucleotides (5 μM) were added to the culture medium (in the presence of serum) for 3 days with repeated additions at 24-hour intervals. Water was added to the control group. Culture medium was replaced only once, to prevent augmentation of Hsp60 expression due to the stress caused by medium replacement. Experiments were terminated 24 hours after the last addition. Cells were washed with phosphate buffered saline (PBS) and further incubated at room temperature for 3 hours with PBS containing 0.1 nM VIP. After the incubation, the conditioned medium was collected and the cells were harvested and subjected to protein extraction.

The intracellular protein content was analysed by gel electrophoresis and Western blotting using specific antibodies for Hsp60 (SPA-804, StressGen Biotechnologies Corp., Victoria, Canada) and ADNF [9]. The detected signals were analysed by densitometric scan and compared to the values of the untreated controls. Actin content was measured using actin-specific antibodies (anti-rabbit, Sigma), diluted 1:500, as a reference point of total protein content in each sample.

15.2.2. Hsp60 over-expression

Late passage C6 glioma cells (50–60), which exhibit an astrocytic phenotype, were used in this study. Cells (1×10^5 cells/ml) were seeded on tissue culture flasks [41] and were co-transfected either with the mouse Hsp60 expression construct [41] and neomycin expression vector using LipofectAMINE Plus (Life Technologies Inc.) at a ratio of 1:20 or with the neomycin expression vector by itself. Experiments were routinely carried out on a clone of the transfected cells, and all the results were confirmed on a number of individual clones expressing mouse Hsp60 and neomycin resistance and control clones expressing neomycin resistance only. Intracellular protein content of Hsp60 over-expressing clones was analysed by gel electrophoresis and Western blotting [13] [42] using polyclonal rabbit anti-Hsp60 antibody (StressGen, diluted 1:1000); polyclonal rabbit anti-ADNF antibody, diluted 1:250 [20] which was affinity purified against ADNF-9 [9]; and polyclonal rabbit anti-actin antibody (Sigma), diluted 1:500.

15.2.3. Immunoprecipitation

Intracellular protein mixtures and conditioned medium of Hsp60–over-expressing C-6 glioma cell clones were immunoprecipitated. Cells were washed in PBS (Biological Industries) and incubated for 3 hours in PBS. The conditioned medium was harvested, dialysed and lyophilised. The product was dissolved in anti-protease–containing buffer and was immunoprecipitated using the ADNF-9–specific antibodies [9]. Ten micrograms of affinity-purified antibody were added to each milligram of protein and allowed to conjugate in 4 °C for 1 hour. The conjugates were then precipitated by incubating at 4 °C for 2 hours with Protein A/G Plus-Agarose Beads (Santa Cruz Biotechnology). The washed precipitate was boiled, and the supernatant was collected and analysed by gel electrophoresis and Western blotting as described previously [14, 20].

15.3. The relationship of intracellular Hsp60 to ADNF

15.3.1. Hsp60 anti-sense oligodeoxynucleotides reduce ADNF-like expression

Although VIP treatment appeared to increase intracellular Hsp60 in astrocytes, albeit insignificantly, an apparent decrease in intracellular ADNF-like immunoreactivity was observed. As expected, anti-sense oligodeoxynucleotide treatment (Hsp60-specific) reduced Hsp60 expression, and this was paralleled by an even more pronounced reduction in ADNF-like immunoreactivity. Actin levels remained constant or were increased (Figure 15.1). Hence, the results suggest that Hsp60 anti-sense oligodeoxynucleotides reduce ADNF-like expression.

15.3.2. Over-expression of Hsp60 increases intracellular ADNF expression: Western blot analysis

Co-transfection generated 18 Hsp60 over-expressing clones, all of which exhibited an enhanced immunoreactivity in the 14-kDa ADNF-like band. Seven additional clones showed only enhanced ADNF-9–like 14-kDa immunoreactivity. Some of these clones were analysed further. In clone 6, a seven-fold increase in Hsp60 immunoreactivity was observed (60 kDa), compared to a

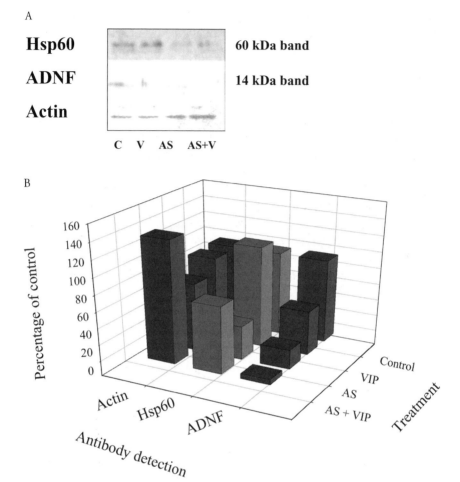

Figure 15.1. (A) Hsp60 anti-sense oligodeoxynucleotides reduce ADNF-like expression: Western blot. Antibodies used were anti-Hsp60 (Hsp60), ADNF-9 and actin. C = control, V = VIP treated, AS = anti-sense oligodeoxynuclotide (Hsp60), AS + V = anti-sense oligodeoxynuclotide + VIP. (B) Changes were quantified by densitometry.

control neo1-clone (which lacks the Hsp60 vector). Clone 13 exhibited a 30-fold greater immunoreactivity with the ADNF-14–specific antibody (14 kDa [9, 20]; Figure 15.2). In clone 14 there was a 30-fold greater immunoreactivity with anti-Hsp60 antibodies and a 258-fold greater reactivity with ADNF-14 antibodies. All of these analyses were performed on the same Western blot. Similar relationships between Hsp60 and ADNF were observed in all the clones analysed.

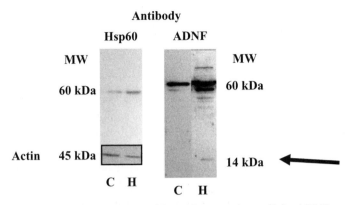

Figure 15.2. Over-expression of Hsp60 increases intracellular ADNF expression: Western blot analysis. MW = molecular weight, C = control, H = transfected with Hsp60 cDNA. Arrow indicates 14-kDa ADNF-like immunoreactivity.

15.3.3. Over-expression of Hsp60 increases intracellular ADNF expression: immunoprecipitation

A specific 14-kDa band was detected by Western blotting following the immunoprecipitation experiments only in the Hsp60-transfected clone lanes. No clear band of this size could be detected in the control lanes (Figure 15.3). The antibody used for both immunoprecipitation and immunodetection by Western blotting was the antibody recognising ADNF. The immunoprecipitation results

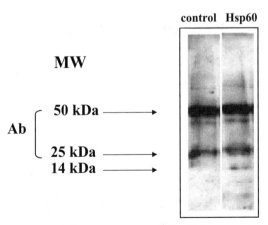

Figure 15.3. Over-expression of Hsp60 increases intracellular ADNF expression: immunoprecipitation experiments. A Western blot is shown and abbreviations are as in Figures 15.1 and 15.2.

of proteins in the extracellular milieu showed several ADNF-like bands at the lower molecular weight range (data not shown).

15.4. Discussion

The aforementioned results suggest a direct association between the expression of Hsp60 and of ADNF. ADNF exhibits a very close structural homology with Hsp60 which implies a chaperone-like activity [21]. ADNF was originally discovered as a 14-kDa potent neuroprotective protein secreted from glial cells in the presence of VIP [6]. Hsp60 is also secreted in the presence of VIP [43]. Hsp60 antibodies induce neuronal cell death which is inhibited by ADNF [6]. Furthermore, VIP-induced secretion of ADNF is associated with enhanced synapse formation [16]. Other studies have shown that over-expression of Hsp60/10 protects against ischaemia/reperfusion injury [44]. ADNF is a secreted protein. Hsp60 might also be secreted and the possibility that peptides derived from these proteins provide extracellular chaperonin activity is intriguing. The possibility of functional–structural interactions between Hsp60 and ADNF is also intriguing.

All the studies just described refer to Hsp60 and ADNF intracellular relationships and emanate from our previous findings of extracellular ADNF and Hsp60 sequences [6, 16, 43]. Because ADNF acts at femtomolar concentrations, high secretion levels are not required and hence the detection of ADNF in the extracellular milieu requires high quantities of conditioned medium. Our preliminary results (not shown) suggest that extracellular immunoreactive ADNF-like proteins in the lower molecular weight range are related to Hsp60. Interestingly, ADNF-9 (SALLRSIPA) is structurally homologous to NAP (NAPVSIPQ), an eight–amino acid peptide derived from ADNP [9]. ADNP is essential for brain formation [45] and NAP, which provides a broad range of neuroprotection [37, 46, 47], has been shown to inhibit the aggregation of the β-amyloid peptide, an aggregation that is associated with extracellular toxicity in AD [48]. Thus, NAP and related peptides may act as extracellular chaperones.

Acknowledgements

We thank Drs. Douglas E. Brenneman, Haya Brody and Ohad Birk for their help. Specifically, we thank Dr. Douglas E. Brenneman for his continuous support, Dr. Birk for the Hsp60 construct and Dr. Brody for her help with the glioblastoma Hsp60 over-expression colonies. These studies were supported by the United States–Israel Binational Science Foundation, the Israel Science Foundation, the Lily and Avraham Gildor Chair for the Investigation of Growth Factors, and the Institute for the Study of Aging and Allon Therapeutics, Inc.

NOTES ADDED IN PROOF: Several relevant new findings have been published or are in press since the original preparation of this chapter.

Divinski I, Mittelman L and Gozes I. A femtomolar acting octapeptide interacts with tubulin and protects astrocytes against zinc intoxication. J Biol Chem 2004, 279: 28531–28538

Gozes I and Divinski I. The femtomolar-acting NAP interacts with microtubules: Novel aspects of astrocyte protection. J Alzheimers Dis 2004, 6: S37–S41.

Furman S, Steingart R A, Mandel S, Hauser J M, Brenneman D E and Gozes I. Subcellular localization and secretion of activity-dependent neuroprotective protein in astrocytes. Neuron Glia Biology 2005, in press.

Brenneman D E, Spong C Y, Hauser J M, Abebe D, Pinhasov A, Golian T and Gozes I. Protective peptides that are orally active and mechanistically nonchiral. J Pharmacol Exp Ther 2004, 309: 1190–1197.

Wilkemeyer M F, Chen S Y, Menkari C E, Sulik KK and Charness M E. Ethanol antagonist peptides: structural specificity without stereospecificity. J Pharmacol Exp Ther 2004, 309: 1183–1189.

Zhou F C, Sari Y, Powrozek T A and Spong C Y. A neuroprotective peptide antagonizes fetal alcohol exposure-compromised brain growth. J Mol Neurosci 2004, 24: 189–199.

Chiba T, Hashimoto Y, Tajima H, Yamada M, Kato R, Niikura T, Terashita K, Schulman H, Aiso S, Kita Y, Matsuoka M and Nishimoto I. Neuroprotective effect of activity-dependent neurotrophic factor against toxicity from familial amyotrophic lateral sclerosis-linked mutant SOD1 in vitro and in vivo. J Neurosci Res 2004, 78: 542–552.

REFERENCES

1. Gozes I and Brenneman D E. VIP: molecular biology and neurobiological function. Mol Neurobiol 1989, 3: 201–236.
2. Gozes I, Fridkin M, Hill J M and Brenneman D E. Pharmaceutical VIP: prospects and problems. Cur Med Chem 1999, 6: 1019–1034.
3. Gozes I, Bardea A, Reshef A, Zamostiano R, Zhukovsky S, Rubinraut S, Fridkin M and Brenneman D E. Neuroprotective strategy for Alzheimer disease: intranasal administration of a fatty neuropeptide. Proc Natl Acad Sci USA 1996, 93: 427–432.
4. Gozes I, Bachar M, Bardea A, Davidson A, Rubinraut S, Fridkin M and Giladi E. Protection against developmental retardation in apolipoprotein E-deficient mice by a fatty neuropeptide: implications for early treatment of Alzheimer's disease. J Neurobiol 1997, 33: 329–342.
5. Gozes I, Perl O, Giladi E, Davidson A, Ashur-Fabian O, Rubinraut S and Fridkin M. Mapping the active site in vasoactive intestinal peptide to a core of four amino acids: neuroprotective drug design. Proc Natl Acad Sci USA 1999, 96: 4143–4148.
6. Brenneman D E and Gozes I. A femtomolar-acting neuroprotective peptide. J Clin Invest 1996, 97: 2299–2307.
7. Brenneman D E, Hauser J, Neale E, Rubinraut S, Fridkin M, Davidson A and Gozes I. Activity-dependent neurotrophic factor: structure-activity relationships of femtomolar-acting peptides. J Pharmacol Exp Therap 1998, 285: 619–627.

8. Guo Q, Sebastian L, Sopher B, Miller M W, Glazner G W, Ware C B, Martin G M and Mattson M. Neurotrophic factors [activity-dependent neurotrophic factor (ADNF) and basic fibroblast growth factor (bFGF)] interrupt excitotoxic neurodegenerative cascades promoted by a PS1 mutation. Proc Natl Acad Sci USA 1999, 96: 4125–4130.

9. Bassan M, Zamostiano R, Davidson A, Pinhasov A, Giladi E, Perl O, Bassan H, Blat C, Gibney G, Glazner G, Brenneman D E and Gozes I. Complete sequence of a novel protein containing a femtomolar-activity-dependent neuroprotective peptide. J Neurochem 1999, 72: 1283–1293.

10. Glazner G W, Boland A, Dresse A E, Brenneman D E, Gozes I and Mattson M P. Activity-dependent neurotrophic factor peptide (ADNF9) protects neurons against oxidative stress-induced death. J Neurochem 1999, 73: 2341–2347.

11. Glazner G W, Camandola S and Mattson M P. Nuclear factor-kappaB mediates the cell survival-promoting action of activity-dependent neurotrophic factor peptide-9. J Neurochem 2000, 75: 101–108.

12. White D M, Walker S, Brenneman D E and Gozes I. CREB contributes to the increased neurite outgrowth of sensory neurons induced by vasoactive intestinal polypeptide and activity-dependent neurotrophic factor. Brain Res 2000, 868: 31–38.

13. Zamostiano R, Pinhasov A, Bassan M, Perl O, Steingart R A, Atlas R, Brenneman D E and Gozes I. A femtomolar-acting neuroprotective peptide induces increased levels of heat shock protein 60 in rat cortical neurons: a potential neuroprotective mechanism. Neurosci Lett 1999, 264: 9–12.

14. Gressens P, Marret S, Bodenant C, Schwendimann L and Evrard P. Activity-dependent neurotrophic factor-14 requires protein kinase C and mitogen-associated protein kinase kinase activation to protect the developing mouse brain against excitotoxicity. J Mol Neurosci 1999, 13: 199–210.

15. Guo Z H and Mattson M P. Neurotrophic factors protect cortical synaptic terminals against amyloid and oxidative stress-induced impairment of glucose transport, glutamate transport and mitochondrial function. Cereb Cortex 2000, 10: 50–57.

16. Blondel O, Collin C, McCarran W J, Zhu S, Zamostiano R, Gozes I, Brenneman D E and McKay R D. A glia-derived signal regulating neuronal differentiation. J Neurosci 2000, 20: 8012–8020.

17. Gozes I, Giladi E, Pinhasov A, Golian T, Romano J and Brenneman D E. Activity-dependent neurotrophic factor: comparison of intranasal and oral administration of femtomolar-acting L and D peptides to improve memory. Soc Neurosci Abstract 2000: 223.

18. Steingart R A, Solomon B, Brenneman D E, Fridkin M and Gozes I. VIP and peptides related to activity-dependent neurotrophic factor protect PC12 cells against oxidative stress. J Mol Neurosci 2000, 15: 137–145.

19. Brenneman D E, Hauser J and Gozes I. Synergistic and non-chiral characteristics in dissociated cerebral cortical test cultures. Soc Neurosci Abstract 2000 223–224.

20. Gozes I, Davidson A, Gozes Y, Mascolo R, Barth R, Warren D, Hauser J and Brenneman D E. Antiserum to activity-dependent neurotrophic factor produces neuronal cell death in CNS cultures: immunological and biological specificity. Brain Res Dev Brain Res 1997, 99: 167–175.

21. Gozes I and Brenneman D E. Activity-dependent neurotrophic factor (ADNF). An extracellular neuroprotective chaperonin? J Mol Neurosci 1996, 7: 235–244.

22. Hashimoto Y, Niikura T, Ito Y, Sudo H, Hata M, Arakawa E, Abe Y, Kita Y and Nishimoto I. Detailed characterization of neuroprotection by a rescue factor humanin against various Alzheimer's disease-relevant insults. J Neurosci 2001, 21: 9235–9245.

23. Ramirez S H, Sanchez J F, Dimitri C A, Gelbard H A, Dewhurst S and Maggirwar S B. Neurotrophins prevent HIV Tat-induced neuronal apoptosis via a nuclear factor-kappaB (NF-kappaB)-dependent mechanism. J Neurochem 2001, 78: 874–889.

24. Zamostiano R, Pinhasov A, Gelber E, Steingart R A, Seroussi E, Giladi E, Bassan M, Wollman Y, Eyre H J, Mulley J C, Brenneman D E and Gozes I. Cloning and characterization of the human activity-dependent neuroprotective protein. J Biol Chem 2001, 276: 708–714.

25. Sigalov E, Fridkin M, Brenneman D E and Gozes I. VIP-Related protection against Iodoacetate toxicity in pheochromocytoma (PC12) cells: a model for ischemic/hypoxic injury. J Mol Neurosci 2000, 15: 147–154.

26. Offen D, Sherki Y, Melamed E, Fridkin M, Brenneman D E and Gozes I. Vasoactive intestinal peptide (VIP) prevents neurotoxicity in neuronal cultures: relevance to neuroprotection in Parkinson's disease. Brain Res 2000, 854: 257–262.

27. Zemlyak I, Furman S, Brenneman D E and Gozes I. A novel peptide prevents death in enriched neuronal cultures. Reg Peptides 2000, 96: 39–43.

28. Gozes I and Brenneman D E. A new concept in the pharmacology of neuroprotection. J Mol Neurosci 2000, 14: 61–68.

29. Gozes I, Alcalay R, Giladi E, Pinhasov A, Furman S and Brenneman D E. NAP accelerates the performance of normal rats in the water maze. J Mol Neurosci 2002, 19: 167–170.

30. Beni-Adani L, Gozes I, Cohen Y, Assaf Y, Steingart R A, Brenneman D E, Eizenberg O, Trembolver V and Shohami E. A peptide derived from activity-dependent neuroprotective protein (ADNP) ameliorates injury response in closed head injury in mice. J Pharmacol Exp Ther 2001, 296: 57–63.

31. Romano J, Beni-Adani L, Nissenbaum O L, Brenneman D E, Shohami E and Gozes I. A single administration of the peptide NAP induces long-term protective changes against the consequences of head injury: gene Atlas array analysis. J Mol Neurosci 2002, 18: 37–45.

32. Spong C Y, Abebe D T, Gozes I, Brenneman D E and Hill J M. Prevention of fetal demise and growth restriction in a mouse model of fetal alcohol syndrome. J Pharmacol Exp Therap 2001, 297: 774–779.

33. Newton P E, Brenneman D E and Gozes I. 30-day intranasal toxicity studies of NAP in rats and dogs. J Mol Neurosci 2001, 16: 61.

34. Pelsman A, Fernanandez G, Gozes I, Brenneman D E and Busciglio J. In vitro degeneration of Down syndrome neurons is prevented by activity-dependent neurotrophic factor-derived peptides. Soc Neurosci Abstracts 1998, 24: 1044.

35. Leker R R, Teichner A, Grigoriadis N, Ovadia H, Brenneman D E, Fridkin M, Giladi E, Romano J and Gozes I. NAP, a femtomolar-acting peptide, protects the brain against ischemic injury by reducing apoptotic death. Stroke 2002, 33: 1085–1092.

36. Smith-Swintosky V L, Gozes I, Brenneman D E and Plata-Salaman C R. Activity dependent neurotrophic factor-9 and NAP promote neurite outgrowth in rat hippocampal and cortical cultures. Soc Neurosci Abstracts 2000, 26: 843.

37. Gozes I, Divinsky I, Pilzer I, Fridkin M, Brenneman D E and Spier A D. From vasoactive intestinal peptide (VIP) through activity-dependent neuroprotective protein (ADNP) to NAP: a view of neuroprotection and cell division. J Mol Neurosci 2003, 20: 315–322.
38. Divinski I, Spier A D and Gozes I. NAP, a peptide derivative of the VIP-regulated gene ADNP, confers neuroprotection through microtubule dynamics. Reg Peptides 2003, 115: 42.
39. Ashur-Fabian O, Giladi E, Furman S, Steingart R A, Wollman Y, Fridkin M, Brenneman D E and Gozes I. Vasoactive intestinal peptide and related molecules induce nitrite accumulation in the extracellular milieu of rat cerebral cortical cultures. Neurosci Lett 2001, 307: 167–170.
40. Steinhoff U, Zugel U, Wand-Wurttenberger A, Hengel H, Rosch R, Munk M E and Kaufmann S H E. Prevention of autoimmune lysis by T cells with specificity for a heat shock protein by antisense oligonucleotide treatment. Proc Natl Acad Sci USA 1994, 91: 5085–5088.
41. Birk O S, Douek D C, Elias D, Takacs K, Dewchand H, Gur S L, Walker M D, van der Zee R, Cohen I R and Altmann D M. A role of hsp60 in autoimmune diabetes: analysis in a transgenic model. Proc Natl Acad Sci USA 1996, 93: 1032–1037.
42. Kurek J B, Bennett T M, Bower J J, Muldoon C M and Austin L. Leukaemia inhibitory factor (LIF) production in a mouse model of spinal trauma. Neurosci Lett 1998, 249: 1–4.
43. Bassan M, Zamostiano R, Giladi E, Davidson A, Wollman Y, Pitman J, Hauser J, Brenneman D E and Gozes I. The identification of secreted heat shock 60-like protein from rat glial cells and a human neuroblastoma cell line. Neurosci Lett 1998, 250: 37–40.
44. Hollander J M, Lin K M, Scott B T and Dillmann W H. Overexpression of PHGPx and HSP60/10 protects against ischemia/reoxygenation injury. Free Radic Biol Med 2003, 35: 742–751.
45. Pinhasov A, Mandel S, Torchinsky A, Giladi E, Pittel Z, Goldsweig A M, Servoss S J, Brenneman D E and Gozes I. Activity-dependent neuroprotective protein: a novel gene essential for brain formation. Brain Res Dev Brain Res 2003, 144: 83–90.
46. Zaltzman R, Beni S M, Giladi E, Pinhasov A, Steingart R A, Romano J, Shohami E and Gozes I. Injections of the neuroprotective peptide NAP to newborn mice attenuate head-injury-related dysfunction in adults. Neuroreport 2003, 14: 481–484.
47. Alcalay R N, Giladi E, Pick C G and Gozes I. Intranasal administration of NAP, a neuroprotective peptide, decreases anxiety-like behavior in aging mice in the elevated plus maze. Neurosci Lett 2004, 361: 128–131.
48. Ashur-Fabian O, Segal-Ruder Y, Skutelsky E, Brenneman D E, Steingart R A, Giladi E and Gozes I. The neuroprotective peptide NAP inhibits the aggregation of the beta-amyloid peptide. Peptides 2003, 24: 1413–1423.

16

Heat Shock Proteins Regulate Inflammation by Both Molecular and Network Cross-Reactivity

Francisco J. Quintana and Irun R. Cohen

16.1. Introduction

Heat shock proteins were initially identified as heterogeneous families of stress-induced proteins characterised by their chaperone activity [1]. Subsequently, they were identified as immunodominant antigens recognised by the host immune system following microbial infection [2] or during the course of autoimmune disease [3–6]. Recently, the role of heat shock proteins as endogenous activators of the innate and adaptive immune system has been unveiled [7]. In this chapter we discuss the relevance of heat shock proteins and their immune activities to the regulation of inflammation and autoimmune disease. We shall see that the regulatory activities of heat shock proteins on inflammation involve two types of cross-reactivity: *molecular* cross-reactivity exists between microbial and self-heat shock proteins and *network* cross-reactivity exists between different self-heat shock proteins.

16.2. Inflammation activates heat shock protein–specific T cells

Although the injection of incomplete Freund's adjuvant (IFA) to BALB/c mice induces local inflammation, Anderton and colleagues demonstrated that the injection of IFA also induces T cells reactive with the mammalian 60-kDa heat shock protein (Hsp60) [8]. These Hsp60-reactive T cells were TCR$\alpha\beta^+$, CD4$^+$ and major histocompatibility complex (MHC) class II-restricted [8]. Notably, Hsp60-specific cells could only be found in the local lymph nodes draining the site of IFA injection, and they were not present in distant lymph nodes. Hsp60-specific T cells are not only induced but also recruited to the site of inflammation [8].

The pro-inflammatory response which drives autoimmune disorders has also been shown to lead to an up-regulation of heat shock protein expression and the recruitment of heat shock protein–specific T cells to the target organ. Mor and colleagues have described that, along with myelin-specific T cells, T cells specific for the mycobacterial 65-kDa (Hsp65) or 71-kDa (Hsp71) heat shock proteins are recruited to the central nervous system (CNS) in rats undergoing experimental autoimmune encephalomyelitis (EAE) [9]. This initial observation was subsequently extended to include self-heat shock proteins and T cells reactive to them, in both EAE and human multiple sclerosis [10–12]. Finally, transplanted organs undergoing rejection show increased levels of expression of endogenous heat shock proteins and are infiltrated by heat shock protein–specific T cells (reviewed in [13]).

In short, heat shock protein–specific T cells are induced by inflammation and are recruited to the sites of inflammation. In this chapter, we will discuss experimental data that support a regulatory role for heat shock proteins and heat shock protein–specific T cells in the control of inflammation.

16.3. Heat shock proteins control inflammation

Adjuvant arthritis (AA) in the Lewis rat [14] and spontaneous autoimmune diabetes in the non-obese diabetic (NOD) mouse [15] are experimental models for two of the most prevalent human autoimmune diseases: rheumatoid arthritis [16] and type 1 diabetes mellitus (T1DM) [17]. Although the clinical signs of the models are naturally different, both experimental diseases are linked by the observation that heat shock proteins can halt the autoimmune attack. We have used these experimental models to study the role of heat shock proteins in the control of autoimmune disease.

16.3.1. Adjuvant arthritis

AA is induced in Lewis rats by a subcutaneous injection of heat-killed *Mycobacterium tuberculosis* in IFA [14]. T cells specific for mycobacterial Hsp65 can both drive and inhibit AA. Although Hsp65-specific CD4$^+$ T cell clones cross-react with cartilage components and transfer AA [18], Hsp65 administered as a protein [19], encoded in a recombinant vaccinia virus [20] or administered as a DNA vaccine [21] can inhibit AA. The administration of Hsp65 can also regulate experimental arthritis triggered by the lipoidal amine CP20961 [22] or by pristane [23].

Inhibition of AA by Hsp65 is thought to involve cross-reactivity with self-Hsp60 [24]. We have studied the specificity of the regulatory immune response that controls AA using DNA vaccines coding for either human Hsp60 (pHsp60) or mycobacterial Hsp65 (pHsp65) [25]. Although both pHsp60 and pHsp65 protect against AA, pHsp60 is significantly more effective [25]. Using DNA vaccines encoding fragments of Hsp60 to identify immunoregulatory regions within Hsp60, the anti-arthritogenic effects of the pHsp60 construct have been shown to reside in the amino acid (aa) 1–260 region of Hsp60 [26]. Using Hsp60-derived overlapping peptides, peptide Hu3 (aa 31–50 of Hsp60) is specifically recognised by T cells of rats protected from AA by DNA vaccination [26]. Vaccination with Hu3, or transfer of splenocytes from Hu3-vaccinated rats, prevents the development of AA, whereas vaccination with the mycobacterial homologue of Hu3 has no effect [26]. Prevention of AA by vaccination with pHsp60, DNA vaccines encoding the N-terminus of Hsp60, or Hu3 was associated with the induction of T cells that secrete IFN-γ, IL-10 and TGF-β1 upon stimulation with Hsp60 [25, 26]. Thus, Hsp60-specific T cells can control the progression of AA. However, what influence do T cells reactive with other heat shock proteins have on such processes?

T cell responses to the mycobacterial 10-kDa heat shock protein (Hsp10) [27] or mycobacterial Hsp71 have also been shown to control the progression of AA [28–30]. We studied whether self-heat shock proteins other than Hsp60 could inhibit AA using DNA vaccines encoding human 70-kDa heat shock protein (Hsp70) or the human 90-kDa heat shock protein (Hsp90). DNA vaccination with Hsp70 or Hsp90 shifted the specific arthritogenic T-cell response from a Th1 to a Th2/3 phenotype and inhibited AA [31]. Thus, Hsp70 and Hsp90 can also modulate arthritogenic T cell responses in AA.

Hsp60-specific responses in patients with rheumatoid arthritis [32, 33] or juvenile chronic arthritis [34] are associated with milder arthritis and a better prognosis. Although no information is yet available on T cell responses to Hsp70 or Hsp90 in human arthritis, these observations suggest that heat shock protein-specific T cells might also have a regulatory role in human autoimmune arthritis. The role of the 70-kDa heat shock protein BiP as a modulator of rheumatoid arthritis is described in detail in Chapter 14.

16.3.2. NOD diabetes

NOD mice spontaneously develop diabetes as a consequence of a T cell–mediated autoimmune process that destroys the insulin-producing β cells of the pancreas [17]. NOD mice have a high frequency of self-reactive T cells [35], which is reflected by a highly self-reactive B-cell repertoire [36]. Several antigens are targeted by diabetogenic T cells, including insulin [37] and glutamic acid decarboxylase (GAD) [38]. Similar to the situation found in AA, T cell reactivity to Hsp65 is a double-edged sword. A peak of Hsp65-specific

T cell reactivity precedes the onset of diabetes [39], and immunisation with Hsp65 can induce a transient hyperglycaemia [39]. However, vaccination with Hsp65 can also inhibit the development of diabetes [39]. These initial reports may be explained by cross-reactivity between mycobacterial Hsp65 and self-Hsp60.

We have shown that self-Hsp60 is targeted by the diabetogenic attack; T cells reactive with the Hsp60 peptide p277 (aa 437–460) can induce diabetes in irradiated NOD recipients [40]. On the other hand, vaccination of NOD mice with peptide p277 has been shown to arrest the development of diabetes [40] and can even induce remission of overt hyperglycaemia [41]. Successful p277 treatment leads to the down-regulation of spontaneous T cell proliferation to p277 and to the induction of a Th1-to-Th2 switch in the immune response to p277 [42]. Other peptides of Hsp60 can also inhibit the development of spontaneous diabetes in NOD mice [43].

NOD mice can also develop a more robust form of diabetes induced by the administration of cyclophosphamide, termed cyclophosphamide-accelerated diabetes (CAD) [44]. Cyclophosphamide is thought to specifically deplete regulatory T cells [44], thereby unleashing a Th1 response which is rich in IFN-γ secreting cells and leads to overt diabetes [45].

We have studied the effect of DNA vaccination with pHsp60 or pHsp65 on CAD. Vaccination with pHsp60, but not with pHsp65, protects NOD mice from CAD [46]. Thus, the efficacy of the pHsp60 DNA vaccine in this situation can be explained by regulatory Hsp60 epitopes that are not shared with Hsp65; indeed well-characterised regulatory epitopes from Hsp60 are not conserved in the sequence of Hsp65 [46]. Vaccination with pHsp60 modulates the T cell responses to Hsp60 and also to GAD and insulin. T cell proliferative responses are significantly reduced, and the cytokine profile induced by stimulation with Hsp60, GAD or insulin revealed an increased secretion of IL-10 and IL-5 and a decreased secretion of IFN-γ, a finding which is compatible with a Th1-to-Th2 shift in the autoimmune response [46].

In conclusion, the administration of Hsp60 peptides, or of whole Hsp60 as a recombinant protein or a DNA vaccine, can halt autoimmune NOD diabetes. Several antigens are targeted during the progression of diabetes [17] and it is therefore remarkable that the immunoregulatory networks triggered by Hsp60 can control diabetogenic T cells that are directed to a range of other antigens, such as insulin and GAD.

B and T cell responses to Hsp70 [47], Hsp60 and p277 [6, 48] have also been described in patients with T1DM. Indeed, a double-blind, phase II clinical trial was designed to study the effects of p277 therapy on newly diagnosed patients

[49]. The administration of p277 after the onset of clinical diabetes preserved the endogenous levels of C-peptide (which fell in the placebo group) and was associated with lower requirements for exogenous insulin, thereby revealing an arrest of β cell destruction [49]. Treatment with p277 led to enhanced Th2 responses to Hsp60 and p277 [49]. Thus, like NOD diabetes, human T1DM appears to be susceptible to immunomodulation by Hsp60 therapy.

Taken together it appears that heat shock proteins can control the progression of inflammation and, in particular, self-heat shock proteins seem to be quite efficient in doing so. However, do we need exogenous heat shock proteins to trigger heat shock protein–based regulatory mechanisms?

16.4. Triggering of heat shock protein–based immunoregulation by innate immune activation

Bacterial DNA stimulates the innate immune system via Toll-like receptor 9 (TLR9) [50] due to the presence of DNA motifs consisting of a central unmethylated CpG dinucleotide flanked by two 5′ purines and two 3′ pyrimidines [51]. Such a sequence is referred to as a CpG motif. We have demonstrated that bacterial CpG motifs can inhibit spontaneous diabetes in NOD mice [52], but not CAD [46]. The prevention of diabetes was characterised by a decreased insulitis [52]. Moreover, we have detected a decrease in the spontaneous proliferative responses of T cells to Hsp60 and its p277 peptide, concomitant with the induction of Th2-like antibodies of the same specificity, thereby revealing a Th1-to-Th2 shift in the autoimmune response of the treated mice [52].

To investigate the mechanisms involved in the regulation of spontaneous NOD diabetes by CpG motifs, we studied the expression of Hsp60 in splenocytes from NOD mice stimulated with a synthetic oligonoucleotide containing CpG motifs (CpG). In vitro stimulation with CpG led to a dose-dependent upregulation of intracellular Hsp60 levels, as demonstrated by Western blot analysis, and also to the release of Hsp60 into the supernatant. A control oligonucleotide containing an inverted CpG motif (GpC) had no significant effect on the intracellular levels of Hsp60 or on Hsp60 secretion [Quintana and Cohen, manuscript submitted].

CpG also affected the responses of T cell clones specific for the Hsp60 peptides p12 (aa 166–185) or p277 (aa 437–460). In the presence of irradiated antigen-presenting cells (APCs), CpG triggered the dose-dependent proliferation of both Hsp60-specific T cell clones, but not of an anti-ovalbumin T cell line [Quintana and Cohen, manuscript submitted]. All the T cells were activated by their target antigen, but not by lipopolysaccharide (LPS), thereby ruling out the possibility that some of the observed proliferation was due to the presence of contaminating B cells [Quintana and Cohen, manuscript submitted]. The analysis of cytokine secretion revealed that CpG stimulation triggered the secretion of higher amounts of IL-10 and lower amounts of IFN-γ than did activation with the target Hsp60 peptides (p12 or p277) [Quintana and Cohen, manuscript submitted]. The Hsp60-specific T cell lines were not activated by CpG in the absence of APCs, and CpG-induced proliferation was inhibited by anti-MHC class II antibodies [Quintana and Cohen, manuscript submitted]. Thus, CpG activates Hsp60-specific T cells by stimulating the presentation of peptides derived from endogenous Hsp60 in the MHC class II molecules of the APC. Because IL-10 is known to have suppressor effects on immune responses [53], the relative increase in IL-10 secretion by Hsp60-specific T cells might explain the protective effect of CpG on NOD diabetes. The reader should refer to Chapters 13 and 14 for a discussion of chaperones that selectively induce IL-10 over IL-1/tumour necrosis factor synthesis.

Figure 16.1 depicts our model for the action of CpG on spontaneous NOD diabetes. The activation of APCs or of other cell types via TLR9 leads to the up-regulation of intracellular levels of Hsp60 and eventually to its secretion. Hsp60 is then presented on the surface of the APC via MHC class II molecules. Hsp60-specific regulatory T cells are therefore activated, halting the progression of NOD diabetes.

A paper by Kumaraguru and colleagues reports that CpG triggers the up-regulation and release of Hsp70 from macrophages; however, the effects of CpG on Hsp70-specific T cell lines were not studied [54]. Based on the APC function of macrophages, it is likely that Hsp70 peptides presented in the MHC molecules of CpG-treated macrophages can modulate Hsp70-specific immunity.

TLR9-mediated activation has been shown to control several experimental models of autoimmune disease including EAE [55], colitis [56] and arthritis [57, 58]. Ligands for other TLRs, such as poly I:C [59] or LPS [60–62], have also been reported to inhibit experimental autoimmunity. Whether the activation of heat shock protein–based immunoregulatory mechanisms is a feature shared by several TLR-dependent signalling cascades remains to be seen. Nevertheless, our results suggest that regulatory Hsp60-specific T cell responses can be triggered by the activation of innate networks that lead to the release of endogenous heat shock proteins leading, in turn, to the activation of the specific T cell populations. Could we use these innate networks to diversify the heat shock protein–specific immune response? In other words, could we administer a particular heat shock protein and induce T cell responses directed to a different heat shock protein?

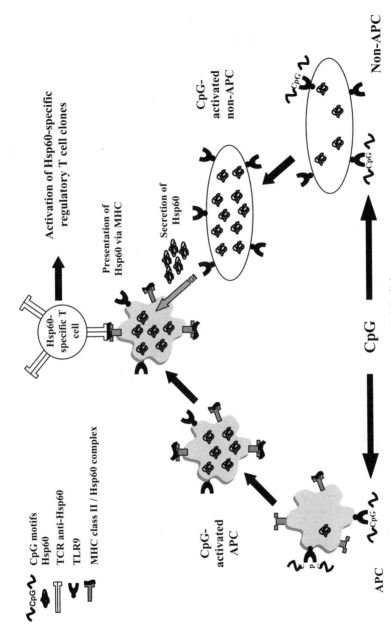

Figure 16.1. Model for the action of CpG on spontaneous NOD diabetes.

CpG motifs
Hsp60
TCR anti-Hsp60
TLR9
MHC class II / Hsp60 complex

APC

CpG

Non-APC

CpG-activated APC

CpG-activated non-APC

Secretion of Hsp60

Presentation of Hsp60 via MHC

Hsp60-specific T cell

Activation of Hsp60-specific regulatory T cell clones

16.5. Connectivity between different heat shock protein–specific immune responses

We have demonstrated that DNA vaccines coding for Hsp60 (pHsp60), Hsp70 (pHsp70) or Hsp90 (pHsp90) can inhibit AA [25, 26, 30]. Moreover, DNA vaccines coding for Hsp70 or Hsp90 can modulate the Hsp65-specific T cell response which drives AA, in a similar manner to that previously demonstrated for Hsp60 [25, 26, 30]. Hsp60, Hsp70 and Hsp90 bear no significant sequence homology or immune cross-reactivity. However, might immunisation with an exogenous heat shock protein trigger the presentation of a different endogenous heat shock protein, leading to the diversification of the immune response induced by vaccination with a particular heat shock protein?

DNA vaccination with pHsp70 or pHsp90 induces antigen-specific proliferative responses: pHsp70-vaccinated rats manifest T cell responses to Hsp70, and pHsp90-vaccinated rats manifest T cell responses to Hsp90 [31]. However, DNA vaccination with pHsp70 or pHsp90 could also induce T cells that proliferated and secreted IFN-γ, TGF-β1 and IL-10 upon stimulation with Hsp60 [31]. Thus different heat shock protein molecules are linked immunologically.

To characterise this connection, we compared the epitope specificity of the Hsp60-specific T cell response induced by pHsp60 with that induced by pHsp70 using a panel of overlapping peptides derived from the human Hsp60 sequence [31]. We had previously found that pHsp60 DNA-vaccination-induced regulatory T cells were reactive with a single Hsp60 peptide epitope, Hu3 (aa 31–50) [26]. However, lymph node cells (LNCs) from pHsp70-vaccinated rats responded to several other Hsp60 peptides: Hu19 (aa 271–290), Hu24 (aa 346–365), Hu25 (aa 361–380), Hu27 (aa 391–410), Hu28 (aa 406–425), Hu30 (aa 436–455), Hu32 (aa 466–485), Hu33 (aa 481–500) and Hu34 (aa 271–290) [31]. Thus, although both pHsp60 and pHsp70 can induce Hsp60-specific T cells, the fine specificities of the T cell responses induced are different. The crosstalk between the Hsp60- and the Hsp70-specific T cell responses is reciprocal, in that pHsp60-vaccinated rats showed significant T cell responses upon stimulation with Hsp70 [31]. These findings are schematically represented in Figure 16.2.

Hsp60, Hsp70 and Hsp90 share no sequence homology and are not immunologically cross-reactive. One possible explanation for the induction of Hsp60-specific T cell responses by pHsp70 or pHsp90 is self-vaccination with endogenous self-Hsp60 which is induced and/or released as a result of the DNA vaccinations. Indeed, we could detect increased levels of circulating Hsp60 in pHsp70-vaccinated rats [Quintana et al., manuscript submitted]. The upregulation of Hsp60 levels in the circulation was dependent on the presence

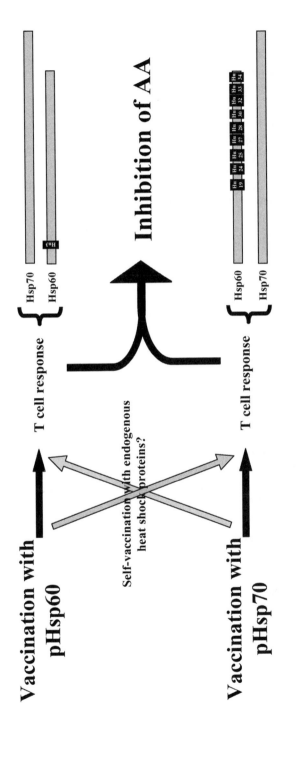

Figure 16.2. Connectivity between Hsp60- and Hsp70-specific T cell responses.

of the *hsp70* gene on the pHsp70 DNA construct; as a control, empty plasmid had no effect on circulating Hsp60 levels [Quintana et al., manuscript submitted]. Reciprocally, vaccination with pHsp60 induced a T cell response to Hsp70; however, we have not yet been able to measure the levels of Hsp70 in the blood of pHsp60-vaccinated rats. Although the molecular mechanisms require further study, the present findings demonstrate that heat shock protein–specific responses are inter-regulated and highlight the multiple immune signalling activities of these molecules.

16.6. Immunoregulatory mechanisms triggered by heat shock proteins

Earlier, we have seen that inflammation induces heat shock protein–specific T cells and that, despite a lack of immunological cross-reactivity between the molecules, T cell responses to different heat shock proteins are connected. We have shown that heat shock proteins can control autoimmune disease, and we have referred to experimental data suggesting that heat shock protein–based regulatory networks can be triggered in the absence of exogenous immunisation with heat shock proteins. We are therefore left with the question of how heat shock protein molecules might control autoimmunity.

The autoimmune attack which leads to overt autoimmune disease is a process that simultaneously engages several cell types and molecular mechanisms that are not restricted to the immune system. Faced by such a multi-front attack, it is not surprising that heat shock proteins bear several features that might prove helpful for the control of the autoimmune response. Heat shock proteins, as described in Chapter 1, are intracellular chaperones which facilitate the correct folding of newly synthesised proteins [1]. Moreover, they also improve antigen processing and presentation by APCs [63]. Heat shock proteins facilitate the induction of T cell responses to free peptide epitopes which are bound by circulating heat shock proteins and taken up by APCs through heat shock protein–specific receptors [63, 64]. Circulating heat shock proteins, not loaded with any peptide, can directly activate several cell types via innate receptors [7]. Heat shock proteins can activate immune system cells, such as dendritic cells [65–67], and also non-immune cells, such as endothelial cells [68]. Finally, heat shock proteins bear regulatory T cell epitopes [25, 69]. However, as we have seen, heat shock proteins can also be targeted by the pathogenic T cells that characterise autoimmune diseases such as T1DM [6].

The sites at which heat shock proteins are expressed can influence their immunoregulatory functions. The intracellular levels of heat shock proteins are increased upon cellular stress. Viral or bacterial infections up-regulate heat shock protein expression [70–73], and necrotic cells release heat shock proteins [74]. Inflammation is a source of cellular stress, and heat shock proteins are over-expressed at the sites of inflammation, such as in the synovium in arthritis [75]. Strikingly, heat shock proteins are also up-regulated in activated macrophages [76] and T cells [77]. Thus, heat shock proteins simultaneously mark the cells targeted by the autoimmune attack and the pathogenic immune cells that carry out the attack.

Based on the intra- and extra-cellular functions of heat shock protein and their localisation, several mutually non-exclusive mechanisms, involving adaptive and non-adaptive immunity, can contribute to the immunoregulatory properties of heat shock proteins.

16.6.1. Adaptive immunity

16.6.1.1. Environmental regulation of heat shock protein–specific immunity

Heat shock proteins are immunodominant bacterial antigens [78]. Because mucosal immunisation is known to induce antigen-specific regulatory responses [79], exposure to bacterial heat shock proteins from the intestinal flora might be a source of heat shock protein–specific regulatory T cells. Indeed, Moudgil and colleagues have demonstrated that environmental microbes can induce Hsp65-specific T cells directed to regulatory epitopes that are cross-reactive with self-Hsp60 [69, 80]. Vaccination with heat shock proteins or their peptides might simply amplify this naturally acquired regulation. However, based on this mechanism, any cross-reactive protein conserved through evolution from bacteria to mammals should be immunoregulatory. This is not always the case, as recently reported by Prakken and colleagues [81].

16.6.1.2. Boost of regulatory T cell responses

Heat shock proteins can bind free peptides and induce peptide-specific immune responses, even in the presence of low amounts of the target peptide [82]. Thus, heat shock protein molecules could be loaded *in vivo* with regulatory self-peptides and subsequently boost or amplify specific regulatory T cell responses. Indeed, Chandawarkar and colleagues have reported that gp96 can both induce and down-regulate tumour-specific immune responses [83]. Furthermore, heat shock proteins purified from the inflamed CNS of EAE rats (and not from naïve rats) can vaccinate naïve rats against EAE [84]. Thus, Hsp70–peptide complexes synthesised at the sites of active inflammation can trigger tissue-specific anti-inflammatory T cell responses [84]. Nevertheless, this mechanism does not explain the immunomodulatory effects of heat shock protein–derived fragments or peptides. See Chapters 17 and 18 for more information on chaperones and peptide-specific immune responses.

16.6.1.3. Cytokine-mediated bystander inhibition

Inflammation leads to the local up-regulation of heat shock proteins. Heat shock protein–specific T cells might therefore be recruited to sites of inflammation, where they could control pathogenic T cell clones by the secretion of regulatory cytokines. Heat shock protein–specific T cells induced by vaccination with immunoregulatory DNA vaccines or peptides secrete regulatory cytokines (IL-10 and TGF-β1) [25, 29, 30, 46].

16.6.1.4. Anti-ergotypic regulation

T cells reactive to activated T cells (but not to resting T cells) can control experimental autoimmune disease [85–87]. The T cell receptor expressed by these regulatory T cells recognises peptides derived from activation markers (ergotopes), such as the α-chain of the IL-2 receptor [86, 87] or the TNF-α receptor [87]. These cells are termed anti-ergotypic [85]. Now it has been reported that mRNAs encoding for heat shock proteins are up-regulated upon T cell activation [77]. Thus, Hsp60 too might serve as an ergotope. We studied whether vaccination with DNA vaccines encoding Hsp60, or with the regulatory peptide Hu3, might induce anti-ergotypic responses. To serve as an ergotope, Hsp60 would have to fulfil two requirements. Firstly, Hsp60 must be up-regulated in activated T cells. Secondly, activated T cells must present Hsp60-derived peptides to Hsp60-specific T cells.

The activation of T cells by the mitogen Concanavalin A, or by specific antigen, up-regulates intracellular levels of Hsp60 [Quintana et al., manuscript submitted]. Thus, the first condition is fulfilled: T cell activation triggers Hsp60 expression. Moreover, activated T cells can present Hsp60. Hsp60-specific T cells proliferate to activated T cells and secrete both IFN-γ and TGF-β1 [Quintana et al., manuscript submitted]. The activation of Hsp60-specific T cells was MHC class II (RT1.B) restricted, since it could be inhibited with the OX6 monoclonal antibody [Quintana et al., manuscript submitted]. Thus, Hsp60 can function as an ergotope *in vitro*; however, can functional Hsp60-specific anti-ergotypic responses be induced *in vivo*?

DNA vaccination with pHsp60 has been found to induce anti-ergotypic T cell responses that are MHC class II (RT1.B) and MHC class I restricted [Quintana et al., manuscript submitted]. In contrast, vaccination with Hu3 induced only an MHC class II restricted (RT1.B) anti-ergotypic T cell response [Quintana et al., manuscript submitted]. Thus, Hsp60-specific $CD4^+$ and $CD8^+$ anti-ergotypic T cells can be induced *in vivo*.

LNCs from rats with AA stimulated with the immunodominant 180–88 T cell epitope of Hsp65 (mt180) secrete high levels of IFN-γ [25]. Since T cells specific for this epitope have been shown to transfer AA [18, 88], the reactivity of LNCs of AA to mt180 is thought to reflect the behaviour of the arthritogenic T cells. LNCs of AA rats stimulated with mt180 in the presence of Hsp60-specific anti-ergotypic T cells (but not with a control anti-myelin bask protein (MBP) line) secrete significantly less IFN-γ [Quintana et al., manuscript submitted]. Thus, anti-ergotypic responses can control the arthritogenic response *in vitro*. Our model for the role of Hsp60-specific T cells in anti-ergotypic response is depicted in Figure 16.3. However, the contribution of the anti-ergotypic response to the regulatory functions of heat shock protein–specific T cells in AA and other autoimmune disorders *in vivo* is still unknown.

16.6.2. Innate immunity

16.6.2.1. Innate activation of regulatory T cells
Heat shock proteins are endogenous ligands for innate receptors. Hsp60 and Hsp70 activate TLR4 and TLR2 [89]; Hsp70 and Hsp90 have also been reported to signal via CD40 and CD91 [90, 91]. See Chapters 7, 8 and 10 for more details of the receptors for chaperones. Caramalho and colleagues have reported that regulatory $CD25^+$ T cells are activated via TLR4 [92]. Thus, it is possible that self-heat shock proteins directly activate regulatory cells via innate receptors. This hypothesis is partially supported by the findings made by Dr. Gabriel Nussbaum in our laboratory, who has generated NOD mice lacking a functional TLR4. NOD mice carrying a non-functional *tlr4* allele show an early onset and an increase in the incidence of spontaneous diabetes. Interestingly, the sensitivity of those NOD mice to CAD remains unchanged (Dr. Gabriel Nussbaum, personal communication). Cyclophosphamide is thought to deplete regulatory cells [44]; thus, these findings suggest that TLR4-mediated signals triggered by self-ligands do activate regulatory cells involved in the control of autoimmune diabetes.

16.6.3. Hsp60 triggers anti-inflammatory activities in T cells via TLR2

Hsp60 and p277 can directly inhibit chemotaxis and activate anti-inflammatory activities in human T cells, via TLR2 [93]. Human T cells activated by mitogen in the presence of Hsp60 or p277 also show a decreased secretion of IFN-γ and an increased secretion of IL-10 (unpublished observations). Thus, soluble Hsp60

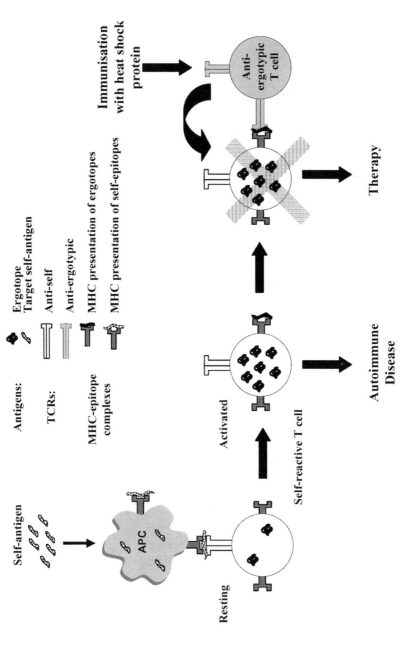

Figure 16.3. Anti-ergotypic response mediated by Hsp60-specific T cells.

or p277, acting via TLR2, can modulate T cells involved in the progression of inflammation.

16.7. Heat shock proteins: physiological modulators of inflammation

Inflammation is physiological [94]; it plays a role in processes ranging from wound healing [95] to neuroprotection [96]. However, uncontrolled inflammation can lead to disease and, as a consequence, precise mechanisms have been selected through evolution for the tight control of inflammation.

Inflammation induces heat shock proteins and heat shock protein–specific immune responses. However, heat shock proteins and the immune responses directed against them can both promote and inhibit inflammation. Heat shock proteins are central nodes in physiological networks that control inflammation; they integrate the intra-cellular response to stress with the inter-cellular signals that spread a cascade of pro- or anti-inflammatory responses (Figure 16.4A). The variety of the anti-inflammatory responses co-ordinated by heat shock proteins is as diverse as the biological activities of heat shock proteins. In the short term, heat shock proteins can activate regulatory mechanisms via innate receptors. In the long term, heat shock proteins can also trigger adaptive immunoregulatory T cell responses directed against heat shock proteins or other self-proteins. Heat shock proteins bridge the innate and adaptive immune responses involved in the physiological control of inflammation.

Regulatory networks centred on heat shock proteins can be boosted by several methods to treat autoimmune disease (Figure 16.4B). Indeed, these therapies could operate by simply mimicking the effects that the environment has on the immune system. The rise in the standard of living achieved during the past century in the developed world seems to have diminished the microbial stimulation of the immunoregulatory functions that keep immune balance. This reduction in immune stimulation might contribute to the increased incidence of autoimmune diseases observed in developed countries, as we have discussed elsewhere [97].

16.7.1. Network cross-reactivity

It is striking that immunisation with defined heat shock proteins leads to the induction of T cell responses directed to other structurally unrelated heat shock protein molecules [31]. The definition of immunological cross-reactivity, usually found in immunology textbooks, would not account for this unexpected finding. Herein, we would like to propose a new definition for cross-reactivity.

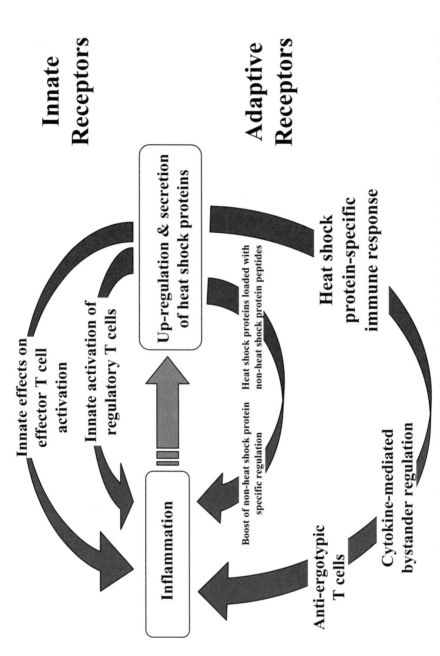

Figure 16.4. Heat shock proteins as regulators of inflammation: (A) physiological regulation of inflammation by heat shock proteins and (B) therapeutic/environmental regulation of inflammation by heat shock proteins.

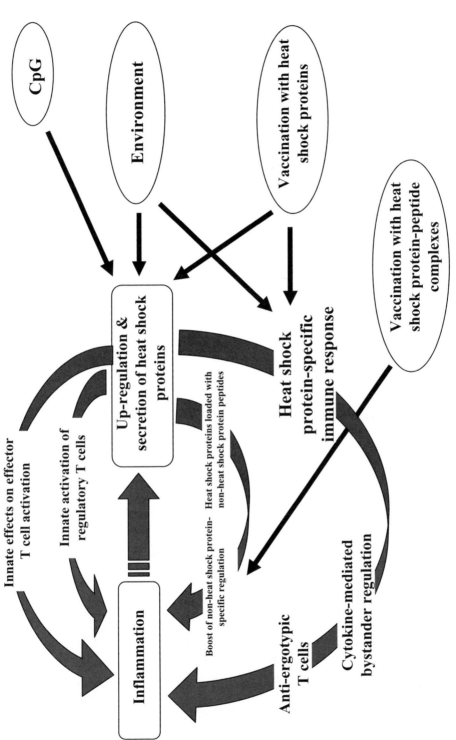

Figure 16.4. (*continued*)

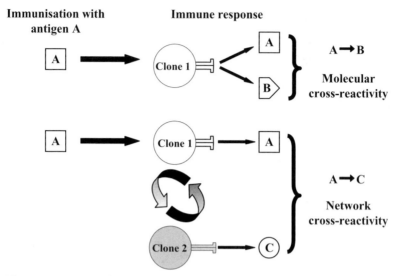

Figure 16.5. Immunological cross-reactivity.

We define *molecular* cross-reactivity as the classical cross-reactivity that exists between antigens that share sequence homology, leading to their recognition by the same T or B cell clones (Figure 16.5). We define *network* cross-reactivity as the immune connection existing between molecules that bear no sequence homology, like Hsp60 and Hsp70, but whose specific immune responses are somehow interconnected (by self-vaccination or another mechanism). Thus, the organisation of the immune network is such that immunisation with antigen 1 can induce an immune response that targets not only antigen 1, but also antigen 2, in the absence of any single T or B cell clone that recognises both antigens (Figure 16.5). The regulatory properties of heat shock proteins might result from the molecular cross-reactivity existing between self and microbial heat shock proteins and the network cross-reactivity that exists between different endogenous heat shock proteins.

The study of the pro- and anti-inflammatory mechanisms mediated by heat shock proteins could lead to the design of novel therapies for autoimmunity: therapies aimed at reinforcing the built-in mechanisms that are based on the physiological control of the immune function by heat shock proteins [94, 98]. The initial success of the Hsp60 peptide p277 in treating human T1DM shows the feasibility of this approach [49]. A deeper understanding of the multiple activities of heat shock proteins on the immune system and body homeostasis might allow us to extend these findings to other autoimmune disorders.

Acknowledgements

Professor Irun R. Cohen is the incumbent of the Mauerberger Chair in Immunology. We thank Ms. Danielle Sabah-Israel for excellent secretarial assistance.

REFERENCES

1. Hartl F U and Hayer-Hartl M. Molecular chaperones in the cytosol: from nascent chain to folded protein. Science 2002, 295: 1852–1858.
2. Young D B, Ivanyi J, Cox J H and Lamb J R. The 65kDa antigen of mycobacteria – a common bacterial protein? Immunol Today 1987, 8: 215–219.
3. Gaston J S. Heat shock proteins and arthritis – new readers start here. Autoimmunity 1997, 26: 33–42.
4. Dhillon V, Latchman D and Isenberg D. Heat shock proteins and systemic lupus erythematosus. Lupus 1991, 1: 3–8.
5. Lehner T. The role of heat shock protein, microbial and autoimmune agents in the aetiology of Behcet's disease. Int Rev Immunol 1997, 14: 21–32.
6. Abulafia-Lapid R, Elias D, Raz I, Keren-Zur Y, Atlan H and Cohen I R. T cell proliferative responses to type 1 diabetes patients and healthy individuals to human Hsp60 and its peptides. J Autoimmunity 1999, 12: 121–129.
7. Beg A A. Endogenous ligands of Toll-like receptors: implications for regulating inflammatory and immune responses. Trends Immunol 2002, 23: 509–512.
8. Anderton S M, van der Zee R and Goodacre J A. Inflammation activates self hsp60-specific T cells. Eur J Immunol 1993, 23: 33–38.
9. Mor F and Cohen I R. T cells in the lesion of experimental autoimmune encephalomyelitis. Enrichment for reactivities to myelin basic protein and to heat shock proteins. J Clin Invest 1992, 90: 2447–2455.
10. Birnbaum G and Kotilinek L. Heat shock or stress proteins and their role as autoantigens in multiple sclerosis. Ann NY Acad Sci 1997, 835: 157–167.
11. Birnbaum G. Stress proteins: their role in the normal central nervous system and in disease states, especially multiple sclerosis. Springer Semin Immunopathol 1995, 17: 107–118.
12. Gao Y L, Brosnan C F and Raine C S. Experimental autoimmune encephalomyelitis. Qualitative and semiquantitative differences in heat shock protein 60 expression in the central nervous system. J Immunol 1995, 154: 3548–3556.
13. Pockley A G. Heat shock proteins, heat shock protein reactivity and allograft rejection. Transplantation 2001, 71: 1503–1507.
14. Wauben M H M, Wagenaar-Hilbers J P A and van Eden W. Adjuvant arthritis. In Cohen, I. R. and Miller, A. (Eds.) *Autoimmune Disease Models*. Academic Press, Inc., New York 1994, pp 201–216.
15. Elias D. The NOD mouse: a model for autoimmune insulin-dependent diabetes. In Cohen, I. R. and Miller, A. (Eds.) *Autoimmune Disease Models*. Academic Press, Inc., New York 1994, pp 147–161.
16. 16. Feldmann M, Brennan F M and Maini R N. Rheumatoid arthritis. Cell 1996, 85: 307–310.
17. Tisch R and McDevitt H. Insulin-dependent diabetes mellitus. Cell Biol Int 1996, 85: 291–297.

18. van Eden W, Holoshitz J, Nevo Z, Frenkel A, Klajman A and Cohen I R. Arthritis induced by a T-lymphocyte clone that responds to *Mycobacterium tuberculosis* and to cartilage proteoglycans. Proc Natl Acad Sci USA 1985, 82: 5117–5120.

19. Billingham M E J, Carney S, Butler R and Colston M J. A mycobacterial 65-kd heat shock protein induces antigen-specific suppression of adjuvant arthritis, but is not itself arthritogenic. J Exp Med 1990, 171: 339–344.

20. Hogervorst E J, Schouls L, Wagenaar J P, Boog C J, Spaan W J, van Embden J D and van Eden W. Modulation of experimental autoimmunity: treatment of adjuvant arthritis by immunization with a recombinant vaccinia virus. Infect Immun 1991, 59: 2029–2035.

21. Ragno S, Colston M J, Lowrie D B, Winrow V R, Blake D R and Tascon R. Protection of rats from adjuvant arthritis by immunization with naked DNA encoding for mycobacterial heat shock protein 65. Arthritis Rheum 1997, 40: 277–283.

22. Anderton S M, van der Zee R, Prakken B, Noordzij A and van Eden W. Activation of T cells recognizing self 60-kD heat shock protein can protect against experimental arthritis. J Exp Med 1995, 181: 943–952.

23. Thompson S J, Francis J N, Siew L K, Webb G R, Jenner P J, Colston M J and Elson C J. An immunodominant epitope from mycobacterial 65-kDa heat shock protein protects against pristane-induced arthritis. J Immunol 1998, 160: 4628–4634.

24. van Eden W, van der Zee R, Paul A G A, Prakken B J, Wendling U, Anderton S M and Wauben M H M. Do heat shock proteins control the balance of T-cell regulation in inflammatory diseases? Immunol Today 1998, 19: 303–307.

25. Quintana F J, Carmi P, Mor F and Cohen I R. Inhibition of adjuvant arthritis by a DNA vaccine encoding human heat shock protein 60. J Immunol 2002, 169: 3422–3428.

26. Quintana F J, Carmi P, Mor F and Cohen I R. DNA fragments of the human 60-kDa heat shock protein (HSP60) vaccinate against adjuvant arthritis: identification of a regulatory HSP60 peptide. J Immunol 2003, 171: 3533–3541.

27. Ragno S, Winrow V R, Mascagni P, Lucietto P, Di Pierro F, Morris C J and Blake D R. A synthetic 10-kD heat shock protein (hsp10) from *Mycobacterium tuberculosis* modulates adjuvant arthritis. Clin Exp Immunol 1996, 103: 384–390.

28. Kingston A E, Hicks C A, Colston M J and Billingham M E J. A 71-kD heat shock protein (hsp) from *Mycobacterium tuberculosis* has modulatory effects on experimental rat arthritis. Clin Exp Immunol 1996, 103: 77–82.

29. Wendling U, Paul L, van der Zee R, Prakken B, Singh M and van Eden W. A conserved mycobacterial heat shock protein (Hsp) 70 sequence prevents adjuvant arthritis upon nasal administration and induces IL-10-producing T cells that cross-react with the mammalian self-hsp70 homologue. J Immunol 2000, 164: 2711–2717.

30. Tanaka S, Kimura Y, Mitani A, Yamamoto G, Nishimura H, Spallek R, Singh M, Noguchi T and Yoshikai Y. Activation of T cells recognizing an epitope of heat-shock protein 70 can protect against rat adjuvant arthritis. J Immunol 1999, 163: 5560–5565.

31. Quintana F J, Carmi P, Mor F and Cohen I R. Inhibition of adjuvant-unduced as arthritis by DNA vaccination with the 70-kd or the 90-kd human heat-shock protein: immune cross-regulation with the 60-kd heat-shock protein. Arth Rheum 2004, 50: 3712–3720.

32. Macht L M, Elson C J, Kirwan J R, Gaston J S H, Lamont A G, Thompson J M and Thompson S J. Relationship between disease severity and responses by blood mononuclear cells from patients with rheumatoid arthritis to human heat-shock protein 60. Immunology 2000, 99: 208–214.

33. van Roon J A G, van Eden W, van Roy J L A M, Lafeber F J P G and Bijlsma J W J. Stimulation of suppressive T cell responses by human but not bacterial 60-kD heat shock protein in synovial fluid of patients with rheumatoid arthritis. J Clin Invest 1997, 100: 459–463.

34. Prakken A B, van Eden W, Rijkers G T, Kuis W, Toebes E A, de Graeff-Meeder E R, van der Zee R and Zegers B J. Autoreactivity to human Hsp60 predicts disease remission in oligoarticular juvenile rheumatoid arthritis. Arthritis Rheum 1996, 39: 1826–1832.

35. Kanagawa O, Martin S M, Vaupel B A, Carrasco-Marin E and Unanue E R. Autoreactivity of T cells from nonobese diabetic mice: an I-Ag7-dependent reaction. Proc Natl Acad Sci USA 1998, 95: 1721–1724.

36. Quintana F J and Cohen I R. Autoantibody patterns in diabetes-prone NOD mice and in standard C57BL/6 mice. J Autoimmunity 2001, 17: 191–197.

37. Wegmann D R, Norbury-Glaser M and Daniel D. Insulin-specific T cells are a predominant component of islet infiltrates in pre-diabetic NOD mice. Eur J Immunol 1994, 24: 1853–1857.

38. Tisch R, Yang X D, Singer S M, Liblau R S, Fugger L and McDevitt H O. Immune response to glutamic acid decarboxylase correlates with insulitis in non-obese diabetic mice. Nature 1993, 366: 72–75.

39. Elias D, Markovits D, Reshef T, van der Zee R and Cohen I R. Induction and therapy of autoimmune diabetes in the non-obese diabetic mouse by a 65-kDa heat shock protein. Proc Natl Acad Sci USA 1990, 87: 1576–1580.

40. Elias D, Reshef T, Birk O S, van der Zee R, Walker M D and Cohen I R. Vaccination against autoimmune mouse diabetes with a T cell epitope of the human 65-kDa heat shock protein. Proc Natl Acad Sci USA 1991, 88: 3088–3091.

41. Elias D and Cohen I R. Peptide therapy for diabetes in NOD mice. Lancet 1994, 343: 704–706.

42. Elias D, Meilin A, Ablamunits V, Birk O S, Carmi P, Konen-Waisman S and Cohen I R. Hsp60 peptide therapy of NOD mouse diabetes induces a Th2 cytokine burst and downregulates autoimmunity to various β-cell antigens. Diabetes 1997, 46: 758–764.

43. Bockova J, Elias D and Cohen I R. Treatment of NOD diabetes with a novel peptide of the Hsp60 molecule induces Th2-type antibodies. J Autoimmunity 1997, 10: 323–329.

44. Yasunami R and Bach J F. Anti-suppressor effect of cyclophosphamide on the development of spontaneous diabetes in NOD mice. Eur J Immunol 1988, 18: 481–484.

45. Ablamunits V, Quintana F, Reshef T, Elias D and Cohen I R. Acceleration of autoimmune diabetes by cyclophosphamide is associated with an enhanced IFN-γ secretion pathway. J Autoimmunity 1999, 13: 383–392.

46. Quintana F J, Carmi P and Cohen I R. DNA vaccination with heat shock protein 60 inhibits cyclophosphamide-accelerated diabetes. J Immunol 2002, 169: 6030–6035.

47. Abulafia-Lapid R, Gillis D, Yosef O, Atlan H and Cohen I R. T cells and autoantibodies to human HSP70 in Type 1 diabetes in children. J Autoimmunity 2003: 313–321.

48. Horváth L, Cervenak L, Oroszlán M, Proháska Z, Uray K, Hudecz F, Baranyi É, Madácsy L, Singh M, Romics L, Füst G and Pánczél P. Antibodies against different epitopes of heat shock protein 60 in children with type 1 diabetes mellitus. Immunol Lett 2002, 80: 155–162.

49. Raz I, Elias D, Avron A, Tamir M, Metzger M and Cohen I R. Beta-cell function in new-onset type 1 diabetes and immunomodulation with a heat-shock protein peptide (DiaPep277): a randomised, double-blind, phase II trial. Lancet 2001, 358: 1749–1753.

50. Hemmi H, Takeuchi O, Kawai T, Kaisho T, Sato S, Sanjo H, Matsumoto M, Hoshino K, Wagner H, Takeda K and Akira S. A Toll-like receptor recognizes bacterial DNA. Nature 2000, 408: 740–745.

51. Krieg A M. CpG motifs in bacterial DNA and their immune effects. Ann Rev Immunol 2002, 20: 709–760.

52. Quintana F J, Rotem A, Carmi P and Cohen I R. Vaccination with empty plasmid DNA or CpG oligonucleotide inhibits diabetes in nonobese diabetic mice: modulation of spontaneous 60-kDa heat shock protein autoimmunity. J Immunol 2000, 165: 6148–6155.

53. Akdis C A and Blaser K. Mechanisms of interleukin-10-mediated immune suppression. Immunology 2001, 103: 131–136.

54. Kumaraguru U, Pack C D and Rouse B T. Toll-like receptor ligand links innate and adaptive immune responses by the production of heat-shock proteins. J Leuk Biol 2003, 73: 574–583.

55. Boccaccio G L, Mor F and Steinman L. Non-coding plasmid DNA induces IFN-γ *in vivo* and suppresses autoimmune encephalomyelitis. Int Immunol 1999, 11: 289–296.

56. Rachmilewitz D, Karmeli F, Takabayashi K, Hayashi T, Leider-Trejo L, Lee J, Leoni L M and Raz E. Immunostimulatory DNA ameliorates experimental and spontaneous murine colitis. Gastroenterology 2002, 122: 1428–1441.

57. Zeuner R A, Ishii K J, Lizak M J, Gursel I, Yamada H, Klinman D M and Verthelyi D. Reduction of CpG-induced arthritis by suppressive oligodeoxynucleotides. Arthritis Rheum 2002, 46: 2219–2224.

58. Zeuner R A, Verthelyi D, Gursel M, Ishii K J and Klinman D M. Influence of stimulatory and suppressive DNA motifs on host susceptibility to inflammatory arthritis. Arthritis Rheum 2003, 48: 1701–1707.

59. Serreze D V, Hamaguchi K and Leiter E H. Immunostimulation circumvents diabetes in NOD/Lt mice. J Autoimmunity 1989, 2: 759–776.

60. Tian J, Zekzer D, Hanssen L, Lu Y, Olcott A and Kaufman D L. Lipopolysaccharide-activated B cells down-regulate Th1 immunity and prevent autoimmune diabetes in nonobese diabetic mice. J Immunol 2001, 167: 1081–1089.

61. Sai P and Rivereau A S. Prevention of diabetes in the nonobese diabetic mouse by oral immunological treatments. Comparative efficiency of human insulin and two bacterial antigens, lipopolysaccharide from *Escherichia coli* and glycoprotein extract from *Klebsiella pneumoniae*. Diabetes Metab 1996, 22: 341–348.

62. Iguchi M, Inagawa H, Nishizawa T, Okutomi T, Morikawa A, Soma G I and Mizuno D. Homeostasis as regulated by activated macrophage. V. Suppression of diabetes mellitus in non-obese diabetic mice by LPSw (a lipopolysaccharide from wheat flour). Chem Pharm Bull (Tokyo) 1992, 40: 1004–1006.

63. Li Z, Ménoret A and Srivastava P. Roles of heat-shock proteins in antigen presentation and cross-presentation. Cur Opin Immunol 2002, 14: 45–51.

64. Srivastava P. Roles of heat-shock proteins in innate and adaptive immunity. Nat Rev Immunol 2002, 2: 185–194.

65. Vabulas R M, Braedel S, Hilf N, Singh-Jasuja H, Herter S, Ahmad-Nejad P, Kirschning C J, Da Costa C, Rammensee H G, Wagner H and Schild H. The endoplasmic reticulum-resident heat shock protein Gp96 activates dendritic cells via the Toll-like receptor 2/4 pathway. J Biol Chem 2002, 277: 20847–20853.

66. Bethke K, Staib F, Distler M, Schmitt U, Jonuleit H, Enk A H, Galle P R and Heike M. Different efficiency of heat shock proteins to activate human monocytes and dendritic cells: Superiority of HSP60. J Immunol 2002, 169: 6141–6148.

67. Flohé S B, Bruggemann J, Lendemans S, Nikulina M, Meierhoff G, Flohé S and Kolb H. Human heat shock protein 60 induces maturation of dendritic cells versus a Th1-promoting phenotype. J Immunol 2003, 170: 2340–2348.

68. Bulut Y, Faure E, Thomas L, Karahashi H, Michelsen K S, Equils O, Morrison S G, Morrison R P and Arditi M. Chlamydial heat shock protein 60 activates macrophages and endothelial cells through Toll-like receptor 4 and MD2 in a MyD88-dependent pathway. J Immunol 2002, 168: 1435–1440.

69. Moudgil K D, Chang T T, Eradat H, Chen A M, Gupta R S, Brahn E and Sercarz E E. Diversification of T cell responses to carboxy-terminal determinants within the 65-kD heat-shock protein is involved in regulation of autoimmune arthritis. J Exp Med 1997, 185: 1307–1316.

70. Hirono S, Dibrov E, Hurtado C, Kostenuk A, Ducas R and Pierce G N. Chlamydia pneumoniae stimulates proliferation of vascular smooth muscle cells through induction of endogenous heat shock protein 60. Circ Res 2003, 93: 710–716.

71. Beimnet K, Soderstrom K, Jindal S, Gronberg A, Frommel D and Kiessling R R. Induction of heat shock protein 60 expression in human monocytic cell lines infected with Mycobacterium leprae. Infect Immun 1996, 64: 4356–4358.

72. Wainberg Z, Oliveira M, Lerner S, Tao Y and Brenner B G. Modulation of stress protein (hsp27 and hsp70) expression in CD4+ lymphocytic cells following acute infection with human immunodeficiency virus type-1. Virology. 1997, 233: 364–373.

73. Saito K, Katsuragi H, Mikami M, Kato C, Miyamaru M and Nagaso K. Increase of heat-shock protein and induction of γ/δ T cells in peritoneal exudate of mice after injection of live Fusobacterium nucleatum. Immunology 1997, 90: 229–235.

74. Basu S, Binder R J, Suto R, Anderson K M and Srivastava P K. Necrotic but not apoptotic cell death releases heat shock proteins, which deliver a partial maturation signal to dendritic cells and activates the NF-κB pathway. Int Immunol 2000, 12: 1539–1546.

75. Boog C J P, de Graeff-Meeder E R, Lucassen M A, van der Zee R, Voorhorst Ogink M M, van Kooten P J S, Geuze H J and van Eden W. Two monoclonal antibodies

generated against human hsp60 show reactivity with synovial membranes of patients with juvenile arthritis. J Exp Med 1992, 175: 1805–1810.

76. Teshima S, Rokutan K, Takahashi M, Nikawa T and Kishi K. Induction of heat shock proteins and their possible roles in macrophages during activation by macrophage colony-stimulating factor. Biochem J 1996, 315: 497–504.

77. Ferris D K, HarelBellan A, Morimoto R I, Welch W J and Farrar W L. Mitogen and lymphokine stimulation of heat shock proteins in T lymphocytes. Proc Natl Acad Sci USA 1988, 85: 3850–3854.

78. Zugel U and Kaufmann S H. Immune response against heat shock proteins in infectious diseases. Immunobiology 1999, 201: 22–35.

79. Chen Y, Kuchroo V K, Inobe J, Hafler D A and Weiner H L. Regulatory T cell clones induced by oral tolerance: suppression of autoimmune encephalomyelitis. Science 1994, 265: 1237–1240.

80. Moudgil K D, Kim E, Yun O J, Chi H H, Brahn E and Sercarz E E. Environmental modulation of autoimmune arthritis involves the spontaneous microbial induction of T cell responses to regulatory determinants within heat shock protein 65. J Immunol 2001, 166: 4237–4243.

81. Prakken B J, Wendling U, van der Zee R, Rutten V P M, Kuis W and van Eden W. Induction of IL-10 and inhibition of experimental arthritis are specific features of microbial heat shock proteins that are absent for other evolutionarily conserved immunodominant proteins. J Immunol 2001, 167: 4147–4153.

82. Blachere N E, Li Z L, Chandawarkar R Y, Suto R, Jaikaria N S, Basu S, Udono H and Srivastava P K. Heat shock protein-peptide complexes, reconstituted in vitro, elicit peptide-specific cytotoxic T lymphocyte response and tumor immunity. J Exp Med 1997, 186: 1315–1322.

83. Chandawarkar R Y, Wagh M S and Srivastava P K. The dual nature of specific immunological activity of tumour-derived gp96 preparations. J Exp Med 1999, 189: 1437–1442.

84. Galazka G, Walczak A, Berkowicz T and Selmaj K. Effect of Hsp70-peptide complexes generated in vivo on modulation EAE. Adv Exp Med Biol 2001, 495: 227–230.

85. Lohse A W, Mor F, Karin N and Cohen I R. Control of experimental autoimmune encephalomyelitis by T cells responding to activated T cells. Science 1989, 244: 820–822.

86. Mimran A, Mor F, Carmi P, Quintana F J, Rotter V and Cohen I R. DNA vaccination with CD25 protects rats from adjuvant arthritis and induces an antiergotypic response. J Clin Invest 2004, 113: 924–932.

87. Mor F, Reizis B, Cohen I R and Steinman L. IL-2 and TNF receptors as targets of regulatory T-T interactions: isolation and characterization of cytokine receptor-reactive T cell lines in the Lewis rat. J Immunol. 1996, 157: 4855–4861.

88. van Eden W, Thole J E R, van der Zee R, Noordzij A, van Embden J D A, Hensen E J and Cohen I R. Cloning of the mycobacterial epitope recognized by T lymphocytes in adjuvant arthritis. Nature 1988, 331: 171–173.

89. Vabulas R M, Ahmad-Nejad P, da Costa C, Miethke T, Kirschning C J, Hacker H and Wagner H. Endocytosed HSP60s use Toll-like receptor 2 (TLR2) and TLR4 to activate the Toll/interleukin-1 receptor signaling pathway in innate immune cells. J Biol Chem 2001, 276: 31332–31339.

90. Becker T, Hartl F U and Wieland F. CD40, an extracellular receptor for binding and uptake of Hsp70-peptide complexes. J Cell Biol 2002, 158: 1277–1285.

91. Basu S, Binder R J, Ramalingam T and Srivastava P K. CD91 is a common receptor for heat shock proteins gp96, hsp90, hsp70 and calreticulin. Immunity 2001, 14: 303–313.

92. Caramalho I, Lopes-Carvalho T, Ostler D, Zelenay S, Haury M and Demengeot J. Regulatory T cells selectively express Toll-like receptors and are activated by lipopolysaccharide. J Exp Med 2003, 197: 403–411.

93. Zanin-Zhorov A, Nussbaum G, Franitza S, Cohen I R and Lider O. T cells respond to heat shock protein 60 via TLR2: activation of adhesion and inhibition of chemokine receptors. FASEB J 2003, 17: 1567–1569.

94. Cohen I R, Quintana F J, Nussbaum G, Cohen M, Zanin A and Lider O. HSP60 and the regulation of inflammation: physiological and pathological. In van Eden, W. (Ed.) Heat Shock Proteins and Inflammation. Birkhauser Verlag A G, Basel 2004, pp 1–13.

95. Werner S and Grose R. Regulation of wound healing by growth factors and cytokines. Physiol Rev 2003, 83: 835–870.

96. Cohen I R and Schwartz M. Autoimmune maintenance and neuroprotection of the central nervous system. J Neuroimmunol 1999, 100: 111–114.

97. Quintana F J and Cohen I R. Type I diabetes mellitus, infection and Toll-like receptors. In Shoenfeld, Y. and Rose, N. (Eds.) Infection and Autoimmunity. Elsevier, Amsterdam 2004.

98. Cohen I R. Tending Adam's Garden: Evolving the Cognitive Immune Self. Academic Press, London: 2000.

17

Heat Shock Protein Fusions: A Platform for the Induction of Antigen-Specific Immunity

Lee Mizzen and John Neefe

17.1. Introduction

The unusual immunogenicity of heat shock proteins (also known as stress proteins) was discovered in studies of the immune response to microbial infection, in which a large proportion of the humoral and cellular immune response to diverse microbial pathogens was found to be specific for pathogen-derived heat shock protein [1]. These studies demonstrated that immune recognition of pathogen-derived heat shock proteins occurs in natural and experimental settings in animals and man. This is discussed in detail in Chapter 16. Immune responses elicited to mycobacterial heat shock proteins have been particularly well studied. In man, recognition of mycobacterial heat shock protein by CD4$^+$ T cells occurs in the context of numerous human leukocyte antigen (HLA) alleles, and epitopes have been identified that are presented by multiple HLA molecules [2]. The promiscuous recognition of mycobacterial heat shock proteins supports their utility as 'universal' immunogens for the genetically diverse human population. The immunogenic properties of microbial heat shock proteins have accordingly led to their application in a variety of immunisation formats as prophylactic and therapeutic agents in models of infectious disease and cancer [3]. In these studies, heat shock proteins have been delivered as subunit vaccines, carrier proteins in chemical conjugates, recombinant fusion proteins and DNA expression vectors for induction of humoral and cellular immunity.

To explain the disproportionate focus of the immune response on a small subset of pathogen antigens, heat shock proteins were proposed to act as 'red flags' – alerting the immune system to the presence of a foreign invader [4]. Given their ubiquity in microbial pathogens and their over-production by microbes in response to 'stressful' conditions experienced within the infected host [5], pathogen-derived heat shock proteins represent highly effective targets for sensing infection. However, evidence is accumulating that the immune system may also respond to endogenous (i.e., self) heat shock proteins during various pathophysiological states, such as inflammation or necrotic cell death. Indeed, in this context, heat shock proteins are postulated to function as universal 'danger signals', alerting the immune system to the presence of stressed, infected or diseased tissue [6].

Given the central role of professional antigen-presenting cells (APCs) in initiating immune responses, the induction of antigen-specific immunity by heat shock proteins may be explained, in part, by the identification of candidate heat shock protein receptors on dendritic cells (DCs) and macrophages. On murine and/or human monocytes, receptors for Hsp90, Hsp70 and Hsp60 homologues include CD14, members of the Toll-like receptor (TLR) family, TLR4 and TLR2, scavenger receptors (SRs) CD91 and LOX-1, and the co-stimulatory molecule CD40 [7–14]. In addition, CD94 is implicated as a receptor for Hsp70 on human NK cells [15]. The receptors for chaperones are discussed in detail in Chapters 7, 8 and 10. A common function of TLRs and SRs is the binding of exogenous or endogenous 'pattern recognition' ligands, respectively, which triggers innate immune responses [16, 17]. Similarly, stimulation of natural killer (NK) cells via CD94 can activate their innate tumour-killing function, as observed with Hsp70 [18]. Finally, co-stimulation of DCs through CD40 is a critical activation signal for initiating adaptive T cell responses [19]. Hence, receptor molecules implicated in heat shock protein recognition share the property of transmitting activation signals from the innate to the adaptive immune system.

Specific heat shock protein recognition and uptake by APCs is also supported by *in vitro* studies demonstrating activation and maturation of DCs exposed to exogenous heat shock proteins [20, 21] as well as the 'cross-priming' of CD8$^+$ T cell responses to antigens associated with heat shock proteins [22, 23]. At present, direct biochemical confirmation of receptor binding by heat shock proteins is lacking; the more prevalent view is that identified molecules are part of signalling pathways in APCs that may involve as yet unidentified co-receptors.

17.2. Heat shock protein fusions: a new approach to immunisation

Given the ability of heat shock proteins to target APCs and to enhance immune responses to associated antigens, a practical approach for the design and manufacture of new antigen-specific vaccine formulations is the construction of heat shock protein fusions by recombinant DNA technology. Here, DNA sequences for heat shock proteins and target antigens are spliced together in-frame ('fused') using standard techniques, and the resultant chimaeric genes or encoded proteins are employed as immunogens. As reviewed next, immunisation with heat shock protein fusions in protein, RNA, DNA or viral vector formats has yielded promising results in animal models of infectious disease and cancer. Significantly, this approach has progressed to the first human testing of a heat shock protein fusion protein as a viral immunotherapeutic.

17.2.1. Heat shock protein fusion proteins

In the first published example of this concept, immunisation with an HIV p24-*Mycobacterium tuberculosis* Hsp70 fusion protein was shown to induce p24-specific immune responses in mice, as characterised by the production of Th1 and Th2 cytokines and the induction of a significant IgG titre that persisted for over a year [24]. This work alerted researchers to the potential of heat shock protein fusions for the induction of immune responses and highlighted some important features that would be confirmed in subsequent research: lack of dependence on adjuvants, induction of antigen-specific B and T cell responses and the significant enhancement of antigen-specific immune responses by covalent linkage (fusion) of heat shock proteins to the antigen.

The ability of heat shock protein fusion proteins to enhance antibody responses was extended by studies employing *Leishmania infantum* Hsp70 or Hsp83 fusions with *Escherichia coli* maltose-binding protein (MBP) [25–27]. A consistent finding in these studies was the induction of a Th1 immune response to MBP, as measured by cytokine secretion and IgG isotype profile. Notably, immunisation of athymic *nu/nu* mice with the fusion protein generated anti-MBP IgG$_{2a}$ antibodies, which is consistent with a T cell–dependent response [26]. As noted in another study, the induction of IgG antibodies by heat shock proteins can possess unusual features [28].

Although heat shock protein fusions can elicit markedly enhanced antigen-specific humoral responses, the DC targeting property of heat shock proteins has been exploited by a majority of researchers seeking to augment cellular immunity to fused antigens. In the first example of cross-priming by a heat shock protein fusion protein, immunisation of mice with an ovalbumin–*M. tuberculosis* Hsp70 fusion induced ovalbumin-specific CD8$^+$ cytotoxic T lymphocytes (CTLs) [29]. These CTLs recognised the H-2^b-restricted ovalbumin SIINFEKL peptide (amino acids (aa) 257–264), indicating that heat shock protein fusion immunisation led to 'natural' antigen processing *in vivo*. Mice immunised with the fusion were protected against challenge with an ovalbumin-expressing tumour cell line, implicating ovalbumin-specific CTLs in tumour rejection.

In another study concerning *in vivo* antigen processing, *Trypanosoma cruzi* Hsp70 was chosen as a vehicle for the discovery of potential human CTL epitopes in a target antigen from the same parasite. In this study, immunisation of HLA-A2/K^b transgenic mice with a kinetoplastid membrane protein-11

(KMP11)–Hsp70 fusion permitted the identification of two HLA-A2–restricted CTL epitopes in KMP11 [30].

In the first example of CTL induction by a fusion protein containing an Hsp60 homologue, peptide-specific CD8$^+$ CTLs were elicited in mice of H-2^b and H-2^d haplotypes immunised with *Mycobacterium bovis* BCG Hsp65-influenza nucleo-protein fusions [31]. In this study, significant CTL activity appeared following a single immunisation. This persisted for a minimum of four months and was boosted by a second immunisation, thereby suggesting the presence of memory T cells. This study also demonstrated that, as observed for Hsp70 fusions, Hsp65 fusion immunisation elicits CTLs specific for defined immunodominant epitopes present within fused antigens.

With these features in mind, a tumour 'vaccine' using *M. bovis* BCG Hsp65 was created and tested in a mouse model of human cervical cancer [32]. Here, prophylactic and therapeutic immunisation with an Hsp65–human papillomavirus (HPV)-type 16 E7 fusion (HspE7) led to the rejection of an HPV16 E7-expressing tumour, TC-1. HspE7 immunisation was associated with induction of a Th1-like cell-mediated immune response, based on the cytokine secretion profile, and the presence of cytolytic activity against TC-1 cells. Tumour rejection following therapeutic immunisation with HspE7 was dependent on the presence of CD8$^+$ T cells during priming and was associated with long-term survival (over 253 days) in 80 percent of treated animals.

A number of studies have provided insight into the mechanisms by which heat shock protein fusion proteins elicit CTLs. A particularly revealing observation is that CD8$^+$ CTLs are induced by mycobacterial Hsp70 and Hsp65 fusions in CD4$^+$ T cell-deficient mice [32–36]. Induction of CTLs without CD4$^+$ T cell 'help' can occur by direct activation of APCs, as shown by CD40 ligation or acute viral infection [37, 38]. Consistent with mechanisms by which APCs 'licence' CD8$^+$ T cell responses, mycobacterial Hsp65 fusion proteins have been shown to directly activate murine bone-marrow-derived DCs *in vitro* whereas, *in vivo*, myeloid DCs recovered from fusion immunised mice displayed an activated phenotype [33]. To obtain more quantitative measurements on the potency of CTL induction by heat shock protein fusions, an ovalbumin–*M. tuberculosis* Hsp70 fusion protein was used to immunise OT-1 mice transgenic for a T cell receptor recognising the ovalbumin SIINFEKL peptide [36]. By comparison with animals receiving the SIINFEKL peptide in complete Freund's adjuvant (CFA), immunisation of mice with the fusion protein induced stronger and more durable proliferation of adoptively transferred OT-1 cells *in vivo*. On a molar basis, delivery of the SIINFEKL peptide within an ovalbumin protein fragment fused to Hsp70 was several hundred-fold more effective in eliciting CTL responses than peptide in CFA. These observations illustrate, in this transgenic model, the quantitative and qualitative superiority of a heat shock protein fusion over the gold-standard adjuvant CFA as a vehicle for CTL induction.

To map the region of Hsp70 involved in CTL induction to fused antigen, mice were immunised with heat shock protein fusions composed of an ovalbumin fragment joined to different portions of the *M. tuberculosis* Hsp70 protein [34]. This study revealed that the ability to elicit SIINFEKL-specific CTLs resided within a region of Hsp70 located in the ATPase domain (aa 161–370). It is noteworthy that the amino acid sequences in Hsp70 associated with induction of antibody or CTL responses to fused antigens have been mapped to the conserved ATPase domain [25, 34, 35]. The region of human Hsp70 responsible for CD40 binding has also been mapped to the ATPase domain [13], in contrast to findings with mycobacterial Hsp70 [12]. However, this discrepancy might be related to the conformation of Hsp70 induced by nucleotide and peptide binding [13]. The structure–function relationship of Hsp70 with respect to binding to CD40 is described in detail in Chapter 10.

The high degree of sequence and functional conservation among heat shock proteins in evolution and predictions from the 'danger hypothesis' [6] suggest that the immunogenic properties of heat shock proteins will extend from microbes to higher eukaryotes, including mammals. In the context of a heat shock protein fusion, evidence for this has been provided in studies in mice that have demonstrated CTL induction to antigens fused to murine (i.e., self-) Hsp70 proteins [34, 35]. In one study, CTLs have been elicited to five different major histocompatibility complex (MHC)-restricted peptide sequences fused individually to murine Hsc70 as minimal epitopes without flanking sequences [35].

17.2.2. Heat shock protein gene fusions

As noted earlier, current research indicates that the induction of antigen-specific CTLs by heat shock protein fusion proteins derives largely from their ability to act upon myeloid APCs in at least two ways: firstly, to efficiently deliver fused antigen sequences into the MHC class I pathway for presentation to CD8+ T cells, and secondly, to deliver activation signals that upregulate co-stimulatory functions and the secretion of cytokines that promote cellular immunity. At present, the intracellular pathways traversed by heat shock protein fusion proteins in APCs are largely unknown. Because endogenous expression of proteins by DNA vaccines can enhance the induction of CTLs [39], it is instructive to compare immune responses induced by heat shock protein fusions encountered by APCs as exogenous proteins versus endogenously expressed proteins.

To enhance cellular immune responses to HPV16 E7, prototype vaccines have been tested in mice based on gene fusions between HPV16 E7 sequences and *M. tuberculosis* Hsp70, delivered in plasmid DNA [40, 41], adeno-associated virus

[42], Sindbis virus RNA replicon [43, 44] and vaccinia virus vectors [41]. In all of these studies, E7-specific CD8$^+$ T cell responses were induced, and when tested these were superior to those that were induced by the E7 gene alone or other vector controls. Moreover, these studies demonstrated the prophylactic or therapeutic activity of the various E7–Hsp70 fusion constructs against TC-1 tumour cells, including, in one instance, a TC-1 variant with reduced class I expression, a phenotype common to many human cancers [41]. These studies contained another important observation: that immunisation with E7–Hsp70 fusions in plasmid DNA or RNA replicon formats induces CD8$^+$ T cells/CTLs and/or anti-tumour responses that are wholly or predominantly independent of CD4$^+$ T cells [40, 41, 43]. In addition to CTL effectors, a subset of these studies also demonstrated the essential contribution of NK cells and IFN-γ in tumour rejection following the administration of RNA or DNA vectors expressing E7–Hsp70 [41, 43].

In a direct extension of previous work using a corresponding heat shock protein fusion protein, immunisation of mice with plasmid DNA encoding a *T. cruzi* KMP11-Hsp70 fusion elicited production of anti-KMP11 IgG$_{2a}$ antibodies and CD8$^+$ CTL recognising predicted HLA-A2 epitopes in KMP11 and conferred prophylactic immunity against challenge with *T. cruzi* [45]. Extending the utility of a DNA vaccine approach to a mammalian heat shock protein fusion partner, a rabbit calreticulin–E7 plasmid DNA vaccine has been shown to induce high levels of E7-specific antibodies and CD8$^+$ T cells in mice and provides prophylactic and therapeutic immunity against TC-1 tumour cells [46]. Notably, CTL induction with the CRT–E7 DNA vaccine does not require CD4$^+$ T cells during the priming phase of immunisation. Therefore, one feature of CTL induction shared by many heat shock protein fusions delivered as protein or nucleic acid–based immunogens is the relative lack of dependence on CD4$^+$ T cells. The induction of both CTL and antibody responses by heat shock protein fusions, as surrogates for delivery of antigen into the MHC class I and class II pathways, respectively, presumably reflects the multiple pathways of heat shock protein trafficking within APCs [47, 48].

17.3. Clinical testing of a heat shock protein fusion protein, HspE7

Based on immunotherapeutic activity in a pre-clinical model of HPV-associated cancer, an *M. bovis* BCG Hsp65–HPV16 E7 fusion protein, HspE7 [32], has progressed into clinical testing in humans. HspE7 (formal designation SGN-00101) is being developed by Stressgen Biotechnologies Inc. (San Diego, CA, U.S.A.) for the treatment of diseases caused by HPV, including genital warts (GW), recurrent respiratory (or laryngeal) papillomatosis (RRP or warts of the respiratory tract), the cancer precursors known as cervical and anal intraepithelial neoplasia

(CIN and AIN respectively) and cervical cancer. HPV is ubiquitous in humans and is estimated to infect over 70 percent of the sexually active population [49]. Based on nucleotide sequence information, there are over one hundred genotypes of HPV [50]. Infection with a subset of 'high-risk' oncogenic HPV types, such as type 16, is associated with CIN, AIN and cancer, whereas infection with 'low-risk' HPV types 6 and 11 is commonly associated with GW and RRP [51].

In the clinical trials, HspE7 is being administered to patients by subcutaneous injection in a buffered saline vehicle. HspE7 treatment is well tolerated, with the most common adverse experience being an injection site reaction, typical of vaccines, which resolves without treatment. An active treatment regimen for HspE7 has been identified: 500 µg given three times at monthly intervals. To date, the trial results in the HPV indications suggest that therapeutic immunisation with HspE7 is active in AIN, GW [52] and RRP. Clinical activity, measured as reduction or complete resolution of indicated HPV lesions, is observed in a majority of patients receiving HspE7 immunisation and, when followed, is durable in a majority of patients for follow-up periods ranging from several months to two years. For patients with high-grade AIN and RRP, the response to HspE7 therapy means a potential reduction or avoidance of invasive surgical procedures. For GW patients, the durability of response to HspE7 therapy is potentially superior to that typically observed with use of caustic/ablative techniques or topical immunodulators.

The activity of HspE7, which contains HPV16 E7, against GW and RRP, which are caused primarily by HPV types 6 and 11, suggests induction of cross-reactive immunity to HPV E7 sequences present in non-HPV16 genotypes. This evidence supports the use of HspE7 as a broad-spectrum immunotherapeutic for diseases caused by multiple HPV genotypes. Other notable clinical findings include the activity of HspE7 in the HPV-infected adult (GW, AIN) and paediatric (RRP) populations, and activity in treating HPV-associated lesions of the anogenital skin and mucosa, and mucosa of the upper respiratory tract. Future clinical trials plan to test HspE7 in HIV$^+$ patients with anogenital HPV disease. The rationale for testing HspE7 in this population is directly supported by the observed activity of HspE7 in CD4-deficient mice [32].

17.4. Conclusions

Heat shock protein fusions, delivered as exogenous protein immunogens or as endogenously expressing nucleic acid and viral-based vectors, engender potent humoral and cellular immune responses. The literature demonstrates that immune responses can be induced with heat shock protein fusions containing antigen sequences that are 1) of varying lengths and character, 2) contained within natural flanking residues or as minimal epitopes and 3) fused to the N- or C-termini of heat shock protein sequences. In addition, antigen-specific immune responses have been demonstrated with heat shock protein fusion partners derived from 1) different heat shock protein families (e.g., Hsp90, Hsp70, Hsp60) and 2) prokaryotic and eukaryotic species, including the same species as the immunised recipient (i.e., 'self'). These observations speak to the breadth and reproducibility of this approach as a robust immunisation platform.

The induction of immune responses and, in particular, CD8$^+$ CTLs by heat shock protein fusions is an important phenomenon from both theoretical and practical standpoints. Firstly, it demonstrates efficient CTL priming to antigens by mechanisms apparently independent of the 'natural' chaperone function hypothesised to underlie CTL induction by non-covalent heat shock protein complexes [53]. Secondly, CTL responses, which are paramount for eradication of intracellular pathogens and transformed cells, are routinely achieved by heat shock protein fusions without adjuvant. This represents a new avenue of inquiry for developers of vaccines and immunotherapies, to whom decades of experience have indicated that CTL induction to soluble protein antigens requires adjuvants that pose safety and toxicity risks [54, 55]. Based on the pre-clinical and clinical experience described in this review, heat shock protein fusions offer considerable promise as a platform for future development of therapeutic vaccines to treat chronic viral infections and cancer.

Acknowledgements

We thank Lori Kernaghan for expert assistance in the preparation of this manuscript.

REFERENCES

1. Suzue K and Young R A. Heat shock proteins as immunological carriers and vaccines. In Fiege, U. (Ed.) *Stress-Inducible Cellular Responses*. Birkhauser Verlag, Basel 1996, pp 451–465.
2. Mustafa A S. HLA-restricted immune response to mycobacterial antigens: relevance to vaccine design. Human Immunol 2000, 61: 166–171.
3. Mizzen L. Immune responses to stress proteins: applications to infectious disease and cancer. Biotherapy 1998, 10: 173–189.
4. Murray P and Young R A. Stress and immunological recognition in host-pathogen interactions. J Bacteriol 1992, 174: 4193–4196.
5. Garbe T R. Heat shock proteins and infection: interactions of pathogen and host. Experientia 1992, 48: 635–639.
6. Gallucci S and Matzinger P. Danger signals: SOS to the immune system. Cur Opin Immunol 2001, 13: 114–119.

7. Asea A, Kraeft S-K, Kurt-Jones E A, Stevenson M A, Chen L B, Finberg R W, Koo G C and Calderwood S K. Hsp70 stimulates cytokine production through a CD14-dependent pathway, demonstrating its dual role as a chaperone and cytokine. Nat Med 2000, 6: 435–442.

8. Kol A, Lichtman A H, Finberg R W, Libby P and Kurt-Jones E A. Heat shock protein (HSP) 60 activates the innate immune response: CD14 is an essential receptor for HSP60 activation of mononuclear cells. J Immunol 2000, 164: 13–17.

9. Ohashi K, Burkart V, Flohé S and Kolb H. Heat shock protein 60 is a putative endogenous ligand of the Toll-like receptor-4 complex. J Immunol 2000, 164: 558–561.

10. Basu S, Binder R J, Ramalingam T and Srivastava P K. CD91 is a common receptor for heat shock proteins gp96, hsp90, hsp70 and calreticulin. Immunity 2001, 14: 303–313.

11. Vabulas R M, Ahmad-Nejad P, da Costa C, Miethke T, Kirschning C J, Hacker H and Wagner H. Endocytosed HSP60s use Toll-like receptor 2 (TLR2) and TLR4 to activate the toll/interleukin-1 receptor signaling pathway in innate immune cells. J Biol Chem 2001, 276: 31332–31339.

12. Wang Y, Kelly C G, Karttunen T, Whittall T, Lehner P J, Duncan L, MacAry P, Younson J S, Singh M, Oehlmann W, Cheng G, Bergmeier L and Lehner T. CD40 is a cellular receptor mediating mycobacterial heat shock protein 70 stimulation of CC-chemokines. Immunity 2001, 15: 971–983.

13. Becker T, Hartl F U and Wieland F. CD40, an extracellular receptor for binding and uptake of Hsp70-peptide complexes. J Cell Biol 2002, 158: 1277–1285.

14. Delneste Y, Magistrelli G, Gauchat J, Haeuw J, Aubry J, Nakamura K, Kawakami-Honda N, Goetsch L, Sawamura T, Bonnefoy J and Jeannin P. Involvement of LOX-1 in dendritic cell-mediated antigen cross-presentation. Immunity 2002, 17: 353–362.

15. Gross C, Hansch D, Gastpar R and Multhoff G. Interaction of heat shock protein 70 peptide with NK cells involves the NK receptor CD94. Biol Chem 2003, 384: 267–279.

16. Krieger M. The other side of scavenger receptors: pattern recognition for host defense. Cur Opin Lipidol 1997, 8: 275–280.

17. Janeway C A J and Medzhitov R. Innate immune recognition. Ann Rev Immunol 2002, 20: 197–216.

18. Multhoff G, Mizzen L, Winchester C C, Milner C M, Wenk S, Eissner G, Kampinga H H, Laumbacher B and Johnson J. Heat shock protein 70 (Hsp70) stimulates proliferation and cytolytic activity of natural killer cells. Exp Hematol 1999, 27: 1627–1636.

19. Yang Y and Wilson J M. CD40 ligand-dependent T cell activation: requirement of B7-CD28 signalling through CD40. Science 1996, 273: 1862–1864.

20. Kuppner M C, Gastpar R, Gelwer S, Nossner E, Ochmann O, Scharner A and Issels R D. The role of heat shock protein (hsp70) in dendritic cell maturation: hsp70 induces the maturation of immature dendritic cells but reduces DC differentiation from monocyte precursors. Eur J Immunol 2001, 31: 1602–1609.

21. Bethke K, Staib F, Distler M, Schmitt U, Jonuleit H, Enk A H, Galle P R and Heike M. Different efficiency of heat shock proteins to activate human monocytes and dendritic cells: Superiority of HSP60. J Immunol 2002, 169: 6141–6148.

22. Arnold D, Faath S, Rammensee H-G and Schild H. Cross-priming of minor histocompatibility antigen-specific cytotoxic T cells upon immunization with the heat shock protein gp96. J Exp Med 1995, 182: 885–889.

23. Suto R and Srivastava P K. A mechanism for the specific immunogenicity of heat shock protein-chaperoned peptides. Science 1995, 269: 1585–1588.

24. Suzue K and Young R A. Adjuvant-free hsp70 fusion protein system elicits humoral and cellular immune responses to HIV-1 p24. J Immunol 1996, 156: 873–879.

25. Rico A I, Angel S O, Alonso C and Requena J M. Immunostimulatory properties of the *Leishmania infantum* heat shock proteins hsp70 and hsp83. Mol Immunol 1999, 36: 1131–1139.

26. Rico A I, Del Real G, Soto M, Quijada L, Martinez-A C, Alonso C and Requena J M. Characterization of the immunostimulatory properties of *Leishmania infantum* Hsp70 by fusion to the *Escherichia coli* maltose-binding protein in normal and nu/nu BALB/c mice. Infect Immun 1998, 66: 347–352.

27. Echeverria P, Dran G, Pereda G, Rico A I, Requena J M, Alonso C, Guarnera E and Angel S O. Analysis of the adjuvant effect of recombinant *Leishmania infantum* Hsp83 protein as a tool for vaccination. Immunol Lett 2001, 76: 107–110.

28. Bonorino C, Nardi N B, Zhang X and Wysocki L J. Characteristics of the strong antibody response to mycobacterial hsp70: a primary, T cell-dependent IgG response with no evidence of natural priming or γδ T cell involvement. J Immunol 1998, 161: 5210–5216.

29. Suzue K, Zhou X, Eisen H N and Young R A. Heat shock fusion proteins as vehicles for antigen delivery into the major histocompatibility complex class I presentation pathway. Proc Nat Acad Sci USA 1997, 94: 13146–13151.

30. Maranon C, Thomas M C, Planelles L and Lopez M C. The immunization of A2/K^b transgenic mice with the kmp11-hsp70 fusion protein induces CTL response against human cells expressing the T. cruzi kmp11 antigen: identification of A2-restricted epitopes. Mol Immunol 2001, 38: 279–287.

31. Anthony L S D, Wu H, Sweet H, Turnnir C, Boux L and Mizzen L A. Priming of CD8$^+$ CTL effector cells in mice by immunization with a stress protein – influenza virus nucleoprotein fusion molecule. Vaccine 1999, 17: 373–383.

32. Chu N R, Wu H B, Boux L J, Siegel M I and Mizzen L A. Immunotherapy of a human papillomavirus (HPV) type 16 E7-expressing tumour by administration of fusion protein comprising *Mycobacterium bovis* bacille Calmette-Guerin (BCG) hsp65 and HPV16 E7. Clin Exp Immunol 2000, 121: 216–225.

33. Cho B K, Palliser D, Guillen E, Wisniewski J, Young R A, Chen J and Eisen H N. A proposed mechanism for the induction of cytotoxic T lymphocyte production by heat shock fusion proteins. Immunity 2000, 12: 263–272.

34. Huang Q, Richmond J F, Suzue K, Eisen H N and Young R A. *In vivo* cytotoxic T lymphocyte elicitation by mycobacterial heat shock protein 70 fusion proteins maps to a discrete domain and is CD4$^+$ T cell independent. J Exp Med 2000, 191: 403–408.

35. Udono H, Yamano T, Kawabata Y, Ueda M and Yui K. Generation of cytotoxic T lymphocytes by MHC class I ligands fused to heat shock cognate protein 70. Int Immunol 2001, 13: 1233–1242.

36. Harmala L A E, Ingulli E G, Curtsinger J M, Lucido M M, Schmidt C S, Weigel B J, Blazar B R, Mescher M F and Pennell C A. The adjuvant effects of *Mycobacterium*

tuberculosis heat shock protein 70 result from the rapid and prolonged activation of antigen-specific CD8$^+$ T cells *in vivo*. J Immunol 2002, 169: 5622–5629.

37. Rahemtulla A, Fung-Leung W P, Schillham M W, Kundig T M, Sambhara S R, Narendran A, Arabian A, Wakeham A, Paige C J, Zinkernagel R M, Miller R G and Mak T W. Normal development and function of CD8$^+$ cells but markedly decreased helper cell activity in mice lacking CD4. Nature 1991, 353: 180–184.

38. Ridge J P, Di Rosa F and Matzinger P. A conditioned dendritic cell can be a temporal bridge between a CD4$^+$ T-helper and a T-killer cell. Nature 1998, 393: 474–478.

39. Gurunathan S, Klinman D M and Seder R A. DNA vaccines: immunology, application, and optimization. Ann Rev Immunol 2000, 18: 927–974.

40. Chen C-H, Wang T-L, Hung C-F, Yang Y, Young R A, Pardoll D M and Wu T-C. Enhancement of DNA vaccine potency by linkage of antigen gene to an hsp70 gene. Cancer Res 2000, 60: 1035–1042.

41. Cheng W F, Hung C F, Lin K Y, Juang J, He L, Lin C T and Wu T-C. CD8$^+$ T cells, NK cells and IFN-γ are important for control of tumor with downregulated MHC class I expression by DNA vaccination. Gene Therapy 2003, 10: 1311–1320.

42. Liu D-W, Tsao Y-P, Kung J T, Ding Y-A, Sytwu H-K, Xiao X and Chen S-L. Recombinant adeno-associated virus expressing human papillomavirus type 16 E7 peptide DNA fused with heat shock protein DNA as a potential vaccine for cervical cancer. J Virol 2000, 74: 2888–2894.

43. Cheng W-F, Hung C-F, Chai C-Y, Hsu K-F, He L, Rice C M, Ling M and Wu T-C. Enhancement of Sindbis virus self-replicating RNA vaccine potency by linkage of *Mycobacterium tuberculosis* heat shock protein 70 gene to an antigen gene. J Immunol 2001, 166: 6218–6226.

44. Hsu K-F, Hung C-F, Cheng W-F, He L, Slater L A, Ling M and Wu T-C. Enhancement of suicidal DNA vaccine potency by linking *Mycobacterium tuberculosis* heat shock protein 70 to an antigen. Gene Therapy 2001, 8: 376–383.

45. Plannelles L, Thomas M C, Alonso C and Lopez M C. DNA immunization with *Trypanosoma cruzi* Hsp70 fused to the kmp11 protein elicits a cytotoxic and humoral immune response against the antigen and leads to protection. Infect Immun 2001, 69: 6558–6563.

46. Cheng W F, Hung C F, Chai C Y, Hsu K F, He L, Ling M and Wu T C. Tumor-specific immunity and antiangiogenesis generated by a DNA vaccine encoding calreticulin linked to a tumor antigen. J Clin Invest 2001, 108: 669–678.

47. Wassenberg J J, Dezfulian C and Nicchitta C V. Receptor mediated and fluid phase pathways for internalization of the ER hsp90 chaperone grp94 in murine macrophages. J Cell Sci 1999, 112: 2167–2175.

48. Castellino F, Boucher P E, Eichelberg K, Mayhew M, Rothman J E, Houghton A N and Germain R N. Receptor-mediated uptake of antigen/heat shock protein complexes results in major histocompatibility complex class I antigen presentation via two distinct pathways. J Exp Med 2000, 191: 1957–1964.

49. Koutsky L. Epidemiology of genital human papillomavirus infection. Am J Med 1997, 102 (suppl 5A): 3–8.

50. Gissmann L, Osen W, Muller M and Jochmus I. Therapeutic vaccines for human papillomaviruses. Intervirology 2001, 44: 167–175.

51. Richart R M, Masood S, Syrjanen K J, Vassilakos P, Kaufman R H, Meisels A, Olszewski W T, Sakamoto A, Stoler M H, Vooijs G P and Wilbur D C. Human papillomavirus. IAC task force summary. Acta Cytologica 1998, 42: 50–58.

52. Goldstone S E, Palefsky J M, Winnett M T and Neefe J R. Activity of HspE7, a novel immunotherapy, in patients with anogenital warts. Dis Colon Rectum 2002, 45: 502–507.

53. Srivastava P K, Udono H, Blachere N E and Li Z. Heat shock proteins transfer peptides during antigen processing and CTL priming. Immunogenetics 1994, 39: 93–98.

54. Raychaudhuri S and Morrow W J W. Can soluble antigens induce CD8$^+$ cytotoxic T-cell responses? A paradox revisited. Immunol Today 1993, 14: 344–348.

55. Gupta R K and Siber G R. Adjuvants for human vaccines – current status, problems and future prospects. Vaccine 1995, 13: 1263–1276.

18

Molecular Chaperones as Inducers of Tumour Immunity

Pinaki P. Banerjee and Zihai Li

18.1. Introduction

Tumour antigens can be broadly classified into four categories: (i) those that are expressed in larger quantities in tumours than their normal counterparts (e.g., tumour-associated carbohydrate antigens) [1], (ii) onco-fetal antigens (e.g., carcinoembryonic antigen) [2], (iii) differentiation antigens (e.g., melanoma differentiation antigen) [3, 4] and (iv) tumour-specific antigens. Tumour antigens in the first three categories could serve as useful markers for diagnostic and prognostic purposes. Although some of these antigens are being used in immunotherapy, none can be called tumour-specific in a true sense. Only the last group includes antigens that are truly specific for tumour cells, in that they contain tumour-specific mutations that are unique for individual tumours such as the tumour-specific point mutation that is found in cyclin-dependent kinase-4. Such a mutation gives rise to a novel antigenic epitope which can be recognised by cytotoxic T lymphocytes (CTLs) [5]. However, for these antigens to be of any value as therapeutic agents, they must be detected in and epitopes isolated from a large range of cancers, and this makes the general use of these antigens difficult.

In the past two decades, evidence has accumulated to support the concept that molecular chaperones or heat shock proteins can be used as a potent source of cancer vaccines [6, 7]. Molecular chaperones, particularly those derived from the Hsp70 and Hsp90 families, are now being tested in the clinical arena for therapeutic efficacy against a range of cancers (Table 18.1). The concept of vaccinating against cancer using molecular chaperones developed from an observation which was made half a century ago, namely that mice immunised with a particular type of irradiated tumour cells were protected against challenge with live tumour cells from the same, but not a different kind, of tumour [8] (Figure 18.1). It was evident that, although tumour A and tumour B were induced by

Table 18.1. Gp96 in clinical trials for different types of cancer

Heat shock protein vaccine	Type of cancer	Status
gp96	Renal cell carcinoma	Phase III
	Melanoma	Phase III
	Colorectal cancer	Phase II
	Gastric cancer	Phase I and II
	Pancreatic cancer	Phase II
	Sarcoma	Phase II
	Ovarian cancer	Phase I

the same carcinogen, were derived from the same histological type or were even present in the same host, the antigenic determinants of each of these tumours appeared to be distinct (see review by Li et al. [9]).

It was postulated that each tumour bears unique tumour-specific transplantation antigens (TSTAs) that could be used to induce protective responses against a particular type of tumour, but not others [10]. A pioneering experiment in which the immunogenic TSTA fraction from tumour lysates was purified led to the identification of gp96 and other heat shock proteins as the protective agents against autologous tumours [11]. However, gp96 is not immunogenic *per se*; rather the immunogenicity of the preparation is defined by novel tumour-derived peptides that are carried by the gp96 and are co-purified with it [12, 13]. Immunisation with gp96 elicits a potent CTL response against the tumour from which gp96 is isolated, the specificity of which is defined by the gp96-associated peptides [14–17]. It has since been shown that gp96 devoid of peptides can activate CD8$^+$ T cells in a dose-dependent manner [18] and can provide co-stimulation to CD4$^+$ T cells which results in the generation of a type 2 helper T cell (Th2) immune response [19]. This chapter will review features and mechanisms underlying the capacity of autologous tumour-derived gp96 and other heat shock proteins to elicit tumour-specific immunity and act as anti-cancer vaccines.

18.2. Application of extracellular heat shock protein gp96 in cancer immunity

18.2.1. Autologous, soluble gp96 as a cancer vaccine

Gp96 was first discovered as the glucose-regulated 'stress' protein [20], the primary function of which was to serve the role of a molecular chaperone [21].

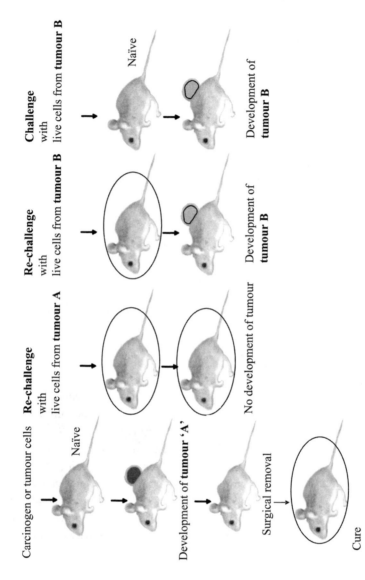

Figure 18.1. When cured surgically, mice bearing tumour A (encircled) became resistant to the challenge from the same tumour cells, but much like the naïve animals remained susceptible to tumour B. This experiment symbolises the existence of tumour antigens and also the generation of immune response against these antigens.

Because this protein is found to be regulated by glucose levels it is, therefore, also known as the 94-kDa glucose-regulated protein (grp94) [22]. This protein is a major glycoprotein resident in the endoplasmic reticulum (ER) and is thus named endoplasmin [23] or 99-kDa endoplasmic reticulum protein (Erp99) [24]. However, under stressful conditions, gp96 tends to redistribute to the Golgi apparatus [25], is found to be enriched to some extent in the nucleus [20] and can also be expressed on the outer surface of the plasma membrane [26]. It has been suggested that the glycosylation pattern of gp96 changes after cellular stress, as denoted by an increased resistance to endoglycosidase H digestion, which depends on the cell types [25]. However, this phenomenon has also been observed in several disease states such as cancer [27], thereby suggesting that cells might be in a state of stress which also results in partial translocation of ER chaperones to the Golgi apparatus.

The protective immunity elicited by gp96 vaccination is exquisitely specific. However, questions have been raised about the polymorphism of the molecule and also there has been doubt about the existence of the mutated form of gp96 in cancer cells, either of which could be a tumour antigen. Molecular cloning and sequencing approaches have rejected these two hypotheses and categorically mapped only one true gene locus, named tra-1 in humans [28]. Sequencing of gp96 from normal and tumour tissues has demonstrated absolute sequence homology between gp96 from the two sources, thereby confirming gp96 as being solely a carrier of immunogenic peptide from the tumour tissues.

Much of the recent studies on the immunologic properties of gp96 have focussed on understanding the interaction between purified gp96 and a variety of immune effector cells including T cells and professional antigen-presenting cells (APCs). APCs such as dendritic cells (DCs) and macrophages carry the receptors of gp96 on the cell surface, one of which is CD91, or α-2 macroglobulin (α-2M) receptor [29]. Binding of tumour-derived gp96 to CD91 leads to a receptor-mediated endocytosis of gp96 complete with its associated tumour-derived peptides (gp96-tumour peptide), and this forms the basis of the initial critical step leading to the induction of tumour peptide-specific immunity [30]. The receptor–ligand interaction activates, matures and leads to the cross-presentation of tumour peptides via major histocompatibility complex (MHC) class I molecules by the DCs. Once activated, DCs from the peripheral tissues migrate to the lymph node in which CD8$^+$ T cell–DC interactions result in the priming and activation of peptide-specific T cells (Figure 18.2). The induction of tumour immunity by gp96 could be a representation of 'danger theory' itself [31]. To clarify this proposal, stressful conditions such as glucose deprivation, hypoxia and acidosis might induce the release of gp96-peptide complexes from the tumour cells, which acts as a danger signal to the immune system. This

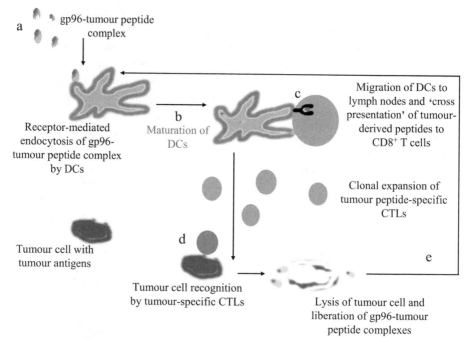

Figure 18.2. Generation of CTL responses by gp96 vaccination. (a) Gp96–tumour peptide complexes can (b) mature DCs and these DCs (c) generate a potent CTL response by cross-presentation of the peptide. Effector CTLs (d) migrate to the site of the tumour to elicit tumour-specific cytolysis. The gp96–tumour peptide from lysed cancer cells (e) can further interact with DCs to prime the anti-tumour response.

extracellular gp96 might activate macrophages and DCs to initiate a cascade of downstream events that result in the induction of an effective anti-tumour immune response.

Intradermal injection of as little as 1 µg tumour-derived gp96 can protect against tumour challenge. The 'mathematical derivation' behind this phenomenon of gp96 is indeed very interesting to note [32]. In brief, such an immunisation incites 10,000 activated DCs to migrate to the lymph node and results in the presentation of an estimated 3,000 MHC tumour-specific peptide complexes to CD8+ cells. As commented by Srivastava, 'one tenth as many DCs presenting one tenth as many antigenic peptides are powerful enough to elicit a potent T cell response' [32]. Other routes of gp96 immunisation such as subcutaneous and intraperitoneal approaches are also effective, however only when 10- to 50-fold higher doses of gp96 are used, respectively [33]. It has been interesting to note that, irrespective of the route of administration, doses of soluble gp96 above or below the optimal do not elicit tumour-protective responses. Higher doses of gp96 can lead to the generation of transferable suppressive T cell subsets that bear CD4$^+$ molecules on the cell surface. A detailed insight into the mechanisms underlying this particular phenomenon has yet to be offered. However, these findings highlight the need for more insight into the immunologic properties of gp96.

18.2.2. Surface expression of gp96 and its role in CTL response

Gp96 contains the C-terminal KDEL (Lys-Asp-Glu-Leu) motif, which is known as the ER retention signal [34]. However, it has been estimated that about 3% of the total amount of gp96 synthesised could be expressed on the surface of the Meth-A tumour cells [35]. Cell surface expression of gp96 has also been observed in a variety of other cell types (see the review by Li et al. [9]). In each case, surface expression was not due to cell death, because surface expression was clearly dependent on the active transport from ER to Golgi. It has been hypothesised that the export of gp96 from the ER to the cell surface is a critical signal to the immune system. To test this concept, we have targeted gp96 to the surfaces of tumour cells by transfection with an engineered gp96 expression vector [36]. The engineered gp96, named 96tm, does not possess the KDEL motif and to the C-terminus of gp96 was attached a transmembrane domain of a platelet-derived growth factor receptor.

Surface expression of 96tm does not lead to changes in the level of endogenous gp96, nor does it have any effect on the folding or transport of MHC class I molecules. 96tm was targeted onto cell surfaces of many tumour types in a type I orientation. A number of important observations were made with these reagents [36, 37]: (i) the immunogenicity of Meth-A fibrosarcomas and CT-26 colon cancers expressing cell surface gp96 is increased; (ii) co-culturing of bone marrow (BM)-derived immature DCs with tumour cells expressing gp96 on their cell surface induces phenotypic maturation of DCs as evidenced by upregulation of CD80, CD86, CD40, MHC class I and MHC class II expression; (iii) DCs activated by cell surface gp96 produce a large amount of pro-inflammatory cytokines such as TNF-α, IL-1β, IL-12 and the chemokine MCP-1; (iv) DC maturation induced by 96tm-expressing tumour cells requires direct cell-to-cell contact; (v) 96tm-expressing tumour cells recruit both CD4$^+$ and CD8$^+$ cells to the site of tumours *in vivo*; (vi) tumour cells expressing cell surface gp96 can cross-prime antigen-specific CD8$^+$ T cells more efficiently and (vii) immunisation with tumour cell expressing cell surface gp96 leads to the development of long-lasting protective CD4$^+$ and CD8$^+$ T cell memory. Thus, over-expression of surface gp96 can also be an effective and alternative strategy for inducing effective cell-mediated anti-tumour immunity against less immunogenic tumours.

Using a similar strategy, the immunological consequence of constitutive expression of gp96 on cell surfaces was examined by generating a gp96tm transgenic mouse model [38]. Interestingly, these animals spontaneously develop a DC-mediated autoimmune disease, which highlights the intrinsic pro-inflammatory properties of gp96. This transgenic murine model might prove invaluable for further mechanistic study of gp96 in immune responses.

18.3. A comparison of the functional mechanism between soluble gp96 and gp96tm

In general, a protective CTL response can be sequentially described by three consecutive events: initiation, maintenance and the induction of immunological memory. Soluble gp96 and membrane-bound gp96 might elicit CTL responses in a subtle, but different mechanism [9, 39]. Important functional differences in immune responses induced by gp96-tm and naturally occurring cell surface gp96 might also exist, and these might be revealed following whole-cell immunisation. Invariably, the chemistry underlying different types of immune responses relies on the nature of the interaction between cells and the gp96 molecule and its associated peptides/proteins in the vaccine preparations. Thus, it would be pertinent to address and also to compare the involvement of various APCs and T cells in the different phases of the anti-tumour immune response which is induced by different forms of gp96-based vaccines.

In the priming phase, the requirement of CD8$^+$ T cells, but not the CD4$^+$ T cells, is important, especially when animals are immunised with the soluble gp96-peptide complex [39]. This phenomenon is found to be largely dependent on the presence of carrageenan-susceptible cells, which could be macrophages and other, as yet to be identified APCs. In contrast, the priming of CTLs following immunisation with irradiated cancer cell requires the presence of CD4$^+$ T cells. During the effector phase, both T cell subsets are important for effective immunisation in response to tumour-derived gp96 or irradiated tumour cells. Much like the priming phase, macrophages play a pivotal role in the CTL effector response induced by purified gp96.

An interesting observation is that although immunisation with soluble protein generally initiates the CD4$^+$ T cell response, the depletion of CD4$^+$ T cells during gp96 administration does not affect the generation of tumour immunity in the host. This might be due to the fact that gp96-tumour peptide is targeted directly to macrophages and DCs [30], which cross-present the 'antigen' to CD8$^+$ cells. Thus, depletion of macrophages from the immune milieu compromises the antigenic response. Both soluble gp96 [40] and membrane-bound 96tm [37] have been shown to elicit memory anti-tumour immune responses.

In the case of 96tm immunisation, both CD4$^+$ and CD8$^+$ T cells are required for tumour rejection in the priming as well as memory phase. The requirement for CD4$^+$ T cells in the generation of CD8$^+$ T cell memory to soluble gp96 immunisation, *in vivo*, has yet to be carefully studied. Suffice to say, both the membrane-bound and the soluble gp96 can elicit a potent anti-tumour immunity. However, the mechanisms under which immune responses are generated are differentially regulated and the mechanistic differences have yet to be elucidated.

18.4. Immunotherapy with other forms of gp96

In addition to the soluble and membrane-bound forms of gp96, gp96-Ig, the secreted form of gp96 from tumours, has also attracted considerable interest for its potency in cancer therapy [41, 42]. A gp96-Ig fusion protein has been made by replacing the KDEL sequence with CH2 and CH3 domains of the murine IgG$_1$ molecule [43]. This construct was transfected to the ovalbumin-expressing E.G7 lymphoma cells, and this tumour model was used as a source of vaccine. Gp96-Ig secreting E.G7 cells, but not gp96-Ig negative control cells, were able to prime adoptively transferred T cell receptor transgenic CD8$^+$ T cells specific for the MHC class I–restricted SIINFEKL peptide sequence of ovalbumin. To determine whether the N terminus or the C terminus are crucial for gp96-mediated peptide presentation, the C terminus of gp96 has been fused with a CTL epitope, and this had a limited effects on the capacity to prime CTL responses [44]. Covalent fusion of gp96 with peptides might inhibit the presentation of these peptides via MHC class I molecules, because heat shock protein-tumour peptides require enzymatic processing before being presented via proteosome-dependent or -independent pathways [45, 46]. Some believe that the N-terminal domain of gp96 does not have peptide binding capability, but that it might induce cross-protection in various tumours [47], whereas others believe that the N-terminal domain of gp96 can indeed carry tumour peptide [13].

Gp96-Ig can efficiently elicit anti-tumour responses in the absence of CD4$^+$ T cells and macrophages [43], and gp96-Ig and gp96-tm are more effective than soluble gp96 when mice are challenged with 5×10^5 live Meth A tumour cells, a dose which is five-fold greater than the maximum number of cells up to which immunity mediated by soluble gp96 can be retained. Gp96-tm has been shown to be slightly more effective than gp96-Ig in a tumour rejection experiment [37].

To increase the efficiency of gp96 in tumour rejection, it was thought that it was necessary to enhance the macrophage/DC population *in vivo* at the time of gp96 vaccine administration. Thus, the administration of Lewis lung cancer (LLC) cells transduced with granulocyte macrophage colony stimulating factor (GM-CSF) in combination with 1 μg tumour-derived gp96 was found to be more effective than either one administered separately [48]. This strategy of combination vaccination can mature, activate and increase the number of the DCs in the draining lymph node within 36 hours after vaccination. Further, this LLC-GM-gp96 therapy also depends on the optimum amount of gp96. DC-mediated activation of CD8$^+$ T lymphocytes is also CD4$^+$ T cell– and natural killer (NK) cell–dependent. Markers of DC maturation (MHC class II, B7.2) are up-regulated by gp96 with no significant change in CD40 expression level. However, the expression of gp96 receptor (CD91) is regarded as being constitutively expressed and unrelated to the presence or absence of gp96 immunisation [48].

18.5. Role of CD91 and gp96 in innate and adaptive immunity

From the preceding discussion, the question might be asked how gp96-mediated anti-tumour immunity is regulated, especially given that its receptor is constitutively expressed on DCs. Attempts to elucidate this fact revealed that cell lysates prepared from necrotic cells, but not apoptotic cells, can deliver a maturation signal to DCs via the NF-κB pathway, and that it is through this route that gp96 delivers it signals to the DCs [49]. Apoptotic cells are known to bear apoptotic markers such as phosphatidylserine (PS) on their cell surface and for which DCs bear the receptor (PSR). Engagement of PS with PSR down-regulates the expression of co-stimulatory molecules by DCs and concomitantly increases the secretion of anti-inflammatory cytokine transforming growth factor (TGF)-β [50]. Other receptors such as CD14 and CD36 recognise apoptotic cell-associated ligands and also transmit similar tolerogenic signals to macrophages and DCs. However, during necrosis the binding of gp96 to its receptor on DCs elicits the secretion of IL-12 and TNF-α, and also up-regulates the expression of co-stimulatory molecules. In the peripheral circulation, any potential effects of heat shock protein–peptide complexes can be neutralised by the presence of α-2M, another ligand for CD91. However, in the tissues, in which α-2M is not present, CD91 becomes accessible to gp96 released from damaged or stressed cells [51].

It is important to point out that CD91 is not the only receptor which is proposed to bind to heat shock proteins. A CD91-independent pathway has been reported to mediate cross-presentation of gp96-chaperoned peptides [52]. Toll-like receptors (TLRs) [53] and scavenger receptor-A [54] are such candidate receptors for gp96. The receptors for heat shock proteins are discussed in detail in Chapters 7, 8 and 10. Furthermore, the observation that gp96 devoid of peptides can activate DCs strongly indicates the co-existence of innate immunity in tumour rejection [55]. Binding of gp96 to macrophages and DCs can elicit the secretion of IL-12, TNF-α, type-1 interferon, IL-1β and GM-CSF. IL-12 activates NK cells which lyse tumour cells directly. On tumour cell lysis, gp96 and its associated peptides are released, and this binds to CD91 on neighbouring APCs, thereby generating an adaptive immune response. Thus, as is the case with other heat shock proteins, gp96 bridges innate and adaptive immune responses, and this capacity probably developed as early as the emergence of vertebrates [7, 41, 56–58]. To exemplify, autologous tumour-derived gp96- and Hsp70-based immunisation induces significant protection against a transplantable tumour in *Xenopus*, the earliest vertebrate to possess an adaptive immune system [59].

18.6. The role of other heat shock protein family members in anti-cancer immunity and a comparison with gp96

Soon after the establishment of gp96 as a tumour 'antigen', attention focussed on the possibility that other molecular chaperones exhibited similar properties. Two of the Hsp90 family members, named 84- and 86-kDa TSTA (i.e., p84 and p86), are also immunogenic for cancer [60]. Unlike gp96, these molecules are devoid of sugar moieties and do not bind to lectins; however, much like gp96 these molecules do not bear tumour-specific polymorphic DNA sequences and are found to protect only against autologous tumours. The overall homology between p84/86 and gp96 is 49%. Interestingly, at the lowest effective dose the immunogenicity of these molecules is dependent on the purification process. Thus, fractions purified using Mono-Q columns offer comparatively better protection against a lethal challenge with Meth A cells than those purified using hydroxyapatite. However, in contrast to gp96, both the purified fractions of p84/86 offered a similar, almost 80%, protection when used in higher dose (i.e., 10–40 µg). Much like p84/86, the immunogenicity of Hsp70 also depends on its preparation protocol. Thus, the replacement of ADP-agarose from the ATP-agarose in the chromatography column completely restores the ability of the Hsp70(-peptide) to elicit anti-tumour immunity [61, 62]. ATP binding and hydrolysis have been shown to release Hsp70-associated peptides [63] and the loss of immunogenicity after ATP treatment provides strong evidence that the peptides chaperoned by Hsp70 are the immunogens [62] as, for example, Hsp70 chaperoned tyrosinase peptide [64].

18.7. Mechanisms underlying Hsp70-mediated tumour regression

To determine which portion of the Hsp70 is essential for the generation of the most potent CTL response against the peptide linked to it, five different class I restricted CTL epitopes have been covalently linked to either the N or C terminus of Hsc70 (the constitutive member of the Hsp70 family) and these have been immunised to mice via different routes [44]. In contrast to gp96-tumour peptide immunisation, intravenous rather than intradermal injection of Hsp70-tumour peptide is the most effective route of immunisation. When compared to a dose of 1 μg, the intravenous administration of 10 μg chaperone–peptide complex generates a much more vigorous CTL response, as determined by the number of IFN-γ–producing cells. The N-terminal region and the C-terminal region flanking peptides are equally effective as far as intravenous immunisation is concerned. However, in an overall 'score' in the CTL assay, and depending on different routes of immunisation, the C-terminal-bound flanking peptides are more efficient than those bound at the N-terminal end.

Studies attempting to elucidate the mechanism resulting in Hsp70-tumour peptide-mediated CTL responses have shown that, much like gp96-peptide, carragenan-susceptible cells but not CD4$^+$ T cells are essential for the cytolytic activity and have demonstrated that Hsp70 binds to newly processed peptides within as little as 1 hour at 25 °C [62]. An elegant study to identify the CTL epitopes of HLA-B46 positive cancer patients resulted in the discovery of two oligopeptides, nine amino acids long, derived from a truncated form of Hsc70 (81% homologous). The sequence of these epitopes varied to a small extent from the wild-type form and was mapped to a position outside of the peptide binding domain [65]. Further, the recognition of surface Hsp72 (a family member of Hsp70) by a distinct population of NK cells has also been documented in the literature [66–69]. Taken together, these findings confirm the probable role of Hsp70 in eliciting the innate and adaptive immunity in tumour regression.

18.8. The recipient of TAP-transported peptides as another group of chaperone vaccines

From the data discussed so far, it is evident that any of the three heat shock proteins (gp96, Hsp90 or Hsp70) could be used as a source of cancer vaccine against autologous tumours. However, to choose the best, a comparison between these three heat shock proteins has been undertaken using administration by the subcutaneous route [70]. This demonstrated that Hsp70 and gp96 are of equal efficiency, whereas Hsp90 is only 10% as effective as its counterparts. Thus, whereas 9 μg gp96 or Hsp70–peptide complex protects mice from tumour challenge, the dose of Hsp90 required for similar protection is 90 μg. Interestingly all three of these heat shock proteins bind equal amounts of peptide. The molecular and cellular bases for the differential immunogenicity of these proteins are unclear.

It is proposed that Hsp70 and Hsp90 act as peptide transporters, in that they carry their peptide cargo to the transporter associated with antigen processing

(TAP), whereas gp96 and calreticulin facilitate the binding of the MHC class I-β2 microglobulin peptide complex in the ER. This reasoning suggests that, much like gp96, other recipients of the TAP-transported peptides such as calreticulin (CRT) and the 170-kDa glucose regulated protein grp170 might also act as tumour vaccines. Grp170 has been tested as a vaccine in a metastatic tumour model, in which it was found to be effective in terms of reducing the number of metastatic colonies and growth of further tumour challenge [71].

CRT has also been found to bind tumour peptides and upon subcutaneous immunisation CRT and gp96 are comparable in their protective efficiency against a tumour challenge [72]. As CRT-tumour peptides can be cross-presented by DCs to generate a CTL response, it is therefore suggested to be the key mechanism behind CRT-induced anti-tumour immunity. Interestingly, DNA vaccination of CRT, or CRT-E7 (CRT DNA linked to HPV-16 DNA) equally reduces pulmonary tumour nodules due to an inhibition of bFGF-induced angiogenesis. This 'T cell–independent tumour regression' has been further proven in a nude mice model in which a slight advantage of CRT-E7 DNA vaccination over the CRT vaccinated mice has been observed [73].

CRT, Hsp70, Hsp90 and gp96 bind to the same receptor on DCs, CD91. CRT-tumour peptides exhibit the greatest capacity to induce IFN-γ secretion from antigen-specific CTLs in $vitro$, and this is almost twice that of gp96 [74]. Taken together, these data clarify the involvement of innate and adaptive immunity in CRT-mediated tumour rejection. The capacity of CRT in the cross-presentation experiment is comparable to Hsp70, whereas the efficiency of Hsp90 lies in between that of gp96 and Hsp70 [70]. Further, a comparative analysis of heat shock protein binding to CD11b$^+$ cells has also shown that, even at the very high concentration (200 µg/ml), Hsp70 binding to CD11b$^+$ cells is not saturated, whereas Hsp90 and gp96 were saturated at this concentration [75]. This observation suggests either a high affinity of Hsp70 for CD91 or the existence of another receptor for Hsp70 on CD11b$^+$ cells. Indeed, a new receptor for Hsp70 (LOX-1) from the group of the scavenger receptor family has now been documented in the literature [76]. The existence of other receptors for Hsp70 argue against the observations that an anti-CD91 antibody can completely abrogate the re-presentation of Hsp70–peptide complex [74] and that Toll-like receptor (TLR-2 and TLR-4) pathways are involved in gp96-mediated DC activation which is preceded by endocytosis [30, 53, 77].

The in $vivo$ data provide insight into the relative merits of different heat shock proteins as anti-tumour vaccines and it would appear that gp96 has a number of advantages given the low doses required to induce CTL responses and tumour rejection. Indeed, gp96 is currently being evaluated in a number of clinical trials (Table 18.1). Hsp70 is also being evaluated in a phase II clinical trial. Studies into the therapeutic potential of other heat shock proteins such as Hsp60, Hsp40 and Hsp27 have also been initiated; however, the intense focus on gp96 has to date overshadowed their contribution [78].

18.9. Conclusion

It is evident that gp96 is not a tumour antigen, nor 'probably' are any of the heat shock proteins; rather the nature of the immune response induced by them is dictated by the peptides with which they are associated. In general, heat shock protein–tumour peptide complexes elicit CTL responses against a wide variety of cancers [15, 79, 80]. However, the mechanisms of action of each heat shock protein and their various forms (membrane bound, soluble or secreted) have yet to be elucidated in any great detail. For example, the mechanistic basis for the distinct CD4$^+$ T cell requirement by soluble and membrane-bound gp96 is unclear. Does each heat shock protein engage with same or different sets of receptors? Do heat shock proteins differ in their abilities to activate innate and adaptive immunity? Despite these uncertainties, these chaperone–peptide complexes exhibit unique advantages over other vaccines as follows:

- They represent the entire repertoire of peptides generated in a tumour cell, thereby abolishing the need to identify and isolate immunogenic tumour epitopes.
- Their use is not restricted to a particular type of cancer, or to patients with a particular MHC haplotype.
- The risks that are associated with the use of transforming DNA, attenuated organisms or immunosuppressive factors such as TGF-β are eliminated.

Finally, with little or no treatment-related toxicity, tumour-derived heat shock proteins likely fulfil all of the immunological criteria to justify their clinical evaluation as cancer vaccines.

REFERENCES

1. Livingston P. Ganglioside vaccines with emphasis on GM2. Semin Oncol 1998, 25: 636–645.
2. Greiner J W, Zeytin H, Anver M R and Schlom J. Vaccine-based therapy directed against carcinoembryonic antigen demonstrates antitumor activity on spontaneous intestinal tumors in the absence of autoimmunity. Cancer Res 2002, 62: 6944–6951.
3. Boon T, Cerottini J C, van den Eynde B, van der Bruggen P and van Pel A. Tumor antigens recognized by T lymphocytes. Ann Rev Immunol 1994, 12: 337–365.

4. Denkberg G, Lev A, Eisenbach L, Benhar I and Reiter Y. Selective targeting of melanoma and APCs using a recombinant antibody with TCR-like specificity directed toward a melanoma differentiation antigen. J Immunol 2003, 171: 2197–2207.
5. Wolfel T, Hauer M, Schneider J, Serrano M, Wolfel C, Klehmann-Hieb E, De Plaen E, Hankeln T, Meyer zum Buschenfelde K H and Beach D. A p16INK4a-insensitive CDK4 mutant targeted by cytolytic T lymphocytes in a human melanoma. Science 1995, 269: 1281–1284.
6. Srivastava P K. Immunotherapy of human cancer: lessons from mice. Nat Immunol 2000, 1: 363–366.
7. Srivastava P. Roles of heat-shock proteins in innate and adaptive immunity. Nat Rev Immunol 2002, 2: 185–194.
8. Foley E J. Antigenic properties of methylcholanthrene-induced tumors in mice of the strain of origin. Cancer Res 1953, 13: 835–837.
9. Li Z, Dai J, Zheng H, Liu B and Caudill M. An integrated view of the roles and mechanisms of heat shock protein gp96-peptide complex in immune response. Frontiers Biosci 2002, 7: 731–751.
10. Prehn R T and Main J M. Immunity to methylcholanthrene-induced sarcomas. J Nat Cancer Inst 1957, 18: 769–778.
11. Srivastava P K, DeLeo A B and Old L J. Tumor rejection antigens of chemically induced sarcomas of inbred mice. Proc Natl Acad Sci USA 1986, 83: 3407–3411.
12. Li Z and Srivastava P K. Tumor rejection antigen gp96/grp94 is an ATPse: implications for protein folding and antigen presentation. EMBO J 1993, 12: 3143–3151.
13. Linderoth N A, Popowicz A and Sastry S. Identification of the peptide-binding site in the heat shock chaperone/tumor rejection antigen gp96 (Grp94). J Biol Chem 2000, 275: 5472–5477.
14. Arnold D, Faath S, Rammensee H-G and Schild H. Cross-priming of minor histocompatibility antigen-specific cytotoxic T cells upon immunization with the heat shock protein gp96. J Exp Med 1995, 182: 885–889.
15. Blachere N E, Li Z L, Chandawarkar R Y, Suto R, Jaikaria N S, Basu S, Udono H and Srivastava P K. Heat shock protein-peptide complexes, reconstituted in vitro, elicit peptide-specific cytotoxic T lymphocyte response and tumor immunity. J Exp Med 1997, 186: 1315–1322.
16. Blachere N E and Srivastava P K. Heat shock protein-based cancer vaccines and related thoughts on immunogenicity of human tumors. Semin Cancer Biol 1995, 6: 349–355.
17. Nieland T J F, Tan M C A A, Monee-van Muijen M, Koning F, Kruisbeek A M and van Bleek G M. Isolation of an immunodominant viral peptide that is endogenously bound to the stress protein GP96/GRP94. Proc Natl Acad Sci USA 1996, 93: 6135–6139.
18. Breloer M, Fleischer B and von Bonin A. In vivo and in vitro activation of T cells after administration of Ag-negative heat shock proteins. J Immunol 1999, 162: 3141–3147.
19. Banerjee P P, Vinay D S, Mathew A, Raje M, Parekh V, Prasad D V, Kumar A, Mitra D and Mishra G C. Evidence that glycoprotein 96 (B2), a stress protein, functions as a Th2-specific costimulatory molecule. J Immunol 2002, 169: 3507–3518.
20. Welch W J, Garrels J I, Thomas G P, Lin J J and Feramisco J R. Biochemical characterization of the mammalian stress proteins and identification of two stress proteins as glucose- and Ca^{2+}-ionophore-regulated proteins. J Biol Chem 1983, 258: 7102–7111.

21. Csermely P, Schnaider T, C. S, Prohászka Z and Nardai G. The 90-kDa molecular chaperone family: structure, function, and clinical applications. A comprehensive review. Pharmacol Ther 1998, 79: 129–168.

22. Lee A S. The accumulation of three specific proteins related to glucose-regulated proteins in a temperature-sensitive hamster mutant cell line K12. J Cell Physiol 1981, 106: 119–125.

23. Koch G, Smith M, Macer D, Webster P and Mortara R. Endoplasmic reticulum contains a common, abundant calcium-binding glycoprotein, endoplasmin. J Cell Sci 1986, 86: 217–222.

24. Lewis M J, Mazzarella R A and Green M. Structure and assembly of the endoplasmic reticulum. The synthesis of three major endoplasmic reticulum proteins during lipopolysaccharide-induced differentiation of murine lymphocytes. J Biol Chem 1985, 260: 3050–3057.

25. Booth C and Koch G L. Perturbation of cellular calcium induces secretion of luminal ER proteins. Cell Biol Int 1989, 59: 729–737.

26. Teriukova N P, Tiuriaeva I I, Grandilevskaia A B and Ivanov V A. The detection of membrane tumor-associated antigens of Zajdela's hepatoma on the surface of cultured rat cells. Tsitologiia 1997, 39: 577–581.

27. Feldweg A M and Srivastava P K. Molecular heterogeneity of tumor rejection antigen/heat shock protein GP96. Int J Cancer 1995, 63: 310–314.

28. Maki R G, Eddy R L J, Byers M, Shows T B and Srivastava P K. Mapping of the genes for human endoplasmic reticular heat shock protein gp96/grp94. Somat Cell Mol Genetics 1993, 19: 73–81.

29. Binder R J, Han D K and Srivastava P K. CD91: a receptor for heat shock protein gp96. Nat Immunol 2000, 1: 151–155.

30. Singh-Jasuja H, Toes R E M, Spee P, Münz C, Hilf N, Schoenberger S P, Ricciardi-Castagnoli P, Neefjes J, Rammensee H-G, Arnold-Schild D and Schild H. Cross-presentation of glycoprotein 96-associated antigens on major histocompatibility complex molecules requires receptor-mediated endocytosis. J Exp Med 2000, 191: 1965–1974.

31. Matzinger P. Tolerance, danger, and the extended family. Ann Rev Immunol 1994, 12: 991–1045.

32. Srivastava P. Interaction of heat shock proteins with peptides and antigen presenting cells: chaperoning of the innate and adaptive immune responses. Ann Rev Immunol 2002, 20: 395–425.

33. Chandawarkar R Y, Wagh M S and Srivastava P K. The dual nature of specific immunological activity of tumour-derived gp96 preparations. J Exp Med 1999, 189: 1437–1442.

34. Munro S and Pelham H R. A C-terminal signal prevents secretion of luminal ER proteins. Cell Biol Int 1987, 48: 899–907.

35. Altmeyer A, Maki R G, Feldweg A M, Heike M, Protopopov V P, Masur S K and Srivastava P K. Tumor-specific cell surface expression of the KDEL containing, endoplasmic reticular heat shock protein gp96. Int J Cancer 1996, 69: 340–349.

36. Zheng H, Dai J, Stoilova D and Li Z. Cell surface targeting of heat shock protein gp96 induces dendritic cell maturation and antitumor immunity. J Immunol 2001, 167: 6731–6735.

37. Dai J, Liu B, Caudill M, Zheng H, Qiao Y, Podack E R and Li Z. Cell surface expression of heat shock protein gp96 enhances cross-presentation of cellular antigens and the generation of tumor-specific T cell memory. Cancer Immun 2003, 3: 1–5.

38. Liu B, Dai J, Zheng H, Stoilova D, Sun S and Li Z. Cell surface expression of an endoplasmic reticulum resident heat shock protein gp96 triggers MyD88-dependent systemic autoimmune diseases. Proc Nat Acad Sci USA 2003, 100: 15824–15829.

39. Udono H, Levey D L and Srivastava P K. Cellular requirements for tumor-specific immunity elicited by heat shock proteins: tumor rejection antigen gp96 primes CD8[+] T cells *in vivo*. Proc Natl Acad Sci USA 1994, 91: 3077–3081.

40. Janetzki S, Blachere N E and Srivastava P K. Generation of tumor-specific cytotoxic T lymphocytes and memory T cells by immunization with tumor-derived heat shock protein gp96. J Immunotherapy 1998, 21: 269–276.

41. Strbo N, Oizumi S, Sotosek-Tokmadzic V and Podack E R. Perforin is required for innate and adaptive immunity induced by heat shock protein gp96. Immunity 2003, 18: 381–390.

42. Strbo N, Yamazaki K, Lee K, Rukavina D and Podack E R. Heat shock fusion protein gp96-Ig mediates strong CD8 CTL expansion *in vivo*. Am J Reprod Immunol 2002, 48: 220–225.

43. Yamazaki K, Nguyen T and Podack E R. Tumour secreted heat shock-fusion protein elicits CD8 cells for rejection. J Immunol 1999, 163: 5178–5182.

44. Udono H, Yamano T, Kawabata Y, Ueda M and Yui K. Generation of cytotoxic T lymphocytes by MHC class I ligands fused to heat shock cognate protein 70. Int Immunol 2001, 13: 1233–1242.

45. Binder R J, Blachere N E and Srivastava P K. Heat shock protein-chaperoned peptides but not free peptides introduced into the cytosol are presented efficiently by major histocompatibility complex I molecules. J Biol Chem 2001, 276: 17163–17171.

46. Castellino F, Boucher P E, Eichelberg K, Mayhew M, Rothman J E, Houghton A N and Germain R N. Receptor-mediated uptake of antigen/heat shock protein complexes results in major histocompatibility complex class I antigen presentation via two distinct pathways. J Exp Med 2000, 191: 1957–1964.

47. Baker-LePain J C, Sarzotti M, Fields T A, Li C Y and Nicchitta C V. GRP94 (gp96) and GRP94 N-terminal geldanamycin binding domain elicit tissue nonrestricted tumor suppression. J Exp Med 2002, 196: 1447–1459.

48. Kojima T, Yamazaki K, Tamura Y, Ogura S, Tani K, Konishi J, Shinagawa N, Kinoshita I, Hizawa N, Yamaguchi E, Dosaka-Akita H and Nishimura M. Granulocyte-macrophage colony-stimulating factor gene-transduced tumor cells combined with tumor-derived gp96 inhibit tumor growth in mice. Human Gene Therapy 2003, 14: 715–728.

49. Basu S, Binder R J, Suto R, Anderson K M and Srivastava P K. Necrotic but not apoptotic cell death releases heat shock proteins, which deliver a partial maturation signal to dendritic cells and activates the NF-κB pathway. Int Immunol 2000, 12: 1539–1546.

50. Huynh M L, Fadok V A and Henson P M. Phosphatidylserine-dependent ingestion of apoptotic cells promotes TGF-β1 secretion and the resolution of inflammation. J Clin Invest 2002, 109: 41–50.

51. Schild H and Rammensee H-G. gp96 – the immune system's Swiss army knife. Nat Immunology 2000, 1: 100–101.

52. Berwin B, Hart J P, Pizzo S V and Nicchitta C V. CD91-independent cross-presentation of grp94(gp96)-associated peptides. J Immunol 2002, 168: 4282–4286.

53. Vabulas R M, Braedel S, Hilf N, Singh-Jasuja H, Herter S, Ahmad-Nejad P, Kirschning C J, Da Costa C, Rammensee H G, Wagner H and Schild H. The endoplasmic reticulum-resident heat shock protein Gp96 activates dendritic cells via the Toll-like receptor 2/4 pathway. J Biol Chem 2002, 277: 20847–20853.

54. Berwin B, Hart J P, Rice S, Gass C, Pizzo S V, Post S R and Nicchitta C V. Scavenger receptor-A mediates gp96/GRP94 and calreticulin internalization by antigen-presenting cells. EMBO J 2003, 22: 6127–6136.

55. Nicchitta C V. Re-evaluating the role of heat-shock protein-peptide interactions in tumour immunity. Nat Rev Immunol 2003, 3: 427–432.

56. Bausinger H, Lipsker D and Hanau D. Heat-shock proteins as activators of the innate immune system. Trends Immunol 2002, 23: 342–343.

57. 57. Gaston J S H. Heat shock proteins and innate immunity. Clin Exp Immunol 2002, 127: 1–3.

58. Wallin R P A, Lundqvist A, Moré S H, von Bonin A, Kiessling R and Ljunggren H-G. Heat-shock proteins as activators of the innate immune system. Trends Immunol 2002, 23: 130–135.

59. Robert J, Gantress J, Rau L, Bell A and Cohen N. Minor histocompatibility antigen-specific MHC-restricted CD8 T cell responses elicited by heat shock proteins. J Immunol 2002, 168: 1697–1703.

60. Ullrich S J, Robinson E A, Law L W, Willingham M and Appella E. A mouse tumor-specific transplantation antigen is a heat shock-related protein. Proc Natl Acad Sci USA 1986, 83: 3121–3125.

61. Peng P, Ménoret A and Srivastava P K. Purification of immunogenic heat shock protein 70-peptide complexes by ADP-affinity chromatography. J Immunol Methods 1997, 204: 13–21.

62. Udono H and Srivastava P K. Heat shock protein 70-associated peptides elicit specific cancer immunity. J Exp Med 1993, 178: 1391–1396.

63. Flynn G C, Chappell T G and Rothman J E. Peptide binding and release by proteins implicated as catalysts of protein assembly. Science 1989, 245: 385–390.

64. Noessner E, Gastpar R, Milani V, Brandl A, Hutzler P J, Kuppner M C, Roos M, Kremmer E, Asea A, Calderwood S K and Issels R D. Tumor-derived heat shock protein 70 peptide complexes are cross-presented by human dendritic cells. J Immunol 2002, 169: 5424–5432.

65. Azuma K, Shichijo S, Takedatsu H, Komatsu N, Sawamizu H and Itoh K. Heat shock cognate protein 70 encodes antigenic epitopes recognised by HLA-B4601-restricted cytotoxic T lymphocytes from cancer patients. Brit J Cancer 2003, 89: 1079–1085.

66. Gehrmann M, Schmetzer H, Eissner G, Haferlach T, Hiddemann W and Multhoff G. Membrane-bound heat shock protein 70 (Hsp70) in acute myeloid leukemia: a tumor specific recognition structure for the cytolytic activity of autologous NK cells. Haematologica 2003, 88: 474–476.

67. Gross C, Koelch W, DeMaio A, Arispe N and Multhoff G. Cell surface-bound heat shock protein 70 (Hsp70) mediates perforin-independent apoptosis by specific binding and uptake of granzyme B. J Biol Chem 2003, 278: 41173–41181.

68. Moser C, Schmidbauer C, Gurtler U, Gross C, Gehrmann M, Thonigs G, Pfister K and Multhoff G. Inhibition of tumor growth in mice with severe combined

immunodeficiency is mediated by heat shock protein 70 (Hsp70)-peptide-activated, CD94 positive natural killer cells. Cell Stress Chaperon 2002, 7: 365–373.

69. Multhoff G. Activation of natural killer cells by heat shock protein 70. Int J Hyperthermia 2002, 18: 576–585.

70. Udono H and Srivastava P K. Comparison of tumor-specific immunogenicities of stress-induced proteins gp96, hsp90 and hsp70. J Immunol 1994, 152: 5398–5403.

71. Wang X Y, Kazim L, Repasky E A and Subjeck J R. Immunization with tumor-derived ER chaperone grp170 elicits tumor-specific CD8$^+$ T-cell responses and reduces pulmonary metastatic disease. Int J Cancer 2003, 105: 226–231.

72. Basu S and Srivastava P. Calreticulin, a peptide-binding chaperone of the endoplasmic reticulum, elicits tumor- and peptide-specific immunity. J Exp Med 1999, 189: 797–802.

73. Cheng W F, Hung C F, Chai C Y, Hsu K F, He L, Ling M and Wu T C. Tumor-specific immunity and antiangiogenesis generated by a DNA vaccine encoding calreticulin linked to a tumor antigen. J Clin Invest 2001, 108: 669–678.

74. Basu S, Binder R J, Ramalingam T and Srivastava P K. CD91 is a common receptor for heat shock proteins gp96, hsp90, hsp70 and calreticulin. Immunity 2001, 14: 303–313.

75. Binder R J, Harris M L, Ménoret A and Srivastava P K. Saturation, competition, and specificity in interaction of heat shock proteins (hsp) gp96, hsp90, and hsp70 with CD11b$^+$ cells. J Immunol 2000, 165: 2582–2587.

76. Delneste Y, Magistrelli G, Gauchat J, Haeuw J, Aubry J, Nakamura K, Kawakami-Honda N, Goetsch L, Sawamura T, Bonnefoy J and Jeannin P. Involvement of LOX-1 in dendritic cell-mediated antigen cross-presentation. Immunity 2002, 17: 353–362.

77. Vabulas R M, Wagner H and Schild H. Heat shock proteins as ligands of toll-like receptors. Cur Topics Microbiol Immunol 2002, 270: 169–184.

78. Ménoret A and Bell G. Purification of multiple heat shock proteins from a single tumor sample. J Immunol Methods 2000, 237: 119–130.

79. Suto R and Srivastava P K. A mechanism for the specific immunogenicity of heat shock protein-chaperoned peptides. Science 1995, 269: 1585–1588.

80. Tamura Y, Peng P, Liu K, Daou M and Srivastava P K. Immunotherapy of tumors with autologous tumor-derived heat shock protein preparations. Science 1997, 278: 117–120.

Extracellular Biology of Molecular Chaperones: What Does the Future Hold?

19

Gazing into the Crystal Ball: The Unfolding Future of Molecular Chaperones

Lawrence E. Hightower

Predicting the future of the exobiology of molecular chaperones is bound to be risky business: after all, unravelling the intracellular lives of the chaperones has become legendary for its unexpected twists and turns. Can we expect differently for their extracellular capers? I cannot claim the clearest crystal, but I do have a unique perspective on the field from my perch as Editor-in-Chief of the major specialty journal in the field, *Cell Stress & Chaperones*. I will refer to papers in recent issues that will lead interested readers to other papers in key areas that I believe provide insights into the future as well. Perhaps we can begin to illuminate the crystal ball by listing major unsolved problems and by identifying the disciplines of the investigators that these problems are now attracting into the field.

One of the exciting and renewing aspects of the heat shock field, as it was known historically, has been the succession of colleagues from different disciplines that have entered and moved the field forward. The chance initial finding of the heat shock response in *Drosophila* by Ritossa in 1962 [1] was pursued by a small group of *Drosophila* biologists until about 1978 when the response was discovered in a variety of other organisms. Molecular geneticists were attracted to the heat shock genes as models of inducible eukaryotic gene expression, and the field took on a more global interest. The 1982 Cold Spring Harbor Heat Shock Meeting: From Bacteria to Man was dominated by molecular geneticists describing gene organisation and transcription, chromatin structure and regulation in several systems besides *Drosophila* [2]. There were a few talks from biochemists, cell biologists and a smattering of physiologists, the vanguard of many more to come who would work out the function of the heat shock proteins, create the names 'molecular chaperones', 'chaperonins' and 'co-chaperones', and establish the physiological setting of cellular stress responses, as the heat shock response came to be known. Progress by the biochemists in the purification and characterisation of the substantial amounts of molecular chaperones attracted structural biologists, and we got our first three-dimensional look at chaperones at the atomic level (see Chapter 1).

Joining this rich interdisciplinary mix of investigators who continue to build on a now solid foundation of basic science, we now find investigators interested in translational research including medical scientists focused on specific diseases along with a few clinicians and M.D./Ph.D. colleagues who are beginning to take the field from the laboratory bench to the patient bedside. Cellular immunologists in particular are having a major impact. So, here is the first light in our crystal ball: the future will bring more research on molecular chaperones in human biology and disease, the effects of environmental stress on human populations, and the use of animal model systems, particularly those with sequenced genomes, to study the complex biology of stress response physiology and cytoprotection, the morphed sibling of thermotolerance and *Drosophila* phenocopy protection.

Our crystal brightens more, because colleagues with translational interests in our field work in biotechnology companies as well as in academia, and my second prediction is that they will succeed in producing vaccines against cancers and dysplasias of viral origin that are both therapeutic and preventative. Some of these vaccines will be directed against the tumour cells of individual patients; that is, our field will spawn one of the first really dramatic successes in personalised medicine (see Chapters 17 and 18). In addition, the process of stress-conditioning to induce tissue cytoprotection will become part of pre-operative patient care, at least for elective surgeries, and new drugs will be marketed to stimulate in some cases and turn off in others the stress response in humans. Perhaps some of the new drugs needed to accomplish these tasks will be based on the cytokine activities of molecular chaperones, knowledge of their extracellular receptors, and the signal transduction pathways to which they are linked. My confidence in these predictions is high and we have opened for 2004 a new section in *Cell Stress & Chaperones* entitled 'Stress Response Translational Research'.

The path to translational research on molecular chaperones is not quite the linear one just outlined. Two of the talks on thermotolerance and heat shock proteins at the 1982 meeting were given by colleagues crossing over from the discipline of radiation oncology. The clinical use of hyperthermia in cancer treatment to kill cancer cells is a line of translational research which has paralleled and sometimes crossed paths with the cellular stress response field. Researchers on both paths have come to view thermotolerance or cytoprotection as a key cellular state that needs to be understood and controlled, and both have come to appreciate the central roles of heat shock proteins in this altered state of cellular physiology. There is renewed interest in the clinical application of mild hyperthermia and fever-like responses, and both groups share a fascination with apoptosis. What will the future hold? I note that the Cell Stress Society International and the North American Hyperthermia Society held a joint meeting in Québec in 2003, The First International Congress on Stress Responses in Biology and Medicine. My third prediction is that more than a few productive collaborations on translational research in our field will trace back to this seminal international meeting.

Perhaps the major unanswered questions are: How do stress proteins egress from cells, and how are they taken up by other cells? Path-breaking work in the 1980s initially identified heat shock proteins among the set known as glia-axon transfer proteins, demonstrating that at least a couple of the heat shock proteins can be transferred cell-to-cell and showed that these same heat shock proteins are released from cultured mammalian cells by a non-ER–Golgi mechanism [3, 4]. Hsp70, Hsc70 and Hsp110 were released from heat-stressed newborn rat cell cultures, and Hsc70 was released from non-heat-stressed cultures, possibly stimulated by medium changes during the experimental protocol [4]. Actin was also released into the medium and transferred cell-to-cell, raising the possibility of a microfilament-aided release or transfer mechanism. The release of actin by a non-ER–Golgi pathway was reported independently by Rubenstein and colleagues [5].

There appear to be at least three general ways in which a nucleocytoplasmic protein like Hsp/Hsc70 might become extracellular. *In vivo*, it might be released locally by tissue damage during wounding (i.e., tissue trauma). If the injury is large enough, sufficient amounts could be released or taken into blood vessels to create a system distribution. Secondly, there appears to be a release mechanism via a non-ER–Golgi pathway which might be very sensitive to triggering by breaks in tissue homeostasis. And finally, there appears to be a cell-to-cell transfer mechanism which might be particularly important in nervous tissue in which glial cells capable of producing heat shock proteins might transfer them to neuronal cells that do not during physiological stress. Tytell and colleagues have pursued a possible therapeutic application for the transfer of protection by introducing purified Hsp70 into neuronal cells as a model for treating spinal cord injury [6].

A portion of Hsp70 is associated with plasma membranes, as early localisation studies in *Drosophila* showed. In fact, the propensity of Hsp/Hsc70 to bind fatty acids and to interact with lipid bi-layers in interesting ways, to produce channels for example, expands to possibilities for egress and uptake mechanisms. It will no doubt take a while longer to sort out these mechanisms, even using model cells such as yeast, in which non-ER–Golgi release has also been observed (See Chapter 3). What is very obvious now is that once heat shock proteins are released from cells, they find receptors on other cells, particularly cells of the immune system and tumour cells, on which they have biological effects. In addition to nucleocytoplasmic heat shock proteins, their relatives in the endoplasmic reticulum, members of the glucose-regulated protein set, can escape, presumably by an ER–Golgi secretory pathway, to contribute to these biological effects.

Another approach to illuminating our crystal is to try to predict some of the keywords that might come to dominate our field. For homeothermic vertebrates, I would include inflammation, vascular endothelium and innate or natural immunity among the key search terms of the future. Whereas external temperature cues appear to dominate the thermal induction of the heat shock response of poikilotherms, it is more likely that the heat generated during localised tissue inflammation, along with a host of other inducers, acts in concert and probably synergistically to activate the heat shock or cellular stress response most frequently in homeotherms. We can also include systematic responses such as fever and the effects of elevated blood levels of pro-inflammatory cytokines on the vascular endothelium. It has been clear since some of the earliest work on the induction of heat shock proteins in the tissues of heat-shocked rats by F. P. White [7] that cells associated with blood vessels are among the most responsive in a broad range of tissues. Innate immunity is considered to be the most ancient of the arms of the immune response, and it appears that a very ancient and evolutionarily well-conserved set of proteins, the heat shock proteins, has become intertwined in this rather non-specific first line of defence against microbes.

The components of innate immunity include physical barriers such as skin, mucosal epithelia and the anti-microbial molecules that they produce such as defensins, the complement system, macrophages, neutrophils and natural killer cells. Some potentially interesting overlaps in our keywords are now apparent. Components of innate immunity contribute to inflammatory responses, and the recruitment of neutrophils into areas of tissue damage requires a regulated response from the vascular endothelium in the vicinity of the wounded or inflamed tissue. In these venues, extracellular molecular chaperones have ample opportunity to exercise their activities as chemokines to regulate tissue-level inflammatory processes. This will be a rich hunting ground for additional responsive cells, more receptors and signalling pathways regulated by extracellular chaperones.

Coincidentally, the issue of *Cell Stress & Chaperones* which is most current to the writing of this chapter contains three articles in succession that would have appeared in our futuristic keyword search. The search term 'inflammation' would have brought up an article by Barton and colleagues providing immuno-histochemical evidence of Hsp32 (hemoxygenase 1) in normal and inflamed human stomach and colon [8]. Hsp32 was detected in inflammatory cells and gastric epithelial cells of normal human gastric and colonic mucosa. It was expressed at higher levels in inflamed gastric mucosa, independent of *Helicobacter pylori* infection, and was particularly high in inflamed colon samples from patients with active ulcerative colitis [8]. In rats, increased expression of Hsp32 has been correlated with decreased inflammation in chemically induced colitis [9]. Whether or not Hsp32 plays a similar regulatory role in the types of chronic human inflammatory diseases described in this study remains to be determined.

The search term 'vascular endothelium' would identify the next paper in this issue, a study by Kabakov and colleagues on the cytoprotective effects of over-expressing either Hsp70 or Hsp27 in human endothelial cells [10]. The authors used an *in vitro* model of ischaemia-reperfusion injury, as may occur during myocardial infarctions and strokes, to show that the expression of either of these heat shock proteins within the first six hours of post-hypoxic re-oxygenation results in significant reductions in endothelial cell apoptosis [10]. This is a more realistic test of the potential for therapeutic intervention than previous animal models in which stress conditioning has been applied prior to the experimental induction of a myocardial infarct or stroke, since these are not predictable events that lead to the initial ischaemia. I fully agree with the authors' contention that protection of the vascular endothelium from oxidative damage and apoptosis is likely one of the most important considerations in facilitating recovery with minimal tissue damage.

The third paper in this sequence describes the finding that Hsp70 reactivity independent of major histocompatibility complex (MHC) class I was associated with increased densities of the C-type lectin receptor CD94 and the neuronal adhesion molecule CD56 on the surface of human primary natural killer (NK) cells following stimulation by the peptide terminal localised Hsp70 sequence 'TKDnnllgrfelsg' (TKD, 99 450–463) [11]. This paper by Multhoff and colleagues provides a glimpse into the exobiology of the human molecular chaperone Hsp70. The cell surface receptors CD94 and CD56 in primary NK cells were selectively upregulated following treatment with the Hsp70 peptide TKD. Elevated densities of these two receptors were correlated with an increased cytolytic response of these activated NK cells against target tumour cells displaying Hsp70 on their surface membranes. An important point to note about all three of these papers is that they employ human cells to address possible links between heat shock proteins and human diseases – they represent translational research contributions. This trend in the molecular chaperone field toward translational research will likely expand in the near future.

REFERENCES

1. Ritossa F A. A new puffing pattern induced by temperature shock and DNP in *Drosophila*. Experientia 1962, 18: 571–573.
2. Schlesinger M J, Tissières A and Ashburner M (Eds.). *Heat Shock, from Bacteria to Man*. Cold Spring Harbor Laboratory Press 1982.
3. Tytell M, Greenberg S G and Lasek R J. Heat shock-like protein is transferred from glia to axon. Brain Res 1986, 363: 161–164.
4. Hightower L E and Guidon P T. Selective release from cultured mammalian cells of heat-shock (stress) proteins that resemble glia-axon transfer proteins. J Cell Physiol 1989, 138: 257–266.
5. Rubenstein P, Ruppert T and Sandra A. Selective isoactin release from cultured embryonic skeletal muscle cells. J. Cell Biol 1982, 92: 164–169.
6. Tidwell J L, Houenou L J and Tytell M. Administration of Hsp70 *in vivo* inhibits motor and sensory neuron degeneration. Cell Stress Chaperon 2004, 9: 88–98.
7. Currie R W and White F P. Characterization of the synthesis and accumulation of a 71-kilodalton protein induced in rat tissues after hyperthermia. Can J Biochem Cell Biol 1983, 61: 438–446.
8. Barton S G R G, Rampton D S, Winrow V R, Domizio P and Feakins R M. Expression of heat shock protein 32 (hemoxygenase-1) in the normal and inflamed human stomach and colon: an immunohistochemical study. Cell Stress Chaperon 2003, 8: 329–334.
9. Wang W P, Guo X, Koo M W, Wong B C, Lam S K, Ye Y N and Cho C H. Protective role of heme oxygenase-1 on trinitrobenzene sulfonic acid-induced colitis in rats. Am J Physiol Gastrointest Liver Physiol 2001, 281: G586–594.
10. Kabakov A E, Budagova K R, Bryantsev A L and Latchman D S. Heat shock protein 70 or heat shock protein 27 overexpressed in human endothelial cells during posthypoxic reoxygenation can protect from delayed apoptosis. Cell Stress Chaperon 2003, 8: 335–347.
11. Gross C, Schmidt-Wolf I G H, Nagaraj S, Gastpar R, Ellwart J, Kunz-Schughart L A and Multhoff G. Heat shock protein 70-reactivity is associated with increased cell surface density of CD94/CD56 on primary natural killer cells. Cell Stress Chaperon 2003, 8: 348–360.

Index

Forgive Me

LESLEY PEARSE

PENGUIN BOOKS

PENGUIN BOOKS

Published by the Penguin Group
Penguin Books Ltd, 80 Strand, London WC2R ORL, England
Penguin Group (USA) Inc., 375 Hudson Street, New York, New York 10014, USA
Penguin Group (Canada), 90 Eglinton Avenue East, Suite 700, Toronto, Ontario, Canada M4P 2Y3
(a division of Pearson Penguin Canada Inc.)
Penguin Ireland, 25 St Stephen's Green, Dublin 2, Ireland (a division of Penguin Books Ltd)
Penguin Group (Australia), 707 Collins Street, Melbourne, Victoria 3008,
Australia (a division of Pearson Australia Group Pty Ltd)
Penguin Books India Pvt Ltd, 11 Community Centre,
Panchsheel Park, New Delhi – 110 017, India
Penguin Group (NZ), 67 Apollo Drive, Rosedale, Auckland 0632, New Zealand
(a division of Pearson New Zealand Ltd)
Penguin Books (South Africa) (Pty) Ltd, Block D, Rosebank Office Park, 181 Jan Smuts Avenue,
Parktown North, Gauteng 2193, South Africa

Penguin Books Ltd, Registered Offices: 80 Strand, London WC2R ORL, England

www.penguin.com
First published by Michael Joseph 2013
Published in Penguin Books 2013

001

Copyright © Lesley Pearse, 2013
All rights reserved

The moral right of the author has been asserted

Set in Garamond
Typeset by Palimpsest Book Production Limited, Falkirk, Stirlingshire
Printed in Great Britain by Clays Ltd, St Ives plc

ISBN: 978-0-241-96149-0

www.greenpenguin.co.uk

To my brother, Dr Michael Sargent, and his lovely wife, Jean.

Thank you, Michael, for all the times I've picked your brains about DNA, poison and goodness knows what else. Even as a small child I suspected you'd come in useful one day!

I am so proud of you too.

Prologue

Flora kicked off her shoes, pulled her dress over her head and tossed it on to the bed. She was about to remove her underwear too, when a glance in the gilt-framed cheval mirror stopped her.

Dressed, she still looked quite trim for a woman of forty-eight, but naked she was flabby and her skin pale. She couldn't bear the thought of anyone seeing her like that. Not even in death.

She opened a drawer, took out the ivory silk slip which matched her bra and knickers and put it on. 'That's better,' she murmured.

Removing the band holding her hair back, she ran her fingers through it till it tumbled down over her bare shoulders. Her Titian-red wavy hair had always been her best feature, and even now, as desperate as she felt, she was proud of it.

The bath was already run in the en-suite bathroom, she was primed with a couple of sleeping pills and some brandy, and no one was due home for at least three hours. She was entirely resolved upon what she intended to do, yet it hadn't occurred to her until now that it would have been kinder to the children if she'd checked into a hotel room so that a stranger found her.

It was the bedroom which prompted this thought. From the expensive red and gold wallpaper that Andrew had raged over to the French gilded bed and sumptuous carpet and

1

curtains, it reflected her true character. It was the only room in the entire house which really did, as Andrew despised what he called 'bordello' style. Everywhere else was muted shades of cream and taupe, as befitted a Georgian country house.

But she wanted to die here in this room which she'd fought long and hard to keep as she planned it. He'd driven her to this point by forcing her to bend to his will about everything else. He claimed he loved her, that everything he'd done was for her, but in reality he'd stifled her true personality and creativity to the point where she could barely remember who she'd once been.

In her early twenties she'd claimed that suicides were cowards. She'd loved life so much then that she despised anyone who didn't embrace it as she did. But she didn't know then what heartache could do, or that a bad choice in a weak moment could change the whole course of your life.

But it was too late for regrets now; she was feeling woozy, and Andrew would be home first so it would be he who found her. As she went over to the dressing table to take one last look at the framed pictures of her children, she was very unsteady.

Sophie and Ben, seventeen and eighteen respectively, grinned cheerfully back at her. The picture of the two of them had been taken on Boxing Day, at the pre-lunch drinks party they had every Christmas for neighbours and friends. They were very alike: tall, slender and dark-haired. They had inherited Andrew's looks, but she hoped they would never become mean-spirited control freaks like him.

In a separate frame was one of Eva. It had been taken on Boxing Day too, but it was not a very flattering picture. She was smaller than the other two, curvy and pretty with lovely blue eyes, but the purple dress overwhelmed her delicate colouring and made her look plump and closer to thirty than only twenty. It pricked Flora's conscience.

'I should've picked a dress out for you,' she sighed. 'Pink or pale blue – that would've done you justice. I also should've told you never to try to be what you think other people want of you. I'm a good example of where that leads. Be true to yourself, and remember I loved you.'

She kissed each one of their faces, biting back tears. Time was running out; she could feel her head swirling, and she still had to write a note for them. She picked up the pen and notepad she'd left by the bedside, but could no longer remember the words she'd planned to say.

'Forgive me,' she began. But nothing more came to her, and in some strange way that seemed enough.

She left the note on the bedside cabinet and went into the bathroom. The new sharp craft knife was ready on the side of the bath. She climbed into the hot water, lay back for a few moments to brace herself and then picked up the knife.

She hesitated. The steel knife felt cold and heavy in her hand. Could she really do it? It was the pain she was afraid of, and of not cutting deep enough to open her veins.

'No more guilt,' she murmured. 'No more pretending. It will all be gone for ever very soon.'

With the knife in her left hand, she quickly drew the blade sharply across her right wrist, then changed hands and cut the left one before the pain could stop her. Surprisingly, it didn't hurt; and the way the blood began to pump out, she knew she'd cut them deeply enough.

She let her arms sink into the hot water and watched the water turn red.

It was done.

Chapter One

As Bette Midler's 'From A Distance' came on, Eva turned up her car radio and sang along with it. It was cold and raining, but she was feeling happy because Olive, her boss, had let everyone go early when the heating broke down. As Eva had arranged to go to a make-up party this evening with some of the girls from work, she would have more time to wash her hair and get ready.

She turned into the drive, but had to slam her brakes on because the wrought-iron gates were unexpectedly shut. She stopped only a whisker from them. 'Damn,' she exclaimed. Not only had she nearly hit the gates, but now she would get soaked opening them up.

As she could see her mother's red Polo parked at the end of the drive by the house, Eva felt irritated. Why had she closed the gates if she was in?

Despite her pique at having to get out in the rain and push open the heavy gates, Eva noticed that the borders of daffodils and other spring flowers around the huge expanse of front lawn looked bright and beautiful. However inconsistent her mother was in so many ways, she lavished care on the garden – in fact, had it not been raining quite so hard she would be out there now.

Hopping quickly back into her car, Eva drove up the drive and parked just behind the Polo, then hurried through the arch which led to the old stable block. A few years ago her parents had converted the yard into a courtyard garden, and the stables to an indoor swimming pool. The courtyard was a real suntrap; the surrounding walls kept off the wind, and at the start of March it had been so mild they had all sat out here for a couple of hours after Sunday lunch.

The back door was unlocked. Eva hung up her coat on a peg, then went into the kitchen, expecting to find her mother there, preparing the evening meal. But she wasn't. The kitchen was so polished and neat, with a carefully arranged bowl of fruit and a vase of daffodils on the black marble worktop, that it looked like it was about to star in a feature in *Homes & Gardens*.

This was rather unusual, as her mother wasn't tidy by nature. When Dad was away on business for a few days she always let things slide. Sometimes Eva would get home in the evening to find the breakfast things still where they'd been that morning. But Dad was fussy, and he liked everything to be immaculate; mostly when Eva got home first she'd find her mother frantically rushing around, putting things straight, polishing and tidying before he got in.

Eva thought today's extreme tidiness must mean Dad was expected early or they were having visitors, as there wasn't so much as a dirty cup or glass anywhere.

'I'm home, Mum,' Eva called out. 'Where are you?'

Getting no answering call, she glanced into the sitting room and the conservatory beyond, then into the study and the dining room. She wasn't there and, like the kitchen, they were immaculate. It was also ominously quiet – usually, the radio was on.

Puzzled, Eva stood at the bottom of the stairs for a moment. Her mother might be unpredictable: some days she made several different kinds of cakes and cooked meals to stow away in the freezer, and on others she was barely motivated to use a tin opener. Yet one thing was constant – and that was, she always welcomed her family home.

Normally just the sound of Eva or Dad's car on the drive was enough for her to break off whatever she was doing to come and greet them.

Like many Georgian houses the hall was large and impressive with the oak staircase rising from the middle, then curving gracefully around to meet a gallery on the first floor. There was a skylight window above, and on a sunny day the staircase was flooded with natural light. Today the light was murky and the rain was drumming against the glass.

Eva went halfway up the stairs, and called out again. When there was still no reply she wondered if her mother had got one of her migraines and gone for a lie-down, so she decided against calling again for fear of disturbing her.

All the five bedrooms and two bathrooms led off from the gallery. But as Eva reached the gallery and saw that her parents' bedroom door was open, she doubted her mother was sleeping. She peeped in; there was a dress on the bed, which suggested she'd changed, perhaps to go out for a walk. Yet that seemed unlikely when it was pouring with rain.

Eva was puzzled as she looked in each of the bedrooms, remembering that the back door hadn't been locked. Mum wouldn't leave that open even to nip quickly to a neighbour's house.

A door with a plain wooden staircase behind it led to three tiny attic rooms. Once servants' quarters, two were now spare rooms, rarely used except at Christmas or other special occasions when someone came to stay; the third one was used for storage. Although it was unlikely her mother was up there, Eva checked anyway. But she wasn't there.

For the past few weeks her mother had been somewhat withdrawn and distant. On several occasions Eva had found her just staring into space, in a world of her own. A couple of days ago Eva had talked about it to Ben, her younger

7

brother. He'd been of the opinion it was her age, because he'd heard that all women got a bit odd in their forties. But now, as Eva began to feel anxious, she wished she'd risked Dad scoffing at her and told him what she'd noticed.

Hoping that her parents' room might offer up a clue, Eva went in there. The dress on the bed was the one her mother had been wearing at breakfast. Dad had been sarcastic, asking if she was going to a tea dance because it was a vintage dress from the 1940s, emerald-green wool crêpe with a small corsage of lighter green velvet flowers on the bodice.

Flora liked vintage clothes. She said they belonged to a gentler period when women looked like women. Her wardrobe was full of old velvet, chiffon and crêpe. Dad was always sarcastic about the way she dressed. To him they were just second-hand clothes, and he thought the wife of Head of Sales for one of the largest paper product companies in Europe should dress the part.

But although Dad got his way about almost everything, he had given in on this point because absolutely everyone else agreed Flora suited vintage clothes. Her red curly hair, curvy body and pale skin could be likened to many of the film stars of the 1940s. Crêpe dresses cut on the cross, beaded boleros and peplum-waisted jackets went with both her shape and her character. Maybe they weren't too practical, but then practicality wasn't exactly Flora Patterson's strong suit.

As Eva stood in her parents' bedroom, she remembered the terrific row her parents had had when Dad arrived home from a business trip to find Mum had completely redecorated this room. Eva had never been quite sure whether she liked the shock of coming from the muted decor elsewhere into this red and gold, grandiose and decadent room. But she really admired her mother for not only decorating it herself, but getting the curtains, carpet and French walnut furniture in while Dad was away, and sticking to her guns when he went mad about it. She had insisted that she was entitled to have one room in the house that was just for her.

Although it was unlikely Mum had braved the rain to do something in the garden or the garage, it would account for her changing her clothes, so Eva looked out of the window.

People assumed by looking at the grand gates, and the sweeping drive, that the back garden must be huge. It had been, but the house was in such a bad state when her parents bought it that they sold all of the land at the back of the house to pay for the renovations. The development company who bought it built a small estate of executive houses there.

There was just a narrow strip of patio at the back now, and an eight-foot wall to give them some privacy. But here in the front of the house it was still possible to imagine how The Beeches had looked when it was first built two hundred years ago because the trees and bushes surrounding the lawn shut out even a glimpse of the newer neighbouring houses. Eva couldn't see the garage from the window, as it was joined to the side of the house, but it was possible that if the door was shut and her mother was engrossed in something, she might not have heard Eva's car.

But as she turned from the window she noticed that the door of the en-suite bathroom was closed. Like the extremely tidy kitchen, that was uncharacteristic. Except for Tuesdays – when Rose, the cleaning lady, came – there was often a trail of dropped clothes from the bed to the bath, with doors and drawers left open.

'Mum, are you in there?' she called out.

There was no reply but she went over to the door and banged loudly on it, just in case Flora had her Walkman on in the bath. She turned the handle and opened the door just wide enough to peep in.

To her relief she could see the top of her mother's head just above the end of the claw-foot bath.

'Oh, here you are! So sorry to intrude. I was getting worried –'

She stopped short, suddenly noticing the bathwater was as red as her mother's hair.

'Mum!' she screamed as she rushed in. 'Mum! What's happened?'

But one look at the pallor of her mother's face and her wide-open, yet vacant eyes was enough for Eva to know she was dead; and the craft knife covered in blood, dropped on the floor beside the bath, told her how it had happened.

Nothing in Eva's entire life had prepared her for such a shocking sight, and she screamed involuntarily, running out on to the landing in fright.

It took her a few moments to pull herself together enough to go back into the bedroom, pick up the phone and dial 999. But as soon as she'd stammered out to the operator what she'd just found and given the address, she went back to the landing and slumped down onto the floor, too shocked and terrified by what she'd seen to go downstairs.

The waiting for someone to come seemed endless. The only sounds were the rain thudding down on the skylight and her heart beating too fast. She wrapped her arms around her knees and sobbed.

Nothing had happened that morning to make Eva suspect something was badly wrong. Breakfast had been utterly normal and, aside from Dad's sarcasm in asking Mum if she was going to a tea dance, nothing unusual had been said. Mum had made a pot of tea as usual, and just sat there drinking hers as Sophie and Ben got themselves cereal. She'd said all the usual stuff. Had Sophie got her games things? Then reminded Ben he must have a proper lunch at school, not just a packet of crisps. She'd kissed them all as they left the house, even asked Dad to pick up his best suit from the dry cleaners. Did she know she was going to do this even then? And why did she tidy the whole house? Did she think her death would be less distressing for everyone if the house was looking perfect?

When Eva heard the siren in the distance, she felt unable to move. She didn't think she'd even be able to speak to the police or ambulance men.

Suddenly the silence was broken by the sound of tyres on the gravel drive and loud male voices. One of them was her father's; he must have arrived along with the emergency services. Knowing he would have Ben and Sophie with him, Eva felt she had to protect her brother and sister from what she'd seen, and so she hauled herself up.

But before she even got to the stairs she heard Dad speaking in the hall below. He must have opened the front door to let the police in that way.

'There must be some mistake,' he was saying with indignation. 'Are you sure it wasn't a hoax call? Yes, that is our elder daughter's car, but did the person who made the call say she was Eva Patterson?'

'Daddy!' Eva called out, clinging to the gallery rail. 'It was me. Don't let Sophie and Ben up here.'

All at once what seemed like a dozen people were all speaking at the same time. There were heavy footsteps and Sophie was yelling that she wanted to know what was going on.

Eva felt as if she was in the middle of a terrible nightmare. But she knew she wasn't going to wake up and find it wasn't

11

real. She really had seen the bathwater bright red with blood. She really had lifted Mum's arm and seen the slash across her wrist. And she hadn't imagined the bloodstained knife lying on the floor.

As the ambulance men came up the stairs she turned to point to the bedroom. But the dark red carpet in there looked to her like a pool of blood, and her stomach heaved. She could hear Sophie screaming downstairs, and Ben's voice too, shrill with anxiety, then Dad's voice above theirs, telling them to be quiet as they were making the situation even worse. She felt herself growing dizzy, and she must have fainted, because the next thing she knew she was on the floor and a policewoman was kneeling beside her.

'There now,' she said soothingly. 'You've had a terrible shock, but come downstairs with me and I'll make you a cup of tea.'

WPC Sandra Markham was thirty-eight and had been in the police force in Cheltenham for twelve years. She knew she had a reputation as being good at weighing up the dynamics in domestics – which one of a warring couple was the vicious one, the liar or the bully. Her opinion was valued because she was very observant, could read body language well, and she also had a knack of getting people to talk.

She had been called upon, hundreds of times, to be present when it was necessary to break the news of a death or serious accident. Each time the reaction was different: some people couldn't take it in, while some guessed what was coming as soon as they saw a police uniform. Some remained dry-eyed and silent, others screamed and wailed, and there were many other variations between the two extremes. But in every other case where children had lost a mother or father, she had never known the remaining parent, however shocked and grief-stricken they were, not rally enough to try and comfort them.

In the three hours Markham had been at the Pattersons' home, she hadn't once seen Andrew Patterson attempt to comfort Eva.

He had arrived at The Beeches with his two younger children at the same time as the police. He'd gone up the stairs right behind the two male officers and Markham had followed him. He didn't even glance at his elder daughter, crumpled up on the landing, as he rushed into the bedroom.

That of course was understandable, given the circumstances. Yet when he came out of the bedroom just a few minutes later, when Markham was trying to get Eva on to her feet to take her downstairs, she cried out to him, and he ignored her.

Once Markham had got Eva down to the kitchen she questioned her, trying to ascertain the girl's exact movements after she returned home from work until the moment she found her mother.

Finding your mother dead in a bath of blood had to be one of the most terrible things for anyone to experience, especially someone so young. Yet Patterson didn't once come over to Eva, put his arm around her, or show any concern for her.

Sometimes in cases like this people appeared vacant, too shocked to really take in what was happening around them. But Patterson was listening hard, and when Eva said how the kitchen had looked like a show house, he interrupted. He curtly asked why Eva found that strange, implying that she was lying.

The house was immaculate, and it looked to Markham as though it was always kept that way. But she didn't think Eva would make any reference to it unless this wasn't always the case. Was Patterson trying to conceal his wife's failings? Could this be a source of conflict which had propelled Flora Patterson to take her own life?

There was no doubt that Andrew Patterson was a very attractive and clearly very successful man: six foot two, athletic build, dark hair with just a sprinkling of grey at the temples, good teeth and very dark eyes. His shock and horror at his wife's death seemed heartfelt, yet his lack of compassion towards his elder daughter was suspicious.

There was only a year between Ben and Sophie, the two younger children, and they could easily have passed for twins, as they were so alike – both tall, slender, with their father's glossy dark hair and eyes. As Markham hadn't seen the mother, she assumed Eva must take after her, because she was much shorter, with blue eyes and light brown hair.

Because Patterson interrupted her questioning several times, and also because Sophie kept rushing in and out of the room wailing and screeching, Markham took Eva into the sitting room to get the whole story.

Distraught as she was, it was obvious Eva was a caring, level-headed girl. She managed to tell her story clearly and showed a protective anxiety for her younger siblings that was very laudable. While she wasn't as strikingly beautiful as her younger sister, she had a sweet face and there was something about her that made Markham want to take her in her arms and cuddle her.

Part of it was because she looked a bit prim and old-fashioned. Her hair was tied back at the nape of her neck, and her navy-blue suit, white shirt and plain court shoes were far too frumpy for a girl of almost twenty-one. Yet despite that, Markham felt she was more worldly than her appearance would suggest.

Usually when Markham interviewed young girls after

something horrific, they were unable to get beyond their own feelings. Eva related her irritation at the gates being closed, and her bewilderment that the back door was unlocked with no sign of her mother, just as any other girl would. She broke down several times too, becoming so upset when she described the moment when she found her mother that Markham felt she might have to halt the interview. But Eva visibly made the effort to pull herself together, and her real concern was not for herself but for what had driven her mother to do it. She was also desperate to go to comfort Sophie, who was by then hysterical.

'Would you say your parents' marriage was a happy one?' Markham asked gently. The house was beautiful and luxurious and it was hard to imagine any woman not being happy there. But she knew from experience that appearances could be deceptive.

Eva nodded tearfully. 'I think so. But they were very different kinds of people. Dad's very ordered and calm; he likes everything just so. Mum could be quite chaotic and disorganized.'

'Did you notice anything, even something quite small, that was different about her recently? Did she seem worried or nervy? Had she been ill?'

'Not really. She had seemed sort of distant for a while, but then she often had periods like that.'

It was at that point Markham looked around and saw Patterson hovering by the doorway, listening. His expression wasn't one of anxiety for Eva; it was more like he was checking on what she was saying. Markham not only wondered why that was important to him, but also why her colleagues hadn't made sure he stayed in the kitchen with the other two children.

There were no grounds to find Flora's death suspicious.

The way she was lying, the absence of any signs of a struggle and the knife dropped over the edge of the bath made it clear it was suicide. The fact she was wearing cream silk underwear and the stark note left in the bedroom, saying only 'Forgive me', suggested she had planned it in advance.

Yet there had to be a reason why a woman who appeared to have everything – a beautiful home, three children and no financial worries – would choose to end her life. Debt, disgrace, terminal illness, an unbearable marriage or an illicit love affair were all possibilities, and perhaps something would come to light later. Yet Markham felt certain Andrew Patterson already knew the reason, or at least could guess at it, but he wasn't the kind to reveal anything which might reflect badly on him.

As for Eva, her total bewilderment proved she knew nothing. Markham could only hope the post mortem or the inquest might throw up some answers for all three children. To be left wondering why would be torture.

Much later that evening, after Flora's body had been taken away to the mortuary and the police had left, Eva sat at the kitchen table nursing a cup of tea that had long since grown cold. She felt completely numb.

Ben was next to her in much the same state, still wearing his navy-blue school blazer, not speaking, his eyes red-rimmed and swollen, and now and again he reached out silently for her hand. Dad was across the table from them, grimly drinking whiskey and only uttering a few questioning words now and then which didn't appear to need answers.

Sophie was the only one who hadn't kept still; she had paced around the kitchen, one minute sobbing loudly, the next angrily demanding to know why their mother had done this. When she got no real answer she would then flounce out of the room, picking up the telephone to cry to one of her friends.

Eva looked at the clock at one point and felt surprised that it was only eleven thirty; it seemed to her that she'd been sitting here for a whole night. She wanted to go to her room, not to sleep – she doubted she'd be able to – but just to escape the atmosphere of brooding intensity that was pressing down on her.

All the images of what had taken place earlier seemed confused now, and out of sequence. There had been so many policemen coming and going, so much noise and confusion. She recollected someone, she presumed it was a doctor, saying that Flora had been dead for around two hours when Eva found her. She wondered why she remembered that when everything else seemed a jumble.

Dad had cried earlier. She went to him to try to comfort him, but he pushed her away, almost as if he held her responsible. Another horrible moment was when the men carried Mum's body down the stairs on a stretcher. Sophie shrieked like a mad thing, saying they couldn't take her away, and when Eva had tried to calm her down and explain that the police had to take her, Sophie accused her of not caring.

WPC Markham had been very kind to her. She'd said people often said and did hurtful things at such times and she mustn't take it to heart. Eva found it odd that much of the detail of what had happened earlier was fading; the only part that was still crystal clear in her mind was her mother's white face above the bloody bathwater. That image played and replayed in her head over and over again.

Was it true that the police had found a note which just said 'Forgive me'?

How could Mum say goodbye, kiss Dad and each of them

that morning, then clean and tidy the house, yet go on to do that in the afternoon?

Why? What could have been so terrible in her life that she couldn't bear it a minute longer?

Earlier she had heard Dad talking to one of the policemen. 'I gave Flora everything she wanted,' he said. 'This house, holidays, she could buy what she liked and go where she liked. She loved her children. How could she do this to us?'

'There isn't always an explanation for why people do this,' the policeman had replied.

But an explanation was needed; they were all distraught. If it was because Mum was terminally ill, if she'd gone mad or had huge debts she'd been hiding, that at least would make some kind of sense of it.

Eva had never felt as helpless as she did now. As the eldest she had always been the one who acted as peacemaker in squabbles between Sophie and Ben. If they were in trouble with Mum or Dad she took their part. She wanted to try to comfort them both now, and to reassure them they would get through this. But she couldn't; she didn't have the words, or the will. Dad, Ben and Sophie – they all seemed like strangers, not her family.

She had never known Dad be anything other than self-assured, calm and in charge in any situation. Her friends always said he looked like Pierce Brosnan, and was tasty for a middle-aged man, but to Eva he was just her dad, officious and controlling, lacking a sense of humour, but always reliable. He had never been demonstrative, nor was he the kind you could have a heart-to-heart talk with. Mum had often accused him of being emotionless.

Yet now, watching him nursing yet another large glass of whiskey, a five o'clock shadow on his cheeks, muttering 'for-

18

give me' over and over again, he bore no resemblance to the man who had always been so controlled and as steady as a rock.

Sophie and Ben both took after him; Ben's hair was as Dad's had been – thick, dark and wavy, flopping over his eyes. At eighteen he was as skinny as a runner bean, and even though everyone told him he would fill out before long, he despaired of ever having the kind of muscular body some of his friends had.

Sophie was seventeen, and very pretty – five foot nine, with fabulous shapely legs, glossy dark hair and a perfect size ten figure. Recently she'd decided she was going to become an actress. In moments of irritation Eva had retorted that she was already a drama queen.

She certainly had been a drama queen tonight. Screaming, wailing, flouncing around saying she felt like killing herself, even when the police were still here. And she kept going over and over what had happened, almost as if she was in a feeding frenzy over the drama of it. She'd even gone into the sitting room and telephoned some of her school friends to tell them all about it.

Eva felt Dad should have asserted himself then and told her she had no right to divulge such a personal thing, because by tomorrow it would be all over Cheltenham. But he didn't seem to notice what Sophie was doing. Yet what upset Eva most was that her sister was only reacting to how this tragedy would affect her. 'What will people think of me?' Sophie had said, just before spreading the story even further.

'How could Mum be so selfish when I needed her to find out about drama colleges?' she said later, seeming totally unaware of how self-centred that remark was.

Eva loved Sophie, but she had always been a spoiled brat. Whatever she wanted, she got. At seven she wanted ballet

19

lessons, and she'd only been going six months when she threw a tantrum because she wasn't picked to be in a show. Dad tried to reason with her and explain she just wasn't good enough yet, and that by next year she would be, but she wouldn't see reason and refused to go to dancing any more.

Next she wanted a pony, and she went on and on about it till she got Pepper. Within two months she was refusing to even feed her, let alone ride her. She said Pepper smelled.

Eva had wanted a pony too, and she asked if she could look after Pepper. She'd never had riding lessons like Sophie because the lesson time on Saturdays coincided with activities Ben and Sophie went to, but she felt she could learn quite easily.

'I'm not throwing more good money after bad,' Dad said in that voice he had when his mind was made up. 'I'm selling Pepper and that's the end of it.'

Eva could see Mum thought this was unfair. 'Give Eva a chance, she's far more responsible than Sophie,' she argued. 'Besides, all three of them need to learn that caring for an animal should be taken very seriously.'

Dad had just cast a scathing glance at Flora, as if he held her accountable for Sophie losing interest in the pony. 'I've made my decision. Pepper is going, and we'll have no more talk about it.'

To this day Eva could still remember the triumphant smirk on Sophie's face. She didn't want Pepper herself, but she didn't want her elder sister to have him either.

Eva wasn't one for resurrecting past hurts but earlier, when Sophie had claimed that it was Eva's fault their mother had killed herself, she'd nearly slapped her.

'How can it be my fault?' she asked. 'I'm the only one who ever helped her around the house. I never demanded anything of her.'

'You did! You've kept on about having a twenty-first birthday party,' Sophie retorted.

Eva's birthday was the twenty-sixth of April – a little less than a month away – and she could hardly believe her sister would claim such a thing, as she'd barely mentioned it at all. She looked to Ben and her father for support. But they just sat there and said nothing.

'It was Dad who suggested I had one,' Eva pointed out. 'If you remember, I said I didn't want a party.'

'You pressured Mum to get a marquee put up in the garden.'

Eva had been incredulous at that. 'That was Mum's suggestion. Tell her, Dad!'

He didn't answer, just gulped down the rest of his drink and filled the glass again.

It was Ben who put an end to the argument. He banged his fist on the table and said it wasn't decent to argue at such a time.

He was right of course, and as much as Eva had wanted to point out that it was Sophie who hassled their mother every single day about something, she knew this wasn't the time for it and had lapsed into silence.

As the chiming clock in the sitting room struck midnight, Eva felt someone had to make a move. 'We can't make sense of anything sitting here,' she said, getting up and looking to Ben and Sophie. 'Perhaps you two should try to get some rest too?'

'I'm not leaving Dad,' Sophie said, sticking out her lip. 'He needs me.'

Eva shrugged; their dad was in a world of his own, and she doubted he needed Sophie's prattling and hysteria. 'If any of you need me, you know where I am.'

Up in her bedroom, Eva lay down on her bed and sobbed. She desperately needed someone to put their arms around her and tell her the misery she was feeling would go away in time. While she knew it was awful for everyone, she'd had the worst shock in finding Mum, and she'd been the one who had been questioned the most. So surely Dad could have put his own feelings to one side for a moment and thought of her? He'd cuddled Sophie and Ben, and even reminded them they still had him, but he'd ignored her.

She really didn't want to dwell on it now, but the truth was she was always the one who was ignored by Dad. Right back when she was only seven or eight years old, she had felt he cared only about Ben and Sophie and she was virtually invisible. Even Granny and Grandpa, his parents, had been the same. They talked to her, bought her presents, and yet the two little ones got the lion's share of their attention.

Mostly she thought it was because she wasn't pretty like Sophie, or clever like Ben. Sophie demanded a centre-stage position and always got it; Ben charmed people and made them laugh.

Maybe that was why she became rebellious at fourteen. She truanted from school, hung around with rough kids from the council estate, and allowed herself to be led into trouble and to dress like a goth. While she knew she was alienating herself from her parents, at least outside the home she felt she was somebody; she was even admired by her new friends because she didn't act like the 'posh' girls they knew.

Unfortunately, when she left school her appearance made things very difficult for her. The only work she could get was in fast-food outlets, and that incensed her parents even more.

A horrible incident when she was nearly eighteen had finally brought her to her senses. Yet even though she had admitted to her mother then that she was ashamed of how she had been, Dad never praised her for changing her ways. Even when she got her present, good job in the mail-order company, dropped the goth look, let the black and purple dye in her hair grow out and wore suits and smart dresses, he still acted as though she was an embarrassment.

Recently she'd been promoted to Head of Customer Services, with a big pay rise, but Dad hadn't once asked what the job entailed, or shown an interest in the people she worked with.

As for the twenty-first birthday party, she had never wanted one. The people she would have liked to celebrate with were the ones she worked with, and they would be uncomfortable at the kind of posh show-off do Mum and Dad wanted.

What would happen to the family now? She couldn't imagine how they could hold together without Mum. She might have been erratic, disorganized and given to being distant sometimes, but she had been the hub of all their lives.

Was she severely depressed, and none of them had ever realized?

Eva didn't know very much about depression, but she had read in a magazine that artistic and sensitive people tended to be more prone to it. Flora was artistic: she'd been at art school when she was young, and Eva remembered her drawing pictures for all three of them when they were small, making lovely Christmas decorations and cards, and she was always called upon to design posters for school events. Even her vintage clothes were part of that. Could she have become depressed because she had no outlet for that side of her personality?

It occurred to Eva then that she really didn't know anything much about her mother. Flora rarely spoke about her youth – what ambitions she'd had, who her friends were – or even how she felt about anything. Eva knew plenty of trivial stuff – that she'd rather have a bar of Cadbury's chocolate than a posh box of chocolates, or that green was her favourite colour and peonies her favourite flowers – but not serious stuff like what made her really angry, or what her worst fear was.

But now she came to think about it, they'd never really talked, not the way Eva talked to other women at work. They told Eva stories about when they were young, about their families, and sometimes they spoke about the mistakes they'd made along the way. Each little confidence brought them closer as friends, but Mum never opened up about anything. It was as if she held up an invisible shield to stop anyone getting close.

It was clear enough that something, or someone, had caused her to be so unhappy that she had been pushed over the edge.

But such things didn't erupt out of nowhere in one day. So why didn't she tell anyone what was wrong?

Chapter Two

Olive Oakley rested her head in her hands, so stunned by the phone call from Eva Patterson that she wasn't even sure she'd managed to offer her sympathy and support.

Olive was a partner in Oakley and Smithson, a fast-expanding mail-order fashion company, and Eva was one of her most promising employees. A statuesque and glamorous blonde in her forties, Olive had worked her way up in the rag trade, from machinist to running her own company, by sheer tenacity and force of personality. Someone in the trade once described her as 'the kind of woman who would eat her own young'. That had amused her; she had retorted that was why she'd never had any children.

Yet however hard-headed she usually was, she had a real soft spot for young Eva. The girl acted like she was tough, but Olive knew that was the armour she hid behind, and underneath she was very vulnerable and unsure of herself. Nothing could be much worse than being the first on the scene of a suicide. And if it was your own mother it was hard to imagine how anyone could recover from such a trauma.

It was nearly three years ago that Eva had arrived here for an interview, yet Olive remembered it as if it was yesterday.

On her way to her office to prepare for the interviews she was holding that day, she had glanced at the three girls waiting in reception and was appalled to see that one of them was a gothic horror. She had thick black eye make-up, black and purple hair like a rat's nest, and was wearing a long scruffy black dress and Doc Marten boots.

Up in the office she looked at all the girls' application forms and decided, based on where each of them lived and the school they'd been to, that the horror was Sharon Oates.

She decided she would interview the other two, then tell Sharon the position was filled.

On paper the one called Eva Patterson looked ideal. Good handwriting, school and address, she could type, and she'd had work experience in telephone sales. It was a little worrying that the only real jobs she'd had were in fast-food outlets, but at least that proved she had a work ethic. She also liked the fact that the girl listed her interests as reading, fashion and sewing.

She buzzed through to reception and asked that Eva Patterson be sent in. To her utter dismay, the girl who came in was the goth.

There was nothing for it but to carry on with the interview.

Yet despite the way the girl looked, she had good manners. She held out her hand and said, 'Good morning, Miss Oakley. I had a look at your catalogue while I was waiting and the clothes you sell are gorgeous. I really want to work here.'

It was even more astounding that such a nice voice came out of such a fright. It was well modulated, clear and with a sparkle to it – all important attributes for someone wanting to work in telephone sales. So she shook the proffered hand, and asked her to sit down.

She began the interview by asking why Eva had had so many previous jobs.

'Because they were all awful places. I don't even like eating that kind of food, let alone serving it,' she said candidly. 'I kept moving on, hoping the next place would be better, but they never were.'

'But you got five Bs in your GCSEs. Couldn't you have aimed higher?' Olive asked.

'I got into the mindset that it was all I could do,' she replied and hung her head. 'And I thought having any kind of job was better than no job.'

'So what finally made you lift your sights a little higher and apply for this job?' Olive asked with a touch of sarcasm.

The girl blushed, visible even through her ghastly thick make-up. 'Because I suddenly saw how low I was sinking, and I was determined to change my life.'

All at once Olive sensed the girl wasn't talking about just the jobs she'd had, but something more. 'Was this getting in with the wrong crowd by any chance?' she asked.

The girl lifted her head and there was a spark of defiance in her blue eyes. 'Yes, it was. I was a fool. I let them lead me around by the nose because I was desperate to have some friends.'

'And now?'

'I woke up and realized they weren't real friends, and if I carried on the way I was going, before long there would be much worse coming to me. I want to turn over a new leaf, to get a job I could love and make something of myself.'

Olive had interviewed dozens of people over the last ten years but she'd never met any other interviewee who was so frank. 'What do your parents think of this idea?'

'I haven't told them about it,' she said. 'I thought action would speak louder than words. Besides, I'm doing this for me, and if it does turn out that it makes them proud of me, then that will be a bonus.'

Olive was reminded of herself at eighteen. She'd been in one sort of trouble or another since she was thirteen, choosing the roughest people in the neighbourhood to pal up with as a protest against parents who ignored her. But the more trouble she got into, the more alienated she became. Finally they threw her out, and but for her aunt who believed there was good in her and took her in, she could well have ended up in prison.

She had a strong feeling that Eva was in much the same place.

'If I was to give you this job,' she said cautiously, 'would you turn up on the first day in normal clothes, without all that hideous gunk on your face, and with your hair neatly brushed?'

She waited, expecting some sort of protest.

But Eva surprised her. 'Yes, I would. You see, when I was waiting downstairs and I saw some of the other girls who work here, I had a bet with myself that you wouldn't even interview me because of the way I look. But you did, and you have looked and sounded interested in me. I appreciate that. Besides, if I'm going to turn over a new leaf, I need a new image too.'

Olive wanted to laugh, but she suppressed it. The girl had spirit, and she liked her straight talking.

'A month's trial then,' she said. 'I do demand a smart appearance, good time-keeping, and politeness and attentiveness to all the customers. Be here at nine sharp on Monday morning.'

Eva arrived ten minutes early in a neat black suit and white blouse. Her hair had been trimmed, the purple tinge toned down, and her only make-up was a little mascara and lipstick. For a moment or two Olive hadn't recognized her as the same girl.

That was nearly three years ago, and Eva had never let her down.

Olive had started her mail-order business ten years earlier in the back room of a dress shop in Cheltenham. Back then she'd sold a limited range of fashionable clothes in larger sizes. Such was the demand for her clothes that she soon had to expand. Now she employed twenty people here in a small industrial park just outside Cheltenham, and they used a factory in Wales to make up their own designs.

Olive knew her success was mainly due to excellent customer relations. Satisfied customers recommended their friends, so she always had to be certain her employees understood this.

Eva grasped it immediately. She had only been working for the company for about six weeks when Olive observed her jotting things down in a notebook after some of the telephone calls. When asked about it she said she made a note of the reasons a garment was being returned. If several people had said it was larger or smaller than standard, or if a colour wasn't quite true to the catalogue colour, she advised customers of this.

Olive was impressed, and when she found there were fewer returns from orders Eva had taken, saving the company money, she implemented a policy that all new lines should be checked for size and colour. This information was now given to all the staff manning the phones.

Eva was also excellent at dealing with difficult customers; she could smooth ruffled feathers, charm the irate and was always diligent in sorting out their problems. As she was also well liked by all the staff, Olive had recently promoted her to be in charge of customer services.

But it wasn't Eva's value to the company that Olive was thinking of now; she was concerned about what the tragedy would do to her protégée. She was likely to be like a ship without a rudder. Drugs, drink, promiscuous behaviour and dropping out of work were all traps she could fall into.

Olive wished she'd been able to vocalize her concern better over the phone. She had expressed her shock, and said that Eva could take as much time off as she needed, but that wasn't quite the same as asking if she had someone to talk it over with or offering a shoulder to cry on.

Olive could only guess at what the girl was going through. Her father, brother and sister would all be in pieces, and Eva was far too young and distraught herself to be able to cope with everyone else's grief. She just hoped that Eva wouldn't start thinking she was in some way to blame.

The phone call from Eva had come on Monday morning, and Olive didn't expect her to ring again until the following week. But on Thursday morning of the same week, Olive arrived at work to find Eva waiting for her in the car park.

She looked as if all the stuffing had been knocked out of her; her shoulders were hunched and she was very pale. She was wearing a black trouser suit, one of the company's best lines, and a pink print blouse underneath it. Olive assumed she'd called round to give her some idea of how long she would be off work, but wondered why she felt she needed to dress so smartly for that.

'You didn't need to come in,' she said. 'You can take as much time off as you need. But how is it at home?'

'Awful,' Eva replied. 'I wondered if it would be alright to come back to work?'

Olive noted the dark circles beneath her eyes and knew she hadn't been sleeping.

'Are you sure you want to do that? Don't worry about losing money, I will make sure you get paid.'

'It's not about the money. I just can't make it any better for my dad or my brother and sister by being there, and at least here I feel useful.' Eva's voice shook, as if she was struggling not to cry.

30

'Come and sit down here.' Olive led her to a bench. 'Tell me all about it?'

'Dad's brooding and drinking,' she admitted. 'Sophie keeps having hysterics. As for Ben, he's just terribly sad and bewildered. I don't know what to say to them, or what to do.' She looked at Olive with haunted eyes. 'Our doctor called round on Monday morning, and I confided in him about it. He said it's often like that until after the funeral. That's been arranged for next Wednesday, to allow time for the post mortem to be done. There will be an inquest too. But I think that will be much later.'

Olive took Eva's hand and rubbed it between hers. 'I sensed three years ago that things weren't great for you at home. But you've been so happy since you've been here, I supposed things had got better. But had they?' she asked.

Eva's eyes filled with tears. 'They had in some ways. Mum and Dad used to always be on at me, but that got better when I stopped going out all the time like I used to, and because I dropped the goth thing. I crept around Mum, doing chores and stuff, so that kept the peace too. But it still wasn't great. I often felt I was a disappointment to them.'

Olive sighed. 'How could anyone be disappointed in you?' she said, and she put her arms around the girl and hugged her to her. 'You are bright, funny, hard-working and you get on with everyone. I'm very glad I took you on. You've certainly never disappointed me.'

Perhaps it was because Eva hadn't expected to be hugged or praised that she burst into tears. Normally Olive couldn't cope with emotional scenes, but her heart went out to Eva and she held her and let her cry. 'I meant it,' she said. 'One of the nicest things about you is that you are completely unaware that you have a great many special qualities.'

'I'm so sorry to burden you with this,' Eva sobbed out,

desperately trying to pull herself together. 'You'll think I'm not fit to be at work now.'

'I'd rather you were here crying than doing it somewhere all on your own,' Olive said.

'It's just that at home Sophie and Dad are making me feel it's my fault,' she said, sniffing back her tears. 'Dad hasn't once put his arms around me or said how awful it must have been for me to find her. It's like I don't count for anything. Why did she do it, Olive? She had everything any woman could want.'

Olive had driven past their house on several occasions and looked at it with envy.

'I don't know, darling,' she said. 'Maybe something will come up in the post mortem to explain it. But even if it doesn't, you mustn't think you are in any way responsible. People do irrational things sometimes and there isn't always a good reason. But as for your dad and your sister, I dare say they are just confused and angry, I believe that's a common reaction to suicide. But if you need someone to talk to, I'm here.'

Eva got up from the bench and attempted a watery smile. 'Thank you for the advice and the kindness. I'll remember them both. But there's work to be done.'

Olive was impressed that, as bad as Eva was feeling, she had kept her dignity and remembered that this was her boss she was talking to, not an aunt or a friend she expected to be able to lean on. So she handed her a tissue and patted her on the shoulder. 'Now go and wash your face, put on some lippy, and get yourself a cup of coffee. Let someone else deal with the difficult customers for the time being. And when you need more time off, let me know. You will get through this.'

Later that morning Olive watched Eva talking to a customer on the phone, and she marvelled at the girl's ability to put aside her own troubles and do her job properly. She was very tempted to phone Mr Patterson and remind him his eldest child needed some support from him. But of course it wasn't her place to interfere.

That evening as Eva drove home she felt a little better for a day at work. It had made her believe she could get through this, and that the sun would shine again before long.

From her first day with the company, she'd loved it. It was only twenty minutes' drive from home, a modern, light and airy two-storey building in pleasant surroundings, and the other staff were all warm, jolly people. Her parents had never taken any interest in her work; they never even looked at the firm's catalogue, and the implication had always been that it was a dead-end job. But it hadn't bothered Eva too much because she was happy there.

She realized Olive must have told the staff what had happened because they all said how sorry they were. But no one had asked how she felt, and she'd been very glad of that. She didn't really know how she felt, or even how she should feel. Was there a proper way to feel about your mother's death?

The horror of the scene in the bathroom was as sharp now as it had been when she found her mother. She suspected it was going to haunt her for ever. Yet she hadn't really cried about it – well, except this morning with Olive. Perhaps that was because she was angry at what had been done to the family. But there was also bewilderment, and anxiety that she may have unwittingly done or said something that had pushed her mother over the edge. But she didn't feel grief as such, at least not the way she'd read about it in magazines. Was that because her emotions were frozen by shock?

She had asked the doctor about grief on that morning when he called round. She was expecting to find she was abnormal in being relatively calm and being able to do normal chores. His response was that grief affected people in many different ways. Drinking and staring into space, like her father was doing, was one way. Ben's silence was another, and Sophie's hysterical outbursts still another. That made Eva feel ashamed that she believed her sister was just milking it for attention, and she resolved to be kinder to her. The doctor had added that some people went into denial and acted as if nothing had happened for a while, but it usually caught up with them sooner or later. Eva didn't seem to fit into any of those camps, and she wondered if she could ask Olive her views on it.

But her overriding distress was the way her dad was treating her. He had never been a warm person; Mum had often said that he lacked empathy. But to all intents and purposes Eva could have been an uninvolved lodger. The day after it happened he had gone to both Sophie and Ben's rooms to talk to them. As she passed the doors she saw him cuddling them and telling them that it would get a little easier every day. But he'd barely said a word to her.

As she drove home she decided that tonight she was going to make him talk to her. If he had some issue with her, she wanted to know about it. He was not going to shut her out; she didn't deserve that.

Eva sighed as she saw the state of the kitchen. She had cleaned it up the previous night but now it was strewn from end to end with dirty dishes, saucepans, food packets and tins. It was almost laughable that the rest of the family were all at home because they were grief-stricken, yet they could still stuff their faces with food.

Music was coming from upstairs, and when she looked in the sitting room there was more mess there – cups, empty Coke cans, plates and crisp packets.

Wearily she went upstairs and found Sophie wearing her

dressing gown, sitting on her bedroom floor drying her hair and listening to a Madonna cassette blaring out. 'I see you found your appetite again,' she said. 'Would it be too much to expect for you to tidy up after yourself?'

Sophie switched off the hair dryer, looking contemptuously at her elder sister standing in the doorway. 'I suppose you think you're in charge now?' she shouted over the music.

Eva went into the room and turned the music off. 'Someone has to act responsibly,' she said. 'Where's Dad?'

'Dunno,' Sophie said sullenly. 'He went out around two. He said something about making arrangements.'

'And Ben?'

'He's in his room.'

'I am not the enemy, neither am I the housekeeper, and we've all got to pull together now to get through this. Now come downstairs and help me clear up.'

'I can't, I'm getting ready to go out,' Sophie retorted.

'Going out where?'

'Meeting my friends, if you must know.'

'Do you think it's appropriate to go out at such a time?' Eva asked.

'You've been to work!' Sophie sounded indignant now.

'That's different, and you know it,' Eva knelt down on the floor beside her sister. On Sophie's bed a short red ra-ra skirt and a skimpy top were laid out. 'Just look at how that would seem to people, you gadding off dressed up for a night out so soon after –' She broke off, unable to actually say, 'Mum's death.'

'Did Mum think of our feelings? Does she deserve any respect?'

There was such hurt in Sophie's voice that Eva took her hand and held it between both of hers. 'No, she didn't consider our feelings and that makes me as sad as it does you.

But we have to behave in the right way, to try to keep some semblance of dignity.'

'She's ruined my life,' Sophie pouted. 'Everyone is talking about it. I hate her now. She was a selfish cow.'

Eva wriggled nearer her sister and drew her into her arms. It was tempting to point out that people only knew about it because Sophie had told them but, as irritating as her sister could be, she was only seventeen and she hadn't stopped to think before she spread the story around.

'Yes, she was selfish, and I don't understand it any more than you,' Eva said, smoothing back the younger girl's dark hair from her face. 'But don't say you hate her; she may not have been able to help herself. We might find out that she had a good reason, and then you'll feel terrible that you said such a thing. You've still got me, and Ben and Dad. I'll cook us some dinner and maybe we can all talk about stuff, decide what we're going to do.'

Sophie clung on to her, crying softly. 'Didn't she care about us?' she said brokenly.

'Of course she did,' Eva said soothingly. 'I've heard that sometimes the verdict at inquests is that they "took their life while the balance of their mind was disturbed". That's like being crazy for a short while. It doesn't mean she couldn't bear us any more. Her note said "Forgive me". I think we should.'

'You'd forgive anybody for anything,' Sophie said. But for once there was no scorn in her voice.

'I won't forgive you if I come home tomorrow night and find such a mess,' Eva said teasingly. 'Now finish drying your hair and come and help me get the dinner.'

Eva was just mashing the potatoes when her father came in. She was pleased to see he looked the way he used to before

36

this happened, in a navy-blue suit with striped shirt and tie, no stubble and his hair combed.

'Did you go to the office today?' she asked.

'Fleetingly,' he said. 'Amongst other more pressing things.'

His curt tone made her wary. She decided not to comment on his appearance.

'I've made sausages and mash,' she said. 'I hope that's OK. I'll need to do some food shopping tomorrow. Would you like a cup of tea?'

He didn't reply and walked through the kitchen to the sitting room. She heard him pouring himself a drink.

A few minutes later she heard him pour more into the glass. She looked at Sophie, who was laying the table; Sophie shrugged, as if to say, 'Here we go again.'

'Will you go and tell Ben that dinner's ready, please?' Eva asked her.

Ben came down with Sophie, and Dad came back into the kitchen. He'd taken off his jacket and tie, and his glass was filled to the brim with whiskey. He sat down at the table and Eva dished up the food.

Nothing was said by anyone for some little time. Ben and Sophie were eating eagerly, but Dad only took a few mouthfuls of his dinner between gulps of whiskey.

Suddenly he put down his knife and fork and looked pointedly at Eva. 'Can you tell me why your mother would leave you her studio?'

'Studio?' she asked, frowning in puzzlement.

'Don't play the innocent,' he said sharply.

'I don't know what you are talking about,' she said truthfully. 'Please explain, Dad. And don't be nasty, I've done nothing to deserve that.'

'You've done nothing to deserve being left a studio in central London that must be worth a small fortune, that's for sure.'

Eva's mouth dropped open. Ben and Sophie looked equally shocked.

'I really don't know anything about any studio. Are you saying Mum owned this?'

'Well, of course I am,' he snapped. 'She lived there with you before we got married.'

Eva could only stare at him in consternation. Instinct and the spiteful look in his eyes told her he was out to hurt her. A cold shudder went down her spine.

'You mean, you, Mum and me?'

He sneered at her. 'Your mother was living there with you when I met her.'

She understood the implication in what he'd said, but she couldn't really believe it was true. Or that he was cruel enough to say such a thing just because he was angry.

'I thought I was born a year after you married,' she said in a small voice.

'That's just what your mother wanted everyone to believe. She never did like the truth too much.'

Eva looked into his dark eyes and saw utter contempt for her. She had a gut feeling he'd been waiting a long time to drop this bombshell.

'Loads of people have a baby before getting married.' Ben spoke out defensively, clearly not really grasping what his father meant. 'Don't be mean to Eva, Dad. It's not her fault.'

'It *is* her fault. I know she was in cahoots with Flora over the studio.'

Eva was shocked and bewildered. She knew nothing about a studio, and she couldn't imagine why her dad believed she did.

'Dad, if Mum had a studio, I promise you I knew nothing about it. Are you saying she's left a will with this in?'

'Yes, I bloody well am,' he said, his voice rising. 'I went to see the solicitor this afternoon. It was bad enough having to explain to him about Flora, but then I found she'd betrayed me. We wrote wills years ago, both of us leaving everything to the surviving partner. But the sneaky bitch had another one drawn up for herself, and in it she's not only left that studio to you, but she also left her half of this house to Sophie and Ben. That means I can't even bloody well sell it and move on if I want to.'

All three siblings looked at each other anxiously. None of them really understood legal matters, but the fact that their normally calm father was so angry told them this was something really serious.

Sophie broke the silence first. 'Does that mean we have to leave here?'

'Of course it doesn't,' Ben said, reaching out to pat his sister's shoulder.

'I've worked my socks off for this house,' Dad raged, growing flushed in the face. 'Your mother wanted for nothing, and never did a day's work. I even took her kid on and brought her up as my own. This is how she repays me. I can't even claim on the life insurance because she topped herself.'

Only one line of that bitter tirade really registered with Eva: *I even took her kid on and brought her up as my own.*

'Are you saying I'm not your daughter?' she asked in a shaky voice, hoping against hope he'd only said it in the heat of the moment.

'Are you stupid along with being conniving? Of course you bloody well aren't,' he said, taking a long swig of his whiskey. Then, putting the glass down, he glared at her balefully. 'Anyone with only three brain cells would've worked that out years ago.'

Suddenly the reason she looked so different from her brother and sister was clear. It had been commented on by other people, but Mum had said Eva took after her side of the family.

The enormity of it, and to be told in such a spiteful way, felled Eva. All she could do was flee, running out of the kitchen into the courtyard and then on down the drive and out into the road beyond.

Her mother had gone and now she was just a worthless stepchild, only there on sufferance.

She kept on running until she came to fields. Seeing a farm gate, she climbed over it and slumped down on to the grass behind the hedge, crying her heart out.

Earlier, she'd told Sophie they should forgive their mother. But how could she forgive this? How many times had they looked through photo albums together? Always Mum had said stuff like, 'Look at you, Daddy's girl,' when she was in his arms or on his lap. Taking her first few steps, or on a climbing frame or riding a tricycle, Dad was almost always there with her. In later pictures, when Ben and then Sophie had arrived, it was still the same. Maybe she was now too big to be in his arms – the new baby had that place – but they were happy family pictures, and she looked as right in them as the other two did.

She had often asked why Sophie and Ben were taller, darker and thinner than she was. But Mum always said that was how it was in families sometimes. Perhaps that was true, but by the time she was six or seven she was old enough to be told she had a different father.

Eva really didn't know anything about a studio. She knew from old photographs that they lived somewhere else before Ben was born, but Mum had never said where it was, just as she'd never said anything much about her own childhood, or her parents.

Eva had asked her about them once. Granny, Dad's mum,

was ill in hospital, and though Eva was only nine she sensed Granny was going to die soon by the way Mum and Dad talked. By then she knew most children had two sets of grandparents, and she asked where her other set were.

'They died before you were born,' Mum said. 'They lived in Cornwall.'

That was it really. Scanty information which, if today's revelations were anything to go by, might not even be the truth. All she really knew for sure about her mother was that she had gone to art college in London during the 1960s. There was one of her paintings in the sitting room – a view of a beach which Mum said was in Cornwall. Perhaps it was close to where she had lived as a child, but she never said.

It had been dusk when Eva went into the field, but now it was pitch dark. She only moved because her teeth were chattering with the cold; she didn't want to go home, but she had no money on her. And in just the sweatshirt and jeans she'd changed into before she cooked the tea, she'd be frozen stiff by morning. She hoped that Dad would apologize and talk to her about things in an adult way. But she didn't hold out much hope of that.

She had only walked about a hundred yards when she saw headlights coming towards her, and as it got closer she recognized the car as hers. Ben was driving it. He'd passed his driving test a short while ago, but he wasn't insured to drive. Spotting her, he did a U-turn in the road, and jumped out.

'Where have you been? I've searched all over for you,' he said frantically. 'I've been so worried. Dad was horrible, I'm ashamed of him.'

Ben had always been sensitive and caring about others, and Eva was so touched by his anxiety for her. She felt she had to make it better for him. 'Yes, he was horrible. But I suppose he's hurting and needed to lash out at someone. I just wish it hadn't been me.'

'I shouted at him and said he should be ashamed of himself.' Ben put his arm around her awkwardly. 'I said if he felt he had to tell you, then he should've picked a better time.'

'Did you and Sophie know all along?'

'Of course we didn't. It was as much of a shock to us as it was to you. But as far as I'm concerned, nothing will change between us. You're still my sister,' he said as he wiped his damp eyes with the back of his hand.

'Thank you, Ben, that means a lot,' she said. 'I don't know what I'm going to do now, though. I think it would be better if I moved out. But I really don't know anything about this studio. Do you?'

Ben shook his head. 'Nothing at all. But if it helps, I think Dad was ashamed when you ran out. He said he'd been telling Mum for years she should tell you the truth. Sophie said he was cruel too. She said it was bad enough losing Mum and she didn't want to lose you too.'

'Did she?'

'Yes, she did, and she meant it. But come on home now, you're like a block of ice.'

Eva got into the car, and as they drove she told him what she'd been thinking while she was in the field. 'What else is there we don't know about? It's scary, thinking you know someone well and then you find out you don't know anything.'

'I wish I'd never found out Dad could be like that,' Ben said tartly. 'It's as if everything we believed in has collapsed. I'd better warn you that Sophie is freaking out again too. She thinks Dad wants to move away. She's an imbecile sometimes; Dad never said he wanted to sell the house, only that he couldn't. He isn't going to dump us.'

'It looks like he wants to dump me,' Eva said glumly. 'We haven't even got through the funeral yet. I dread to think what else will come out of the woodwork then.'

Chapter Three

On the morning of the funeral Eva woke to the sound of heavy rain.

She got up and went to the window. Her room looked out over the garden wall on to the estate of new houses. She could see that the patio on the nearest one was awash.

The last few days had been really spring-like with warm sun, and she'd hoped it would remain that way for the funeral. Now the thirty or more people coming back here afterwards would have to be indoors; the house was big enough for that, but it would have been less stressful if some of them could have been out in the courtyard.

Flora's post mortem had shown she had no medical problems. Eva had been hoping against hope there would be something, as at least that would make sense of why her mother chose to end her life. If there was another man in her life, that hadn't come to light either. But then Eva had known that was never a possibility because Flora rarely went out on her own. She was always at home.

Her death had been the most terrible thing. Just thinking about it made Eva's chest so tight she could hardly breathe, and she wondered how she would ever get over it. Then for Andrew to tell her he wasn't her father in such a nasty way had crushed her even more. He had made an apology of sorts the following night, his excuse being that he was upset and had drunk too much. But however much she wished she could forgive him, she found herself running his words over and over in her head, just as she was constantly dwelling on how she'd found her mother.

She had never had the kind of affectionate, jovial relationship with him that she'd observed other people seemed to have with their fathers. He had always been stern and critical, and had never invited any kind of confidences. Her mother had often rolled her eyes at his lack of compassion and sense of humour and told Eva that she must make sure before she got married that her man had both those important qualities.

It was funny how she kept recalling similar remarks her mother had made. Had she been trying to tell her that the marriage wasn't a happy one?

Yet Eva felt Andrew was being honest when he insisted he'd tried to make Flora tell her the truth for years. He said he had been afraid that if she ever asked for her birth certificate, she would see her mother's maiden name Foyle was on it, and a gap left where her father's name should have been. He said Flora had always promised she would tell Eva the truth at an appropriate time.

'Each time you reached a milestone – your sixteenth birthday, then your eighteenth – I insisted she told you,' he said. 'But she always said, "Not now. I'll know when the time is right." But I was always afraid that you would need your birth certificate at some stage. Do you remember just before Christmas when you said you'd like money for your twenty-first, rather than a party, because you wanted to go to Thailand with a friend at work? Well, that made Flora panic; she thought you'd need a visa, and for that you have to produce a birth certificate.'

Eva did remember talking about wanting to go to Thailand. She also remembered that her mother got very uptight about it. She even said it was selfish to go away with a friend rather than have a nice family party they could all enjoy.

'You aren't trying to say that was the reason she killed herself?' she asked him incredulously.

He was looking at her accusingly. 'Well, I think it certainly played its part,' he said. 'She was terrified of how you would react.'

'Then why didn't you take over and tell me yourself?' she snapped at him. 'And don't you dare say that it was because you were afraid to, because you didn't have any problem spitting it out the minute she was dead!'

'Oh grow up, Eva,' he said scornfully. 'You know the truth now, so deal with it.'

That cold dismissive response was like a knife through her heart. She was certain that any other man who had brought up a child as his own from babyhood would have reassured her that he'd always loved her, even if she wasn't his biological daughter. But she didn't have the words to say how deeply wounded she felt, and she was also certain that even if she did, it wouldn't make any difference to him.

'I suppose you don't want me in the house any more then?' she said in an attempt to get him to say he still had some feelings for her.

'I certainly think it would be best if you moved out after the funeral,' he replied, turning away from her as if he couldn't bear to look at her. 'After all, you do have a place to go to now.'

Since that night she'd stopped calling him Dad. The word stuck in her throat.

It had been very tempting to leave the house immediately. She had a little money saved – enough to stay in a bed and breakfast for a few weeks – but she didn't go, because of Ben and Sophie. They were bewildered and hurting, and right now they needed her.

A few days ago Andrew had gone out to dinner with a colleague straight from work. Eva made spaghetti Bolognese for Ben, Sophie and herself, and Sophie began talking about the dressing-up clothes they used to keep in one of the rooms in the attic.

'I used to think it was magic that there was always something different and new in there,' she said. 'Remember, Eva, when we found the two princesses' dresses?'

Eva did remember; she was about eleven, and Sophie eight. They had gone up to the attic to play and the two dresses were hanging up – one gold to fit Sophie, and a midnight-blue one for her. Eva knew Mum had made them because she'd seen her come back from the market with the satin, and Mum had put her finger to her lips when Eva asked what it was for. But she hadn't seen the finished dresses before. And they were marvellous, each with a train and an Alice band headdress decorated with jewels to go with them. There was a purple cloak for Ben too.

'She was so good at making us surprises,' Ben said wistfully. 'We played in those outfits so much. Remember the play you wrote for us, Eva? You made us rehearse it nearly every day before we put it on for Mum and Dad.'

Eva laughed. 'The two princesses were competing for the hand in marriage of the prince. Sophie and I had to do all kinds of tasks to show how accomplished we were.'

Ben laughed then. 'And I had to do quick changes to be your servant and test your skills. All I really wanted to do was strut around in the cloak being the prince.'

'Mum clapped so hard when we finally performed it for her and Dad,' Sophie said with a thoughtful smile. 'She was the best actress, she made out she had no idea what we'd been doing up in the attic for weeks.'

'We used to have a lot of fun playing together,' Ben said wistfully. 'She once said to me, "Stick close to your sisters as you grow older, Ben. You three will need each other when I'm gone." Do you think she knew then that she wasn't going to grow old with us?'

'I don't think so,' Eva said, seeing that Ben's lower lip was trembling with emotion. 'I think she only meant that she wished she'd had brothers and sisters to share things with.'

'But she wasn't always fun and happy,' Sophie reminded them. 'What about that time we tried to make plaster rabbits with that rubber mould? She flew right off the handle, and all we'd done was spill a bit of plaster in the kitchen.'

'And that time you collected flower petals to make perfume.' Ben grinned at Sophie. 'I thought she was going to kill you.'

'Well, to be fair, you had made a mess in the kitchen with the plaster. And you did pick off all the flower heads in the garden, Sophie. Anyone would get mad about that,' Eva pointed out.

'She was a bit irrational sometimes, though,' Ben admitted. 'Remember her throwing that dish of lasagne at the wall because Dad said he was bored with it? I could've understood it if she'd done it while he was there watching her, but she waited until he'd gone out.'

'She was afraid to throw it in front of him,' Sophie said. 'He'd have gone mental.'

Sophie's incisive remark surprised Eva. She'd always thought her younger sister wasn't aware that her father had a nasty side. But she wasn't going to agree with her, as Sophie was quite likely to tell tales later. She thought she'd better change the subject.

'Getting back to what Ben said earlier about Mum wanting us to stay close. We will, won't we?'

'Of course,' both Sophie and Ben agreed.

Eva smiled at them. If they felt like that, perhaps there was even hope that Andrew had only reacted the way he had because of his grief and that in time he'd come round too.

But there was no evidence of it yet. He wasn't drinking as much now, and he hadn't been nasty to her again, but an atmosphere hung around the house, heavy with unspoken recriminations on both their parts.

She wanted to talk about it all to clear the air. She thought of offering him this studio she knew nothing about, to convince him she was not guilty of any kind of conspiracy with her mother. But he avoided being alone with her, and she felt he was just waiting for the funeral to be over so that he could tell her to go.

That was what made her confide in Olive. She didn't want to – it was bad enough that Olive knew her mother had taken her own life – but she had to tell someone. It was all whirling around in her mind till it reached the point where she felt she might go mad with it. As it was, it turned out to be the best thing she could have done. Olive explained about wills and what 'probate' meant properly. She said it might be months before it was all settled and the property came to her, so the best thing was to move out right after the funeral, and at that point to go and speak to the solicitor.

Eva had decided that was what she must do, but she couldn't help wondering how Sophie and Ben would cope when she left. They had never done anything around the house; she doubted they even knew how to work the washing machine. In the last couple of weeks she had tried to get them to help her out, because she was doing all the cooking, shopping, washing and tidying up. But although Ben tried,

Sophie refused point blank, and Andrew took the view it was women's work. Eva was afraid everything would fall apart without her there.

'That isn't for you to worry about,' she told herself as she went to have a shower. Ben would be going to university in the autumn anyway, and it might make Sophie less self-centred if she didn't have someone looking after her all the time.

It rained remorselessly all day. When they got back to the house after the funeral, Ben directed the parking on the drive, Sophie took umbrellas and coats, and Eva offered drinks.

'Your children do you great credit, Andrew,' Eva heard a large woman in a very theatrical black hat remark. 'I wouldn't have blamed them if they'd retreated upstairs. It must be awful for them.'

Eva wished she could go and hide. Most of the people who had come back to the house were strangers to her; there were only a handful of family friends and neighbours. She wondered who all the strangers were, and how they knew about the funeral, as Andrew hadn't said he'd contacted people.

The clergyman was a stranger too, and though Eva knew he'd called at the house to talk to Andrew about Flora, it was all too obvious that he didn't know her personally and that he was uncomfortable holding a religious service for a suicide.

Sophie cried constantly before, during and after the service, and both Andrew and Ben kept wiping their eyes. But Eva remained dry-eyed because the hymns, prayers and words about her mother didn't seem to have anything to do with her. The clergyman did mention that she had been very artistic, but that didn't cover how at Christmas Flora would transform the whole house into fairyland with beautiful handmade decorations. She could do amazing arrangements with a few twigs and leaves and whatever flowers she could find in the garden. When she made any of them birthday cakes they were always something fantastic: monsters for Ben when he was little, and Cinderella's coach for Sophie. Eva remembered she had once had a Little Red Riding Hood cake on her birthday, with a forest of marzipan trees, Grandma's cottage in the middle and the wicked wolf spying on Red Riding Hood.

Eva recalled how Mum used to get the three of them to dance with her to old rock 'n' roll records, the wonderful picnics she used to make, and how every birthday she painted them a card. And it always reflected what they were currently interested in, from whales to dinosaurs.

Even when Eva had gone out into the garden earlier in the morning in the rain to pick a bunch of spring flowers, she didn't feel the expected surge of emotion. Mum had loved the garden; in good weather she would be pottering out there all day, and it was beautiful in every season because of her care. Eva had thought she could make a lovely flower arrangement for the top of the coffin. But although she'd found the right shallow container and put oasis in it, the way Mum always did when she made table decorations, when she began to put the flowers in, it looked like something a six-year-old had put together. She did cry then, because she felt she was letting her mother down. She left it in the kitchen; she didn't want Andrew thinking she couldn't even get that right.

The only thing that made her want to cry during the service was when she suddenly realized she hadn't really known her mother. She had always believed that she had; she knew what made Flora laugh or cry, her favourite music, television programmes and types of food. She had even been very good at picking out clothes her mother would love. But now that seemed so very superficial – the way icing on a cake gave you no indication as to what lay beneath it.

It was impossible to imagine her mother's curvy small body was inside the pale wood coffin. Flora had often joked that when she died she wanted her body to be put in a boat, surrounded by flowers and then floated down a river like the Lady of Shalott. Eva knew the famous painting by Waterhouse, and the model even looked like Flora with her long red wavy hair and very pale skin.

She didn't think she ever wanted to see that picture again.

As they came out of the chapel at the crematorium, there was a far larger group of people waiting rather impatiently to go in for the next funeral. That was a further reminder that Flora was only special to her children and husband. Eva supposed the other group, who had lost a loved one through illness or an accident, wouldn't hold out much sympathy for a suicide.

Everyone walked very slowly past the part of the Garden of Remembrance where the undertaker had put the flowers. That too seemed pointless – a waste of money, as they were destined to die by the next day. Eva gathered up the cards with the flowers because she thought they should all read the kind messages. But the rain had made the ink run and most were illegible.

Back at the house, two friends of Rose, the cleaning lady, had put out all the food Eva had bought the day before, and made pots of tea. Eva busied herself taking round a large plate of canapés, but she watched Andrew talking to the guests, a large glass of whiskey in his hand. He had forced a smile at the compliment about his children. She heard him telling Sophie that the lady owned an antique shop in Montpellier, and that Flora was always popping in there and buying things when they first moved here.

It was that comment which made Eva suddenly aware she

might be able to find out more about her mother from some of these people.

Taking courage into both hands, Eva made her way towards the oddest couple in the room because she was certain they had never been friends with Andrew.

The man was tall and thin with lank hair straggling over his collar and John Lennon glasses. He looked like he'd borrowed his dark suit; it didn't fit him anywhere. The woman he was with had coal-black dyed hair, bright green eye shadow and a too short and too tight navy-blue dress for someone plump and past fifty. But it was plain to see she had been a beauty: her green eyes were lovely and her cheekbones sharp, and she had an air about her of someone well used to being admired.

'Hello, I thought I'd introduce myself and ask where you fit into Mum's past,' Eva said, holding out her hand to shake theirs. 'I don't think we've met before.'

'We've met you, Eva,' the woman said with a warm smile. 'We often minded you as a baby. I'm Lauren Calder and this is Jack Willow. We were at art college in London with Flora. We shared a house together as students.'

Jack stepped forward and, instead of taking her proffered hand, he kissed both her cheeks, his hands resting on her shoulders. 'We are both so sorry about your mum, Eva. It must have been a terrible shock to you all.'

The genuine sorrow and sympathy in his voice was soothing. 'It was,' Eva agreed. 'It's very difficult to get your head around such a thing. It's also made me realize that Mum hid a great deal from me. I know nothing about her past, not even about her student days.'

She noticed the way the couple looked at one another. It was the kind of look that said they weren't sure if they should be the ones to divulge anything.

'I'm not looking for a complete biography,' Eva added quickly. 'Just a few little stories. There're so many people here that I don't know. Did Andrew call you all about Mum?'

'He called me,' Lauren said. 'I think I'm the only one of our student group that Flora kept in touch with. Even that wasn't much, just a few words on a card at Christmas really. I rang around the other people, and we all came today because Flora had a special place in our hearts.'

'That's a nice thing to say,' Eva said. 'But why?'

'For many reasons; because she was such fun, so very talented, and because she gave us so much encouragement when she became successful.'

Eva was puzzled at that. 'Successful?'

'Surely you know your mother was a very good artist?' Jack sounded surprised that she didn't appear to know. 'She was selling her work when the rest of us were just dreaming about it.'

'She was?'

'My goodness, she really did keep you in the dark,' Lauren said with a nervous giggle. 'I know she gave up painting, and I never understood why, but I didn't imagine she wouldn't tell her kids about those days.'

'I was surprised to see only one of her paintings here,' Jack said. He pointed to the Cornish beach scene on the wall. 'That's a very early one. I remember her working on it, she said it was a beach near to where she grew up. Are there any more in other rooms?'

'No. That's the only one,' Eva said, looking round at the picture. It had been hanging there in the alcove by the chimney for as long as she could remember – just sea, beach and rocks, nothing in it that had ever made her ask questions about it. But looking at it now, as if for the first time, she could see that it really was a very good painting. The light, clouds and the texture of the rocks were so realistic it could almost have been a photograph.

'She would paint birthday cards in watercolours for friends and for us, but she never used oils or did any big pictures. But if she was so good, why did she give it up?' she asked.

Lauren reached out and took Eva's hand in hers. 'Maybe it was because when she had you three children, she didn't feel the need to paint any more.'

Eva nodded. 'I can understand her not keeping it up when we were small, but it seems strange she didn't start again once we were all at school. She never worked, you know, and it wasn't as if she was that house-proud.'

Jack smiled. 'She used to be the untidiest person I knew,' he said. 'She always claimed she was born to be waited on.'

'She said that to me once too.' Eva smiled back at him. 'I think she only kept the house perfect because Andrew insisted that she must. She didn't do it when he was away –' She stopped short, suddenly aware she shouldn't tell people such things.

Lauren took her hand again, perhaps guessing what had cut her short. 'It's OK, Eva, you can talk about it, especially to us. I bet the last couple of weeks have been a very lonely time for you? My mother died when I was just a bit older than you. I felt so confused, angry, sad, every kind of emotion, and I had no one I could talk it over with. How's your dad been? He was very curt on the phone, I didn't dare ask him anything more.'

'He's been struggling with it,' Eva admitted. 'But then he would, just like all of us. We didn't see it coming. But there is one thing I'd like to ask you. Mum's studio, do you know about it?'

'Well, yes, of course I do,' Lauren replied. 'She bought it

while we were all still sharing a house. That's where I babysat you.'

'She never said anything about it to me, and apparently she's left it to me. That's caused some hard feeling,' Eva said carefully. 'Where is it?'

'In West London. Holland Park.'

Eva didn't know London at all, but she had heard of Holland Park being a smart area.

Jack must have read her expression because he smiled. 'Parts of that area were good even back in the sixties, but not where the studio is, that was virtually a slum. We all said she was mad buying it. But she'd inherited some money – from her father, I think – and she was determined. It wasn't as if it was a real artist's studio, only a little terraced house. I haven't seen it since you were small, but I should imagine it's been tarted up since then.'

Ever since Eva had been told about this 'studio', she'd imagined it was just one big room, perhaps with an adjoining bathroom, because that was what estate agents called such places. She was very surprised to discover it was a house.

Her earlier caution left her. 'Do you know who my real father is?' she asked. 'I thought it was Andrew until Mum's death, but he said he isn't my father.'

Jack and Lauren exchanged glances again.

'She didn't tell you before?' Lauren said, and looked very uneasily at Jack. 'I wonder if that's why she never invited us here?'

'You think she was afraid you'd let the cat out of the bag?' Eva prompted.

Lauren hesitated. 'Maybe, dear. Flora always did play her cards close to her chest.'

'So who is my father?' Eva asked, keeping her voice down as she could see Sophie hovering close by.

'Well, we've always assumed it was Patrick O'Donnell, the illustrator. He was part of our group and they were together for several years.'

'You only imagine! Flora was your friend, surely you know for sure?' Eva said a little sharply.

'Look, Eva,' Jack took over. 'It was a strange time, with lots of things going on, and we weren't always part of it. Flora was with Patrick for a few years, but she left him and later on took legal action to get him out of her place. Pat went off to Canada, tail between his legs, and Flora didn't tell us anything, not even that she was expecting you. By the time we caught up with her again, you were a couple of months old.'

Chapter Four

Eva switched on the light and looked at the clock. It was after two, but she couldn't get off to sleep. She could hear Andrew snoring – just a soft, distant rumbling because his bedroom door was shut. She'd listened to the sound a thousand times in the past and found it comforting that he was close by. But now she knew he wasn't her real father she found it irritating.

She almost wished she hadn't spoken to Jack and Lauren today at the funeral. She'd hoped for some new understanding about her mother, but all she'd got was more puzzles. First, that she'd been a successful artist; Eva knew little about the art world but she did know that only a handful of artists made any real money from it. So why on earth hadn't Mum ever told her that she was one of those few?

Then there was the news about an illustrator called Patrick O'Donnell who might be her father. Jack had said he knew he was living back in England now, and had suggested she look him up. But how could she? If he didn't want her as a baby he wasn't likely to care about seeing her now.

She had spoken to some of Flora's other old friends too, and although she didn't get as much from them as she did from Jack and Lauren, they had created a picture for her of the young Flora they knew. They all said how much of a party animal she'd been, the last one to leave, always up for anything. Someone said how she had mad ideas – camping in midwinter, skinny-dipping in the Thames – and she got people joining her with sheer force of personality. Yet none of this fitted the woman Eva knew; she'd always seemed rather reclusive, and certainly not bold or impulsive.

There were also some pointed little remarks from a couple of people that hinted at Flora being mercenary, hard and devious. Eva thought that might be because as students they'd have all been on their uppers, and they were probably jealous of Flora's success. It was clear from the clothes and the cars of these old student friends that they were still poor, and coming to The Beeches to discover Flora had never had to struggle financially, as they had, might have resurrected that envy.

But why had Flora cut herself off from them? And why when she had been a successful artist had she given it up?

Was it Andrew's doing?

Eva had watched him as these people tried to talk to him during the day and she could see by his strained expression and body language that he was struggling to be polite and had absolutely no interest in any of them. But then Andrew was a businessman through and through – his interests were the stock market, politics and sport, not art. Maybe when he found Lauren's phone number in her mother's address book he'd thought she was a more recent friend.

Eva doubted he'd ever grown his hair long, worn tie-dye T-shirts or patched jeans. He'd never gone to a rock concert, was appalled at drug taking or even smoking cigarettes, and he sneered at New Age people, alternative lifestyles, astrology and vegetarians. In fact he was probably appalled that by contacting Lauren about Flora's death he had unwittingly given an open invitation to a bunch of people he saw as just cranks.

But if these people had been Flora's friends, Eva wondered how and why she ever got together with Andrew. It was odd for a woman who had apparently seen life in technicolour to settle down with a man who only saw black and white and who lived his life through spreadsheets.

Eva knew that Andrew hadn't been wealthy when he married Flora; that came later, when they moved to Cheltenham. This house had been dilapidated at the time. It was selling off the land at the back of it which had enabled them to turn it into what it was today. Yet Flora was the creative one, so why did she always bow to Andrew's taste?

Eva was about sixteen when Andrew first came up with the idea of putting a swimming pool in the old stables. She remembered Flora gently pointing out that pools cost a lot to maintain. Nothing more was said about it for months, then one night Eva walked in the front door and overheard them in the sitting room having a row about it.

She went halfway up the stairs, but stayed there to listen.

'It's just showing off,' Flora insisted. 'The stables aren't big enough for a decent size one that you can really swim in. You just want the neighbours to be impressed. But they won't be forking out for the heating bills, will they?'

'It's me that brings the money in, so I can decide what to spend it on,' Andrew argued. 'The kids will love it.'

'They might at first. But they'll be bored with it in no time. They aren't that keen on swimming, and you've never been interested.'

'I would be if it was right here,' he said. 'Besides, it's a statement that I'm doing well.'

'As I thought, you just want to pose,' Flora snapped back at him. 'It's a waste of money.'

'My money,' he said, and with that he opened the sitting room door to walk out.

Eva had no alternative but to flee up to her room before he caught her eavesdropping.

They went on rowing for some time that night. Eva couldn't hear what they were saying but at one point she heard something smash, then it went quiet.

Flora was very silent and brooding for the next few days, and although Eva asked her what was wrong, she refused to say. As nothing further happened for a few weeks about the swimming pool, Eva assumed Andrew's plan had been abandoned.

When the conversion of the old stables finally got started, Flora didn't protest, but Eva was aware she was still against the idea, because of her tight-lipped false smile. Eva, Ben and Sophie were all thrilled with the pool when it was finished. But, as Mum had predicted, it was a nine-day wonder. They had a few weeks of going in there every night after school and at the weekends, but gradually their enthusiasm tapered off, as did Andrew's. Eva couldn't remember when she'd last seen him use it.

Was that what was wrong between them? Did Flora feel trapped in a middle-class world with a control freak – a man who liked to impress the neighbours with his ride-on lawn mower, his swimming pool and a new top-of-the-range car every year? He played squash with other men, occasionally went to watch cricket or rugby with someone, but Eva didn't think he had even one really close friend. She remembered once, when he and her mother were planning a dinner party, Flora had complained that one of the couples he'd chosen were very dull. Andrew's reply had been that they were 'well connected'. Eva supposed that meant he thought they could be useful to him.

None of these things had fully registered with Eva before today. About the only thing she'd really noticed was that Flora was at her happiest when she was gardening or being creative. She wished so much that she'd thought to ask her mother how she felt about things – deeper questions that might have given her some insight into what made her mother tick.

Maybe that was part of the reason why Flora killed herself; because she felt her family took no interest in her as a person? It must have been very demoralizing to be thought of as just a mother and housewife, especially if she'd once been a successful artist.

Only one certainty had come out of the events of today, and that was that Eva must find a place of her own as quickly as possible. Around six, after everyone had left, Andrew had totally ignored her as she was clearing up. She'd heard him praise Ben and Sophie for holding themselves together and acting with dignity, yet she didn't even get a thank-you for buying and preparing the food.

She wasn't going to stay on here as an unappreciated skivvy. Tomorrow she'd make an appointment with the solicitor, and she'd start looking for a flat.

On Monday afternoon, five days after the funeral, Eva left work early for her appointment with Mr Bailey, the solicitor. After seeing him she was going to view a bedsitter. She would take it, whatever it was like, as the atmosphere at home had become poisonous since the funeral.

It was like walking on eggshells with Andrew. He snapped at her about everything – from moving his piles of paperwork from the kitchen to his study, to asking what he'd like for an evening meal. He kept saying the house was a tip, but he was as much to blame as Sophie and Ben. She was trying so hard to run the house, to keep up with the washing, ironing, shopping and cooking while working full time too. But all he did was complain and criticize.

Sophie sucked up to him constantly, and continued to do nothing to help around the house. Ben escaped as often as he could.

On Saturday morning Eva was just going past her parents' bedroom when she saw Andrew pulling all their mother's clothes out of the wardrobes and drawers and stuffing them into black bin liners. She was so shocked she couldn't stop herself from asking what he was going to do with them.

'I'm taking them to a charity shop,' he snapped.

'Isn't it a bit soon?' she ventured. 'And some of her clothes were very expensive.'

'I know that, I paid for them,' he retorted, not even looking up from stuffing a beautiful brown velvet jacket into the bag.

'What if I sorted them out and took the best vintage ones to sell back to that shop Mum bought them from?' she suggested.

'So you can have the money?' he said with a nasty sneer. 'My God, Eva, you are a piece of work!'

She burst into tears, because nothing had been further from her mind. What she wanted was to see him treating her mother's belongings, whether that was clothes, jewellery or other things, with respect because he had loved her. Shovelling them into bin liners without any thought for the memories they held was so cold-hearted. It was as if he hated Flora now.

'That's right, cry and make a big drama out of it,' he said scornfully. 'Your mother always did that too. She took her own life, Eva! I knew she was a self-centred bitch. But I never thought she'd put herself before the needs of her family. She didn't give a toss for any of our feelings. So you tell me what possible reason could I have for holding on to this lot?'

'Because it's too soon to get rid of it all,' Eva ventured through her tears. 'You might be sorry later.'

'The quicker I get everything of hers out of this house, the better I'll feel,' he said, stuffing more things in.

'Including me, I suppose,' she said and turned away, not wanting to hear his response.

Yesterday she had cooked Sunday lunch for them all: roast

beef, Yorkshire puddings and all the trimmings. Ben didn't come back, Andrew put his on a tray and took it into the sitting room to watch TV, and Sophie ate hers in silence.

Eva went up to her room after she'd cleared up, and she hadn't been there long when the phone rang. She opened the door, intending to go and answer it if no one else did, but Andrew picked it up down in the hall.

'I can't talk now,' he said in the kind of half-whisper that Eva had used in the past when speaking to people her parents wouldn't approve of. 'The kids are all here.' There was silence for a few moments before he spoke again. 'I know, but it won't be long now. The wait is nearly over. I'll ring you tomorrow night.'

Eva closed her bedroom door very quietly. No one did that lowered voice thing unless they were afraid of being overheard and feeling guilty. She was sure it had to be a woman he was speaking to. So was he having an affair and Mum found out? Was that what drove her to suicide? And if it was, how could Andrew put on that huge display of grief?

She stayed in her room until bedtime. No one came to see her, and she felt so terribly alone and uncertain about everything that she cried herself to sleep.

She'd woken this morning feeling tougher and determined. She got the local paper on the way to work, saw the bedsitter advertised and rang to make an appointment to view it at six o'clock. Now as she drove into the car park of the solicitor's, she told herself that even if one door was closing behind her, there was freedom behind the door in front of her.

Mr Bailey was just as she imagined a solicitor to be – old, small, slightly stooped and with half-glasses perched precariously on the end of his nose. His office was lined with thick leather-bound books.

'Do come in and sit down, Miss Patterson,' he said after shaking her hand and offering his commiserations on the death of her mother. 'It had been my intention to contact you right after your mother's funeral, but you pre-empted that by calling me.'

Eva suddenly felt she might cry, but took a deep breath and explained that Andrew had told her about a studio she was to inherit. 'He seemed very cross about it,' she added.

'He had no right to be, or to be surprised by it. I drew up a will for your mother when they first moved to Cheltenham and the studio was left to you even then. He was here with her then, and she made the position quite clear. On that occasion she also changed your name by deed poll to Patterson.'

'Until the night Andrew told me about her will, I didn't even know he wasn't my father,' Eva admitted.

'Oh dear!' Bailey exclaimed, taking off his glasses and cleaning them on a handkerchief. 'To have that revealed so soon after your mother's death must have been very distressing for you.'

'It was, but he seemed to think I had influenced Mum in giving me the studio. He didn't believe that I didn't even know Mum owned one.'

'He shouldn't have taken out his pique on you at such a time, but I dare say it was because of the nature of her death and because I had to reveal to him that your mother had recently changed her will without his knowledge. He was angry with me about that.'

'You mean leaving her half of the house to Sophie and Ben?'

'Yes, my dear. But she had every right to decide what was to become of her assets. These days it is becoming much more common for people to ensure that the remaining partner doesn't have total control of them. Usually they are afraid their other half will remarry and the children of the new husband or wife will inherit the marital home.'

Eva nodded and hung her head. She could feel tears welling up, and she'd promised herself she wouldn't cry.

'You poor child,' Bailey said gently. 'I only met your mother a few times, with a gap of many years in between the first and last couple of times, so I can't claim I knew her well. But I did notice a difference in her the last time she came here. That was just before Christmas. She seemed to have lost the vibrancy I remembered; I wondered then if she was ill, and if that was why she wanted to change her will. In fact I asked if something was wrong, but she smiled and said there was nothing, and that she was just making sure all three of her children would be looked after, rather than just you. That was entirely reasonable in my view.'

'My stepfather doesn't see it that way,' Eva said glumly.

'I can imagine. But in fairness to him, he'd already had the shock of his wife's death to contend with, and it must have been distressing to find that she didn't consider what effect her new will would have on him, and on his security. He has worked very hard for years to keep you all in comfort, and now with your mother's half of the family home being bequeathed to your brother and sister it means his finances are restricted. But even if he should choose to contest her will, it wouldn't change anything. He still owns half the house, and any judge would see that as adequate for his needs.'

Eva thought it served Andrew right, if he had been cheating on her mother. But she couldn't say that without proof. 'I can understand him being upset about that,' she said. 'But if Mum had always said that the studio was to go to me, and he never had any stake in it, why be mean to me about it now?'

66

Mr Bailey made a shrugging gesture with his hands. 'At times like this people don't always think logically, my dear. I hope in time you can heal the rift between you, as I'm sure your mother wouldn't have wanted you to fall out over it. Getting back to the studio, I have no idea of the condition it is in. Your mother did have an agent who took care of letting it, but apparently she dismissed him a few years ago. She had a building society account which rent money was paid into. I need to look into that for you because I have a note here that your name was also on the account. That was another thing she did to make certain it went straight to you.'

Eva looked up at him in shocked surprise.

He smiled at her expression. 'I have no idea now how much money that will be – she may have been drawing it out as fast as it went in – but we'll see in a few days. Meanwhile, it would be prudent to say nothing of this to anyone.'

'I won't, Mr Bailey,' she said. 'I'm planning to leave home. In fact I am going to see a place after I leave here. I can't stay at the house any longer, my stepfather is making me feel very unwelcome.'

'I'm so sorry to hear that,' Bailey said, looking at her with concern. 'I'm afraid suicide always has far-reaching effects on families. Sometimes people act irrationally because they feel unable to grieve in the way they would if the loved one had died of natural causes.'

'I think Andrew ought to remember that I am grieving too.'

'Quite so. All things considered, I think moving out is the best thing for you. But you do understand that you won't get either the property or the money immediately?'

'Of course, I didn't come here expecting that.' She felt embarrassed now that he would think it was only the money she cared about. 'I came just to find out the legal position

67

and get some advice from you. I'm told it has to go through probate, and that can take months.'

'Well, not in your case, Miss Patterson. You see, your mother took the precaution of putting the property in trust for you until you are twenty-one, which I understand is very soon. She also made sure any money in the building society account was payable to you on her death. But there are still papers to be drawn up and signed.'

She nodded, wanting to know how long it would take, but afraid to ask.

'I will make a start on that tomorrow,' he said. He pushed his glasses further up his nose and looked at her gravely. 'I see it as my duty to point out why your mother made these arrangements. She came to me initially because she wanted to make sure you had financial security which wasn't dependent on her husband. She had concerns that if she had other children you might not all be treated equally. As it turned out, she was very astute, and she has made it possible for you to make a new start and to be independent. So you must use it wisely.'

He paused, looking at Eva over his glasses.

'Try to forgive your mother for not telling you the truth about your birth. Such things become more and more difficult the older a child becomes. And however hurt and let down you feel about it, and about her taking her own life, this legacy is proof of her love for you. She took special care and planned ahead for you.'

Mr Bailey's words were like soothing ointment on a sore place. Eva's eyes filled up with emotional tears, and for the first time since her mother's death she felt comforted.

'Thank you,' she said. 'I will remember that.'

'Good.' He smiled at her. 'No fast cars, wild parties or other extravagant silliness. I have no idea how much the property is worth, but it is probably the only legacy you will ever get. Remember that, if you are tempted to fritter it away. And keep your own counsel about it. Sadly, there are a great many people in this life who will befriend you just for a slice of your inheritance.'

Eva nodded in agreement. Olive had said much the same thing. She got up to leave. 'I'll ring you in a day or two and confirm my new address. Thank you for the advice.'

Mr Bailey got up too and took her hand in his. 'The clouds will roll away soon, my dear. You are young and you must look ahead and plan for your future,' he said. 'I'll be in touch with you soon, but if you have any further questions or need any advice, just call me.'

Chapter Five

Eva sniffed back tears as she hauled two large bin liners stuffed with clothes and bedding out of her car. Number 44 Crail Road didn't look any better in the early evening sunshine than it had when she'd viewed the room five days earlier under grey skies. But then she'd supposed few houses were ever going to look as nice as The Beeches.

It must have been a smart address back in Victorian times, a tree-lined avenue of big semi-detached three-storey houses, each with a basement for the servants. But now most of the houses in the road were converted into flats, with front gardens paved over for parking, and they all exhibited a general lack of care.

'It could be far worse,' she murmured to herself. 'And it's not for ever.'

As she struggled up the path between two scrubby areas of overgrown grass, a young man came bounding down the steps from the front door.

'Hi! Moving in? Let me help you,' he said and came forward to take the two bags from her hands.

'Tod!' she exclaimed, recognizing him as a temporary driver who had been taken on at Oakley and Smithson for the Christmas rush. All the girls at work had drooled over him, as he was both charming and handsome – mid-twenties, tall, slender with floppy fair hair and deep blue eyes. Eva had been as guilty as everyone else of fancying him like mad. 'Do you live here too or are you just visiting? I've taken the room at the front on the ground floor.'

He frowned as if trying to place her, then came a warm smile of recognition. 'Oh yes, Eva! I always remember the pretty ones. Welcome to "The Ritz". I've got one of the small hovels at the back. How much more stuff have you got?'

'A suitcase, a TV and a few boxes,' she said, blushing furiously at the compliment. 'But I don't want to impose on you if you're going somewhere.'

'I'm only going down the pub,' he said. 'And I can see you've been crying, so I'll help you in and try to make you think you've found paradise.'

'I got a bit teary saying goodbye. I've never been away from home before,' she admitted, wishing she'd checked her face before she got out of the car. 'It all feels a bit overwhelming. But I'll be fine once I've settled in.'

It took only a few minutes to get the rest of her belongings and dump them on the floor of her room. 'It will soon look like home when you've put all your things around,' he said cheerfully. 'Girls are always good at that. I think homemaking is inbred.'

She had never had an opportunity to say more than a few words to him while he was at Oakley and Smithson, but she'd heard from the girls who went out of their way to talk to him that he made each of them feel like they were the most interesting person in the world. Eva was desperately in need of some kindness and sympathy and it was tempting to blurt out the reason she'd left home. But she stopped herself; she didn't want to frighten him away. 'Would you like a cup of tea?' she said instead. 'I've brought all the stuff with me.'

'That's an offer I won't refuse. I haven't got the nerve to invite you into my room because I left it like a tip. What made you leave home?'

'I just thought it was time I became independent, and I wanted a bit more freedom.'

The landlord had called it a studio apartment, but in reality it was just a bedsitter, with the kitchen part of the room divided off by a breakfast bar. There was a shower in a cupboard and she would share the lavatory with the other ground-floor tenants.

Although it looked shabby and battered it did have all the basic equipment. And there was a new shiny kettle on the breakfast bar which was an addition since she'd viewed the room. Eva filled it up, and as she got out her bag of groceries she asked Tod where he was working now.

'Care in the community,' he replied and pulled a face.

'Old people and stuff like that?'

'Sort of. I drive the disabled or old people to hospital for appointments, but I also do a few shifts helping out vulnerable young people living in sheltered housing. You know the kind of thing.'

Eva didn't really, but she nodded anyway. 'That's a kind thing to do,' she said.

He shrugged. 'I'm just gaining general experience. I want to be a counsellor.'

'How interesting,' she said.

He laughed. 'People always say that, then start telling me their problems.'

'Well, I'm not going to. I don't have any, except how to make this place a home.'

'Then tell me the truth about why you've been crying?'

There was a good reason for the girls at work drooling over him; aside from his lovely blue eyes and soft full mouth there was something very sexy about him. Part of it was that he obviously came from an upper-class background yet seemed very comfortable with and interested in ordinary people. At work he'd always been in company navy-blue overalls, but now in a worn T-shirt and ragged jeans, with untidy hair as if he'd just got out of bed, he looked even better.

'Like I told you, just saying my goodbyes,' she said. 'My younger brother didn't want me to go. I suppose I started to worry that I'd be lonely here.'

'You won't be.' He smiled reassuringly at her. 'This is a quiet night in here, not a sound because everyone's out. But it gets quite rowdy and they are a nice bunch who all pitch in together. You might feel like moving on somewhere quieter in a few weeks.'

The kettle boiled, she made the tea and opened a packet of biscuits.

'Marks and Sparks biscuits!' he exclaimed. 'If it gets around that you buy those, you'll never get a moment's peace.'

They chatted while they drank the tea – about ordinary stuff like their taste in music, clubs in Cheltenham, and preferred takeaways. He asked too about some of the people who worked in the packing department. He said he'd enjoyed working there, as it was always a good laugh.

'Speaking of packing,' he said with a laugh, 'would you like some help with unpacking?'

She wanted him to stay but sensed he was anxious to get to the pub. So she thanked him and said she'd rather do it alone.

'Well, I won't hold you up,' he said, getting up off the sofa. 'But I'm in number six, down the passage at the back, if you need help with anything tomorrow.'

'You've been marvellous, thank you,' she said, hardly daring to look into those blue eyes in case he sensed that she fancied him. 'I think I'm going to like it here.'

She watched him walk down the front garden, admiring his graceful lope and the suggestion of muscles under the shabby T-shirt. If he hadn't been around, she thought she might have thrown herself on the sofa and sobbed for hours. But she wasn't going to do that, however hurt and sad she felt. She must keep in mind that this was a new start, and make it work for her.

Turning back to look at her room, she thought that it didn't look quite so bleak now either. It was big, very light, and at least the walls were neutral, even if they were a bit grubby. The grey cord carpet was worn thin in places, but she could jolly it up with a rug. She pulled out the sofa bed and winced at how lumpy the mattress was, but she hadn't expected it to be anything else.

It would be fun to buy things to pretty the room up: a few pictures, a cloth on the small table, books on the shelves. And at least she had her own shower. Sophie had always hogged the one at home.

As she started to unpack her clothes and hang them up in the wardrobe, she wondered what had been said after she left home this evening. She had told Andrew she was leaving two days ago. At the time he'd seemed indifferent, but tonight he'd been very nasty, saying things that cut her to the quick.

He was standing at the door of the sitting room when she came downstairs with the first two bin liners of clothes. 'How quickly the rats leave the sinking ship,' he said in a sneering tone. 'Just like your mother. She never appreciated all I did for her either. That car, for instance. I went to a great deal of trouble to get you that for your nineteenth birthday.'

She was afraid he was going to take the car, and as she didn't know where she stood legally, she felt she must be careful with him. But she couldn't just ignore what he'd said; to do so would just be spineless.

'I'm only leaving because you've made it impossible for me to stay,' she said carefully. 'I do appreciate you gave me a good childhood, and the car was a lovely present. But it's a shame you had to spoil all that by informing me you weren't my father practically the minute Mum was dead!'

'I wouldn't have had to if she'd told you herself,' he snarled.

Normally when he used that tone of voice with her she got frightened, but she was determined not to be intimidated by him. 'Yes, she should've told me. But you let yourself down by telling me in the way you did. It was pure spite, and you know it.'

'How dare you?' he said, taking a threatening step towards her. 'I fed and clothed you for years, paid for you to go to a good school.'

'And I stupidly thought you did that because you loved me,' she said, squaring up to him. 'How wrong could I be?' She flounced out to her car then, and stayed outside until he'd gone back into the sitting room.

Andrew remained in the sitting room right up until she came downstairs with the last box. Then he appeared in the hall again.

'I hope you haven't taken anything that doesn't belong to you,' he said.

'Of course I haven't. Unless you count the Christmas presents you gave me?'

'If I find you've been in my room and taken any of your mother's jewellery, I'll call the police,' he warned her.

She couldn't believe he'd say such a thing. 'Sophie's the one who has been eyeing that up,' she said icily. 'You'd better check her room if you find anything missing.'

'Don't think you can just walk back in here when you get tired of fending for yourself,' he spat at her. 'You go, and that's it.'

Something snapped inside her. She put down the box and glared at him. 'What would there be for me to come back for?' she asked. 'You've made your feelings about me quite plain since Mum died. Not one word of consolation, and no praise that I tried to keep everything together.'

'Have you got any idea of what it felt like to arrive home here and find my wife had killed herself?' he shouted at her.

'Of course I have. I found her, remember? Lying in a bath of blood. Is finding your mother dead less traumatic than a wife? But I didn't turn on you.'

'It was you who drove her to it,' he said, walking menacingly towards her, his handsome face suddenly twisted and ugly. 'Truanting from school, hanging around with guttersnipes, dressing like some gothic tart, and taking drugs. All those times you stayed out and we hadn't a clue where you were.'

She became scared then, afraid he might hit her. But she had to stand her ground. 'That was all over years ago, and you know it,' she retorted. 'I couldn't keep up with my school work, and I was bullied too. I only ever smoked a bit of dope, nothing worse, almost everyone does that. I expect even your precious Sophie does. Maybe you ought to look to yourself to find out why Mum didn't want to live any more.'

She didn't wait for his reaction to that, but picked up her last box and hurried out. She half expected him to come out after her, but he didn't.

Ben came running out just as she was starting up her car, and he looked really upset. 'I heard all that,' he admitted, leaning into her car window. 'He's been so mean to you, and he deserved what you said. But I'm not like him, Eva. You're my sister, and I love you.'

'Nothing will change between us,' she said, reaching up and caressing his cheek. 'You've got my address and work phone number, so ring me if you want to come round, you'll be welcome any time. I love you too, Ben. But I won't come back here, not ever.'

As she drove down the drive for the last time, tears were pouring down her face. All her memories were here: leaving for her first day at school holding her mother's hand, learning to ride a bike, and to swim. She'd pushed both Ben and Sophie around the garden in their prams, thinking they were real-life dolls. So many happy Christmases and birthdays. In barely a week's time she would be twenty-one, and the only person that date meant anything to, other than her, was gone. Ben might keep in touch for a while, but once he got to university he'd forget about her. Sophie was probably eyeing up her old room even now, because it was bigger than hers. The only time she would give a thought to her elder sister was when Andrew expected her to iron his shirts and tidy up.

'Don't think about it any more,' she murmured to herself now as she put her cassette player on the table, and plugged her television in.

She would make new friends here, and she had the ones at work too. She could go where she wanted, be with whoever she wanted. She didn't need anything from Andrew Patterson.

A week later, on the 26th of April, a rap on the door quickly followed by Tod shouting out that it was 'wine o'clock' made Eva swiftly zip up her new dress and rush to open the door.

'Gosh, you look gorgeous,' he said, waving a bottle of wine at her. 'Don't let me down by telling me you've got a date.'

It was an auspicious day for her. It was her twenty-first birthday, and this morning she'd received notification that the balance in the building society account her mother had used for the rent on the London studio had been transferred to her and she only had to go into the local branch to get her passbook. She'd gone there in her lunch hour to find there was £6,040 in the account.

She couldn't believe it was such a large sum. She had never expected to get anything more than a few hundred, and it was all she could do not to shout it from the rooftops.

Mr Bailey had sent her a birthday card. With it was a letter explaining that the last tenant in the studio had left after not paying the rent for several months, and she should be prepared to find the place in a bad state. He had added that if she decided to sell it, he would be glad to do the legal work for her.

Yesterday she'd been to a hairdresser's that opened late and had blonde highlights and a new cut. She had been wearing her hair long and straight for the last couple of years, but in a moment of wanting an entirely new image, partly because of Tod, she'd decided a jaw-length bob and some highlights might suit her better.

She was thrilled with her new look; her hair felt marvellous, so much thicker and bouncier. It made her brave enough today to try on the slinky turquoise sleeveless dress she'd admired in the catalogue at work. To her amazement it looked really good on her, as she appeared to have lost a bit of weight since her mother died.

'No date. But I was hoping someone might ask me to go down the pub with them, as it's my birthday.'

'Oh shit, why didn't you give me some warning so I could get you a card?' he said as he came into the room.

'I'm not so sad that I have to announce my birthday in advance.' She laughed.

'Fortuitous that I bought some vino then,' Tod said, but stopped short when he saw the cards on the mantelpiece. 'Double shit, it's your twenty-first!'

There were only three cards: the one from Mr Bailey, one from Ben and Sophie, and a third large one signed by everyone at work. They'd had a whip-round and bought her a

Chanel No. 19 gift set. Olive had given her a lovely tan leather handbag.

'It's just another day in paradise,' Eva giggled. 'Now pour that wine.'

She liked Tod so much. The day after she moved in, he'd knocked and said he was going to the launderette and asked if she would like to go with him so she'd know where it was. She had never expected that something as mundane as going to a launderette could be so much fun. Tod could talk about anything; he observed stuff about people and made her laugh with it.

She could see from the small mountain of washing that he only tackled it when he'd got nothing clean left to wear. As she carefully smoothed out and folded his clean clothes, he watched her with amusement.

'I bung everything in the bag as it comes out of the dryer,' he said.

'If you do it like this, nothing will need ironing,' she said.

'Ironing!' he exclaimed. 'What's that?'

Since then Tod had knocked on her door several times, mostly to cadge something – tea bags or some milk – but he usually stayed a little while for a chat. His visits shortened the evenings, and they took her mind off her mother's death and Andrew being so nasty.

She'd kept her resolve not to tell him about the recent events, portraying herself as a happy, independent career girl, and hoping that in time she would actually feel that was what she was.

'Then I'd better take you down the pub, not just because it's such an important day, but because you look so gorgeous,' he said as he looked at her cards. 'There isn't one from your mum and dad. Why's that? Surely they haven't forgotten?'

She said the first thing that came into her head. 'I'm going there tomorrow for lunch. And I'd love to go to the pub with you. Now, about that wine? Are you going to pour me a glass?'

It was a lovely evening. First, the pub where Tod announced to all and sundry that it was her twenty-first, and everyone bought her drinks and made a fuss of her. Then, when the pub closed, they went on to a club around the corner where the music was so loud it was impossible to talk to anyone and so crowded there was barely room to dance either. Eva was happily drunk, content to watch her new friends being silly together, yet feeling protected from approaches by predatory lone males, because she was part of a group.

When a drunken man became very insistent that she dance with him, Tod stepped in.

'Sorry, mate, she's spoken for,' he said and, putting his arm around her, he drew her on to the dance floor.

He had been dancing ever since they got into the club. As he put both his arms around her and drew her to his chest, it felt like he was on fire. 'Phew, you're hot!' she exclaimed.

'And so are you,' he said. 'But not in the sweaty way, like me.'

Eva giggled at the compliment.

'You don't know how lovely you are, do you?' he said, catching hold of her face with both hands and looking into her eyes.

No boy had ever said anything like that to her, but she assumed he'd only said it because he was drunk. Yet his hands on her face, lips so close to hers, were making her heart beat faster. It was so tempting to just sink into it, to let him kiss her, but she knew where that would lead.

In the past she hadn't had a place of her own to take anyone. But she'd been with boys to their flats, or in their cars, and afterwards it was always the same, they gave her that

'What was I thinking of?' look. The muttered 'I'll ring you', which they never did. From sixteen to eighteen there had been so many shaming times like that. The last one, just before her eighteenth birthday, was the worst. He'd been so rough with her, virtually rape, and then after he'd had his way he turned her out of his car and drove off, leaving her to walk home alone.

That was the wake-up call she'd needed and she realized she had to change or she would spend her life being humiliated. She looked at herself long and hard in a mirror, and accepted that she was short, rather plain, overweight, and that it was unlikely any boy was going to want her for anything other than casual sex. She realized too that the so-called friends she hung around with were toxic for her. She had copied their goth look, heavy drinking, drug taking and promiscuous behaviour in an effort to fit in, and if she didn't break away from them she would end up in the gutter.

She got the job at Oakley and Smithson a few weeks later. She had seen Miss Olive Oakley's horrified expression when she walked into her office for the interview, and in that second she knew she had to reinvent herself if she wanted to get anywhere at all. But Olive must have seen something in her to like, because she got the job.

Everything changed for her then. Her first port of call was to a hairdresser, where she got them to re-dye her hair to tone down the heavy black, get rid of the purple streaks and cut off all the straggly bits. Next out was the heavy make-up and the black grungy clothes. She turned up for her first day at work in a suit her mother had bought her, which at the time she had ridiculed as being 'Normal Norah'.

Olive's smile of approval on Monday morning was enough to convince her that her new image was the right one.

It was far easier living with approval than being nagged at

constantly. She loved her job, started having driving lessons, and she found out that staying home at night to watch television with the family and helping around the house made for a more tranquil life.

The girls at work often gently teased her because she hadn't got a boyfriend, and she had learned to make a joke of it herself, claiming she was waiting for Mr Right. She'd been chatted up by the van drivers at work, and on nights out with the girls from work there had been a few blokes who bought her a drink and flirted with her. But she'd never once let it go further than that.

In well over three years, Tod was the first boy she'd met that she really did want. She loved his sense of humour, his interest in people, the way he could chat about anything, and that he was a gentleman. But even now when it seemed that he really liked her, she was too afraid of waking up in her bed with him in the morning and seeing remorse or horror on his face to take a chance.

So she just grinned at him. 'And you are very drunk, Tod,' she said. 'It's been a brilliant evening, but don't get soppy on me now. I want to go home.'

Tod lurched off to his room when they got back, and she went to hers. But once she'd got into bed she found herself crying. It seemed to her that she'd been lonely for most of her life. Not alone, because there had always been her family and other people around, but it was a loneliness that came of having no one to share her thoughts and dreams with, no one she could tell about moments like this, when she didn't feel she belonged anywhere.

She woke at half past nine the next morning with a thumping headache, and remembered she'd told Tod she was going to see her family to celebrate her birthday. She wanted to stay in bed, but she knew she must go out, and stay out all day. So she got up, made tea and took some painkillers, then showered, got dressed and put make-up on.

Driving to Bath to have a look around seemed the best idea. Maybe she could buy herself a piece of jewellery, and when she got back she could pretend it was a birthday present from her parents.

It turned out to be a long, dreary and lonely day. Bath was full of tourists, as it always was in the spring and summer, and though she'd always loved coming here for the day with her mother, wandering around the narrow streets with all the little specialist shops, checking out the vintage clothes shops in Walcot Street, or even going for a walk in Victoria Park, it was no fun on her own. Stopping to have some lunch in a cafe full of couples and friends only heightened the feeling of loneliness still more.

She bought a silver bangle, a pretty tea towel and a tablecloth, and as the shops were closing she drove home. There was a traffic jam at Almondsbury interchange, an accident involving three cars on the slipway on to the M5. It took almost an hour to get past it, and by the time she got back to Crail Road it was after eight o'clock.

Slipping her shoes off, Eva filled the kettle and had just put the new tablecloth on the table, when there was a knock on the door.

It was Tod.

'Good timing,' she said. 'I've just put the kettle on.'

'Did you have a nice day?' he asked, coming in and just standing there, as if he wasn't sure he was welcome.

'Lovely, thanks, but it's nice to be home,' she replied. 'What's up? You look a bit anxious. Or is that just because you're still suffering from last night?'

'Why didn't you tell me your mother died?' he asked

quietly. 'Didn't you feel you could trust me? I thought we were friends?'

Eva was so shocked she nearly dropped the milk she was getting out of the fridge. 'How did you find out?' she asked.

'Your brother called to see you, and I invited him in for some tea as he seemed upset you weren't here.' Tod held out a little parcel to her. 'He brought you this for your birthday.'

Eva took the present from him with shaking hands. 'How did he come to tell you such a thing?' she asked nervously.

It had never occurred to Eva that Ben would turn up without ringing her at work first. And Ben wasn't one to divulge anything to a stranger; he wasn't even very forthcoming with people he knew well, so it was strange that he'd talked to Tod.

'I told him we'd become friends, even told him what a good time we'd had last night, and that you'd gone home to see your folks today. He looked really puzzled, and he said you wouldn't do that because you'd fallen out with your dad since your mum died. Then he told me how she died.'

'I'm sorry,' she whispered, suddenly feeling faint. 'I mean, that I lied and said I was going home today. I only said that because you asked why there was no birthday card from my mum and dad. It's not easy to tell someone that your mother killed herself. I'm shocked Ben told you.'

'I think he needed to talk. He began by just saying he ought to have come round last night to check you were alright and not lonely, because birthdays are always tied up with your mother. But then he kind of blurted it all out. He said it was bad enough for him and your sister, but they didn't find her dead in the bath, like you did, and he was frightened it might have pushed you over the edge.'

Eva couldn't speak; she felt stunned that Ben was mature enough to understand the complexities of how she might feel, and her eyes welled up with tears.

Tod came over to her and pulled her into his arms. 'So I've been worried sick since he left,' he said, his lips against her hair. 'I wish you had told me about it.'

'I couldn't bring myself to, I was afraid I'd seem tragic,' she murmured against his shoulder. It felt so good to be held and to know he'd been worried about her.

'I thought it was a bit odd that you came here to live when your home was so close,' he said. 'Ben hinted that your father was mean to you. Was he?'

'Yes, really horrible. He informed me I wasn't his child.'

'He did what?' Tod sounded incredulous. 'Look, I'll make the tea. You sit down and then you can tell me the whole story.'

'Are you sure you want to be burdened with it?' she asked.

'That's what friends are for,' he said firmly, pushing her down on to the sofa.

Chapter Six

It took Eva some time to tell Tod the whole story, mainly because he wouldn't accept the shortened version and kept stopping her for more details. But finally it was told and he cuddled her to him.

'That's truly awful,' he said, looking really shaken. 'You seemed so together, I would never have guessed you'd been through something so awful. You must be a very strong person.'

'I don't know about that, mostly I feel pretty feeble.' She managed a weak smile. 'I keep telling myself that I'm lucky I've got a job I like. And Mum left me the studio in London, so I've got a nest egg. But it is hard to deal with finding out that your dad, who you loved, and thought loved you, isn't and doesn't.'

'People lash out when they are hurt,' he said gently. 'That doesn't excuse him of course; he's behaved appallingly. But once he's had time to reflect on it, he might very well come round.'

'At the moment I don't care if he never does,' she said. 'But looking back, and that's something I've done a lot of since it happened, I can see that he was never the same to me as he was to Ben and Sophie. He wasn't cruel or neglectful, just uninvolved. But once I got into my teens, and I admit I did start acting up, he was always on at me. I couldn't talk to him, he was so scornful about everything. I think that's what made me want to be anywhere but home, and that sent me on a downward spiral, hanging out with all the wrong people.

But even when I did get back on track, found my present job and stayed home at nights, he wasn't really any nicer to me. I often felt he didn't like me.'

'What about the relationship with your mum?'

Eva sighed. 'When I was little it seemed pretty good, but it certainly had deteriorated by the time I was thirteen or fourteen. I don't know how you judge these things. Most of the girls at school complained about their mums, so I don't know if mine was better or worse. But I did feel she was distant – disappointed in me. In the last three years I tried my best to please her, but it didn't make that much difference.'

Tod nodded. 'I know that feeling!'

'You've had problems with your parents?'

'Yes, I'm a disappointment to them too. But I want to know about you.'

'Why do you think Mum kept so much hidden?' Eva asked. 'The art thing, and about her childhood, family and teenage years? Most mothers do tell their children stuff, even if they claim the kids couldn't care less and aren't even listening. The thing that upset me most at her funeral was feeling I didn't really know her at all.'

'Is Ben like her, or like his father?'

'He looks like Andrew, but he's far more sensitive and feels things deeply. I suppose that comes from Mum.'

'I liked him,' Tod said reflectively. 'I suppose you take after your mum?'

'I suppose so; she was short and had blue eyes. But her hair was red, and she had the pale skin to go with it.'

'I think you should find this guy Patrick O'Donnell and check him out. If he's an illustrator, it should be easy.'

'It's odd that both my parents were artists but I can't paint or draw,' she said.

'My father is a barrister and my mother taught maths.

They produced a child who doesn't like to argue and is useless at maths.'

'You didn't tell me your father was a barrister,' she said in some surprise.

'Well, we didn't really do the "You tell me about your family and I'll tell you about mine", did we?' he laughed.

'No, but now I've told you all my secrets, you'd better share yours.'

'No secrets. Shipped off to boarding school at eleven, hated it and was bullied. So we've got that in common. Dad wanted me to do law, Mum wanted me to be a doctor. All I've done is bum around since leaving uni. They've more or less washed their hands of me.'

'I'd be proud of having a son who cared about other people as you do,' she said.

'They see it as having no drive,' he sighed. 'To them you are a failure if you don't go into the "professions". But it's the way my dad is that put me off. I was about sixteen when I heard him telling Mum about a man he'd defended. The guy was charged with the murder of both his parents. It seems the parents did terrible things to their son, and in the end he set fire to their house and they died in the fire. I was listening hard, and I could feel Dad's anger that he'd lost the case, but I didn't sense any sympathy for his client being sentenced to life imprisonment. I wondered how Dad could be that way. He knew just what that poor guy had gone through during his life, he'd dug up every last piece of evidence to show the jury that there were extenuating circumstances and that the parents had driven him to it. Yet Dad switched off from the guy's plight the minute the jury gave their guilty verdict. I suppose most lawyers are the same; it's the nature of the beast. But it chilled me that my dad had so little compassion. So the thought of law as a profession flew out of the window.'

They discussed things about their parents for a little while and then Eva offered to make some supper for them. 'I can do cheese on toast or egg on toast,' she offered. 'I should've bought some food today.'

Tod chose cheese on toast, and while she was preparing it he asked her more about the studio. 'Is there a tenant in it?'

'No, not now, apparently the last one did a runner. My solicitor told me to be prepared to find it in a mess. It will all be finalized this week. I ought to go down there and look next weekend.'

'Would you like me to come with you?' Tod asked. 'It might be a bit scary on your own.'

'I couldn't put you to all that trouble,' she said.

'I was thinking of it as more of an adventure,' he smiled. 'I'm used to pigsties, remember. We could take sleeping bags with us and doss there, or if it's too vile we could book into a hotel.'

Looking at the eagerness in his face made her heart turn a little somersault. He had nothing to gain by helping her. She hadn't told him about the money she had, because she'd promised Mr Bailey she would keep it quiet. And he'd suddenly made her feel joyful rather than apprehensive.

'That would be wonderful,' she said. 'If we go in my car I can take all the cleaning materials with me. I can tackle anything wearing a pair of rubber gloves.'

'Then I'll buy you a pair and you can clean my room,' he laughed.

They had their supper, and another cup of tea. She kept expecting Tod to say he had to go, but he didn't. They put the TV on, and in one advert there was a weird-looking goth girl. Eva laughingly admitted she'd once gone for a similar look.

He raised both eyebrows in disbelief. 'I can't imagine that, you always look so feminine and neat.'

'Does that equal boring?'

'Certainly not. What right-minded male would want a woman wearing biker boots, with matted hair and piercings all over her face?'

'Men that look like that too, perhaps,' she said. 'I never really liked it, but the crowd I hung around with then were all into it and I just followed like a sheep. I'm glad I was too afraid of needles to have the piercings and tattoos. Also, I knew Mum would freak out.'

'My mum despairs at the way I dress,' he said. 'I tried to tell her "grunge" is in now, but she sniffed and said I looked far better in a suit. As if I could wear a suit for the work I do! But I do have a suit and I don't even mind dressing up for an appropriate occasion.'

'Which would be?'

He looked at her and smiled. 'Taking you out somewhere special.'

She blushed and looked down at her lap to hide her confusion.

He turned to her on the sofa and lifted her chin up. 'Am I wrong in thinking there's something going on with us? I'm not that experienced with girls, but I get the feeling you like me.'

'I do,' she whispered. 'I just –' She broke off.

'You just didn't think I was interested in that way? Or you aren't interested in that way?'

'I didn't think you were,' she whispered.

'Well, I am,' he said and leaned closer and kissed her.

It was the sweetest, gentlest kiss she'd ever known. The kind that said he cared about her, and wanted to take it slowly. But as the tip of his tongue flickered between her lips she felt a surge of wanting, and she relaxed into his embrace for more.

The kissing went on and on, growing deeper and more passionate. And yet he didn't try to get her clothes off, or suggest they went to bed.

He moved away slightly, and she opened her eyes to find him looking down at her in a way she could only think of as loving.

'Tomorrow I'll put on my suit and we'll go out for lunch,' he said. 'Just us, a real date, and we'll take it from there.'

Half of her wanted to pull him back close to her, yet the other half loved that he wanted to do things properly. 'That would be lovely,' she said, her voice choked up with wanting him.

'I'll think of somewhere nice and make a reservation,' he said. 'I'll let you go to bed now and see you about twelve thirty tomorrow afternoon.'

One last kiss and he was gone, leaving Eva wanting to dance and sing; she felt so happy.

Sunday was warm and sunny, and they walked to the restaurant Tod had picked in Montpellier. He said he'd never been there before but had heard it was good.

Eva wore a pink cotton dress she'd bought last summer; the colour really suited her, but she'd hardly worn it because she thought it made her look fat. But it didn't now. In fact when she looked in the mirror she saw it was loose around her waist, because she'd lost weight.

Tod looked completely different in his navy-blue suit, shirt and a tie. He'd even polished up a pair of black shoes. There was something deliciously wonderful about having such a handsome man holding her hand and talking to her as if she was the most important person in his life.

The restaurant was good, busy but not frantically so, with plain wooden tables, lots of modern paintings on the walls, and young, friendly staff. They started on a bottle of white wine while they studied the menu.

'I always find it so hard to choose,' Eva said. 'And when I have, then I always see someone else eating something I think I would've liked better.'

'My dad is a food and wine snob,' Tod confided. 'I used to hate being dragged out for meals with him because he always wanted me to have things he thought I ought to try to widen my experience. All I really wanted was a chicken Kiev.'

'I love those too,' she admitted. 'We never went out to restaurants much. Mum said it was too much of a pain with three kids. I expect it was too. Sophie didn't want anything but egg and chips, and all Ben wanted was pizza. I was a pig, I'd eat anything.'

'Living on my own I've learned to eat anything too. Dad was astounded when I ordered mussels one day. That put him in his place.'

'Where do they live?'

'In Yorkshire, near Harrogate. I hardly ever go home these days. It's always question time. What am I going to do next? Wouldn't it be better if I did this or that? On and on it goes. I just wish I'd had some brothers and sisters to share the load.'

'So I shouldn't ask what your plans for your career are then?'

'I don't mind you asking. I have enrolled on a counselling course for September, here in Cheltenham, and I can fit that in with work.'

'I think you'll make a brilliant counsellor,' Eva said. 'You are so easy to talk to.'

The time flew by: a lovely meal, and yet another bottle of wine, and then Tod looked around them and saw they were alone, except for a waitress laying up tables for dinner that night.

'Everyone's gone! I can't believe it. I didn't notice everyone go!'

Eva giggled. 'Neither did I, but I think we've outstayed our welcome.'

All the way down the road they laughed about the polite yet all too eager way the waitress had taken Tod's credit card for the bill. 'She couldn't get us out of the door fast enough,' Eva said. 'Poor girl, she should've given us a hint earlier if she wanted to close up.'

'Anyone would think we hadn't got a home to go to,' Tod said as he pulled her into his arms and kissed her on the street.

They were both a little drunk, the sun was warm and they ambled home slowly, stopping to kiss every now and then. There were lots of people around, families out for a walk, people with dogs, children on bicycles, but they were so engrossed in one another they barely noticed.

'I thought this street looked quite nasty when I first came here,' Eva said as they reached Crail Road. 'But it's grown on me.'

'I'm glad about that.' Tod stopped walking, turned to face her and took her hands in his. 'You looked so forlorn when you first arrived. But today you look happy and pretty.'

'You've done that,' she said. 'Even if it doesn't work out for us, I'll keep how you've made me feel today inside me for ever.'

As the words came out of her mouth she was afraid she'd sounded too intense. But his eyes were shining, and that lovely full mouth of his was waiting to come down on hers again.

They had barely closed her room door behind them when they fell on each other. All hesitation was gone, their clothes came off and they tumbled on to the sofa, not even thinking to pull it out into a bed.

The touch of his bare chest against her breasts was so good, she forgot that sunshine was streaming through the window, that anyone out in the hall might hear them and even that she'd never been entirely naked with a man in broad daylight before.

It was a roller coaster of sensual delight, and it was only when Tod gasped that he had to get a condom from his jacket that she was pulled up enough to remember that no other man she'd ever had sex with before had worried about contraception.

She didn't feel cheap this time. His touch was loving, he was murmuring lovely things to her, wanting to give her pleasure too. It was that which moved her. With other men she'd always sensed they didn't care about her feelings.

It was over too quickly, but she heard him say her name just as he came, and that was enough for her.

But if the sex had been good, the way he kissed and cuddled her afterwards was even better. 'You are so lovely, Eva. I told you on Friday night that you didn't know how lovely you are, and I know you thought it was just the drink talking, but I meant it then and I do now. I didn't think I'd ever feel like this about anyone.'

She got up to make tea later, and put on his shirt; Tod pulled the bed out properly. Nothing had ever felt so good as getting into it with him beside her, leaning back on his bare chest as they drank the tea, and he talked about them going to London on the following Friday.

They did it again later, and this time it was slower, and very loving. She began to cry when he made her come with his tongue. She'd never had an orgasm before, at least not with a partner.

He asked her why she was crying.

'Because that's never happened before,' she sobbed. She wished she could explain to him how he had made her feel special, that all the past hurts and humiliations that men had dumped on her were washed away now. But some things about her past were best left in the past.

It was only when she got up to take a shower later that she remembered she'd meant to call Ben and thank him for the present. It had been a watch with a lovely blue strap.

'I ought to phone Ben, he'll be worried about me,' she said, wrapping a towel around her. 'But I'm afraid Andrew will answer it, and I really don't want to go out to a phone box.'

'Then don't do it. Ben's bound to ring you tomorrow when you're at work,' Todd said. 'He knows you aren't friendless now, so it can wait.'

She looked at Tod lying there naked on the bed and saw affection and concern in his eyes, and her heart fluttered.

He was right, she wasn't friendless, and this felt like the start of something wonderful.

Chapter Seven

At last Eva knew what the expression 'loved up' meant.

She went off to work on Monday morning in a delicious bubble of happiness and expectation. It was hard to keep her mind on driving or what she would be doing at work, because her insides kept doing little flips as she thought about Tod making love to her. It had been wonderful to be woken early this morning with him caressing her breasts, sleepy kisses that soon grew into fiery ones as his fingers slid into her.

An orgasm exploded inside her even before he entered her; she'd never imagined she could feel such utter rapture or feel so greedy for more.

But it was his tenderness that made her want to cry with joy, his sweet smile when she said how wonderful it was, and the way he folded her in his arms as if he never wanted to let her go. He had made her tea and toast while she was taking a shower, and then promised he was going to make some dinner for them that evening.

Nothing could burst the bubble, not even the first irate customer who rang in that morning, complaining bitterly that she'd been sent the wrong size dress and so had nothing to wear to an important function at the weekend. Eva sailed through it, oozing so much charm that the woman ordered a second dress. By mid-afternoon, when she'd sorted at least a dozen other problems, Olive buzzed her to come into her office.

'You seem remarkably happy today,' she said with a smile. 'I suspect a new man!'

Eva's smile was as wide as a slice of watermelon. 'You

suspect right, he lives in the same house as me. I had the most marvellous birthday weekend.'

'Then I'm very glad for you,' Olive said. 'You deserve some happiness and I hope it works out for both of you.'

It was very tempting to tell Olive much more about Tod – she wanted to shout it to the whole world – but she controlled the urge and thanked Olive, then went on to tell her she'd have the keys to the studio in London by the weekend. 'Everything seems to be coming up roses for me now.'

Olive looked at Eva's glowing face and hadn't the heart to offer any warnings that she should take this new romance slowly.

'Enjoy,' she said.

That evening Tod made a curry for them, and later they drove out to the country for a walk until it grew dark, then back home for more lovemaking.

It was a little disappointing to get a phone call from Mr Bailey later in the week to say that there had been a slight hold-up and it would be another week before she got the keys, but Eva was too happy to care about such incidentals. Tod wanted to be with her every evening, and he had plans for them for the weekend too.

Ben came round on Tuesday. It was only a quick visit, and he was driving Flora's red Polo which he said his father had given him in a rare moment of generosity.

'He was only trying to creep around me because I was angry about the way he was with you. But even the car won't change my plans, I'm going to move to Leeds as soon as I've sat my A levels. I can't stand it at home any longer,' he said fiercely.

'Is that wise?' Eva asked. 'Don't do anything rash. You might regret it later.'

'You don't know what it's like at home,' he said plaintively. 'Dad's definitely got another woman, he goes out every night. The house is a mess, Rose left a note to say unless we keep it tidier she'll have to leave. Sophie is taking full advantage, stopping out late, and sometimes she doesn't come home at all.'

Eva could take no pleasure in being proved right in her belief that everything would fall apart when she left. It made her sad that Ben was worried. 'I really don't think it's a good idea to go to Leeds now,' she said. 'I know you want to go to the university there. But what if you don't get accepted?'

'There's no reason why I shouldn't if my grades are good enough. And I'll make sure they are, Eva. I really want to be there. Besides, I've got a couple of mates in Leeds already. I can get a summer job. Sharing a flat with a bunch of other people has got to be better than the way it is at home now.'

She agreed with him on that point but said he must try to leave on good terms with his father. 'Just because I fell out with him doesn't mean you should too. I just hope he pulls himself together enough to see what Sophie's getting up to. Does he ever say anything about me?'

'He's too busy with whoever he's shagging to think of anyone else,' Ben said bitterly. 'When he is home he's just irritable with us. On Sunday Sophie put a red top in the washing machine with all his shirts. They came out pink. He went ape with her, and when I tried to stick up for her he went to punch me. He keeps banging on about Mum's will too. I've already told him that if he wants to sell the house Sophie and I will agree to it. But he can't sell it even with our agreement. Not until Sophie is eighteen.'

Eva thought it was so wrong that he was taking it out on Sophie and Ben. They had enough to contend with at the loss of their mother. 'He was used to having all the power when Mum was alive,' she said thoughtfully. 'Now he's trying to wield it over you two.'

'I think you might be right there,' Ben said glumly. 'He used to have everything his way. He decided everything – from where we went on holiday, to who got invited round for dinner. I can't remember Mum ever going against him. She was always there, gardening, cooking, making sure everyone had whatever they needed. She didn't really have a life of her own, did she?'

'No, not looking back on it, Ben. She wasn't always like that, though. Her old friends I talked to at the funeral said she was a madcap party person, some of them even hinted at her being selfish too. It had to have been some kind of mental illness – not bad enough for the funny farm, but enough to knock her off-centre. Maybe she just found it easier to do what Andrew said?'

'He can wear you down,' Ben agreed. 'I find myself doing what he says, just for a quiet life. Mum was different when he was away on business, wasn't she? Remember how she put the tent up in the garden that hot summer of 1986 and we all slept out there?'

Eva laughed, suddenly remembering how much fun that had been. Flora had put night lights in jam jars all around the garden, filled up the big paddling pool to pretend it was a lake, and they'd had a moonlight barbecue, before all snuggling up to one another in the tent for ghost stories.

'Andrew would've gone ape if he'd known about that,' she said. 'She took us all to Weston-super-Mare too, we went paddling after it was dark. Remember, we all had to hold hands and jump over the waves? We didn't get home until about three in the morning, you and Sophie were spark out in the back of the car.'

'She made us promise we wouldn't tell Dad too,' Ben said

thoughtfully. 'I never thought that was odd, because she said everyone has to have secret good times. Do you think she had any secret good times without us?'

'You mean like an affair?' Eva asked.

Ben nodded.

'If she did, I never got even an inkling of it,' Eva said. 'I suppose she could've met someone during the day. Maybe she wanted to be with him, but she knew Andrew would never let her take us with her?'

'Do you think that was why she did it? Love gone wrong and all that?'

He looked so forlorn that Eva put her arms around him and hugged him tightly. 'I don't know, Ben, but I really don't think so. But one thing I do know for certain was that she was very proud of how clever you are, as I am too. Don't go against her wishes and make the house over to your dad, whatever he says. Stall him, anything but that. And make sure you get good grades in your exams, and try to stop Sophie going off the rails.'

He smiled weakly. 'I'll do my best. I am glad you seem happy here. Is Tod your boyfriend?'

'Yes, he is now, but only just, up until my birthday he was just a friend.'

Ben beamed then. 'I'm glad. I thought you might be cross that I told him about Mum. But he's a good bloke and I couldn't help coming out with it. He's just what you need.'

Ben was right, Tod was just what she needed. It was so lovely when she got home from work to have him banging on her door within five minutes to ask how her day had been, and offering to make her a cup of tea. She loved sharing a meal with him and planning to go down to the pub later, only to fall into bed and end up not making it there. All those miserable, sad nights back at The Beeches with Andrew being so nasty were all behind her now. It seemed as if her life was getting better and better.

On Saturday morning they went shopping together and he insisted on buying a bunch of flowers for her, which choked her up. She had often looked at couples buying groceries together and felt envious at their togetherness, and she could hardly believe that at last she knew what it felt like.

'You're so easy to please.' He laughed. 'I hope that five or six years down the line you are still the same.'

She knew she loved him and wanted to say so, but she didn't quite dare, not yet. After all, they had only been an item for one whole week, but that remark seemed to confirm he felt a permanence in their relationship too.

Word had got around at Oakley and Smithson that she had a boyfriend, and who he was, and back at work on Monday she was teased about it constantly. Pictures of hats with notes attached from the other girls, saying they'd picked this one out for her wedding, appeared on her desk. She put on her jacket one evening and found the pockets were full of confetti, and in the staff room a silly wedding-present list had been pinned up, with everything from a three-bedroom detached house to a tin opener on it. People had written their names and funny comments beside some of the items.

But Eva didn't tell Tod about any of this; she was afraid he'd feel he was being pushed into a corner.

On the Friday evening Eva picked up the keys for the studio on her way home. She was really excited and expected that Tod would be ready to drive to London. But when she got home he was sunbathing on the grass in the front garden in just a pair of shorts.

'How long is it going to take for you to get ready?' she

asked him. 'I planned to go straight away, just in case it's too awful to stay there and we need to find somewhere else.'

He didn't get up, just grinned at her. 'I don't fancy being in London for the weekend if it's going to be good weather. Let's leave it till next weekend?'

Eva felt she'd just had her balloon popped. 'But I'm really excited to see the studio. We don't have to be in it all weekend, we could go to one of the parks with a picnic,' she pleaded.

'But I've already told the lads we'd be going down the pub tonight. Josh is having a party tomorrow night too. I wouldn't want to miss that.'

Not wanting to look possessive or demanding, Eva swallowed her disappointment and managed to force a smile. 'What if I go to London by train tomorrow on my own then?' she suggested. 'I'm dying to see the place and I don't think I could stand to wait another week.'

'Will you be back for the party?' he asked.

That wasn't the reply she'd expected. She'd thought once he realized that her heart was set on seeing the studio he would change his mind and go with her. 'I don't suppose I'll feel much like it after a long day in London. Especially if there's a lot of cleaning to do. But you go anyway.'

He got up and kissed her. His body was warm, silky and smelled of suntan lotion, and she immediately felt aroused.

'I won't enjoy the party without you there too,' he said, nuzzling at her neck with his lips.

'That's a fib, you'll have far more fun without me.' She laughed, only too happy to forgive him anything when he made her feel dizzy with desire. 'But if you can bear to leave the sun for now, I could show you some fun things to do indoors.'

Eva caught the train to London shortly after seven the following morning. She hadn't slept very well because she was nervous about finding her way around London on her own. She hadn't been there more than five times, and always with her mother. She was very excited to see the studio of course, but it would have been so much more fun if Tod had come with her.

She had hoped he would change his mind at the last minute, but he didn't even stir when she took a shower, so she crept out without waking him. In a small wheeled suitcase she had packed a variety of cleaning materials, tea bags, bin liners and other essentials, plus some old jeans and a T-shirt to work in. She would take some photos before she moved or cleaned anything, so she could show Tod when she got back.

Paddington Station was extremely busy and, as she was apprehensive at finding her way on the tube, she got a taxi to Pottery Lane in Holland Park.

Olive had told her that Holland Park was a smart area with big houses but warned her that the smaller side roads going down towards Ladbroke Grove were very different. She said she had lived near there during the 1970s and parts were quite squalid, so Eva wasn't to expect too much. She added that she thought the studio might be just a couple of rooms above a garage.

As the taxi took her through Notting Hill and on down to tree-lined Holland Park Avenue, Eva was pleased to see that the big houses Olive had spoken of all looked very similar to those in the best parts of Cheltenham. There were restaurants, a couple of trendy gift-type shops and a delicatessen too. As the taxi turned off into Portland Road the houses were slightly smaller and terraced, the front doors opening straight on to the pavement with no gardens, but they were still very smart with black painted railings.

She assumed that the taxi still had quite a way to go, so

when the driver pulled up suddenly by a pub called The Prince of Wales, she was taken by surprise.

'That's Pottery Lane,' he said, glancing at her over his shoulder and pointing straight ahead up a much narrower street which went off at an angle. 'It's difficult to stop there, it's too narrow.'

Eva thought for a second he'd made a mistake, for although the houses in the narrow street were tiny and humble compared with those behind her, they certainly weren't squalid. It had the look of a mews, because some of the houses had been converted with a garage beneath them. She could see they'd been built in the last century as workers' houses, and many of them were still just plain yellow brick, but some had been painted white or pastel shades, and it looked like a very desirable place to live. The street sign confirmed she had been brought to the right place.

After paying the driver she stood for a little while just looking down the street. She hadn't for one moment considered how much her legacy might be worth. It was enough that she'd been left something. But even though she knew nothing about London house prices, she could tell this area was way out of the league of ordinary people. The quality shops she'd seen, the number of BMWs, Mercedes and other smart cars parked on Portland Road, all pointed to this being a yuppie ghetto.

But No. 7 stood out like a rotten tooth in a row of healthy ones. Upstairs there was one huge grimy window, when almost every other house in the row had two or three smaller windows which were in proportion to the building. There was a garage or workshop beneath, a front door to the left of it and a small window too. The doors were battered and the grey paint was peeling off; the little window was so thick with grime she couldn't see through it.

Taking a deep breath she put the key in the lock on the front door and turned it, but there appeared to be something behind the door preventing it from opening. An image of a body lying behind it sprang into her mind. But telling herself that was stupid, she slid her hand around the crack. To her relief it was only a mountain of mail which she was able to push aside.

An appalling fetid smell and the buzzing of flies greeted her, making her stomach turn over. She froze momentarily, nervous of going any further.

With her hand over her nose she went in. It was very dark and she had to open the front door again to see. To her right was the wall of the garage with a connecting door, but to her left and straight ahead appeared to be just one big room. As she edged her way gingerly forward she saw pinpricks of light and realized that the window right at the back was boarded up.

The whole floor area was strewn with rubbish: paper, takeaway food containers, old cardboard boxes, cigarette ends, beer cans and bottles.

She stood on the spot, terrified that the smell was something far worse than decaying food and afraid to take a step further for fear of what she might step on. But as her eyes adjusted to the gloom she saw an open-tread staircase to her right, set against the back wall of the garage. She also saw that all the walls were covered floor to ceiling in black graffiti.

Her initial reaction was to back away. She had never seen or smelled anything so appalling; it even made the seedy rooms of the friends she'd visited during her goth period look like palaces. But she knew if she did back off now, she'd only have to come back some other time and deal with it.

She tried the light switch but no light came on, and her heart sank even further.

It seemed to her that squatters must have got in, for surely no ordinary tenant would leave a place in such a state. The only piece of furniture still intact was an ancient deck chair. Other pieces – chairs, a table and remnants of a chest of drawers – had been roughly chopped up.

The smell made her gag as she edged her way forward through the mess. The kitchen area was in the left-hand corner by the boarded window. To the right was a back door, but the glass was broken in that too and boarded over. The door lock was very stiff, and it took her several attempts to turn it. But as she opened it, and light flooded in, the room looked even more hideous.

The kitchen cupboards had been ripped out, leaving only a filthy sink unit and an equally filthy electric cooker. Eva gingerly turned on the tap and was relieved to find that the water hadn't been turned off.

With a sinking heart, she tried to recall the photographs of herself and her mother that had been taken here some nineteen years ago. But though she did remember one where her mother was wearing a vivid green jump suit, and a matching band around her hair, she couldn't remember what the background of the room looked like.

Taking her courage in both hands, she went upstairs. The big room at the front had clearly been planned and used as an artist's studio because of the huge window, and there were paint splatters everywhere. But there was only one narrow window at the side that opened, and the window frames looked rotten. There was still more rubbish here, including a filthy mattress.

The smaller room at the back, however, was reasonably clear of debris, and it was decorated with a hand-painted frieze of teddy bears.

That cheered her, because she guessed by the age of it that

it had been painted by her mother, and finding a link to her early childhood was something positive.

Finally the bathroom, and she gagged when she saw the toilet was full of excrement. She flushed it, fully expecting to find that it was blocked up. But to her great relief it wasn't, and most of the mess disappeared. She waited till the cistern had refilled and flushed it again, breathing a sigh of relief when she saw the waste was all gone.

The lavatory was still filthy – as were the bath and wash-basin, but she felt she could deal with those.

Olive had told her that she must read the meters, so she went back downstairs to find them. There was no gas, and the electric meter was in a spidery built-in cupboard up by the front door. She jotted down the reading before she forgot.

The walled backyard was as rubbish-strewn as the house, but climbers from the houses on either side were tumbling over a trellis on the top of the wall, and there were plants struggling through the debris too. She didn't think it would take too much effort to make it pretty.

She found a key hanging on the back door, and it fitted the door through to the garage. She braced herself for more squalor but surprisingly found that it was fairly clear: just a few old empty cardboard boxes, a stepladder and an old suit-case with a broken handle.

To gather herself she stood in the fresh air at the front door for some minutes. While she had expected an artist's studio to be dirty and shabby, she had allowed her imagination to build up a romantic picture of a discarded easel and palette, paint brushes in pots and a worn chaise longue where models posed. But the graffiti suggested the last tenants' attempt at art had been fuelled by drugs, and they were filthy people who had no respect for themselves, let alone some-one else's property.

She wondered if these tenants had added to her mother's anxiety. She had been given Flora's old building society pass-book and there had been monthly deposits of £600 up until eighteen months ago, and at that time there had been a bal-ance of over £8,000. Since then there had been no more deposits, and Flora had made one withdrawal of £1,500 pounds in addition to smaller amounts. Eva had no idea what she'd used the large sum of money for, she could only sup-pose it was for repairs on her car or something similar, as there was no sign of the money being spent on the studio. She wondered why Flora hadn't got the tenants evicted and relet it to someone who would pay the rent? Or had she seen what they'd done to it and felt defeated?

Eva could understand that. But she wasn't going to let it defeat her. Yet at the same time she knew the cleaning mat-erials she'd brought with her were not enough. She needed a broom, dustpan and brush, a mop, bucket and toilet brush. And a great many more bin bags. She also thought she would try to get a Calor gas camping stove and a kettle to heat up water.

She needed to go and buy these things but decided that, before she left, she would take photographs to show Tod just how bad it was. She didn't think she could adequately describe how horrible it was with mere words.

Hailing a taxi on Holland Park Avenue, she asked the driver to take her to the nearest hardware shop.

In less than an hour she was back in another taxi. She had everything she needed, including an inflatable mattress and pump. The last purchase had made her spirits rise a little, because it would mean she and Tod would have something to sleep on when he came to see the place.

First, she changed into her old clothes. Then she opened

the front and back doors and all the windows upstairs to let some air in, then put on rubber gloves.

The stink was coming from rotten food in takeaway cartons, and she gagged again when she saw maggots crawling over it. But she shovelled it all up into bin bags, tied them up tightly and put them in the backyard.

Three hours later she had filled sixteen bags with rubbish, stacked the broken furniture and the mattress in the garage, and swept right through the house.

The bathroom had been the hardest thing to clean, and it had turned her stomach imagining the kind of people who had used it. She'd heated up around six kettles of water on the little camping stove to scrub it. But the limescale remover which had been recommended to her was very efficient on both the bath and lavatory, though the fumes nearly knocked her out. She left it to work further while she cleaned the inside of the upstairs windows, and by the time she went back to the bathroom the last of the limescale had dissolved.

She tried hard to see potential in the house while she was working, but apart from it being in a good area, and the rooms being a good size, the scale of what was needed to make it habitable was frightening. A new window was needed downstairs, not just new glass, because the frame was rotten. The cooker was disgusting, and she'd have to get new kitchen cabinets and a fridge. What if the immersion heater upstairs didn't work, or the roof leaked? And who could she go to for advice on these things?

With her savings she had almost £7,000. But although she'd thought before she got here that this sum made her rich, she realized now that it wouldn't go very far. She couldn't even let the place to someone else until she'd made it habitable again.

Yet however horrible and squalid the house was, it must

have been important to her mother, or she would have sold it long ago. It was strange to think of herself living here as a baby: taking her first steps, toilet trained in the bathroom, and playing out in the backyard. Had Flora been happy here?

Later, Eva went along to the cafe in Portland Road for tea and sandwiches, and then took a little walk around the neighbouring streets.

However disheartened she was by the project before her, she couldn't help but be cheered by the area. There was a pretty park nearby, and she found the huge bottle-shaped kiln which gave her road its name. A middle-aged man she questioned told her that it was the only one of its kind left in London. He said that back in the eighteenth century the area had been known as the Pottery and Piggeries as so many people kept pigs here.

She bought a can of lemonade and some apples, then went back to the house and sat down on the floor by the open window in the little bedroom to think while she heated another kettle of water to mop all the floors. The sun had been shining on the front of the house when she arrived here, and now it was shining in the back bedroom and on the backyard. Looking out of the window, she could see into the neighbours' yards. The one on her right was a very pretty courtyard, beautifully paved and with lots of exotic-looking palm-type plants in tubs. The one to her left just had white painted walls, and a table and chairs painted blue. She wished she could see into their houses to see what they'd done with them.

She wanted to feel excited at the challenge of renovating this house, but instead she felt mostly dread as the problems appeared insurmountable. The only real way they could be tackled was by moving here. But how could she do that when she worked in Cheltenham? And then there was Tod. She doubted he'd want to come here with her. But even if he did, it wasn't just decoration that this place needed. It required building skills: carpentry, electrical and plumbing.

Looking out at the backyard, she wondered what it had been like when her mother lived here. She had loved gardening, so it was inconceivable that she hadn't turned it into something beautiful. Had she sat at this very window with Eva in her arms and planned it all?

Did she sleep in here with her? Or did she have a bed in the studio?

Looking up, she noticed there was a trap door in the ceiling, giving access to the attic. She wondered if there was anything in there, and thought she would get the stepladder from the garage later to look.

Her mind turned to Patrick O'Donnell then. In all the excitement of the love affair with Tod, she'd forgotten about him until now. He'd lived here too. If she was to find him, would he want to know her? But as Flora had thrown him out, the chances were he'd slam the door in her face.

It was eight o'clock when Eva finally caught the train back to Cheltenham. She was exhausted, filthy and she ached all over, but she felt very satisfied at what she'd achieved during the day. She'd left the bathroom sparkling and the whole house looked and smelled clean because she'd mopped it all the way through with gallons of hot water, cleaning fluid and disinfectant. She'd even cleaned up all the rubbish in the backyard.

But she didn't know what to do next. Even if she knew what was involved in getting a builder in to do the work needed to make the house saleable or fit to be let, which she didn't, how would she know she could trust him? And she didn't think £7,000 was going to be enough to pay for everything.

She had picked over the mail that was lying behind the door and found bills from the electricity board, as well as water and rates. They were all red ones, addressed to the old tenant. Would she have to pay them? She'd brought them back with her, and she hoped Tod might know what she had to do. She hoped he'd know what to do about everything.

Perhaps it was just because she was so tired, but all at once this legacy looked more of a burden than an asset. London was too big and scary for her. She'd got yet another taxi back to Paddington, because she couldn't face trying to find her way on the tube. But she couldn't do that every time she went there, it would cost a fortune. How was she going to get rid of all the rubbish in the garage? Until she'd done that, she couldn't put her car in there – and there were yellow lines on the street.

When she got back to Crail Road at ten thirty, the house was silent and in darkness, except for one light in the window on the top floor. Tod hadn't folded up the sofa bed when he got up, but he had made it up after a fashion. She drew the curtains and stripped off to have a shower, hoping that would revive her enough to get dressed up and go and find him at the party. But it didn't, and the thought of walking down into the town alone so late at night was too daunting. So she made herself some hot milk, left the door on the latch so that Tod could get in when he came home, and got into bed.

She was woken the next morning by someone talking in the hall. Tod wasn't in bed beside her, and when she looked at her clock she saw it was nearly ten. She guessed Tod had got so drunk last night that, out of habit, he'd staggered into his own room when he got home.

Putting on her dressing gown, she went to his room. She knocked, but there was no reply. She showered, washed and dried her hair, put on jeans and a T-shirt and then tidied up her room and dusted. There was a shared Hoover in a cupboard in the hall, and she really wanted to use it but was afraid she might annoy the other tenants who were still sleeping. So she decided to drive to the supermarket and get some groceries and the Sunday paper.

On leaving the supermarket Eva was still immersed in thoughts about the studio. She was driving on automatic pilot, and found herself turning the wrong way, going towards her old home. When she realized what she was doing, she took a right turn to get back to Crail Road through the back streets.

As she drove down a street with small houses, to her surprise she saw Tod up ahead, standing on a doorstep.

Excited to see him, she automatically slowed right down. But as she did so, she saw he wasn't alone. There was a girl with dark hair, wearing a pink dressing gown and standing in the doorway talking to him. Her first thought was that he'd just called at the house, but as she pulled up some twenty yards away, to her shock he leaned forward to kiss the girl. It was not a kiss of greeting but a full-on goodbye kiss, the kind that followed a night together.

Her heart plummeted, tears sprang into her eyes and she watched in horror as they clung to each other. A flush of rage and nausea rose up inside her and made her grip the steering wheel so tightly it hurt her hands.

Tod took a step back from the girl, then reached out and stroked her cheek in a gesture which, even from a distance, was clearly full of tenderness. The girl stepped forward towards him, flung her arms around him and kissed him again. They were locked there, so wrapped up in each other they were oblivious to anyone watching.

Eva began to tremble. She had put all her trust in him,

believed he felt the same way about her as she did about him. Yet he'd gone to bed with someone else the moment she wasn't around.

Glancing in her rear-view mirror, she saw another car coming up behind her. If she didn't move the driver would beep his horn, and Tod would turn and see her. But there was no parking space to drive into, so she had no choice but to put the car into gear and drive on.

As she passed Tod and the girl she kept looking straight ahead, but out of the corner of her eye she saw Tod look round. She knew he would recognize the car – men always did. She didn't know why she hoped he wouldn't.

At the end of the road she turned left, and then right, tears running down her cheeks. She had no plan of where she was going, she was just fleeing. It reminded her of the evening she'd run out when Andrew told her he wasn't her father. She had the same thumping in her heart, the same sick feeling, and she knew she must park up somewhere before she had an accident.

There was a sense of irony when, for a second time that day, she found herself driving towards her old home. For anyone else that would be the right place to go and be comforted, but she wasn't wanted there either.

All at once she knew she was going to be sick, and that forced her to pull in. She had barely got out of the car and round to the pavement when she vomited into the gutter. She was vaguely aware of a man mowing his lawn on the other side of his fence, but she felt so terrible she didn't care what he thought of her.

Forcing herself to get back into the car, she drove past her old home and out into the countryside. Pulling into a lay-by, she gave way to floods of tears.

Meeting Tod the night she moved into Crail Road had

helped her to feel less lonely and bitter. His friendliness had made her feel she was worth something. Since they became lovers he'd become her whole world, and she'd felt that nothing could ever hurt her again.

But seeing him with that other girl was like stepping on a trapdoor which opened and plunged her into a deep black hole of misery and worthlessness. She had no idea how she could climb out of it, she didn't even believe there was a way.

They might only have been lovers for two weeks, but just the thought of him with another girl was like having a knife twisted in her belly.

How could he do that to her?

Chapter Eight

Putting her head down on the steering wheel, Eva cried great heaving sobs that came from right down in her stomach. She felt such utter despair that she could finally understand why someone would take their own life.

She tried to convince herself that Tod might not have betrayed her trust and that the girl she'd seen him with was just an old girlfriend who he was being kind to. But she knew that was false hope; no one kissed in the street like that unless they had a love hangover from spending the night together. She knew this with utter certainty because she and Tod had been like that.

What was she going to do now? He'd become the axis on which her world spun. Without him she couldn't function, there was nothing left but black emptiness.

She stayed there in her car, crying for what seemed like hours. She tried to reason with herself that being with him for such a short time didn't give her the right to expect fidelity and that she was wildly overreacting. But telling herself that didn't help. She'd fallen in love with him on that first lunch together, and she'd really believed he felt much the same way about her, even if he hadn't said so.

It was only a faint hope that she might have been mistaken. But the possibility that she'd just seen someone who looked like Tod made her drive home. As she drove she imagined he was back at Crail Road worrying about her.

All hope of that vanished when she got in and listened at his door. There was no sound, and no message stuck beneath her door. If he had come back fleetingly, he'd rushed off again to the pub or a friend's house to avoid seeing her.

The rest of the day passed so slowly. Each time she heard someone in the hall, she jumped up, only to hear another tenant's voice or the sound of their footsteps going up the stairs. She couldn't read the paper or watch television because of her tears. She just lay on the sofa, torturing herself with the image of Tod making love to that girl.

One of the things she'd loved most about him was that he was kind. He'd been so sympathetic about her mother's death, he'd cooked for her and made cups of tea. But now even that seemed to be pretence – or surely he'd be concerned about her now? If he hadn't been serious about her, why see her every night for two whole weeks? If he'd said he wanted to meet up with other friends too, that would have shown her that he didn't feel committed.

It was after seven in the evening when she glanced out of the window and saw him coming in through the gate. Without stopping to think, she rushed out and opened the front door to him.

He looked aghast. That confirmed he had spotted her earlier. He also smelled like a brewery.

'How could you do that to me?' she said, and began to cry again.

He took hold of her arm and led her back to her room, shutting the door behind them. 'You don't have a monopoly on me,' he said fiercely. 'And why were you snooping on me?'

'Snooping!' she exclaimed. 'I wasn't. I took the wrong way back from the supermarket and just happened to go down that road. I thought you were still in bed in your room.'

He leaned back against the door, folded his arms and looked at her contemptuously. 'I don't believe you. Far more likely you've been nosing around in my address book.'

'I've never even seen your address book, much less nosed into it,' she retorted with indignation. 'But if that girl is in it, then you've been two-timing both of us.'

'Don't be so dramatic and needy,' he said irritably. 'I've never made a secret of having lots of friends. She just happens to be one of them.'

'I'm not needy,' she said, her voice shaking. 'Neither am I being dramatic. How would you have liked it if I went out and slept with someone else?'

'I wouldn't have minded at all,' he said airily. 'That would've been a whole lot better than thinking I'd got to be with you seven days a week, just to shore you up.'

'Shore me up?' she questioned, not really understanding what he meant by that. 'I never expected you to see me every day. It was you who instigated that.'

'Only because I felt sorry for you.'

Eva felt as if he'd slapped her. She stared at him in horror.

'I asked you out to lunch that day after your birthday because of what you told me,' he went on, not even looking directly at her, as if she was just some passing stranger. 'It was just pity. You'd had such a bad time and I thought I could help you get over it. But you're too buttoned up and prissy for me, you want someone to fill up all the holes in your life, and I can't do that. That's why I said I didn't want to go to London. I hoped you'd become less clingy if you had something else to focus on.'

She looked at his face, and all the warmth and eagerness she loved him for was gone. This man had made passionate love to her again and again; he'd talked about the future as if he intended them to share it. But his eyes were cold now and he was wearing the same scornful expression Andrew had worn when she left his house.

It was unbearable, yet in just the same way that Andrew's

nastiness had stirred up anger in her, so Tod's cruel rejection fired up the remnants of her spirit.

'You flatter yourself, thinking I need you to fill holes in my life,' she retorted, willing herself not to cry again. 'I came back from London really excited by all the new possibilities for me there. In fact I was intending to tell you today that I'm going to move there. I had of course hoped that you would want to share in my good fortune and come and visit now and then. But, silly me, I hadn't realized what an insincere arsehole you are. You belong in provincial Cheltenham with all those sad people who think you care about them. So bugger off and join them.'

There was a slight satisfaction in seeing she'd surprised him. He scuttled out like the rat he was, and she slammed the door behind him.

She leaned back on the door and cried, wishing she could turn off her feelings for him as quickly as she'd made that hasty declaration. She hurt so much inside that she felt she could die from it.

Putting some music on drowned out the sound of her crying, but there was nothing she could do to fix her broken heart. His insulting words kept milling around in her mind, 'dramatic and needy', and what did he mean by her being 'buttoned up and prissy'? Was it the way she looked, the way she was in bed, or did he mean she was dull company?

But even worse was the thought that while she'd been weaving rosy daydreams about them being together for ever, he'd seen her as some fragile loony that had to be watched over. That was so insulting.

She saw him going out again at eight; he had his suit on, and that meant he was going somewhere smart, perhaps with the girl she'd seen earlier. That was the final blow.

It was a little later, almost nine o'clock, when she began to pack up her things. She had no choice but to leave. To stay, seeing his face, hearing his voice, would just be too painful. Besides, she wasn't going to look even more pathetic by not sticking to what she'd told Tod she was going to do.

Her belongings seemed to have multiplied in her time here, and she had to take more care packing them into her car. She also had to be quick – the last thing she wanted was to still be here when he returned.

Stripping the bedding was the worst part; there was still a faint smell of him lingering on the duvet and pillows. But she bundled them into a bin bag and pushed it down hard behind the driving seat.

By ten thirty the car was packed and she returned once more to the room for one last check. As she looked around she saw she'd made no lasting impression on it; the room was just as bleak and forlorn as when she'd first arrived three weeks earlier. Even the love she thought she'd found here was just a mirage.

As she drove away towards the M5 her eyes kept welling up with tears, but she brushed them away angrily.

By the time she reached the M4 there was little traffic on the motorway. She wondered if there was anyone else out there in the darkness, fleeing to another town because of heartbreak. But even if there was, she doubted they were going to such an unwelcoming destination as she was. No electricity or hot water, and if she hadn't bought the inflatable mattress she would be sleeping on the floor. Would she even be able to find her way to the house? She hadn't got a map – all she knew for certain was that the M4 went through West London.

What was Olive going to think when she didn't turn up for work tomorrow? She supposed she'd have to phone her. But what would she say? Somehow she knew her boss wasn't going to think being dumped was a good enough excuse for running away without giving notice.

It was nearly midnight when she saw a sign ahead which said the next junction was for Hammersmith and Shepherd's Bush. Knowing Shepherd's Bush was very close to Holland Park, she turned off there and followed the signs. She went the wrong way at Shepherd's Bush Green and stopped at a garage to ask directions, then turned round and got back to Holland Park Avenue. It all looked so different by night, but after a couple of wrong turns she saw The Prince of Wales and Pottery Lane.

By day Pottery Lane had looked inviting, but now under yellowy street lighting it looked faintly menacing. Pulling up close to the front door, she unloaded the car and then drove down the lane to find somewhere to park without yellow lines.

By the time she got back to the house it was after one, and on realizing that she hadn't had the presence of mind to get candles or a torch, she began to cry again. The house didn't smell evil any longer, but the image of maggots and filth was still in her head and she was scared. Fumbling around in the dark, she eventually found the camping gas ring with the matches beside it. She lit it, but as the blue flames lit up the graffiti-covered walls she felt even more frightened.

Holding the gas ring in one hand, she carried the bag of bedding upstairs to the little bedroom where she'd left the new inflatable mattress, and set to work pumping it up. That made her cry even harder, because when she'd bought it she'd imagined doing this with Tod, laughing at the grimness of the house, and Tod throwing her down on the mattress to make love to her.

She wanted a hot drink, and she realized she hadn't had anything to eat since first thing that morning when she'd had a slice of toast. But unable to face going downstairs again and rummaging through bags of stuff to find biscuits, tea bags, milk and a mug, she made up the bed as best she could, tried to ignore the smell of Tod on the duvet, stripped off down to her underwear and crawled into it.

Tired as she was, sleep wouldn't come. The inflatable mattress felt very strange and smelled of rubber. Her mind flitted from imagining all kinds of creepy-crawlies in the room, to Tod with that other girl. This house might belong to her but it felt like a Dickensian prison, or one of those awful places where glue sniffers gathered, and she felt so terribly alone and afraid.

She wished she'd thought to check into a bed and breakfast in Cheltenham instead of rushing off here. Then she could have gone into work in the morning and asked Olive's advice about what to do. She didn't have the first idea how to get the electricity back on. Or how to go about getting the boards off the windows and new glass put in. Misery overwhelmed her and she sobbed into the darkness, asking why she had been singled out for so much unhappiness.

Rain beating against the window woke her. She looked at her watch and saw it was nearly eight o'clock. She got up and went to the bathroom, but hurried back to bed afterwards as there was nothing to get up for.

She wanted to sleep and sleep, because in that way she could avoid thinking about anything. But a ball of misery was pumping away inside her like a second heart, growing larger and stronger by the minute, and it wasn't going to let her sleep. It wanted to list and enlarge upon her problems, starting with being unlovable.

She was jobless, in a place where she knew no one. And there was a multitude of things to be done to make this place liveable, most of which she had no idea how to do.

As if that wasn't enough, another voice was speaking loudly in her head, telling her she was worthless, stupid and plain.

Olive glanced at the clock on Monday morning. She saw that it was now eleven o'clock and still Eva hadn't rung in with an explanation as to why she wasn't coming to work. Normally when staff failed to turn up she felt only irritation, but in this case she was worried. While she knew there was no phone in the house at Crail Road, she felt certain Eva was the kind to ask someone else to ring in if she wasn't able to do it herself.

By the end of the day, with still no news, Olive drove round to Crail Road on her way home. There was no answer to Eva's bell so she rang some of the others. After what seemed like an interminable time a dark-haired girl wearing jeans and a sweatshirt opened the door. Olive asked if she knew where Eva was.

'She left here last night,' the girl said. 'I'm in the room above hers, and I was looking out of the window about ten and saw her packing stuff into her car.'

Olive was shocked. She explained she was Eva's employer, stressing that she was concerned for Eva.

'She had a row with Tod,' the girl confided. 'I heard their raised voices yesterday, early in the evening, and he went out later on his own. Maybe she's gone home to her parents?'

'She hasn't got any,' Olive said. 'When does Tod get in?'

The girl shrugged. 'For the last couple of weeks since he's been with Eva, he's been coming in at half five, but if he dumped her I expect he's gone straight to the pub.'

Olive's heart sank. She had seen the way Eva was about this young man, and if he'd let her down there was no knowing what she'd do.

'Was Eva's room this one?' Olive pointed to the window nearest the front door.

The girl nodded.

'I'll just get up on the window sill and see if she has taken everything,' Olive said.

The girl looked at Olive's business suit and high heels. 'I think you'd better let me do it,' she said, and she nimbly jumped from the steps at the front door on to the wide window sill to peer over the short net curtains. 'Everything's cleaned out,' she called back. 'Her television, all her cushions and all the other bits and pieces.'

'Is there anyone in the house she might have left a forwarding address with?' Olive asked.

'If she was going to leave one it would've been with me,' the girl said. 'I used to talk to her more than anyone else. I wish I'd come down last night when I saw her loading the car. But, to be honest, I didn't know what to say to her. I wanted to warn her when Tod came on to her that he's a womanizer and not to take him seriously, because he's never with anyone for long. But I didn't – I suppose I thought she might change him. I expect that was what the row was about. She must have found out he'd got someone else.'

Olive was frightened for Eva now. She might only have been seeing this young man for a couple of weeks but Olive had seen the transformation in Eva since she met him. He'd given her hope for the future, taken her mind off her mother's death and her stepfather's nastiness. It wasn't a casual fling to Eva; she'd pinned everything on him. And now that her dreams were shattered, she would be in pieces.

'Could you let me have the landlord's phone number then?' Olive asked.

There was a faint chance Eva might have had a personal

reference from someone she was close to when she took the flat. If she told him why she was worried, he might pass on that person's address.

When Olive got home she rang the landlord. He was pleasant, but said he'd taken only a bank reference and a deposit from Eva. He said that he would relet the room when her advance rent ran out and that if Olive should hear from her, she was to tell the girl to get in touch so he could return her deposit.

Olive wondered what else she could do. She didn't know the exact address of the studio in London; she didn't even know which solicitor had acted for Eva, so she couldn't ask him. Andrew Patterson would know of course, but she was loath to inform him about this.

He was hardly likely to care anyway.

But she supposed it was a good sign that Eva had had enough spirit to take off. A weaker person would have just taken to her bed and stayed there. She had some money and a place to stay, after all.

Yet all the same, Olive couldn't bear the thought of her in a strange place, all alone and hurting.

Eva was hurting. She stayed in bed all day Monday, listening to the rain beating down and wishing she could just die. The teddy bears on the wall were no comfort now, they were just another reminder of betrayal. Flora had ruined her life, and then foisted this hideous dump of a house on to her to create more misery.

Evening came, and as it gradually grew darker she felt panicked that she would soon be plunged into complete darkness again. Later she heard people calling out to one another when the pub closed, car doors banging and the sound of them driving away, then silence fell, broken only occasionally by the tapping of heels on the pavement.

The night seemed endless. The inflatable mattress seemed to be going down and she couldn't get comfortable. At one point she thought of going out to her car and driving somewhere. But where could she go in the middle of the night?

She was still awake at dawn; pigeons were cooing somewhere close by and she knelt up at the window, watching the sky gradually grow lighter. It had stopped raining and she recognized the climbing plant on the back wall of the yard as a clematis, because a few white flowers had opened up. In the half-light it looked pretty and it stirred something inside her, a feeling that if a plant could survive without anyone caring for it, so could she.

When she went downstairs to find tea, milk and sugar, the sight of all her things dumped in that dark dungeon of a room nearly sent her running back upstairs to bed again. But she forced herself to find a mug, some tea bags, and to sniff the milk to check it hadn't gone off. Fortunately it was so cold in the room that it hadn't. She also found some biscuits – which reminded her she hadn't eaten anything for two days.

Two mugs of tea later, she sat up on the bed with a notepad and made a list of things she needed. The first priority was getting the electricity put back on. She dug out one of the old bills addressed to the previous tenant and found a telephone number on the back. She thought she would ring that when the offices opened, and give them the present meter reading.

There was so much she needed. Plates, cutlery, saucepans had all been provided in Crail Road and she hadn't any of her own. She'd have to get some boxes to put her clothes in, and one to keep food in. And then there were candles and a torch, just in case it was a few days before she had power.

But it was the boarded-up window downstairs that worried her most. How much would that cost? How would she know she wasn't being ripped off? Without a job her money wasn't going to last very long. Should she ring Olive and tell her what had happened, and perhaps go back to Cheltenham and stay in a bed and breakfast?

Cheltenham was a small place, though, and someone who knew Tod was bound to see her. Besides, it was a matter of pride not to go running back there, defeated. She would find a job here in London, get the house fixed up somehow and prove to herself she could be strong and manage alone.

Did she have to ring Olive? Why not just write her a letter? She was more than likely to start crying the minute she heard Olive's voice. How would she be able to explain properly then?

But she owed Olive a great deal; she'd been there for her when no one else had. She also knew her boss would think Eva was spineless by not speaking directly to her. She would have to ring, even if she didn't want to.

As it began to get dark on Tuesday evening Eva was feeling a bit better. The electricity company had promised the power would be on again in the morning, and for tonight she had candles and a torch.

Speaking to Olive had helped far more than she had expected. As always, Olive had been very blunt and told her in no uncertain manner that a broken two-week-old romance might hurt, but it wouldn't kill her.

'I know you feel used and forlorn, but better you discovered his true colours now than in a few months' time,' she said crisply. 'Now, stop snivelling and tell me about the house.'

She listened patiently to Eva's story of what a mess it was.

'So what should I do?' Eva said as she finished her tale of woe and wiped her tears away with her sleeve.

'You know what you've got to do. Stay there and fix it up. I know it's a huge challenge, but you are bright and practical. You'll find a way. As you said, you can't let it to anyone in that state. And if you come back to Cheltenham, how are you going to oversee anyone doing the work on it? We are all missing you here, and I certainly didn't want to lose you as an employee, but it seems to me that fate has stepped in and given you a chance to prove yourself. I'd be the last person to try to talk you out of that.'

Eva phoned the electricity board after Olive, and that was surprisingly painless. Instead of going back to No. 7 afterwards she walked up to Notting Hill and then went on to Portobello Road, walking around in a big circle until she came back to Pottery Lane. The fresh air, time to think more calmly, and also having a proper meal in a cafe, put things back into proportion for her. Her heart wasn't broken, just a bit bruised; she had money, and a roof over her head. She'd got to stop feeling sorry for herself.

Later she drove to a branch of B&Q and bought some large plastic boxes and a dress rail for her clothes, two five-gallon tins of white emulsion paint, as well as paint brushes and a roller. While she was in Notting Hill she'd seen a shop that sold all kinds of china, glass, cutlery and other household essentials, and she intended to go back there once the electricity was back on.

Maybe her improved spirits were because she was warming to the idea of the challenge before her.

There was no doubt that this was an interesting area to live in. Beautiful houses which were obviously the homes of the super-rich were cheek by jowl with council flats. As she'd walked down Portobello Road she'd heard many cockney accents, along with West Indian voices, but just as many plummy public school ones too. There were antique shops selling fabulously expensive heirlooms, market stalls piled with rubbishy bric-a-brac outside them. Trendy health-food shops jostled between second-hand clothes and displays of local artists' work. Every colour and creed were there: a group of black men were playing steel drums, South Americans were playing pan pipes, elegant girls who could be top models rubbed shoulders with skinheads, and there were a great many people who looked like they were stuck in the Peace and Love Sixties.

She had found a second-hand furniture shop just off Portobello Road that had everything she needed to turn her house into a home, and the owner had even offered to do her a special price and deliver the furniture free when she was ready for it.

Satisfied she was tired enough to sleep well tonight, she pumped up the inflatable mattress a bit more and made a cup of tea. Tomorrow she would phone around some of the numbers she'd found for window repairs and get some quotes.

It was something of a surprise to wake the next morning and find it was after nine. The last thing Eva remembered was wondering how women back in Victorian times managed to do exquisite needlework by candlelight, as she couldn't see to read even with four candles.

To her irritation the milk had gone off overnight. But as she went to the bathroom to fill the kettle for a wash, to her surprise she saw the glow of an electric light downstairs.

Despite being alone she cheered aloud, and the sour milk was forgotten. It was only when she tried all the other light switches in the other rooms that she found the light bulbs were missing. With trepidation she opened up the cupboard on the landing where the immersion heater was, half expecting that it wouldn't work. But to her delight a little red light came on when she threw the switch, and she cheered again.

The thought of being able to have a bath later lifted her spirits even higher. As she ran around plugging in her hair dryer, a bedside lamp and even the television, and finding they all worked – albeit there wasn't an aerial lead to plug in the television and get a picture – she felt almost ecstatic.

Within half an hour she was washed and dressed in jeans and a pink T-shirt. She even put on some make-up. She thought she would have breakfast in the cafe along the road, then walk up to Notting Hill and get her hair washed and blow-dried because it looked so awful. While she was out she'd ring some of the window companies she'd got the numbers for yesterday, and also look out for some job agencies and see what they had on offer.

In the cafe there was a young couple at another table, looking into each other's eyes and whispering together. It was a sharp reminder that such a short while ago she and Tod had been like that too, and a lump came up in her throat. But she was determined not to get sucked into thinking about him again, so she opened the paper she'd just bought and studied the job vacancies while she waited for her breakfast.

Chrissie, the girl who did her hair later, was warm and chatty and around the same age as her. When Eva told her she was new to London and looking for a job, Chrissie suggested she go to Kensington High Street to have a look around.

'All the big-name shops are there,' she said. 'They always want new people, and there's lots of agencies for office work too. Just cross the road and walk down Kensington Church Street, it's not very far.'

Eva was pleased with the way Chrissie had done her hair; it looked so shiny and bouncy and it instantly made her feel more confident. While she wasn't dressed for job hunting, she thought she would go and take a look anyway. She crossed the road and, seeing a cash machine, drew out fifty pounds to tide her over, then she turned into Kensington Church Street.

It was a busy street of smart restaurants, antique shops, jewellers and art galleries, reminding her of Montpellier in Cheltenham. But the shops here seemed to cater only for very rich people, with eighteenth-century desks and tables, chandeliers, and paintings in ornate gilt frames that looked like Old Masters.

She was looking at a display of antique jewellery in a shop window when suddenly someone banged into her. As she staggered to right herself, she felt her bag being pulled from her shoulder. It was a scruffy-looking young lad trying to take it.

'Get off,' she yelled, clutching at her bag.

There was a brief tussle but he punched her in the stomach, making her double up in pain, and ran off through the crowd towards Notting Hill with her bag.

Despite the pain she still managed to shout out that he'd stolen her bag. When no one made any attempt to stop him, she shouted again and tried to chase after him. People moved out of the way, gawping at her like idiots. But although she yelled out for them to help her, they still didn't react and the distance between her and the thief was widening by the second.

All at once a man in white overalls appeared out of a shop doorway. He looked first at Eva and then at the lad fleeing up the road, and set off after him in pursuit.

Eva had never seen anyone run quite so fast. His legs were

going like pistons, then he lunged at the thief in a flying tackle and knocked him to the ground.

Someone cheered loudly but Eva's view was suddenly obliterated by the crowd. She hobbled nearer, holding her stomach, and saw that the man in overalls had the thief pinned down on the pavement, holding him there with his foot. When he saw her he waved her bag in the air, and grinned jubilantly at her.

As Eva reached them her rescuer lost his hold on the lad, who wriggled away and ran for it.

'Thank you so much,' Eva gasped out. 'Gosh, you were marvellous, and so quick. My keys – and everything else – are in the bag. I don't know what I'd have done without them.'

'Shame I didn't hit him harder, bloody low life,' the man said. 'But are you alright? Did he hit you?'

He was around twenty-five and built like a rugby player. His dark hair was cropped and his stained white overalls suggested he was a decorator.

'Yes, in the stomach,' Eva said, still holding it with one hand. 'It's winded me. But that doesn't matter now, I'm just so grateful to you for getting my bag back. I only just got some money from a cash machine.'

He looked at her with concern and handed back her bag. 'I expect he saw you getting it and followed you. They do that a lot around here.'

'It was very brave of you to tackle him,' she said. 'He might have had a knife.'

He shrugged nonchalantly. 'He was just a druggy opportunist. I should've given him a good kicking, that might have deterred him from doing it again.'

Perhaps it was the thought of what might have been if the thief had got away with her bag, money, door and car keys, her chequebook, and papers that had her address on, but all at once she felt faint.

'Are you alright, love? You've gone white as a sheet.'

The man's voice seemed very far away.

'I think it's shock,' she said weakly. Her knees were buckling under her and she thought she might be sick.

She felt his arm going around her for support. 'Come with me and I'll make you some tea. I'm working just here, I came to the door when I heard the commotion.'

He led her into an empty shop and sat her down on a stool beside some sacks of plaster. 'Take some deep breaths, and I'll get you some water and put the kettle on.'

The wall that ran down the whole length of the shop had just been plastered, and it was still wet and dark brown. But it seemed to be spinning, as if she was drunk.

'Now don't you pass out on me,' she heard the man say as he pushed her head down between her knees. She felt him lift her hair from the back of her neck and put a cold, wet cloth on it. She almost asked him not to mess her hair up, but then realized if she could think of her hair she couldn't be in such a bad way.

After a few minutes he put his hand under her chin and lifted it to look at her. 'Try drinking some water now, your colour's coming back.'

Eva tried to smile. He had a nice open face with dark brown eyes, and his wide full mouth turned up at the corners as if he was permanently smiling. His gallantry and kindness on top of the shock made her eyes fill up with tears. She didn't want to cry; she was afraid if she started again she wouldn't be able to stop. 'I've only been in London two days. Is this what it will be like? Having to be on the alert all the time?' she asked him.

'No, of course not,' he said soothingly, and with the wet

cloth in his hands he wiped her damp eyes. 'You've just been unlucky, that's all. Where are you living?'

'In Holland Park,' she said.

'I've just about finished here so I'll run you home,' he said. 'But first a cup of tea, and I think I've got some chocolate too. That's good for shock.'

Eva *was* trembling with the shock. It was one of those situations she'd heard people talk about, but had never expected would happen to her. Coming on top of all her other troubles, it was all too much; she didn't think she'd ever felt this helpless and afraid. A small voice was whispering in her head that she couldn't let a stranger take her home, yet she knew she wasn't capable of getting there alone.

'You're very kind,' she said when he came back with a mug of tea and the chocolate.

'There isn't much that chocolate can't fix,' he said, breaking her off a piece. 'And who wouldn't help a pretty girl in her hour of need?'

'No one else moved to help me,' she said. 'But you've done more than enough now. I'll just get a taxi home.'

He crouched down on his haunches in front of her and handed her the mug of tea. 'Sorry, love, but I'm taking you back, whether you like the idea or not. Shock can do funny things to people. Anyway, Holland Park is on my way.'

She could see by his determined expression that there was no point in refusing, so she just nodded, and despite her good intentions tears spilled over.

'Don't cry, love,' he said, patting her shoulder. 'You're quite safe now. I promise I'm not a mass murderer.' He looked at the chocolate still in her hand, and smiled. 'Now, are you going to eat that chocolate? If not, shall I have it back?'

There was something about that jocular last remark which

reminded her of Ben. He never wanted to share chocolate bars either.

She put it in her mouth. 'Too late, it's gone,' she said, trying to smile as she wiped her eyes with the back of her hand.

'That's better,' he said with a wide grin which showed very white even teeth. 'Now I'm Phillip Marsh, but only ever called Phil – except by my more loutish mates, who like to call me Swampy. What's yours?'

'Eva Patterson.'

'Well, Eva. You just sit there and drink your tea while I clear up and get my things together.'

He was whistling 'Blue Velvet' as he scraped up fallen lumps of plaster from the floor. Eva observed his movements – they were graceful and fluid like an athlete. When he took off his overalls, revealing jeans and a green T-shirt beneath them, she saw that his body was very taut and muscular. He was nice, really nice, kind, good-looking and capable.

The clearing-up done, and the tools packed away in a box which he took out to his van parked outside the back of the shop, he came back for her and helped her into the passenger seat.

'Just got to lock up,' he said, fastening the seat belt around her as if she was a child. 'You'll be home in five minutes.' As he climbed into the driving seat a few moments later, he asked 'What road?'

She told him it was Pottery Lane, and he nodded as if he knew where that was.

'I put in new damp courses in three of these houses,' he said as they turned into Portland Road minutes later. 'I'm told they go for over a million! You must be paying a very high rent to live here.'

She didn't respond to that question, because although she liked him she didn't want to reveal anything about her situation. 'Just leave me by the pub,' she said. 'My place is only around the corner, and I need to get some milk and light bulbs from the shop before I go in.'

He pulled into a parking space just by the pub, but then he jumped out and came round to her window. 'I'm not leaving you anywhere,' he said very firmly. 'I'll go over there and get you milk and light bulbs, but then I'm taking you right to your door. Stay there. How many light bulbs?'

'Three,' she said. 'But I can't let you get them, I'm fine.'

'You aren't,' he said sharply. 'You look very pale and I need to check there is someone there to keep an eye on you.'

She felt too weak to even attempt to dissuade him, or even admit she was alone. She watched him bound across the road to the shop, and considered getting out of the van and rushing to her front door before he got back.

But he was too quick – he was back to the van before she'd even thought to open the door.

She got out her key as they reached her front door. He took it from her and opened the door.

The darkness at the end of the passage made him turn to her with a puzzled look. 'You aren't squatting here, are you?'

'No, it belongs to me, but there's a lot needs doing to make it habitable.'

'Living here on your own?'

She nodded sheepishly and went in, hoping he'd go. But he followed her and, as she slumped down on to the stairs, he just stood there looking around. She felt such shame at her clothes spilling out of bin bags, the graffiti on the walls, the boarded-up window and back door. She thought he would make his excuses and leave.

But he turned back to her, his face wreathed in concern.

'No one comes to a place like this without good reason. Will you tell me about it?'

Eva fought against bursting into tears. 'It seemed like a good idea at the time,' she said, trying to keep her tone light. 'I know it looks awful, but hey, the electric was put back on this morning, that's a start.'

He grimaced. 'Is it as bad upstairs?'

'I've got an inflatable mattress and tea-making things,' she said. 'And it's brighter.'

'In that case, I'm taking you up there. I'll make you some more tea, and you can lie down and tell me all about it.'

Eva felt she ought to have alarm bells jangling in her head at this suggestion. But she didn't, because she was sure he was a genuine nice guy. After all, would anyone but a good person tackle a thief, comfort her and bring her home?

Upstairs he made her lie down, took her shoes off for her and covered her up with the duvet, then filled the kettle in the bathroom and put it on the gas ring. He sat down on the floor, resting his back against the wall, and then nodded as if he expected her to start spilling the beans.

Eva explained briefly that her mother had left her this place when she died recently, and described how she'd come on Saturday to clean it up, intending then to get some advice on how to proceed. But then something unexpected had made her leave Cheltenham on Sunday night.

'Something unexpected?' He raised one eyebrow.

'I discovered my boyfriend was cheating on me,' she said bluntly. She saw no point in lying; it only complicated matters. 'So I packed my bags and rushed off here. Foolhardy, really, but he lived in the same house as me. And I knew if I stayed there it would be a case of having my nose rubbed in it.'

'I think that was brave, not foolhardy.'

'You wouldn't have said that if you'd seen me stumbling in here in the dark on Sunday night without even a candle or a torch!' She giggled, suddenly seeing the funny side of it. She was liking Phil more and more, and really hoped he wasn't married. That seemed so absurd when just a day ago she had believed she'd never get over Tod. 'But when the electric came on this morning I shook myself out of the doldrums and got my act together. I got my hair done, and I was just going down Kensington Church Street to look for a job when that guy snatched my bag. Talk about one step forward and two steps back. But for you I'd have been locked out of this hovel with not even ten pence to use a pay phone. I think that might have pushed me right over the edge.'

He looked at her appraisingly for a moment. 'Things can only be on the up now then. And by the way, your hair looks very nice,' he said eventually. 'So let's have a cup of tea, and then let's talk about how you can turn this hovel into a home.'

'The main thing is to get the boards taken off the window and door downstairs and the glass replaced,' she said. 'And I need to find out where the council dump is – to take sixteen sacks of stinking rubbish. Once that is done, I think I can paint it all myself.'

He said nothing while he poured the hot water over the tea bags and then squeezed them out. He poured some milk in the tea, then passed her mug to her.

'I don't do windows. Damp-proofing and plastering is my game,' he said. 'But I've got mates that could do the window for you. Would you like me to contact them?'

'Have you got any idea what it's likely to cost?' she asked cautiously. 'I haven't got much money.'

'Not really – like I said, I don't do windows. But you've got no choice but to get the window done, even if you have to

138

borrow the money. Or if you really hate the place, you could sell it.'

She shrugged. 'Who would want to buy it?'

'A property developer would bite your hand off to buy it.'

'Really? Maybe that's what I should do then.'

'Sure, but they'd only give you perhaps two hundred thousand at most. If you got it done up nice, you'd maybe get six for it.'

'Six hundred thousand!' she exclaimed.

He laughed at her surprise. 'You are a little innocent country girl, aren't you? Well, Eva, one of the first things you ought to do when you feel better is go and look at other properties for sale around here, and see for yourself.'

'I don't have enough money to do it up really nicely,' she said glumly. 'I can probably manage to get the window and back door fixed, and maybe get some cheap second-hand furniture. But that's about it. Anything else – like a kitchen – will have to wait until I've got a job and saved up some more money.'

'Surely you could borrow the money to do it up, using this place as security?'

He laughed at her surprised expression. 'You really are an innocent, Eva. The bank would have nothing to lose, lending you the money. And whatever they charge you in interest is going to be a drop in the ocean compared with the extra value you'll put on the house. Property developers don't use their own money; they just borrow, do the place up and sell it on. Say you borrowed fifty thousand to do the work – and that much would turn it into a little palace – bingo, you'd have a place worth double or more what it's worth now.'

Eva felt this should have occurred to her. After all, she knew Andrew had sold off the land at the back of The Beeches to fund doing it up. But he was a businessman who knew about such things. Phil probably was just as wise if he worked in renovation. She felt she could trust him; she'd met enough low-life men in her time to be fairly certain he wasn't one. But she was still wary of laying herself wide open to be fleeced by his cowboy friends.

'I'll think on that,' she said. 'Maybe I could go to my bank and discuss it with them.'

'Are you feeling any better now?' he asked.

'Yes, much better, thanks to you,' she said. 'You've been so kind.'

He smiled. 'Well, like I said before, who wouldn't help a pretty girl in distress? Shock is a funny thing, and I don't think you ought to go out again today. But would you like me to walk around and check everywhere, make a list of stuff that needs doing? You'll need to be able to present your bank with details if you are going to ask them for help.'

'I can't expect you to do that after everything else you've already done for me,' she said. 'I've taken up enough of your time already.'

He got up from the mattress. 'Eva, I wouldn't offer if I couldn't spare the time. I'll go and get a ladder out of my van and check the roof first. You take it easy.'

He was gone for quite some time, and Eva began to worry that she'd been too trusting. She'd heard of con-men claiming a roof needed fixing when there was nothing wrong. He might have rushed to her aid, but then when he saw where she lived he might have got less honourable ideas about her.

She heard him coming back into the house and walking around downstairs. Then he came upstairs again and went into the big room first, and finally the bathroom.

Eva nibbled at her nails and wondered if she should say that, if any work was needed, she'd have to run it by an uncle or someone.

Phil came back into the little room and grinned down at her. 'Well, darlin', good news first. The roof is in good shape, looks like it was redone a few years ago, but the gulleys both back and front need clearing of old leaves and stuff. That's a ten-minute job.'

Eva felt cheered by that.

'The house was rewired just three years ago – there's an electrician's card tucked in by the meter with the date he did it – so that's seriously good news. But the bad news is that all the windows need replacing.'

'Oh no!' she exclaimed.

'You can't put new glass in the one downstairs, the frame's far too rotten. But you really need double glazing anyway. Whatever crank put in that bloody great window in the other bedroom must have frozen in winter and roasted in the summer. If it were me, I'd take it out and put in two smaller windows. That would make it more comfortable, cut down on heating in the winter, and also make the house look a whole lot better from the street.'

'That sounds like a good idea,' she said weakly, because she was somewhat bowled over by the way he appeared to be the proverbial knight on a white charger.

'If it was me, I'd brick up the back door and widen the window to make French doors,' he went on. 'It would be so much lighter, and airy. But getting back to the list: there's no damp, I got out my meter and checked; no rot in the floorboards either.'

Eva hadn't even considered damp or rotting boards, but she was very glad to hear she hadn't got that to sort out too.

'Then there's a new kitchen needed,' he went on. 'That can cost anything from two thousand upwards, but if you didn't mind a second-hand one, I've got mates who are kitchen fitters. They are always ripping perfectly good ones out of big houses and replacing them with top-of-the-range ones. Mostly they just dump the old ones, unless they know some-one like you who wants one. They often get the sinks, the appliances, the whole works. I got my mum one that way, cost me just five hundred for them to fit it in. You just can't be too picky about the colour and stuff. But my mum got a lovely pine one.'

'That sounds marvellous,' she said, suddenly feeling a surge of excitement and hope. 'Would they come and give me a quote? And your window man? Obviously I wouldn't be able to go ahead until I'd got the money sorted.'

'Sure. I'll ring around tonight. Business is quiet at the moment. This slump thing is affecting the whole of the building trade quite badly. Some of them are only working three days a week, so they'll be glad of a job on the side. But that downstairs window should be your first priority. You can't plan or do anything else until that's done.'

He sat down again and stayed talking to her for quite a while. He told her he lived in Acton in a flat he'd bought when he was going to get married. 'Then three months before the wedding I found out she was having it off with her boss,' he said with a rueful grin. 'So I know how you feel. It takes a while to pick yourself up after that. But my brother is sharing with me now, and mostly I think I had a lucky escape.'

Eva felt a little buzz of pleasure to hear he was single.

He asked about her family and, as he'd been so open about himself, Eva felt she had to be honest too. She didn't go into much detail, only saying that her mother had killed herself and that she had to move away as her stepfather was being mean to her. 'Tod said he only went out with me because he felt sorry for me,' she said with a wry smile. 'I think that was the worst thing – to think all he felt was pity, when I believed he loved me.'

'Well, I am sorry about your mum. That's a terrible thing for anyone to go through. But Tod sounds like a right dick-head.'

Eva laughed. She liked Phil's directness. Tod had always wanted to analyse people, and that had fooled her into thinking he was sincere, but she realized now that he hadn't been – not about anything.

'I've been a bit of a dickhead myself,' she said. 'If I'd stopped to think, I would've found another room, carried on at work and seen to this place at weekends until it was habitable. I can't really believe I threw all my toys out of the pram for a man I'd only known a few weeks.'

'I did a bit of that too,' Phil admitted. 'But then I got to thinking I'd show Claire what I was made of. So I got the flat done up, went off on a holiday to Goa with the lads, and made sure she knew how well I was doing without her. That's what you've got to do too, Eva.'

He had to go then; he said he had to see someone to give them a quote, but he'd call round when he'd spoken to his friends about the work she needed doing.

'Would you like me to get you a takeaway?' he said as he got up to leave. 'I can see you've got nothing to cook anything in – or on.'

'You've done more than enough for me already.' She smiled up at him. 'I've got some bits and pieces to eat. Tod may have accused me of being needy, but I'm not.'

'We all have wobbly moments when we need a friend,' he said. He dug in his pocket and pulled out a card. 'So ring me if you need anything or just want to talk. But I'll pop round again to tell you when I've spoken to the lads.'

Chapter Nine

Phil had left Eva with a great deal to think about. That same evening, and much of the next day, she walked about the house, looking, measuring and planning. From outside in the yard she could see how good French doors would look, just as from the street she saw Phil had been right in saying two smaller windows upstairs would make a vast improvement to the house.

In the afternoon she paid a visit to the Notting Hill branch of her bank to arrange to have her account transferred there from Cheltenham, and she also made an appointment to see the manager the following day.

On the way home she went round to the greengrocer's in Clarendon Road, just a couple of streets away from Pottery Lane. Seeing a card in the window of an Italian bistro for a lunchtime waitress, she went in to ask about the job.

A tall dark-haired girl who introduced herself as Marcia was laying up the tables for the evening. 'Antonio, the boss, isn't here right now,' she said. 'But could you come in for a trial tomorrow, about twelve thirty? He'll be here then and he can see how you shape up.'

Eva was glad to agree; the bistro was close to home, and it had a nice relaxed atmosphere. She might have only had experience of Burger King and KFC, but if she could handle working in those places, she was sure Antonio's would be a pushover. As her appointment with the bank manager was at eleven, she could go to the bistro straight afterwards.

*

Eva put on her black suit and white shirt the next morning and slipped some flat black shoes in her bag, as she didn't think she could wait on tables later in her high heels. She arrived at the bank ten minutes early, fired up to convince the manager that she was worthy of a loan if necessary.

Mr Dodds was a plump, bald and genial middle-aged man. Eva told him about the house and explained that she might need a loan later to get it fixed up.

'Where do you work, Miss Patterson?' he asked.

She explained that she'd only been in London for less than a week but that she was starting a job that day. 'I do have six thousand and forty pounds in a building society account,' she said, showing him the passbook. 'And you can see from my current account and the deposit account with you, that I've nearly another thousand there. I am intending to draw on my own funds to get the windows and the kitchen done, but later on I'll need to put in central heating, and I might need a loan then. I thought once I'd got the house straight I could take in a lodger, which will create another income.'

He wanted evidence that the house did indeed belong to her. She showed him letters from Mr Bailey, her solicitor in Cheltenham, which verified this and confirmed he was holding the deeds.

He looked at her very intently for a moment or two. 'Well, Miss Patterson, you appear to be a very level-headed young lady, and I can see from your account with us that you have always acted responsibly in the past. My only concern is your youth, and the fact that you have only been in London for a short time and won't be able to supply a reference yet from your employer. I suggest you get the vital work done with your savings and come back to me in a couple of months if you find you do need more money. I can review the situation then, and possibly arrange a small mortgage for you that you can manage on your salary.'

Eva left the bank feeling elated and proud of herself. She could hardly believe that in just one week she'd been on a white-knuckle ride, from excitement at going to see the house for the first time, to the shock of finding it was a wreck, then the terrible hurt of finding out Tod's true feelings for her, and rushing to London.

All the despair she'd felt a few days ago seemed ridiculous and over the top now. She'd made a friend in Phil, Mr Dodds had taken her seriously, and unless she made an idiot of herself at Antonio's today, she had a job too.

It was just after half past three when Eva got home from Antonio's. Her feet ached from being on them for so long, but she'd got the job, and would start properly on Monday. It wasn't very good money at £3 an hour, and for now Antonio only wanted her three days a week – from twelve to three – but Marcia said she should get at least £5 a day in tips. And as Rose, their cleaner at The Beeches, always used to say: 'That's better than a slap around the face with a wet kipper.'

In the early evening Eva was scrubbing at the old cooker with a Brillo pad, wearing a pair of old baggy shorts and a scruffy T-shirt, when there was a knock at the front door. To her shock, when she answered it Phil was standing there; with him was an older man.

Eva blushed scarlet at being caught looking such a fright. She had a scarf tied around her hair, no make-up, and she thought she must stink of oven cleaner.

'Hi, Eva.' Phil grinned. 'I told you about my kitchen-fitter mate. Well, this is Brian and it just so happens he's got a kitchen in his van that might be perfect for you.'

'Excuse how I look,' she blurted out, peeling off her rubber gloves. 'I've been trying to clean the cooker. But I haven't got any money here.'

'Don't worry about that,' Brian said. 'If you like it, you can have it and pay me later. I'll have to dump it unless I find a new home for it in the next few days.'

Brian had one of those round smiley faces that Eva associated with kindliness and fatherly qualities. She wondered if Phil had laid it on thick to him that she was in need of help, and she felt a bit awkward at being seen as a charity case. But Brian was already opening the back of the van and beckoned her to come and have a look. It was stacked high with kitchen units and wall cupboards all with white matt doors and brass knobs.

'What do you think?' he asked as he pulled out a base unit for her to look at. 'Nice, ain't it?'

'It looks brand new,' she said. She couldn't help but be suspicious of something that seemed too good to be true.

'I know!' Brian grinned at her. 'Some folk 'ave more money than bleedin' sense. Their puppy chewed one door. We ordinary folks would just replace the damaged door, but not these people – they wanted the whole lot ripped out.' He climbed into the van and pulled out a long, pale grey Formica worktop stacked up at the side of the van, and passed it to Phil. 'This came with it. Not a scratch on it and enough of it to fit out a huge kitchen. There's a stainless-steel sink, a fridge and a washing machine too. I can do you a lovely job with these, if you like. All you'll need to get is a new cooker cos they had one of them whopping great range things.'

Eva was stunned. She hadn't really believed that Phil would come up with anything, and at best she'd expected dark brown imitation wood from the early 1970s. She would have been glad of even that, but a lovely white kitchen was beyond her wildest dreams. 'It's marvellous,' she said weakly.

Brian moved the van right up against the garage so cars could pass by, and then the three of them went back into the house. She showed Brian a rough plan she'd made, telling Phil she thought his idea of French doors was what she wanted. Brian measured the space, did a few calculations, then began to draw a plan in chalk on the floorboards.

'You can't have it quite like you've drawn. It would mean two wasted spaces in the corners, and it'll looked cramped,' he said. 'I suggest two lots of units facing each other. The stove, washing machine and sink unit will be at the back, with one other unit there. But the fridge needs to be opposite, under your breakfast bar. And you can have a couple of wall cupboards either side of the cooker.'

'And you've got all that?' she asked.

'Sweetheart, I've got enough units to go right around the whole room, if you wanted them.'

'It sounds great, but how much is it going to cost?' she asked nervously.

'Can you go to six hundred, including the fitting?' he asked. 'That's including the appliances of course.'

After seeing the price of new kitchens in a showroom, and the extra cost of having them fitted, Brian's offer was like a gift from heaven and she wanted to hug him. But she controlled her glee and said that quote was very reasonable. And she asked him when he could start on it, as she had a job now.

'I'll fit around you, love,' he said cheerfully. 'I can't do anything until that window is done, anyway. But I'd be obliged if I could leave the units and the appliances here to free up my van. If you change your mind, I can always come and collect them again.'

She could hardly believe that anyone was so trusting. 'I'm not going to change my mind,' she assured him. 'I didn't even dare hope I'd get such nice units.'

'Right, we'll get them in then.' He nodded to Phil to help him.

The two men took about twenty minutes to bring everything in, stacking the units neatly away from the area where they would eventually go. Brian plugged in the fridge in the corner by the sink so she could use that right away. As soon as she heard it whirring away she ran upstairs to get the milk and other perishables to put in it. She felt she could even cook a meal on the old stove now she had somewhere to store her food.

'I'll be off now,' Brian said, handing her his business card. 'Just give me a bell when you want me to fit it. And you'd better get yourself a new cooker too, I'll need to wire that in when I take the old one out.'

Phil didn't go with him; he said he would catch a bus home. After Eva had seen Brian out, she turned to Phil in some excitement.

'You are a wonder,' she said. 'I can hardly believe you arranged that for me. Would you like a cup of tea before you go home?'

He made a hangdog face. 'I was hoping you might like to come to the pub with me for some dinner. I haven't got anything in at home. And I bet you haven't eaten anything much today?'

'No, I haven't, I've been too busy,' she said. 'But I'll only come if you let me pay. I owe you a dinner for arranging this.'

'If Your Ladyship insists,' he said, making a mock bow. 'To tell the truth, I've been worried about you. This dark room is enough to give anyone nightmares. Were you alright after I left the other day?'

'I was fine,' she said, touched by his concern and a little flustered at finding herself thinking how attractive he was. 'No ill effects at all. But I've been more careful with my bag since, and I've been very busy making plans. I'll tell you about them at the pub. But can I go and change first? I'm sure you wouldn't be seen dead with me looking like this!'

She scurried up the stairs, leaving Phil standing at the back door and looking out into the yard.

It took her just ten minutes to have a quick wash and put on a pair of pink jeans and a white T-shirt. She'd bought the jeans in a sale at work when she embarked on the affair with Tod, but they'd been so tight she'd never worn them. To her astonishment they were perfect now, so she must have lost more weight. She had only a make-up mirror so she couldn't see herself full length. She wished she could.

Her hair was fine with just a brush. She hastily applied some make-up and perfume, then she slipped on a pair of pink high heels with ankle straps.

Phil was perched on a wooden crate in the backyard. He looked up and smiled as she appeared in the doorway. 'You look really nice,' he said.

'Well, thank you, kind sir. Of course I couldn't have looked worse than I did when I opened the door to you earlier.'

'You looked OK to me,' he said lightly. 'Before you came down I was just thinking that someone must have cared for this garden once, it's got a good feeling.'

'I think that was my mum,' she said. 'She loved gardening. Some of the plants like the clematis look old enough to have been planted by her. I removed a lot of rubbish and pulled up lots of weeds, but I want to plant up some tubs, get a table and chairs out here. It's a real little suntrap in the afternoon.'

He looked appraisingly at her. 'You're like a different girl today. Bouncy, smiley and – dare I say it? – happier!'

'I am,' she said with a wide grin. 'You've been like a lucky charm to me. First, getting my bag back, then getting a job, and now the kitchen. So let me feed you as a thank you.'

The Prince of Wales was a friendly pub. It had a very mixed clientele – mostly yuppie types in the thirty to forty age group, but a good proportion of working-class people too – and there was lots of banter between them. Eva mentioned to the landlord, George, that she had just moved into Pottery Lane, and he insisted on giving her and Phil a drink on the house to celebrate.

'I used to eat here a lot when I was working down the street,' Phil said when they'd been given a table for two in a corner. 'The food's not too poncey or expensive. They do a lovely Sunday roast, and the people who drink here regularly aren't toffee-nosed either.'

'If the boards don't come down off that window soon, I might become a regular, propping up the bar every night,' she joked.

'You've got a lot of guts, and I like that,' he said approvingly. 'It must be tough to move to a new town when you don't know anyone. And then to get your bag snatched!'

'I wouldn't have met you but for that,' she said flirtatiously. 'Thanks to your advice, I pulled myself together. I've found myself a job in a bistro – only part time, but it will do for now – and I think I've got my old optimism back. You and Brian have renewed my faith in people.'

They ordered steak, chips and salad, and as they ate Eva told him about the job and visiting the bank.

'I've got enough money for the kitchen and probably the windows, depending of course how much they'll cost. But I thought it was best to see if the bank would lend me some more later so I could put in gas central heating and a new bathroom too. He seemed OK about it, and I can always get someone to share with me to help out, as this waitress job is really only a stopgap until something better comes along.'

'I asked my mate John about doing your windows,' Phil

said. 'I could bring him round tomorrow to have a look, if you like. He hasn't got much work on just now, and I'm sure he'll be happy to do it in stages to suit you. I'll make good all the plaster for you. I'd like to help.'

As the evening progressed Eva found herself becoming more and more attracted to Phil. He had a lovely sense of humour, he was interesting, and he was very interested in her too. He was comfortable to be with, as if she'd known him for a long time. And he had real opinions of his own, not half-chewed-over ideas gathered from other people – the sort she realized now that Tod had. The word she thought best summed up his character was: honest. He told it as it was, and he believed in doing a good day's work for a day's pay. He took pride in his work and had no understanding of people who were lazy, or those who expected something for nothing.

She liked the respectful way he had been with Brian too – she'd sensed a strong bond between them, almost like father and son. 'Tell me about Brian?' she asked. 'Have you known him a long time?'

'He's the salt of the earth,' he said. 'Happily married with two kids he adores, a real craftsman too. I've worked on lots of jobs with him, right since I was a stroppy young lad who thought he knew it all. He's always even-tempered, calm, caring – and a laugh too. Trust him, Eva, ask his advice about stuff, he's a really good man.'

She had thought at first that his respect for Brian might be because he didn't get on well with his own father, but she found that wasn't so. He told her his father worked on the railways, and his mother had a few cleaning jobs, and he was proud of them.

'When I was a little kid, I used to think we were rich just because our house was always neat and tidy,' he chuckled. 'You see, the estate we lived on was a bit rough, and most of my mates' homes were squalid. Their mums had fags hanging out of their mouths, and their dads got drunk a lot. But our mum was always there when we got home from school, in a clean pinny. She baked cakes, knitted us jumpers, and our garden was really pretty with loads of flowers. I never realized that we were better off than others just because Dad did lots of overtime and didn't drink, and Mum did all those cleaning jobs while we were at school. They were careful with what money they had – Dad even had an allotment and grew all our vegetables.'

'They sound lovely,' Eva said.

'They are. I see that now of course, they've got all the right values. But I still went through a stage at fifteen or so of rebelling, wanting to be a hard case like some of my mates. I wanted a motor bike, to hang around on street corners, and I used to bunk off school too, sniffing glue and stuff. If Dad hadn't come down on me like a ton of bricks, I would have ended up in serious trouble. But he talked to me, took me fishing and to football, and he got me an apprenticeship as a plasterer and talked me into playing rugby, going running and stuff.'

'To keep you out of mischief?'

'Partly that, but he also thought I could let off steam that way. I still play rugby for a local team and I still go running. But it was men like Brian that I worked with who really pulled me around. They teased me out of sullen moods, showed by example how to be a real man, and they kept an eye on me too. I found men like them could be a good laugh too, it was them who made me realize how lucky I was to have good parents.'

Eva found it touching that he appreciated what others had done for him. It made her think of Olive and how, by taking her on at Oakley and Smithson, she'd been able to help Eva break free from people who were pulling her down.

'I did my share of rebelling,' she admitted to Phil. 'And for much longer than you. My excuse was that I didn't fit in with the sort of girls that were approved of. I didn't really fit in at home either. I wasn't clever like my brother Ben, or stunning-looking like Sophie. I suspect I became a goth to shock my parents into noticing me.'

'A goth!' He spluttered with laughter. 'I can't imagine that.'

'Thankfully, I've got no photographic evidence of it.' Eva laughed. 'I think it was the happiest day in Mum's life when I bundled up all the black stuff and put it in the dustbin.'

'Will you tell me about your mum?' he asked. 'Or is that a taboo subject?'

'No, it's not taboo.' She went on to explain how she'd never known that Flora had been a successful artist when she was younger, or that Andrew wasn't her father. 'All I know about her past has come from old friends of hers who turned up for her funeral. One of them told me the name of the man who shared the place in Pottery Lane. They think he is my real father. I was thinking of trying to track him down. Do you think I should?'

Phil shrugged. 'I'd want to. Even if it turns out he isn't your dad, he might be able to throw light on things you don't understand.'

Eva felt the conversation was getting a bit too heavy and one-sided, so she lightened it up by asking him if he liked travelling.

'So far I've only been to Ibiza, Benidorm and Goa. But I'd like to go to a great many more places. How about you?'

'Only family holidays in France and Spain,' she admitted. 'I got as far as applying for my own passport when I was eighteen, as I'd been on the family one before. I talked about going off somewhere, but never did. One of my stepdad's lame excuses for telling me he wasn't my real dad was that I might need my birth certificate for a visa and I'd see his name wasn't on it.'

Phil nodded in understanding. 'Well, you're going to show him what you're made of, aren't you?'

Eva smirked. She'd had that same thought many times in the last few days. 'Yes, I intend to. I bet he knew how awful Mum's old studio was and hoped I'd fall flat on my face with it. But thanks to your help and advice, I don't think I will.'

'That's what mates are for.' He grinned. 'We are mates now, aren't we?'

Later that evening Eva sat for a while on the crate in the backyard, just enjoying the fresh air, even though it was dark. She heard the landlord at the pub ring the bell for last orders, and she marvelled that in a little less than a week here, she'd begun to think of it as home.

Phil had caught the bus home after their meal. Outside the pub, as they were saying goodbye, she thought he was going to kiss her. It had been a strange moment – half of her wanted him to, the other half was afraid. But all he did was kiss her cheek and say he'd see her tomorrow when he called with John about the windows.

She turned to watch him walking off down towards the main road. He had a good walk, light on his feet and his back straight. He'd left her with a warm glow inside, a good, secure feeling. He had said nothing to suggest he fancied her, but a sixth sense told her he did. That could be wishful thinking on her part; she certainly wasn't going to make any move on him to find out, because it would just be humiliating if he didn't respond. Maybe that question about being mates was his way of telling her he didn't see her in that light?

She wasn't ready to embark on another relationship, any-way. What was important to her now was to raise her own self-esteem. She was never going to allow anyone to feel sorry for her again.

Until the windows and kitchen were done she couldn't do much in the house, but she could make a start on the backyard. Tomorrow, after Phil and John had been, she would buy some gardening tools, some plants and tubs, and maybe a table and chairs too.

The prospect of transforming the grubby weed-strewn yard into something beautiful was really appealing. First thing tomorrow morning, she thought she would get out here and scrub the paving stones clean.

While she was at it, she'd mentally scrub Tod and Andrew out of her mind too. She couldn't move forward as long as she kept looking over her shoulder.

Chapter Ten

Eva stood in her backyard and admired her new French doors, feeling ridiculously emotional. 'Don't they look wonderful?' she said to Phil, who was cleaning up dropped plaster from the floor. 'I can't believe how they've transformed the room. It looks twice the size, really modern and airy.'

It was Sunday afternoon, and two weeks had passed since Phil first brought John, his window man, round to meet her. Like Brian, John was another middle-aged man, tall and skinny with little to say for himself, but he was a fast worker. He had begun the job with his son Rory on Thursday, knocking out the old boarded-up window and door. Even the old sink, cooker and the graffiti didn't look half as bad with sunshine and fresh air coming in. Eva had watched entranced as he began laying bricks to take the frame for the new doors.

Even the weather was on their side, as June arrived with hot sunshine. On Friday they had set the uPVC frame in, returning on Saturday to put on the doors. Eva had been horrified by all the mess, rubble, old cement, dust and bricks in the backyard, but John and Rory had taken away every last bit of it that evening, including all the bags of rubbish and the mattress in the garage.

Phil had come round today to make good all the plaster, and he'd skimmed the kitchen area and the right-hand wall of the room too. He said he was coming back to do the rest of the room once Brian had installed the kitchen units.

'If you don't look round at that,' Phil said, nodding at

where the units were stacked in front of the graffiti-covered wall, 'you wouldn't think it was the same place.'

All Eva could do was grin with delight. The doors had cost £1,000, which seemed an awfully big chunk of her money, but they were worth it. John was coming back in two weeks to do all the other windows in the house. And Brian had said he would fit the kitchen during the coming week, as he had a couple of spare afternoons.

Everything was going well. She really liked working at the bistro; Marcia, the other waitress, was fun. And the owner Antonio seemed to like her too. The short hours gave her time to work on the garden, and she was getting to know her way around London.

Eva handed Phil a cold beer from the fridge. She hadn't seen him since he'd brought John round to see about doing the windows. He'd been working in Windsor, so it was very nice to have him here all day today. While he'd been plastering, she'd been planting flowers. Then she got fish and chips for them at lunchtime, which they'd eaten sitting at her new little table and chairs outside.

He got hot working, and earlier he'd stripped off his overalls – down to just a pair of khaki shorts. His whole torso, face and hair were now speckled with plaster. Eva had furtively watched him as he was working, turned on a little, not just by his muscles and smooth skin, but by the graceful sweep of his arm as he smoothed the plaster, and the concentration in his face.

'I ought to get going soon,' he said. 'I told Mum I'd pop round to see her this evening, and I'm off early in the morning to Dorset for a job.'

'You haven't told me yet how much you want for the plastering,' she said.

'All you need to pay for is the plaster,' he said with a grin. 'And not now, when I've finished the room will do. Is it

158

alright if I come again next Saturday? Brian should just about be finished by then.'

'What would I have done without you?' she said. 'I'm really grateful to the handbag snatcher now.'

He smiled, and reached out and touched her cheek lightly. 'You would've charmed some other guy,' he said. 'Anyway, it's been nice today. You are so easy to be with.'

His light touch had sent a little shiver down her spine. She wished he could stay, that they could go to the pub, or just sit out in the garden with a few drinks.

'Can I be really cheeky then and ask you to do just one more thing before you go?' she asked. 'Just to get up in the attic and see if there's anything in there? I've tried, but I can't move the hatch.'

'Hoping for some treasure?' he said, finishing off his beer.

'I suppose so. People do put things in attics and forget them, don't they?'

'My mum and dad certainly do. There're old Christmas decs, boxes of stuff they don't even remember putting there, and the cot my brother and I had when we were babies. So you never know what we'll find.'

He took the stepladder up into the back bedroom, and had to move a pile of books to set it up.

'Have you read all these?' he asked.

'Yes, they are all old friends,' she said. 'I love reading and I can't wait until I can put some shelves up for them all. I joined the library the other day. I have to have a book on the go. Do you read?'

'Not much,' he replied. 'The books I own wouldn't even fill one shelf. I'm more of a magazine man, but I'd read if I was lying on a beach.'

Once the stepladder was in place he went up it and pushed hard on the hatch till it finally opened.

'Pass us your torch,' he said to Eva, who was holding the stepladder. 'It's pitch black in there.'

Climbing up a little higher, his top half was then in the loft.

'Can you see anything?' she called out.

'There're a couple of boxes . . . one's got a load of paintings in it,' he shouted down. 'Want me to get them out?'

'Yes, please,' she shouted back, suddenly really excited.

There was a kind of shuffling noise as if he was pulling stuff closer to the hatch, then he moved down the stepladder a couple of rungs.

'Paintings first,' he said and hauled out a couple of big canvases.

Eva took them from him eagerly. One was a woodland scene with the ground carpeted in bluebells. The other was of an old door set in a wall covered in creepers. It reminded her of the book *The Secret Garden*. She let out a squeal of delight when she saw her mother's initials F. F. in the bottom right-hand corner.

Next he handed down a box in which there were about a dozen more smaller canvases. She didn't stop to look at them, as Phil was already heaving out a much larger box which was sealed up with tape.

She had hardly put that on the floor before a second smaller one came down.

'That's it now, nothing else up there,' he said, putting the hatch back again and climbing down. 'But you could do with getting some insulation up there before next winter.'

He picked up the painting of the bluebell wood. 'This is amazing. Not that I know anything about art.'

'It was painted by my mum,' Eva said excitedly, pointing to the initials. 'Her maiden name was Flora Foyle. Isn't it beautiful?'

It was in fact so beautiful that it made all the hairs on her arms stand on end. The sunshine filtering through the trees was remarkable, and the details – not just the bluebells, but the bark on the trees, shiny ivy growing over an old tree stump – took her breath away.

'I'll have to get it framed and hung downstairs,' she said.

'A housewarming present from your mum,' Phil said, putting one big plaster-splattered hand on her shoulder. 'It's a beauty. She was very talented.'

But Eva wanted to see what was in the boxes. She tore off the tape on the first one. Whatever was in there was carefully covered in tissue paper. She folded it back. 'Baby things,' she gasped on seeing a tiny pink jacket. 'Mine?'

'I would think so.' Phil smiled at her stunned expression. 'I bet she packed them away when they got too small for you, and she forgot to get them when she left here. But as much as I'd like to go through this lot with you, I've really got to go. I'll see you next Saturday.'

All that evening, Eva pored over the contents of the boxes. The box of baby clothes appeared to be outgrown things which Flora had packed away and then forgotten, just as Phil had suggested. All that was really notable about them was that they looked rather old-fashioned – hand-knitted jackets and smocked dresses. There were old sprigs of lavender packed amongst them, and a faint hint of it still clung to the clothes.

The collection of paintings was superb, and she was staggered by her mother's talent. The ten smaller ones, all about twelve by fourteen inches, were very varied in subject matter. A couple were of vases of flowers, exquisite in their detail. Then there were three landscapes – all different – one of a baby sleeping in a pram, which she felt certain was her, and another of a rather run-down row of shops. The final three were of gardens: dreamy, sun-filled pictures with statues peeping out from behind voluptuous peonies and roses. She liked those three the best.

But the second smaller box was really intriguing. Eva didn't know if she was being fanciful, but it seemed to her that it had been purposely left here for her to find. She felt there was a meaning in every item, whether that was the old photograph album – with pictures of people who must be her grandparents and aunts and uncles – or snaps of Flora as a young student, many in fancy-dress costumes, press cuttings praising her art, and diaries, some dating back to when Flora was in her early teens.

There was an envelope containing a pencil sketch of a cottage, and with it a photograph of that same run-down row of shops as in one of the oil paintings. They seemed to belong together. Could the owner of one of the shops have lived in the cottage? Or were they both places where Flora had once lived, and so were important to her? Eva wondered why she hadn't attached an explanatory note to them.

Also in the box was Eva's full birth certificate. Just as Andrew had said, there was a dash in the space for her father's details. Eva guessed Flora had hoped it would never come to light that her daughter was illegitimate.

A beautiful silver necklace designed as a series of joined small hearts was tucked into a small box with a card saying simply 'I'll love you for ever' and signed 'P'. Was that from Patrick O'Donnell, the man who might be her father?

There were several invitation cards to exhibitions of Flora's work. They were from various art galleries, mostly in London, dated from the mid to late sixties. There was also an estate agent's leaflet giving details of this house; the asking price was £1,500. Flora had written on it in pencil: 'This is the one.'

A book called *The Prophet* by Khalil Gibran had been inscribed inside to Flora. The message was: 'Books, art and music belong to those who can see and hear true beauty. May your eyes and ears remain sharp for ever.' Sadly, whoever had given it to her hadn't put their name, just the date. April 1968. Eva skimmed through it and was entranced by the author's beautiful, lyrical prose. She intended to read it properly later.

There was also a sketchbook full of pencil drawings of children. Eva felt Flora must once have had the idea of becoming a children's book illustrator, as there was the glimpse of a story in the pictures of untidy, street children reacting to one another in comical ways.

Yet the item that affected her the most was a notebook with a Liberty-print fabric cover in shades of pink and mauve, tied with pink ribbon. On every page was a quick sketch of Eva's head and upper body as a baby, each with a caption beneath that appeared to reflect Flora's thoughts of the day.

It began when she was about two weeks old and sleeping. Beneath it Flora had written: 'So angelic now after screaming for nearly two hours.'

A few sketches further on, she had drawn Eva screaming and had caught perfectly the screwed-up face of an angry baby. Beneath this one Flora had written: 'At times like this I want to walk out of the door for good.'

Eva could see her own progress as she turned the pages, her features becoming more pronounced, her small hands becoming chubbier and her hair starting to grow. She could also sense Flora's exhaustion in her words. 'Will the day ever come again when I'll have the time and energy to paint?' was one comment.

The sketches continued until she was perhaps six months old. In the last one she was smiling, showing two teeth clearly.

Beneath that one was simply 'Precious One'. Eva's eyes filled with tears at this, and a terrible feeling of loss overwhelmed her. All at once she understood what grief really meant, because this was more acute than the pain she'd felt on the day Flora died.

Yet there was comfort too in being able to touch these drawings, almost as if Flora was there in the rooms with her, whispering that her baby had meant everything to her.

Why didn't she give her this book on her eighteenth birthday? It would have been such a perfect gift. But then perhaps Flora had forgotten about it? Eva wondered too if it would have had the same impact on her if she'd been given it when her mother was still alive.

Was it Andrew coming into Flora's life that had stopped the sketch diary? Eva certainly had a sense that it was just mother and baby together at the time Flora had made the sketches. Had he even seen this box of things? Somehow, she doubted it. She sensed Flora had put them up in the attic around the time she met him.

Was that because once she had met Andrew she didn't want reminders of the time when she was alone with her baby, or reminders of the father? But what was it about Andrew that made her turn her back on her art when it had clearly once been so important to her?

Although Eva had been badly hurt by Andrew, and knew him to be something of a control freak, she couldn't believe that he would ever have wantonly suppressed Flora's talent. What reason would he have had for doing so?

Reading all the diaries carefully might throw some light on everything that she found so puzzling. She resolved to read some each night, make notes of any names or places mentioned, and try to piece it all together.

*

'You are just in time,' Brian called out as she came in from work at three thirty on Thursday afternoon. 'Can you come and hold this for me?'

She saw he was struggling to get a wall cupboard on to its fixings.

She grasped the bottom of it and held it up while he clambered up on the stepladder. 'So much easier with two pairs of hands,' he said. 'Fancy becoming my apprentice?'

Eva laughed. 'I wouldn't mind. Maybe then I could do some jobs myself,' she said.

Once he had the wall cupboard fixed, he climbed down again. 'Right then, tomorrow, after I've secured the work surfaces, I'll show you how to use the drill and some other basic things,' he said. 'Now for that last cupboard before I go. I need to get to the dump with the old cooker and sink before the place closes.'

Eva helped him out later with the cooker and sink. As he drove off in his van, honking the horn in farewell, she smiled. He was such a lovely man – funny and fatherly. He'd arrived yesterday to start the kitchen just as she got in from work. He'd been almost as thrilled as she was over her new cooker, and the French doors. He'd worked through till nearly seven, laughing and chatting with her, yet getting an amazing amount done.

She thought his wife was one lucky lady and hoped she appreciated him.

The following morning Eva answered the door but instead of it being Brian as she expected, it was her next door neighbour. She didn't know his name, but he'd nodded at her a week earlier when they both arrived home at the same time.

He was around forty, tall and well built with a ruddy complexion; Eva had seen his wife sunbathing in the garden while she'd been looking out of her bedroom window. She was a good bit younger than her husband, slim, long-legged and looked like a fashion model.

'Hello,' Eva said. 'I'm Eva Patterson. And you live next door, I believe?'

'Yes. Francis, Simon Francis. I should have called when you first moved in, but I thought you were just another tenant. As I understand it, you are the new owner of the house?'

'Yes, that's right,' she said. She didn't like his condescending tone one bit. 'You don't call on tenants then?' she added with faint sarcasm.

'Not if they are like the last ones here,' he said. 'They lived like pigs, rooting around in their own filth. Music blaring out all night, fights and rows all the time. We were delighted when they left.'

'It certainly looked like a pigsty when I arrived,' she said. 'But I'm getting it into shape now.'

'That's what I called about,' he said, running a finger around the collar of his shirt as if he was a little nervous. 'We saw you'd put on new doors at the back.'

'Yes, they are a great improvement on a boarded-up window.'

'Are you planning to replace the front windows and door too?'

'Yes, I am,' she said, a little baffled that he thought this was an appropriate way of welcoming a new neighbour. 'Why do you want to know?'

'I am concerned that you might be intending to put in plastic window frames there too,' he said.

Eva could only stare at him in astonishment for a moment. 'And my window frames are your business, because . . .?'

'Well, plastic does rather lower the tone,' he said in a pompous manner.

She was flabbergasted. 'Well, Mr Francis,' she said in her most icy tone, 'I own this house, and if I want to have window frames made of play dough, plastic or solid gold, it will be my choice and mine alone.'

'I'm only speaking out because you are young and probably don't appreciate the history of these houses,' he retorted.

'Oh, I do,' she smirked. 'They were thrown up for the workers at the pottery, probably funded by the pottery owner who lived in one of the more salubrious Georgian houses nearby and paid them a pittance.'

'There's no need to take that attitude with me,' he said.

'I wouldn't have if you hadn't been so rude and un-neighbourly,' she said. 'Now if you'll excuse me, I have things to do.'

She shut the door in his face, smarting with anger.

A few minutes later there was another knock on the door. Thinking it was her neighbour back again, she wrenched the door open ready to lay into him.

But it was Brian.

'Oh, thank goodness!' she said. 'Am I glad to see you!' She blurted out what had just happened.

'Bloody snob.' Brian sniffed contemptuously. 'These flaming yuppies around here get right up my nose. I agree that a Georgian house should have traditional sash windows with wood frames, but this is just a little Victorian working man's house. And it don't make no sense to stick in windows that need painting every year.'

'I can't believe anyone could be so snotty,' she said, her face flushed red with anger. 'Who the hell does he think he is?'

'You'll have to get used to that sort of crap if you live around here,' Brian said. 'Gentrified areas always attract snobby arseholes. If they had their way, they'd tear down the council houses and the people who live in them would be dumped somewhere else, then they'd build a wall around the whole area to make sure no common folk got in again.'

'What makes some people so mean?' she said, her anger fading and now replaced by hurt. 'Why couldn't he just have welcomed me, asked how I was getting on and if I'd like to come in for a cup of tea?'

Brian patted her shoulder in sympathy. 'Because he's one of those "I am it" prats. He probably made his pile selling insurance and pensions, and he's terrified that his house – which he considers "an investment" rather than a home – might come down in value. My gran was born around here. She's told me how it was in the 1950s. It were a slum area then, all the way from here down to Ladbroke Grove and Westbourne Grove. She talked about the race riots in Notting Hill, and how that bastard slum landlord Peter Rachman stuck half a dozen West Indian families in just one room. It weren't such a desirable place to live in then. Your poncey neighbour is just afraid it might go that way again.'

'Was it still slummy in 1968?' Eva asked. 'I think that's about when my mother bought this house.'

'Yes, it would've been – not as bad as Westbourne Grove and Ladbroke Grove, but still pretty grim. They made that film *Blow-Up*, with David Hemmings and Vanessa Redgrave, here in Pottery Lane in 1966. I think that was the start of "arty" people moving here, because it was much cheaper than Chelsea or Hampstead.'

Brian asked her to hold the end of his measure against the sheet of Formica, while he marked off where he needed to cut it, and as she helped him she told him about the stuff of her mother's that Phil had found in the attic.

'It was so exciting to find her paintings,' she said. 'But the other box of stuff feels like she left it there on purpose for me to find. Everything in it seems to be a clue, like in a treasure hunt, and I've got to work out what it means.'

'What did she die of, Eva?' he asked

'She killed herself,' she said bluntly.

She expected him to be shocked, but he looked as if he already knew what she was going to say.

'Phil told you?' she asked.

Brian shook his head. 'No, he didn't, he wouldn't break a confidence. I just suspected it was something like that, because people who have recently lost someone close do tend to tell you what they died of. Besides, there was something about you that first time I came here with Phil, you looked like a lost and frightened puppy.'

Eva told him the whole story then: about Andrew, and then what happened with Tod. 'So I packed my bags and came here,' she said with a shrug. 'It seemed the only thing to do, even if this place was a tip.'

'If you was my girl, I'd have wrung that lad's neck,' he said stoutly. 'But then if you was mine, you wouldn't have moved away with your mother hardly cold – whether you was my blood or not. You've been through it, no doubt about that. But you're a brave little thing, a real sweetheart, and I think you'll do well for yourself here in London.'

Once Brian had fixed the worktops Eva whooped with delight at the finished result. It was all very simple, just grey and white, but though it was a small kitchen it would be very practical. 'It looks marvellous,' she said. 'Like something out of *House & Garden*.'

Brian laughed. 'Well, I don't look in them hoity-toity magazines,' he said. 'But I like it when the ladies I do kitchens for act like a kid in a Wendy house.'

'Is that what I'm doing?' she asked as she ran her hands over the worktops and opened cupboards and drawers.

'Yes, you are. I bet you'll be down here half the night tonight, arranging knives and forks in the drawers, stacking up your saucepans and stuff.'

'I haven't got much to put in here yet,' she giggled. 'But it is my first home. And I bet your wife was the same with hers?'

He was adjusting the doors so they hung correctly, and he looked up and grinned. 'We had two grotty rooms by Shepherd's Bush market back then. We had mice, just two gas rings, and we had to share the bathroom with four other couples. But we thought it was wonderful, because we'd been living with her mum up till then, and she gave me earache the whole time.'

'Did you do it all up?'

'Not really, we didn't stay in that place long enough. I worked all hours to get a deposit to buy a house. We moved in there two days before the first baby came. The big thrill then was having our own bathroom.'

'I keep wondering how my mum felt about this place when she first bought it,' Eva said. 'It has got a nice vibe about it. I must have been born in a hospital near here, my pram must've stood here somewhere. It's strange thinking about that.'

'Pity she never told you anything,' he said. 'But I guess she must've felt a bit ashamed that she was a single mum. My sister got up the duff in 1968, she was only seventeen, and even if it was all Flower Power and rock concerts then, my dad went ape shit, there was the shotgun wedding an' all.'

'So your niece – or is it a nephew? – is only two years older than me.'

'Niece. Yes, and a right little cracker she is. And her dad and mum are still together. We all wonder what all the fuss was about now. But that's just the way it was back then. Things didn't really change till the mid-seventies.'

'Yes, I can see that's how it was for most people. But Mum was supposed to be a free spirit, a bit wild and stuff, so I don't see why she would be ashamed of me being illegitimate.'

'Then maybe it was your stepdad who had the hang-up about it?'

'You mean it was his idea to pretend I was his baby?'

Brian shrugged. 'Maybe. His parents might have been old-fashioned and wouldn't approve of him marrying a single mother. Have you ever seen pictures of their wedding?'

Eva shook her head. 'I don't think there are any. Mum said it was a very small, quiet one. She never even said where it took place.'

'You might find that out in the diaries. My guess is they got married in secret and then told the family they'd done it a lot earlier than they really did. I expect Andrew wanted to protect her from gossip!'

'So if he loved her enough to do that, and to bring me up as his own, why did he turn nasty later?'

Brian scratched his head. 'Who can say? People change. Sometimes one of the partners loves more than the other one. I know people who are bitter because they didn't get what they expected out of it. There are dozens of reasons for a change of heart. I'd guess that you were just the easiest one to have a pop at. But I promised you some lessons on using an electric drill – and that's going to be a lot more useful to you than finding out what happened between your mum and stepdad.'

That evening Eva went through the box of baby clothes; she thought she might take the best ones to a charity shop. But when she got to the bottom of the box, she found another small box. It was a pretty pink one – the kind a present for a baby might have been put in. Opening it, she found a very tiny pink matinee jacket, matching bonnet and bootees, and a little dress, only big enough for a newborn baby. The dress was nylon, overly frilly, and the hand-knitted jacket, bonnet and bootees were very lacy with satin ribbons. She couldn't imagine that her mother had picked them out – she had always said she loathed fussy baby clothes. But maybe someone dear to her had made them for her and that was why she'd kept them and packed them away so carefully. Maybe Eva's grandmother?

Eva lifted out the tissue paper from the box to repack the clothes, and to her surprise there was a black and white photograph beneath it. It was of a tiny baby, lying in a pram, and Eva was certain it was her. But if it was, why had Flora always claimed there were no pictures of her as a newborn baby, because she didn't have a camera then?

The diaries were proving hard to read. For one thing, the entries were rarely dated, and Flora had an irritating way of using people's initials rather than their names. And in the first diary she hadn't once said where she was. If she hadn't numbered them, Eva wouldn't have even known in which order to read the diaries.

She also wrote in what seemed like riddles. 'Wishing M would stop behaving like I was still six.' Obviously the 'M' was for mother and she was still in Cornwall, as there were many other references to nagging and wanting to go to 'L', which must mean London. Although she wrote 'arrived in L, and it's so huge it's scary', she didn't say what part of London, or whether she was with anyone else. Later on she mentioned The Bistingo, and this appeared to be a restaurant she worked at, as there were many entries referring to possible friends, again only using the initial when describing someone who had come into The Bistingo on different occasions. 'J' was mentioned most, as owing her money, getting on her nerves, or having no talent. So was 'J' male or female? A lover, or just a friend?

She thought that the only way she was going to make any headway was to note down anything she thought might be relevant, even if she didn't understand it, and then try to contact Patrick O'Donnell to see if he could throw any light on it.

But for now she had too much else to do. She needed to buy cooking equipment and crockery. One saucepan, two plates and enough cutlery for one didn't justify even having a kitchen. She needed to get a telephone line installed, arrange for John to come back and do the other windows, get quotes for central heating and a new bathroom suite, and decorate the house. Brian had told her today that she would need to give all the woodwork, doors, skirting boards and the banister a really good rub down before applying undercoat, then at least two coats of gloss paint. Likewise any holes in the walls upstairs would need to be filled and rubbed down before painting. That all sounded very boring and a lot of hard work, but she supposed she would have to do it properly. Until it was done, she couldn't buy furniture or put carpets down.

'You'll have to wait, Mum,' she said, putting the lid back on the pink box. 'I suppose as you kept your secrets for twenty-one years, a bit longer won't make any difference.'

Saturday turned out to be a really good day, still warm and sunny, and fun too because Phil arrived at ten ready to finish skimming the living-room walls.

Eva didn't admit to him that she'd been up at six that morning; she'd made herself scrambled eggs on toast, delighting in the new cooker, the shiny sink and the taps that were easy to turn on. She'd eaten it in the garden looking back through the French doors so she could gloat over the new kitchen. Then she'd put a load in her new washing machine and hung it outside on an airer. She just wished Tod and Andrew could see her now; she felt as if she was putting two fingers up to them.

Almost the first thing Phil said when he arrived was that he'd like to take her out for a meal that night.

'That would be lovely,' she replied.

'Glad you think so,' he said. 'But don't get alarmed that I've brought a bag with me – that's just some smarter clothes to change into later. I thought we could walk through Holland Park, there's a great Chinese up on Kensington High Street.'

Eva liked the fact that he didn't want her to think he expected to stay the night. She really liked him, she kept thinking about him when he wasn't here, and it was nice that he was so gentlemanly with her. She certainly didn't want to be pushed into anything until she was really sure.

Shortly afterwards, Brian turned up with a couple of boxes of tiles and some sheets of hardboard.

'I found all these tiles in the shed last night, and I thought you might like them for a splashback,' he said, opening a box to show her some white tiles with a pale grey motif on them. 'They were left over from a job, and I reckon there're more than enough to do the wall behind the sink.'

Eva was thrilled. They were just what she would have picked herself. 'Gorgeous! Thank you so much, Brian. But you shouldn't be wasting your Saturday doing stuff for me, you should be taking Julie out.'

'She was going shopping with her sister, anyway.' He grinned. 'She don't want me trailing along behind her. And the hardboard is to lay on the floor ready for lino tiles – you can't put them down on floorboards. I brought a bit extra too for the back of the units under the breakfast bar. That'll look OK when it's painted white.'

Eva left the men to get on while she drove up to Notting Hill to buy some kitchen equipment, plates, bowls, and cutlery too. While she was up there she looked in an interior design shop to get ideas for fabric to make some curtains. Everything was terribly expensive, but then she'd already planned to get her fabric from a cheap shop she'd seen in Shepherd's Bush.

Brian had done most of the tiling by the time she got back, and Phil had nearly finished his skimming.

'Goodbye graffiti,' she said as he began to cover the last bit. 'I'm sorry, but I'm not going to miss you one bit.'

Phil laughed. 'What kind of arsehole does it to a place he lives in?' he said. 'I can understand it on empty buildings, or even the thrill of doing it on a bridge over a railway line, but not in your house.'

'I expect there was nothing on the telly,' Brian said. 'I'll have to try it one night and see what Julie thinks of it.'

'I think she might dismember you,' Eva said. 'Did you tell Phil about what that creep next door said about plastic window frames?'

'He did,' Phil said. 'I felt like knocking on his door and marking his card, bloody cheek! But I reckon the reason he stuck his oar in was because he'd hoped to get his hands on this place.'

As Brian hammered down the hardboard on the floor, he glanced at Eva unpacking all the kitchen stuff she'd bought and thought how much better she looked compared with the first time he had called here. Phil hadn't revealed anything much about her – just that the house was a tip, and she needed help.

Brian had more than enough work lined up to do already, but he knew Phil wouldn't have asked him if he hadn't been worried about the girl. As soon as he got here he saw right away why his friend was concerned. It wasn't just that the house was such a mess. It was her: she looked forlorn and scared, only one step away from tears, yet she was desperately trying to act confident.

If it had been anyone else living in that area, he would have charged a couple of thousand at least. But faced with someone who looked like the world was against her, it was all he could do not to give her a hug, and offer to do the kitchen for free.

It turned out to be a pleasure, because she was a little sweetheart. She made him tea, offered her help, and showed so much gratitude and admiration for his work. Then when he got her story out of her, he understood why she'd looked so forlorn. She might have been left a property valuable enough to set her up for life if she used her head, but losing her mother, and then her stepfather turning on her, was enough to crack even the toughest person. And she could easily fall prey to some unscrupulous bastard who would rip her off.

Brian really hoped she and Phil would become an item, because it was obvious they were ideal for one another. But Phil was as bad as Eva; he'd been hurt badly and he was afraid to trust again.

The two of them needed their heads knocking together or they might spend an eternity pussyfooting around, both too scared to make the first move.

By two in the afternoon both men had finished their jobs, and they went out into the garden with a beer.

'You've turned this into a real beauty spot,' Brian said

appreciatively, noting she'd scrubbed the paving stones and planted flowers in every available bit of soil and still more in tubs. It had looked as forlorn as she did the first time he'd seen it, but the warm weather had made everything grow. She'd dug out all the dead plants, trimmed back the straggling climbers, and there wasn't a weed in sight now.

'Did you leave anything for anyone else at the garden centre?' Phil joked, looking at a couple of large empty pots, sacks of compost and trays of still more bedding plants waiting to be planted.

'Don't you scoff, she's a good little homemaker,' Brian said. 'You mark my words, by the time she throws a housewarming party the whole place will be like a palace.'

Eva glowed at the praise. 'I will have one, and you must bring Julie so she can see how lovely the kitchen is. I'll pin the pictures up of when it was a hovel. And then you, Phil and John can all bask in everyone's admiration.'

'Once John's done the windows you ought to get the heating and the bathroom done,' Phil said. 'I've got some numbers for you to ring, Eva. I don't want you getting any cowboys in.'

'If you get the lino tiles for the kitchen, I'll come back and put those down for you,' Brian said. 'And if you want a fitted wardrobe upstairs, just shout.'

'I will,' she said, smiling at him. 'I might need more lessons in DIY too.'

He jotted down something on a scrap of paper and handed it to her. 'That's how many lino tiles you'll need, and my phone number so you can ring me.' He got up and clamped his hand on her shoulder. 'I must go now. It's been a pleasure doing your kitchen, and you keep in touch now.'

Eva went into the house with him to get the money she owed him. He looked as if he didn't want to take it, but she pressed it into his hand and kissed his cheek.

'I'm going to miss you,' she said. 'Your Julie is a lucky lady.'

He looked faintly embarrassed. 'And you are a lovely girl,' he said. 'Now if you're worried about anything, or want to know something, just call me.'

It was a beautiful evening as Eva and Phil walked through Holland Park to the restaurant. She hadn't realized that the tube took its name from this park. In fact it was more of a wood, really – some of the trees must have been planted a couple of hundred years ago.

'I love this park best in May when the leaves are all vivid and new, and the bluebells are out,' Phil said thoughtfully. 'Mum used to bring me and my brother here for picnics sometimes. She liked to look at the posh houses in the streets around here, and me and Lee liked to climb the trees.'

'I wonder if this is where my mum painted that bluebell picture?' Eva said. 'She loved them but said that they were a pest in the garden. I love them too, but maybe I should heed her advice and not plant any.'

'You ought to become a gardener,' he said. 'You've got a real flair for it.'

'I don't know nearly enough,' she replied. 'But I know I ought to think about getting a real career – working part time in a bistro is hardly that.'

'So what would you like to do? I mean, if you could choose anything?'

Eva thought for a moment. 'I'd like to train to be an interior designer. I've been into that posh shop in Notting Hill twice, and I've watched the woman in there making up a board with paint colours, swatches of curtain materials and stuff. I think I could do that. Well, maybe with the right training.'

'Could you go to a college for that?' he asked.

'There was a place in Cheltenham that did courses. So there must be dozens in London. I can sew too, so I could make curtains and cushions. I made the ones in both Sophie's and my own bedroom at home. But I expect it costs a lot for a course – it's always those glossy, far-back sort of women that do it.'

'You could make inquiries,' he said. 'Maybe even get yourself taken on in a shop like the one in Notting Hill. I've been working in some of these big houses where interior designers come in and throw their weight around about colour schemes. They've always struck me as just chancers, anyone with an eye for colour and a big budget could do it.'

'Both you and Brian are doing a fine job on building up my confidence.' She laughed. 'A few weeks ago I knew absolutely nothing about plumbing, plastering . . . or anything, really. I've learned such a lot watching you two. And John too.'

'Now, tell me about the stuff in the attic boxes,' he said. 'Was there anything in there worth anything to you?'

Chapter Eleven

It was almost eleven. Eva's feet were aching, and as she watched the last remaining table of six chatting and laughing she wished they would pay their bill and leave so she could go home.

She had started to do a Friday evening shift, along with lunchtimes at the bistro, the week following her meal with Phil in Kensington. Soon she was doing Thursdays as well, and now it was Saturdays too.

At first she had been glad of the extra work; she got far bigger tips on evening shifts, the time flew by because the bistro was so busy, and she was getting to know people in the neighbourhood. But what she really wanted was time to have some fun, which seemed to be evading her.

It was the end of July, and if she was going to throw a party before the end of the summer, she needed carpets down and some furniture. Not that she had any real friends aside from Brian, John and Phil to invite to a party. But she had thought she could invite a few of the staff here, and some of the more friendly customers, and that way she'd get to know them better.

She needed something to aim for; it seemed to her that all she did was work and sleep. There was no time to laze in the garden with a book, to explore London, or even to go and buy some new clothes to put in the lovely fitted wardrobe that Brian had built in the big bedroom.

She had bought a single bed when the inflatable mattress punctured, and a chest of drawers. A couple of stools for the breakfast bar meant she could at least sit down to eat. But she couldn't order carpets until the decorating was done, and there was no point in buying a sofa before the carpets.

Eva was afraid she'd hurt Phil's feelings too. First, by never being free on Friday and Saturday nights. And then he'd offered to get a couple of his friends in to paint the house throughout, but she'd snapped at him and said she wanted to do it herself.

She really did want to do it herself – that wasn't an excuse to stop him coming round. But she was a little anxious about how he felt about her. He hadn't so much as kissed her yet – well, apart from on the cheek. Yet he had a way of looking at her sometimes, as if he was willing her to make a move on him. Maybe she should; after all, she did fancy him. But she was too afraid of getting hurt again, and that stopped her.

Eva had rung Olive a few days earlier when the phone was finally installed. Her excuse was to give Olive the new telephone number, but in reality she wanted the older woman's advice about Phil. They chatted for a while, and Eva told her how the house was progressing. The new bathroom and central heating had just been completed, and she'd got a low-interest home improvement loan to pay for it.

Olive had lost none of her directness and went straight for the jugular, asking if 'Prince Plasterer' was still around. Eva admitted her quandary.

'Don't be so daft, girl,' Olive said. 'Get a bit tiddly, give him a kiss and if he doesn't respond, apologize and tell him it was just the drink.'

'But what if –' Eva began.

Olive cut her short. 'What if the Moon collides with Earth tomorrow, or the Russians fire a nuclear bomb? Don't waste time on "what ifs", life is too short for that. It's obvious he fancies you madly or he wouldn't do all this stuff for you. Jump in with both feet, girl.'

Eva had been amused and tempted by Olive's advice, and thought she might try it next time Phil came round. If he came round again. Maybe she'd put him off for good?

As she polished glasses and replaced them behind the bar she watched the remaining customers enviously. The men were suave and confident and the three women with them were like so many of the women around here – wearing chic clothes, with perfect hair and make-up. They had an aura about them, as if they'd never had a moment of panic that they weren't pretty or clever enough.

She wondered what it was that gave some people that self-assurance, and why she, who had grown up with so many advantages, didn't have it.

At last the group got up to leave. The three women and one of the men went outside, and the oldest of the three men came forward to pay. But the third man stood just behind him, smiling at Eva.

She smiled back. He was in his late twenties, about five foot ten, very good-looking with bright blue eyes and impossibly long dark eyelashes. She'd seen him in the bistro once before, having a business lunch with some older men.

'Eva, isn't it?' he said as his friend put his wallet away and turned towards the door to leave. 'I heard someone call you that. It's a pretty name.'

'Well, thank you,' she replied, blushing because he was looking at her so intently.

'It suits you. I bet you've been dying for us to leave?'

She laughed. 'Sort of . . . my feet are aching. I hope you enjoyed your meal.'

'I enjoyed looking at you more,' he said.

'Are you coming, Myles?' his older friend asked, holding the door open and looking back. 'The girls are waiting.'

Myles reached out, took her hand, lifted it to his lips and

kissed the tips of her fingers. 'Can I see you tomorrow?' he asked.

The suddenness of this completely threw Eva. 'I can't,' she said instinctively.

'You aren't working here, it's closed on Sundays.'

'I've got stuff to do,' she said. 'And your friends are waiting for you.'

'I'll come round, anyway,' he threw over his shoulder as he walked towards the door. 'Pottery Lane, isn't it?'

She didn't get a chance to ask how he knew that, because the door closed behind him. When she looked out of the window he had his arm around one of the women and they were walking towards Holland Park Avenue.

Eva locked the door, turned the sign round to closed, cleared the table and carried the glasses and coffee cups through to the kitchen where Antonio, the owner, was cleaning the preparation surfaces.

Antonio was only half Italian and had been brought up in England, but he put on an Italian accent for the customers. He'd told Eva in confidence that his real name was Roger. She liked him; he was short, fat, with a sallow complexion and bad teeth, but he was funny, generous and kind-hearted, and he was a fantastic chef.

'Do you know a customer called Myles?' she asked. 'He was in that last group to leave.'

'Good-looking bastard?' he said.

'Well, yes. He asked to see me tomorrow.'

'Don't sound so surprised,' he said, as he rinsed out the cleaning cloth and hung it over a rail to dry. 'You are a pretty girl. Shows he's got good taste.'

'But isn't it a bit weird to ask that when you are already with someone? She was just outside.'

'Well, he's the playboy type – every time he comes in here

he's with a different woman. I thought I might ask him for lessons.'

Eva laughed. 'I'll be off now then. See you on Monday lunchtime.'

As Eva passed by The Prince of Wales the last drunken stragglers were coming out, and one called out to her.

She gave him a wide berth and hurried home, her mind on Myles. He had made her feel fluttery inside, but it wasn't a good feeling – it was troubling. Except in films, good-looking men like him did not go for a very ordinary-looking waitress wearing a green apron. Not when they were already in the company of a glossy, expensively dressed model-type woman. Why had he done it? Was it a wind-up?

Once home she made herself a cup of tea, put a cardigan around her shoulders and went out into the garden. She loved sitting outside on warm nights in the darkness. The white daisies and petunias in the tubs were almost luminous. The honeysuckle on the fence smelled beautiful, and above her the inky black sky was sprinkled with stars. Sometimes her neighbours, the unpleasant Mr Francis and his wife, were out in their garden. They had lights and a barbecue, and they drank a great deal, often speaking so loudly that Eva winced. It amused Eva to know they had no idea she was out here, listening to him running down his work colleagues or arguing about how much his wife had spent on clothes. Sometimes she was tempted to jot down what they'd been saying, and then stick the note through their letter box to shame them into silence in future.

But she had it all to herself tonight; there were no lights either side, and it was very quiet.

She thought about Myles again and decided he was one of those men who just couldn't resist trying to pull a girl, just to prove himself. He probably lived close by and had seen her come out of the house at some time, and that was how he knew which road she lived in. He wouldn't turn up tomorrow; he and his friends had drunk so much wine he probably wouldn't even remember he'd spoken to her.

That was almost the story of her life. She just wasn't memorable to anyone. She guessed that if any of the customers who'd been in tonight were asked to describe their waitress, none of them would be able to. Was that how it was always going to be? Was she the reliable, hard-working girl who would never be remarkable in any way?

Until tonight she had thought what she'd achieved since moving to London *was* remarkable. The house was sorted, she'd got a job, she knew her way around now, and she could laugh at the state she'd got into over Tod. Yet she could also see she hadn't really moved forward at all: she'd made no girlfriends, and she knew no one to have a drink with or take shopping. She hadn't tried to find Patrick O'Donnell yet, and she hadn't gone right through her mother's diaries either.

At the bistro she saw so many girls of her own age having lunch or coffee. From overheard conversations she knew they were really living, taking full advantage of everything London had to offer, going to parties, clubs, cinemas and concerts, buying new clothes, going out with men. But she was just marking time, and feeling lonely for most of it.

What did she have to do to become like those other girls? Should she hurry up and get all the rooms decorated and furnished, and then find someone to share the house? Or find a new job where she'd meet interesting, friendly people and be on the same level as them?

These thoughts made her feel unbearably sad. She seemed to have spent her whole life being on the outside, looking in.

'You're just tired,' she murmured to herself as she got up

to go in. She locked the door and went upstairs to bed. But even as she got ready for bed, she couldn't help but think longingly of her job back in Cheltenham where there was always someone to have a chat to and have a laugh with.

The next morning she got up early. The sadness she'd felt the night before had vanished, because it was another lovely day. She made herself a cup of tea, then put on some old shorts and a T-shirt to finish painting the big bedroom. She'd already finished the ceiling and two white walls; all that was left was the wall where she intended to put the bed. That was going to be turquoise, and she'd already bought curtain material for the two windows – a white background with a dainty turquoise motif.

The sun coming in through the bedroom windows made her feel good as she painted with a roller. She thought she would have a bash at putting up the curtain poles later and make the curtains tonight.

When she'd finished the first coat, she went downstairs to get some breakfast.

By the time she got back to the bedroom, ready to start on the second coat, *Sunday Love Songs* was on the radio. She was singing along with Whitney Houston's 'One Moment In Time' when there was a loud knock on the front door.

She thought it might be Phil. She hoped it was, because she had missed him and wanted to apologize for snapping at him about the decorating.

But when she opened the front door and saw it was Myles she was thrown into confusion, because she hadn't for one moment thought he'd turn up. He was grinning at her, waving a bottle of sparkling wine, and he was dressed in a pale pink polo shirt and jeans.

She could only stare in consternation, very aware she was

speckled with paint. Her hair was held back with a stretchy band and, with no make-up, she knew she looked awful.

'I didn't expect you to come,' she said weakly. 'I'm painting.'

'So I see,' he said. 'But I'll scrub your back if you want to jump in the bath.'

That remark told her his sole purpose for coming round was to get her into bed. But he was over the threshold before she could gather her wits and make it clear that sex wasn't on the menu.

Brushing past her, he walked straight into the living room. 'You've made a nice job of this,' he said, looking around. 'It was a hellhole before.'

'Yes, it was,' she said. 'Awful. But there's still a lot more to do. I really ought to get on with the painting. I only get Sundays to do anything.'

'You can stop for a glass of bubbly and see where that takes us, can't you?' he said. With that he popped the cork, which flew out and hit the wall. 'So where're the glasses?' he asked. He then proceeded to go into the kitchen section of the room and opened cupboards, pulling out two glasses.

'It's a bit early for drinking,' she said uncertainly. He might be undeniably handsome but she didn't like him behaving as if he had a divine right to do whatever he pleased.

'Never too early to drink,' he said and poured out the sparkling wine. 'Let's go out into your garden so I can see what you've done there.'

Eva found herself meekly following him. 'Have you been in here before?' she asked.

He seemed to know his way around.

'Yes, I have – about a year ago. I heard it was coming up for sale, and I was interested.'

He sat down at the table and lit up a cigarette, then offered her one.

'No, thank you. I don't smoke,' she said.

'Don't drink early in the morning and don't smoke! What are your vices then?'

The way he sat with legs astride, the exaggerated way he drew in on the cigarette, and the way he looked at her like she was a piece of meat, was just so arrogant. She didn't want him in her house at all.

'I didn't know it was up for sale a year ago,' she said, ignoring his last question.

He laughed, a humourless sound. 'It wasn't, that was just a rumour, but it was obvious the owner was a crackpot letting it to junkies. Apparently she was living miles away, letting it fall apart. They said she was an artist. I came round here to get the owner's address. I made out I was from the Council, following up a complaint about vermin, and made an inspection.'

'So did you get the owner's address?'

'Yup, I frightened the tenants into it and they handed it over sweet as you like. But their bloody landlady didn't even have the grace to answer my letters.'

'What did you offer her for it?' Eva said.

'A hundred thousand by the last letter, though much less to start with. Told her it was riddled with damp and infested with vermin and that I'd get the tenants out for her too.'

'But it wasn't damp or infested with vermin,' Eva said. 'Fancy telling her that!'

He just laughed. 'All's fair when you are after a bargain property.'

'I don't call it fair to try to intimidate someone in order to get what you want.'

He looked hard at her for a moment. 'You paid the going rate, didn't you? What a chump,' he said scornfully. 'You've got to wise up, girl, or you'll get skinned alive buying property.'

'I didn't have to buy it,' she said. 'I inherited it when my mother died.'

His face tightened.

'Yes, that bloody crackpot landlady was my mother. And I'd like you to leave now.'

'Oh, come on, Eva. How was I to know she was your mother? Lighten up, girl.'

'You are extraordinarily arrogant,' she said. 'I didn't invite you here, and now I want you to leave. I am in the middle of decorating.'

He picked up his drink and gulped it down in one. 'Who the fuck do you think you are, telling me to leave?' he said.

'Because this is my house, and I don't want you in it. So please leave now without saying another word, or I'll call the police.'

He scowled at her. 'Fucking waitress,' he said.

'That's it, get out now,' she said angrily, pointing towards the French doors. 'And take your cheap sparkling wine with you.'

He got up, picked up the wine bottle and began walking into the house. She followed him, her heart thumping, afraid he might damage something just out of spite. But he kept on walking straight towards the front door.

She moved closer, so she could shut the door when he'd gone. But suddenly, without her seeing it coming, he turned and pounced on her, grabbing her by the shoulder and pushing her up against the wall.

'No one speaks to me like that,' he snarled at her, and he smashed the wine bottle at the wall. She screamed and tried to get away as wine and chunks of glass showered down on her, but his hand that had held the bottle was now on her throat, pressing hard on her windpipe. She tried to push him off, but he was cutting off her air supply and she felt powerless.

With his free hand he grabbed her crotch, digging his fingers into her.

'If you didn't want to be fucked, why didn't you say so last night?' he hissed at her. 'Surely even a dumb waitress would know I wasn't coming round for a cup of tea and a chat.'

Eva thought he was going to beat her up and rape her, unless she found the strength to fight him off. His face was contorted with rage, and he bent his head towards her as if he was going to bite her mouth.

She acted out of pure instinct, bringing her knee up with all the force she could muster to hit him squarely between his legs.

He yelped with pain and jerked back involuntarily, letting go of her. Quickly she bent to the floor, picked up a big piece of broken bottle and brandished it. 'Get out, you bastard,' she screamed, and jabbed the glass at his face. He backed away from her towards the still-closed door; he was holding his crotch with one hand, bent over with pain, and blood was trickling down his cheek.

Rage gave her new strength. She reached for the catch on the door and pulled it open. Then, jabbing the glass up to his face again, she kicked at his legs until he had no alternative but to back out of the door.

She slammed it shut, put the chain on too, then ran for the phone to call the police.

As soon as she'd reported what had happened, she slumped down on to the stairs, trembling with shock.

Phil had claimed she was an innocent more than once. She knew that if she told him about this he would ask why she had allowed Myles to come in. He would never believe that she hadn't encouraged the man in any way, and hadn't even told him where she lived. The police were likely to be much the same, and she had no doubt that by the time they went to arrest Myles he would have a plausible story ready, and make out she was some kind of madwoman who attacked him out of spite.

She put her fingers to her neck. It hurt, and it felt as if bruises were coming up. Would that be enough evidence to prove he'd almost throttled her?

The police arrived within ten minutes. By then Eva was crying and unable to stop herself shaking. The woman police officer made her a cup of tea while the policeman questioned her about what had happened.

'So you didn't make a date with him?' he asked, after she'd explained what had happened the night before.

'No. He flirted with me, and said he'd call round today. He already knew where I lived, but I didn't take him seriously. If I had, and really liked him, do you think I'd be dressed like this and painting my bedroom? I'd have been all dolled up with make-up on.'

'But you didn't say he wasn't to come round?'

'Not exactly. But we only spoke for a minute. The woman he was with was outside the bistro, and he went off with her. I felt he was only winding me up. I told Antonio about it, and he said he was a bit of a playboy.'

Eva related everything that had been said between them this morning – how she'd got angry and told him to go – and then she showed them where he'd pinned her to the wall. The evidence was still there, with the broken bottle on the floor and wine dripping down the wall.

'And you cut his face with a piece of glass?' the policeman asked. 'After you'd kicked him in the testicles?'

'You make it sound like I was the attacker!' Eva said angrily. 'I kneed him in the groin, because that was all I could do to get free. What was I supposed to do? Let him throttle me and rape me?'

They put the piece of glass she'd used into a plastic evidence bag, as well as the glass Myles had been drinking from to test for fingerprints. Eva had said she hadn't touched the wine he poured for her, and they could see that was true because it was still on the table in the garden.

'You'd better come with us to the station so we can get a photograph of the bruises on your neck,' the policeman said. 'We'll get you home immediately after we've taken your statement.'

The police drove Eva back home just after three in the afternoon. Her heart sank as they turned into Pottery Lane and she saw Phil knocking on her door.

'Do you know that man?' the police constable asked.

'Yes, he's a friend,' Eva said. 'But I wish he hadn't called now.'

'You need someone with you,' the policeman said. 'But if you think he may give you a hard time, I can ask him to leave.'

'That won't be necessary,' she said. 'Thank you for bringing me home.'

As the car pulled up she saw Phil's surprise. He came over to the car and opened the door. 'What's happened?' he said. 'Are you alright?'

'She's a bit shaken up,' the policeman said, leaning across Eva to speak to Phil. 'She could do with some TLC.'

When Eva opened the front door and saw the spilled wine and broken glass on the floor, she burst into tears. Phil closed the door behind him, put his arms around her and let her cry for a minute.

'I'm going to put the kettle on and get the garden chairs in here. You can sit down while I clear up that mess, then you can tell me all about it,' he said gently.

*

Phil listened carefully, fighting down the desire to find the piece of shit that had done this to her and kick his teeth in. Just the way Eva was dressed, in paint-splattered old shorts and a T-shirt, was all the evidence he needed that she hadn't been expecting this bloke to call on her. The bruising on her neck and the broken wine bottle were proof that she had been in real danger.

From what she'd said the police hadn't been at all sympathetic, and it didn't sound as if they were stirring themselves to find the man. Did they think she deserved such treatment just because she let him in?

'I'm going to stay here tonight,' he said, taking both her hands in his. 'He won't come back of course – even if the police haven't arrested him yet, he'll know better than to risk getting himself in even deeper shit. But you might be scared alone, and I can bunk down here on the floor. I've got an old sleeping bag in my van.'

Eva had a bath and changed while Phil cooked sausages and mash for them both. She felt calmer now she wasn't alone, and they discussed what she should do about her job.

'I can't go back there,' she said. 'I'll be afraid he'll come in.'

'Why don't you just take a month off without trying to find another job,' Phil suggested. 'You've got lots of stuff to do in the house, and perhaps you could go up to Leeds and see your brother too? And how about finding Patrick, the man who might be your dad? I could take some time off too and we could take some day trips to places like Brighton, or just cruise about London and see the sights.'

'That sounds very appealing,' she admitted. 'If I can get the house decorated and furnished, I could advertise for another girl to share it with me. I'd feel much safer with someone else here.'

'You'll have to make sure you get someone who will

become a real friend, not one of those stuck-up know-it-all Sloane's this area is full of,' he said.

Later on, Eva felt much better and gave the bedroom a final coat of paint, while Phil rubbed down the doors upstairs and the banisters ready for painting. She felt cheered to see the bedroom ready for a carpet and furniture. And if she wasn't going to be working for a while, she could spend some time choosing things to decorate the room.

Phil looked a bit apprehensive when she said she wanted to put up the curtain poles.

'I know I can do it,' she insisted, realizing he didn't believe she was capable. 'I've got all the equipment: a power drill, spirit level, Rawlplugs and the tape measure. You just watch and stop me if I go wrong.'

He had that look on his face that men always got when they didn't believe a woman could do something. And once she'd begun, she could see him twitching because she was so slow. Yet he didn't interfere, and he grinned at her encouragingly as she drilled the wall for the brackets.

'Well, I'd take my hat off to you, if I had one,' he said when the poles were finally up. 'That's really good. Most women I know haven't got the strength to get the screws right in.'

'There will be no stopping me now,' she joked. 'If that creep comes back, I'll screw him to the wall!'

'Just make sure you do it in the garage then.' He laughed. 'I don't want the walls I skimmed being messed up.'

The next morning Phil woke her with a cup of tea. 'I've got to go to work now,' he said. 'But I'll be finishing early. If you like, we could go up to that bookshop in Notting Hill and make some inquiries about Patrick O'Donnell. If he illustrates children's books, someone there is bound to know about him.'

'That would be nice,' she agreed, thinking how kind and thoughtful he was. She hadn't met many men who she thought would cheerfully sleep on the floor just to make her feel safe. 'I'll have to phone Antonio this morning and explain why I'm not coming back. Do you think he'll understand?'

'Of course he will. Anyone would. Now go back to sleep for a bit. Yesterday must have drained you.'

The phone ringing about an hour later woke Eva up. It was the police, informing her that Myles Babbington had been arrested that morning, charged with assault and would be appearing in court the following morning. They said he was certain to be bailed pending his trial, but he would be warned that he must not approach her again.

Just the thought of being called as a witness at his trial made her feel frightened all over again. She knew his defence lawyer would try to make it look like she'd led him on.

She phoned Antonio straight away to tell him she didn't feel able to come back to work. He wasn't surprised, as the police had contacted him about it.

'I told them exactly what you told me,' he said. 'And I said you weren't one to flirt with customers, that they should talk to Marcia because she could tell them how the evening had gone in the bistro – she only left about ten minutes before you.'

It was nice that Antonio was sympathetic; he even said she could have a job there again any time she wanted it. He said he would drop her wages round to her. 'I never liked that man,' he said. 'Always bragging about deals he'd made, women he'd pulled. I'm really sorry he hurt and frightened you, and I'm going to miss you.'

Eva spent the rest of the day making her curtains. With no sewing machine she had to sew them by hand, and although she tried hard not to think about Myles, he kept creeping into her head. It was more than likely he'd only get probation, or a suspended sentence, and part of her wondered if the humiliation she'd probably encounter at his trial was worth it.

She was hanging the finished curtains when Phil arrived around four o'clock. 'They look lovely,' he said. 'I am very impressed.'

There was something about Phil that really lifted her spirits. He was so manly. Chasing after the man who stole her handbag, and sleeping on the floor without ever making a big deal of it, was evidence of that. He was also calm, he had a dry sense of humour, and he didn't try to ingratiate himself with her. But, above all, he was kind. She hadn't met many men who had that quality.

He asked if she minded if he had a shower, making a joke about hers being a posh one; he claimed his one at home was just a glorified rubber hose on the taps.

'Shall we have something to eat out after the bookshop?' he yelled out from behind the closed bathroom door.

She shouted back that she'd made some Bolognese sauce and would cook some pasta when they got back.

'Yum yum,' was his reply.

She smiled, as that response appealed to her.

The bookshop Phil took her to in Notting Hill had a very well-stocked children's section.

They wandered around the shop for a while, but the huge selection of books made the likelihood of stumbling upon one illustrated by Patrick very unlikely.

'Is there anyone here who might know about book illustrators?' she asked the woman behind the counter.

'I don't,' the woman said. 'But Mr Temple, the owner, probably does. He's back there,' she said, pointing out a rotund grey-haired man right at the back of the shop.

They walked up to him. 'Hello, Mr Temple,' Eva said, smiling at him. 'I wonder if you can help me? Have you ever heard of an illustrator called Patrick O'Donnell?'

'Indeed, I have, my dear. His illustrations in the Mr Bear books are an absolute delight,' he said. 'The latest one, *Mr Bear Goes Camping*, is number three in the children's book chart right now.'

Eva felt as if someone had just switched on a light inside her. 'Really! He's well known then?'

'One of the best.' Mr Temple beamed. 'Let me show you.'

Being shown O'Donnell's work was as exciting to Eva as finding her mother's paintings in the attic. The Mr Bear books, aimed at under fives, were written by someone called Mabel Brown.

Eva opened the book to look. She read each page, but it was the stunning pictures that brought the simple stories about a family of brown bears to life.

Mr Bear was hapless, and his long-suffering wife was constantly sorting out his mistakes. One picture, where it transpired that Mr Bear hadn't packed the tent poles for their holiday, made Eva laugh out loud. Mr Bear was scratching his head and looking helplessly at the heap of canvas on the ground, and the various little bears were either crying, sheltering under trees or climbing them. Mrs Bear was standing with her hands on her wide hips, with a very grumpy expression on her face, saying: 'I can't trust you to do even the simplest thing, Mr Bear. I asked you before we left home if you'd packed the poles.'

'Lovely, aren't they?' Mr Temple said. 'It's all the detail: one little bear seizing the opportunity in all the confusion to steal an apple from the picnic basket, another one pinching his little sister, and that one trying to snuggle into a blanket. When you read to small children it's good to have stuff like that to point out. But Mr Bear always triumphs in the end. He catches a big fish for their tea, or he chases away a scary eagle or something. Look at the last picture,' he said.

Eva turned to it. The tent canvas was tied to bushes and the whole bear family were snuggled up together under it. She smiled; it gave her a good, safe feeling. She could imagine a small child dropping happily off to sleep at that picture.

'Would you know how I could contact Mr O'Donnell?' she asked. 'He was a great friend of my mother's, and she died recently. I wanted to talk to him about her.'

'I'm so sorry about your mother,' he said. 'I've spoken to Patrick at book events, but I haven't a clue where he lives. The best thing to do is write to him care of the publishers – they will pass the letter on to him.'

Eva bought the book *Mr Bear Goes Camping*, thanked Mr Temple for his help, and then she and Phil left the shop.

'Wow,' Phil said as they got out on to the street. 'If he is your dad, he's someone to be proud of.'

'He might not be,' she said. 'I'm not going to build my hopes up.'

But as Phil drove back home she couldn't help but hope that he was. A man capable of such sensitive and beautiful illustrations was going to be a lovely person. She just hoped he'd want to meet her.

'Are you sure you're going to be alright tonight on your own?' Phil said as he got his things together ready to leave, at about ten.

'Yes, I'll be fine,' she said. 'I'm going to write a letter to Patrick and then I'm going to delve into Mum's diaries a bit more. Thank you so much for staying last night. And for taking me to the bookshop. You are such a good friend.'

'I only came for the shower and the spag bol,' he joked. 'I

suppose you'll be out tomorrow, looking for stuff for the new bedroom?'

'I will,' she said. 'I want to find a Victorian or Edwardian dressing table.'

'Well, don't go looking in Portobello Road, they are silly prices. There's a shop near Shepherd's Bush market that has stuff like that at half the price.'

She gave him a goodbye hug, and kissed his cheek.

'Nigh night,' he said at the door. 'I'll ring you later in the week.'

An hour later, in bed, Eva looked at the Mr Bear book again. She had written to Patrick at the publishers. Because she thought the letter was likely to be read before being passed on to him, she kept it very brief: just that she was Flora's daughter, that her mother had died recently and she hoped he would agree to meet her, as she knew they were old friends.

'I hope you are my dad,' she said, turning the book's pages and poring over the pictures. 'I could really use a dad now.'

Chapter Twelve

At the knock on the door on Saturday morning, Eva came running down the stairs. She paused only a second at the mirror to check her appearance. She was pleased at how good her blue sundress looked against tanned skin, and how flattering her scrunch-dried hair was.

It was mid-August now and London was in the grip of a heatwave. She'd almost given up on Patrick O'Donnell responding to her letter, but last night he had rung her and asked if he could call on her today. She had been so excited after their very brief telephone conversation that she found it impossible to sleep, so she'd got up at five to spring-clean the house. Now he was here at last.

As she opened the door she was surprised to find him looking older than she expected, with saggy bloodhound-type jowls. But though he was portly and probably in his mid to late fifties, he was tall and held himself straight-backed. His grey receding hair was long and tied in a ponytail, and his smile was warm. He had the look of a man who normally wore very casual clothes; she felt that his smart cream linen suit and pink shirt were for her benefit.

'Patrick!' she exclaimed. 'I am so pleased to meet you. Do come in.'

'Sorry I couldn't give you more warning but, as I said on the phone, I've been away and your letter had been lying with other post, unopened,' he said as he stepped in. 'But I was very saddened to hear of Flora's death, and I felt I had to come and see you immediately.'

'I understand,' she said. Their phone call last night had been very brief, as he had said he was already late for an engagement. She'd already told him in her letter that she'd come to live here after her mother's death, that she knew very little about her mother's youth, and that she hoped he'd feel able to tell her a little more, so she didn't add anything to this during the phone call.

She was just touched that her mother still meant enough to him to call round for a chat. 'Shall we sit in the garden? It's such a beautiful day.'

He stopped short in the living room, looking around. 'My word, this all looks so lovely and stylish,' he said. 'A far cry from how it looked when I lived here.'

'It was hideous and disgustingly dirty when I first got here,' she said. 'I sometimes think I never want to look a paint brush in the bristles again!'

Since leaving the bistro she'd been very busy; the whole house was painted now, including the banisters, which had taken many hours. The landing, stairs and living room had grey cord carpet, a bargain that had been salvaged from Earl's Court Exhibition Hall by someone Brian knew. Both the bedrooms had new cream carpet, gas central heating had been put in, and there was a new modern bathroom.

There was still very little furniture downstairs – just a red sofa, her one extravagance, and a small table and chairs which she'd found in a junk shop and painted pale grey.

Flora's paintings, as well as the curtains, which were a fabulous poppy print on a white background, added splashes of colour, and Eva had put a huge vase of red silk flowers in one corner.

'I can imagine how bad it was.' Patrick winced. 'I drove past here one night a few months ago, and it looked very rough. You've even painted the front and garage doors, and put in new windows.'

She was glad he commented on the doors because she'd really laboured over them, rubbing them down for hours and then painting them a glossy French navy. They made her feel happy each time she got home. 'Let's go outside. Would you like tea or coffee, or a cold drink?'

'Something cold,' he said with a warm smile. 'And if you don't mind, I'll take off my jacket. I'm roasting.'

He paused to look at Flora's paintings before going outside and taking a seat in the shade. He didn't comment on the paintings but she sensed he was moved to see them.

She poured two glasses of orange juice and carried them outside.

Patrick smiled at her. 'You've got your mother's green fingers. It takes me right back, she was always out here, deadheading and pottering.'

'The framework was already here,' Eva said, pointing to the honeysuckle, climbing roses and clematis. 'I just tidied it all up and added more flowers and the tubs. I used to buy fashion magazines, now I get gardening ones. I practically live out here.'

'Flora did too. She would paint out here, even when it was quite chilly, she said it made her feel happy. The room upstairs which was supposed to be the studio was freezing in winter and too hot in summer. Ironic, really, as she bought the house because of that room.'

Eva nodded. 'She was the same about the garden at our old house, she would be out there in all weathers. I used to think that was a little peculiar, but I'm getting just like it too. Unless it's raining I come out here the minute I wake up, I have a cup of tea, listen to the birdsong and admire the flowers.'

He smiled, looking at her speculatively. 'I imagined you

with red hair and pale skin like Flora, but you are blonde and suntanned. You must take after your father.'

That remark was evidence that it had never crossed his mind she could be his child, and she realized she must tackle the subject carefully.

'I'm not a natural blonde.' She blushed as she admitted it. 'I had it lightened. And I don't actually know who my father is. That is one of the reasons I wanted to meet you.'

Patrick folded his arms and raised one eyebrow. His stance seemed to confirm she had shocked him.

'You see, I believed Andrew Patterson was my father. There had never been anything to make me doubt that,' she continued, 'but soon after Mum died, he told me he wasn't, and in a very nasty way. It was a total shock. It seems I was already around when he met Flora.'

Patrick's pale blue eyes widened. 'He was that callous? At such a time? That is appalling!'

'It was. I felt as if my whole childhood had been built on a lie. It was bad enough losing Mum, but then to be told I was a kind of cuckoo in the nest, that floored me. I've tried to make excuses for him – he was, after all, in shock at losing Mum. And he'd also just found out that she'd left me this place and her half of the family home to my brother and sister. But that doesn't really excuse him, does it?'

'It certainly doesn't. It sounds to me as if he was making you his scapegoat. Are you sure this is true?'

Eva nodded. 'Oh yes. I found my full birth certificate here, in a box with some of Mum's things. There's just a dash where my father's name should be.'

'You didn't say in your letter what Flora died of,' he said. 'Would that have any bearing on it?'

'Well, yes. It does explain his bitterness. She committed suicide, Patrick.'

He gasped and covered his face with his hands.

'I'm so sorry. I should've found a less blunt way to tell you,' Eva said. 'But it isn't something you can dress up to make it sound less shocking.'

He took his hands from his face and she saw his eyes were swimming with tears. He reached across the table and caught hold of her hand. 'Don't apologize to me, Eva. You are the one who deserves sympathy. I can imagine how terrible it must have been for you!'

'For all of us,' she said. 'We didn't see it coming.'

He was silent for a while; he looked as if he was struggling with the news and what he should say next.

'As shocked and horrified as I am,' he said eventually, his blue eyes now fixed on her, 'in a way it was almost predictable. You see, Eva, the Flora I knew, for all her talent, was slightly off balance. She was complex, at times gregarious, outrageous even, at other times behaving like a hermit, hiding away here. She was also very impetuous and she often acted irrationally.

'I loved her unpredictability, because it was exciting never knowing what was coming next. Yet she also scared me, because she didn't accept boundaries, and often she went too far, too fast and had scant consideration for anyone caught in the fallout.'

'People at her funeral hinted at that too,' Eva said. 'But I never saw that side of her. In fact I didn't recognize the woman they described as being my mother. I was never aware of the wildness they spoke of; I never even knew she'd been a successful artist. While she certainly wasn't conventional, she was a good mother, always there for us.'

'Did anything unusual happen prior to her death?'

'No, nothing at all. My brother and I had noticed that for some weeks she seemed kind of distant and withdrawn, but she often had periods like that and she would never talk about it. She was unpredictable, like you said. Some days I'd arrive home to find she'd made enough cakes for the whole neighbourhood, and on other days she hadn't done a thing all day, not even clearing away the breakfast things. So there wasn't any reason for us to think there was anything seriously wrong. Our home was beautiful, and there were no rows or money worries. It is such a mystery.'

'Then she left no explanation for you?'

'Only a note saying "Forgive me", nothing more. Nothing further has come to light. I'm still as much in the dark about it as I was that day in March.'

Patrick sighed deeply. 'I know how that feels. When she left me there was no real explanation either,' he said sadly.

The ideas that Eva had formed about this man just from his illustrations seemed to be correct. He was caring, sensitive and he had great warmth. She felt he had a strong moral code, and that she could trust him.

As they continued to talk – about her old home in Cheltenham, Sophie and Ben, and the kind of life they'd had before Flora died – Eva found herself liking him even more. He didn't shy away from asking questions and she found it easy to answer them truthfully, because he wasn't judgemental. In no time at all she was telling him how her mother's death had changed everything she had once thought was set in concrete.

'It felt like I'd been pushed out to sea in a small boat without even any oars. I felt totally alone. If Andrew had told me that he wasn't my true father for a good reason, I think I could have accepted that quite easily. It was the maliciousness of it that hurt so much, as if he'd always despised me and I'd only ever been there on sufferance.'

'I really cannot offer any explanation for why he, or any

man, would do that,' Patrick said, shaking his head in bewilderment. 'Grief and loss can make us irrational, but he'd taken you on as a baby, so he must have cared for you. But how did you find out about me?'

'From Jack and Lauren, two old student friends of Mum's that came to the funeral.'

'They came?' He looked astounded. 'I lost touch with them years ago, but they were very good friends at one time. Lauren was really the only close girlfriend Flora ever had, and Jack and I were inseparable as students. I've often regretted that we lost contact.'

'They seemed to think you were my father,' Eva blurted out.

His eyes widened at that. He shook his head and didn't speak for a moment. 'They are very much mistaken, Eva,' he said eventually. 'The way things have been for you, I wish I could lay claim to that honour. But I am not your father, my dear. Like you, I always believed Andrew was. I think I must tell you more detail about Flora and myself to set this straight.'

'I would appreciate that.'

'Flora and I met at Goldsmiths Art College in South London in 1964. She was twenty-one, studying Fine Art, I was twenty-eight and taking a short course in jewellery design. We became friends, and the following year we and some other people, Jack and Lauren included, shared a house together in New Cross. Let me explain how it was in 1965. That was the year the rigidity of the Fifties gave way to what people now call the "Swinging Sixties". London was suddenly the place to be, everything was opening up, there was a real buzz in the air.

'Yet, even so, our lot were a little before our time. It was almost unheard of then for men and women to share accommodation, but we art students considered ourselves "Beats". Beatniks were the forerunners of hippies, really. We wore black baggy jumpers and skintight jeans, and we weren't concerned with society's petty rules.

'I think I loved Flora from the moment I clapped eyes on her, but our relationship was a platonic one at that point and she had several other men in her life. We didn't become lovers until the Christmas of 1965. By then she had her art degree, and I was making jewellery and selling it to boutiques. But Flora was already becoming noticed as an artist. We had so many dreams then, Eva, commune-type visions of living in an old farmhouse and all sharing our possessions and any income from our art . . .' He paused and a wry smile played around his lips.

'Impractical hippy dreams?' Eva said.

'Oh yes, totally impractical, as none of us was doing more than scratching a living.' He chuckled. 'But it was a good time for all of us in that house, we had some wonderful times – that is, when we weren't freezing or starving, and the landlord wasn't getting heavy about the rent we owed. Then, in 1967, quite suddenly everything changed for Flora. She became a minor star in the art world. Galleries exhibited her work, and they sold it hand over fist. Her mother had died of cancer before I met her, but during the latter half of that year her father died too, and he left her some money.'

'Jack and Lauren told me some of that,' Eva said. 'And that she bought this house.'

'Yes, that's right. But what they probably didn't tell you was that all our friends, Jack and Lauren included, were still clinging to the idea of a commune. When Flora bought a house for herself, not for all of them to share, they chose to see it as selling out to Capitalism. That was ridiculous, really. If any of them had come into money, they would have done exactly as she did. But it soured things. Anyway, Flora and I moved here, and she got pregnant. We were thrilled – completely wrapped up in one another – and we intended to get married. But, sadly, she miscarried at six months. It was a little girl.'

'I'm so sorry,' Eva said. She could see the sadness etched into Patrick's face.

'It was never the same afterwards,' he said glumly. 'Flora went into a very dark place. At the time I was very afraid that she might end her life. She didn't of course – she just ended it with me. She just upped and left one day while I was out. Her note said it was over and she'd gone away "to find herself".'

'How awful for you!'

Patrick shrugged. 'I remained here in this house, hoping of course that she'd return before long. I didn't hear anything from her for weeks. I didn't even know where she was, and if she was alright. Then I got a little watercolour of a cottage from her in the post. She said in the accompanying letter that she was in Scotland, and that she was sorry it had all gone wrong for us. But then she went on to say that if I wanted to stay on here, I would have to start paying her rent, as she needed an income, and that her solicitor would be in touch with the terms.'

'That was very cold-hearted!' Eva exclaimed. Yet however unkind she thought it was, it did bear out some of the things her old friends had hinted at about Flora.

'It was fair enough, being asked to pay rent,' Patrick sighed. 'I didn't mind that. What upset me most was that she believed it was she alone who had suffered in losing our baby. I felt the pain just as badly as she did, and she shut me out at a time when we should've been grieving together. I forgave her so much in our time together, but I could never forgive her for that.'

'So what happened then?'

'I duly paid rent to her solicitor for some weeks. Not another word from Flora. Then out of the blue I got an eviction notice. That was the last straw. I was very bitter, and for a while I was tempted to refuse to leave – just to spite her. But I'd been thinking of going to Canada, anyway. And on balance I realized it made more sense to sever all connections with her and go. So I went. Some time later – well over a year, I'd say – I heard from a friend that she was back here in this house with Andrew Patterson and they had a baby girl. But if Patterson came along after your birth, then she must have met your father while she was in Scotland.'

'I hoped it would be you,' Eva said in a small voice. 'You see, I bought *Mr Bear Goes Camping* and got carried away, thinking my father was Mr Bear.'

He smiled. 'Strangely enough, Flora's pet name for me was Mr Bear.' He looked a little embarrassed at admitting that. 'I often wondered if she ever looked at the Mr Bear books and thought about our time here together. She was the successful one. I wasn't doing very well selling my stuff, so I worked as a meter reader for the electricity board to keep my end up. But we were so very happy before she lost our baby. I decorated the little room upstairs with a frieze of bears, though I expect she painted over it.'

'She didn't,' Eva exclaimed. 'It's still there. I've repainted the room, but I left it in place because it's lovely. I thought Mum had painted it for me. But now I know it was done by you, I'll keep it for ever.'

'Oh, Eva,' he sighed. 'I'd have been so proud to call you my daughter, and I'm really glad you like Mr Bear. I also wish I could shine more light on to the time after Flora left me, to make things better for you.'

Eva felt a rush of affection for this nice man, and she

couldn't help but think her mother would have been a great deal happier with him than she ever was with Andrew.

'You've done that just by coming here. After what you've told me, I wouldn't have blamed you if you wanted nothing to do with her daughter.'

He shook his head in denial. 'Flora was the big love of my life, and I'd have been a very odd sort of man if I had no curiosity about what happened to her. When I got back from Canada I asked friends about her and was told she'd married Patterson and moved away. I spoke to one or two gallery owners who had exhibited her work in the past, and they seemed baffled as to why she hadn't been in touch with them. You said she didn't tell you about her earlier success, but did she carry on painting?'

'No, the only things she ever painted were pictures for us kids. I don't understand it either, we only had one oil painting of hers at home. Like I said, she never talked about her past, never even gave us a hint she'd once had exhibitions of her work. The paintings here are ones I found up in the attic. Lauren said that maybe once she had children she felt no further urge to paint.'

'I suppose that could be the explanation, though I can't think of any other artist who gave up painting all together. But enough of Flora . . . tell me what you work at. I can see you're artistic by the decor of the house. Do you paint and draw too?'

'No, I'm useless at it. But I suppose I am artistic in that I've always liked sewing and handicrafts, and now interior design too,' she said with a grin. 'Back in Cheltenham I worked in telephone sales for a mail-order fashion company. I was involved with every aspect of that company, and I really loved it. I was working here in a bistro, but there was an unpleasant incident there and I decided to take a month off and get the house fixed up.'

A little later, Eva got the box of her mother's things down to show him. She lifted out the necklace and asked if he had made it.

'Yes, I did,' he said, his face lighting up. 'Fancy her keeping it!'

'It's beautiful,' Eva said. 'And if it's any consolation to you, it must have meant a lot to Mum too, because all the things in this box appear to have some special significance.'

She showed him the watercolour of the cottage, and he said it was the same cottage she'd sent him a picture of. 'I can't be absolutely certain after all this time but I'm pretty sure she said it was in Pitlochry.'

'What about this row of shops?' she asked him, holding out the photograph. 'There's a painting of it too.'

Patrick shook his head. 'I haven't a clue. Maybe she knew someone who lived there, or stayed there herself for a while. But I really can't imagine why she would paint such a dreary scene, I only ever knew her go for vivid or very beautiful subjects.'

'I've got a feeling that place is important,' Eva said thoughtfully. 'I've been working through her diaries, but so far I haven't found anything that ties it in. But then they are quite hard to read – awful scrawl, and so many initials instead of names.'

He grinned. 'I can imagine. She used to leave notes like that to me, and sometimes I hadn't a clue what she was trying to tell me. I can remember her scribbling in a diary, she even joked that if anyone ever tried to read it they'd be baffled.'

'Well, she's succeeded in baffling me.' Eva laughed.

Patrick picked up the necklace and fastened it around her neck. 'You should wear this. I might not be your father, but I'd like you to keep some room in your life for me. I haven't got any children of my own, and I think Flora would like me to take you under my wing.'

'I'd like that,' she said, smiling up at him. 'And I love the necklace.'

'Good, because it was made with love and therefore a powerful amulet. But let's talk about you and your career. I think you should train in interior design. You've got a flair for it.'

'Really?' It was odd that he'd picked up on the one thing she kept thinking she'd like to do. 'I've really enjoyed doing this place up, and I've got quite good at doing jobs such as putting up curtain poles, shelves and things. But surely you need to know more than that?'

'Having a good eye for colour and design is the main requirement. That's inborn – it can't really be taught. But there are courses on the other skills you need, and how to go about making a career in design. I had a girlfriend back in Canada who was an interior designer. People who have money but very little imagination employ them. They oversee the making of curtains, the choice of wallpaper and furniture, sometimes even do up the whole house.'

Eva liked the sound of that. 'So where would I find a course to take?'

'Many of the art colleges run them, though I imagine most start in September or October. I'll make some inquiries for you. I expect you could get a grant for it too. But from what you've told me, Eva, ever since Flora died you've been pushing yourself too hard. What was this incident at your work?'

She blushed, wishing she hadn't told him that. She was quite over it now and mostly wished she hadn't even called the police, as when the court case came up she knew it would be unpleasant. 'A man made a nuisance of himself,' she said quickly. 'He turned up here and barged his way in, and the upshot was that he assaulted me.'

'You poor girl!' he exclaimed. 'I hope you called the police?'

She felt she had to explain now, and added that she wished she hadn't involved the police.

'You can't let men get away with such things,' he said sternly. 'I think you should take yourself off on a holiday and recharge your batteries.'

'I wouldn't know where to go.' She laughed. 'Besides, I'm more or less having a holiday now, doing nothing but lying in the sun and pottering around the garden.'

'It's a change of scene that rests you,' he said firmly. 'You could go up to Scotland to see if you can find that cottage your mother painted. You might discover your real father there. But more importantly, to quote Flora, you need to "find yourself".'

Eva laughed. 'Very Sixties, "Peace and Love"! But haven't I already found myself by coming to live here?'

'In part. You've become independent, and your creative side has emerged, but you are still being held back by all these questions you have about Flora. Until they are answered to your satisfaction, or you decide they are no longer important to you, you won't be totally your own person.'

'So how will a holiday achieve that?'

'It will give you time to relax, reflect and take stock. Find out what's important to you and what isn't.'

'I'll think about that one,' she said. 'Now let's have some lunch, shall we?'

She made a tuna salad and they ate it in the garden. Patrick had been to Italy recently, and he talked about the works of art he'd seen in Florence and Rome with such enthusiasm and awe that she wished she could see them too and resolved to get some art books from the library to learn more about the artists he mentioned. They discussed fiction too, and although their tastes were mostly very different, they both loved Tom Sharpe's books and reminded each other of the parts they'd found funniest.

Eva could never have had such a conversation with Andrew. He had often sneered at the things she said and belittled all her attempts to try anything new. But she felt totally at ease with Patrick and liked the fact that he was interested in her, what she thought about, her interests and goals. He also seemed to really understand how it felt to be cut off from family.

Sophie hadn't responded to any of her letters. Ben had come down the previous weekend, but it soon became obvious that he saw her house as just somewhere to stay so he could visit old friends, not to spend time with her. On the Saturday night he went off, and didn't come back till midday on Sunday. She had been very hurt and disappointed but she didn't say so, because she didn't want to lose him all together.

She told Patrick a little of this. 'Ben moved up to Leeds a while back, but he's gone back home to Cheltenham for a bit of a holiday. And to try to straighten Sophie out, because apparently she's running wild. But it seems Andrew's got this woman staying at the house most of the time, and Ben couldn't stop ranting about her. He thinks his dad is totally insensitive to his and Sophie's feelings, and that's why Sophie's playing up. He says that once university starts in October he's got no intention of going home ever again.'

'You must let them get on with their lives,' Patrick said calmly. 'The way Andrew treated you was despicable, so you don't have to feel responsible for his children. A pretty girl like you should be out with boyfriends, not worrying about a half-sister whose own father should be doing that. Have you got a boyfriend?'

'There is someone, but he's just a friend.' And she went on

to tell him how she had met Phil. 'He's really nice – fun, interesting and kind.'

He smiled and raised one eyebrow. 'Is he the reason you are reluctant to take a holiday?'

'No,' Eva said, but she felt herself blush. 'It isn't a romance, just a friendship. I make dinner for him, we go to the pictures and to the pub. I really like him, he's such good company, but that's all.'

'That sounds very much like how it was with your mother and me when we first met,' Patrick said. 'I accepted that I had to go along with her terms. But I wanted more.'

'I don't think Phil does.'

'Oh really!' Patrick smirked. 'Men don't usually go for the "just good friends" thing. I think it's more likely he's biding his time, because he's waiting for your bruises to heal. You have been hurt by someone other than Andrew, haven't you?'

She was a little surprised that he'd homed in on that, but then he was very perceptive about everything. So she told him about Tod. 'I'm over it now,' she said quickly. 'Actually, I can't really believe that I reacted the way I did – after all, I'd only known him a short while.'

'Coming so soon after your mum's death, your reaction wasn't surprising,' he said, crossing his arms and looking at her thoughtfully. 'You probably were very needy and intense, and men do get frightened by that. But if you want my honest opinion, Eva, I'd say Tod did you more good than harm. He made you happy and boosted your confidence when you needed it most. Remember that, and forget how it ended. I'm a firm believer that everything happens to us for a purpose. You ran here because of Tod and, as it turned out, that's a good thing.'

'You are very wise,' she said.

'It comes of having made every mistake in the book,' he

admitted ruefully. 'If I could go back and do it all again, the one thing I would do differently is not to hop into bed with people too quickly. You make rash promises, you let lust cloud your vision. There was a lot to be said for old-fashioned courtship. It's good to get to know someone really well before you sleep with them. Perhaps your friend Phil thinks that too.'

She got out the old photograph album that was in Flora's box. As Patrick turned the pages he was able to tell her that the very old couple were her great-grandparents. 'That's your grandmother,' he said, stopping at a very faded picture of a woman standing beneath a tree wearing an apron over her clothes. Next, he paused at a man in a cloth cap and tweed jacket leading a horse. 'This is your grandfather. I think Flora was closer to him than to her mother, who she said was very neurotic.' He turned the pages till he came to one of her grandfather standing by a gate with a very gaunt-looking woman. 'That's your grandfather's sister. I can't remember her name, but Flora said she was very stern. She looks it, doesn't she? Of course they are all dead now. And as Flora was an only child, and her aunt was a spinster, I don't think you have any more relatives.'

They moved on then to the books of sketches of children. Patrick smiled at them. 'We both had the idea of illustrating children's books back then,' he said, 'but it's a hard field to get into. I was lucky in that Mabel, the writer of the Mr Bear books, was a friend and insisted I illustrate them for her. Without her guidance I doubt I'd have become a success. She was able to tell me what small children and their mothers like in illustrations. But Flora wasn't the kind to take any guidance from people. She believed a writer should fit the story around her illustrations, and it doesn't work like that.'

Eva didn't expect Patrick to pore over the diaries with her, but she opened up one to illustrate how confusing they were with no dates, no names and barely a hint of where Flora was at the time of writing.

Patrick put on a pair of reading glasses and frowned as he tried to read. 'I'm none the wiser,' he said. 'I remember the Bistingo. She worked there in the evenings – it was somewhere in Bayswater, I think – but I never went there. These people she mentions must have been casual friends she made there, I don't think she even told me about them. Like I said before, she took a perverse delight in being obscure. I think it was partly because her mother used to pry into her life as a teenager. So good luck with trying to pin her down, Eva – I suspect it would tax a professional code breaker.'

Patrick finally left at five. He only lived in Chiswick but he said he had to get home early, as he had some important work to do. He left a card with his telephone number and address and said she could call him any time.

'I'd like to take you out to dinner next weekend,' he said, giving her a hug. 'We'll go somewhere very swish so you can dress up. You might not be my biological daughter, but I'd have been thrilled if you were. So if you want me to be a stand-in dad, and boss you around, I'm happy to fill the role.'

She stood on tiptoe and kissed his cheek. 'OK, Mr Bear, the job is yours. All I can say is that Mum was a mug to ever leave you.'

As she waved him off at the door she felt warm inside. She now had a far greater sense of who Flora had been, at least in her youth. She might still have a great many questions to find the answers to, but it was a good start.

Chapter Thirteen

Through the rain Eva spotted a road sign up ahead which said it was seven miles to Carlisle. She straightened up her aching back in relief; she should have taken Phil's advice and stopped for the night in Lancashire, or even gone to see Ben in Leeds, instead of pushing on this far in one day.

But even if she was tired and aching from the long drive, she felt good about leaving London to go to Scotland. She didn't wish to run into Myles again, and she needed to think about her future.

Meeting Patrick had been marvellous; she would put him second only to Phil in the list of people she felt lucky to have in her corner. She'd had several chats on the phone with him since their first meeting. And he'd taken her out to dinner, which had been a real treat.

His words about Phil's feelings for her had stayed with her too. He could be right that Phil really did want to be more than just a friend, because he'd called round to see her yesterday evening as she was packing.

'How would you feel if I suggested I joined you up in Scotland?' he asked a little sheepishly.

'I'd tell you to get lost,' she joked. Seeing his face fall, she was quick to tell him that was supposed to be funny. 'I'd love it. But aren't you busy at work?'

'I am right now with this big job in Hampstead, but I think that will be done in another five or six days. I've got holiday due to me and I've never been to Scotland. I really want to see the Highlands.'

He had a funny look on his face, as if he wanted to say something more and couldn't get it out.

'I'd love to see Scotland with you,' she said. 'I doubt it will be much fun roaming around on my own. You could come up by train when you're ready, and I could meet you at the station.'

His face lit up. He moved closer to her and put his hands either side of her face. 'I don't only want to see the scenery. I also want to make sure you don't run off with some wild Scotsman.'

'I'd rather run off with some lovely Londoner,' she said, looking right into his eyes.

He kissed her on the forehead, but she sensed he really wanted to kiss her lips. She didn't know why she didn't just slide her arms around him to give him some encouragement, but she supposed she was afraid of taking the initiative.

'Phone me in the evenings and let me know where you are,' he said. 'Meanwhile, I'll sort it at work.'

Thinking of that moment warmed her, and if he did join her up here, then maybe they could both stop being bashful.

He'd been very happy for her that she'd met Patrick and that they were getting along so well. She wanted Patrick to meet Phil; she was sure they'd like one another. Perhaps after the holiday?

Reaching over to the passenger seat, she picked up the directions to the guest house she had booked into for the night. It was in a place called Wetheral, which didn't appear to be too far from the motorway.

She had never been further north than Blackpool before. Although she'd heard hundreds of times that the scenery in Cumbria was breathtakingly beautiful, she was still astounded by it. She remembered when she was small having a tin box of Lakeland coloured pencils for Christmas one year. The picture on the lid was of mountains with a purple tinge which Flora had said was heather. She'd seen that majestic view today: tiny white dots that were sheep grazing at seemingly precarious heights, and only a few tiny stone cottage houses nestling here and there to prove it wasn't complete wilderness.

She guessed that in sunshine and away from the motorway the scenery would be even more spectacular, and she hoped that if Phil joined her they could explore it together.

The rain stopped and the sun came out just as she got to Wetheral. To lift her spirits even more she was thrilled to find it was a real village, with a village green and pretty cottages all around it. The Briars, the bed and breakfast she'd booked into, was amongst them.

Before checking in she took a short walk to stretch her legs and found a wide, fast-flowing river just a hundred yards below the green. With the late afternoon sun shining on the water, and green hills all around, it looked beautiful.

The Briars was equally lovely in a slightly old-fashioned flowery way, spotlessly clean and smelling of lavender polish. Eva's room overlooking the green was pretty, with peach Laura Ashley wallpaper, a matching quilt on the very comfortable double bed, and frilly curtains. The bathroom was minuscule, but there were fluffy peach towels and an array of mini toiletries.

Mrs Hobbs, the middle-aged landlady, gave her a warm welcome, suggesting that Eva come down after seeing her room to have some tea in the guests' sitting room. There she met a couple in their thirties who had also just arrived. Like Eva, they were breaking their journey before driving on to Scotland in the morning. Mrs Hobbs said they could have dinner, if they wished, and that tonight it was roast chicken.

By ten that evening, Eva was more than ready for her bed.

Dinner had been absolutely delicious, with a choice of three different puddings afterwards. It had been served at one large table, and aside from Eva there were three other couples. It had been a jolly evening, as all the guests were very chatty. One of the things which had worried Eva about coming away on her own was that she would feel lonely, but if this bed and breakfast was anything to go by, she wouldn't be.

After a huge fried breakfast the next morning, Eva packed her bag and paid her bill. It was such a lovely morning that she went for a walk before driving on to Scotland. As she walked along the river bank she thought how lovely it would be to buy a house somewhere like this village to turn into a guest house. She wondered if Mrs Hobbs had enough guests all year round to earn a decent living or if her husband, who Eva hadn't seen, had a job that provided the real income and they only earned pin money from the guests.

All the way to Pitlochry her thoughts kept returning to the idea of owning a guest house. She assumed property would be a great deal cheaper than in London, and it would be fun to do up a big house room by room, each one with a different theme. Perhaps she should forget the idea of becoming an interior designer and instead go in for something in the hotel trade to gain experience? Or get the guest house up and running, and then train to be an interior designer?

But even as these thoughts came to her she smiled, knowing this was only a pleasant daydream. She could bet that half the people who visited Cumbria and Scotland had such thoughts while on holiday. And of the few that actually opened a guest house, most would find it wasn't anywhere near as rewarding as they imagined.

Pitlochry was surrounded by mountains and built above the wide river which ran through a wooded valley, but the town was bigger and busier than she expected. Essentially Victorian, it had a gracious charm with many fine, big houses. The main street was lined with shops and pubs which catered for the thousands of tourist, who passed through it on their way to the Highlands.

Eva thought Flora might have intended to do the same, but ended up staying here because she realized she wasn't actually cut out for isolation.

Sadly, Brae Bank hotel was not what she imagined from its name. Not a pretty hotel perched on a riverbank, but a rather forbidding grey stone building which looked as if it had once been a public house. Extensions had been added without any thought to the overall look of the property. Inside, the tartan carpets were worn and the wood panelling in the reception area was scuffed and dusty-looking.

A coach pulled up just as she was signing the register and disgorged about eighteen elderly people who appeared to be touring around Scotland. Eva guessed that this was the usual clientele, and she wasn't likely to find any soulmates here. Yet the receptionist was very pleasant, giving her the breakfast times and pointing out the bar. Then she asked a lad to escort Eva to her room, and said if she needed anything more she only had to ask.

It was a small drab room on the top floor in the oldest part of the building. The double bed was covered in an orange candlewick bedspread – the kind she remembered Andrew's parents having in their home. But whatever the room's shortcomings, the view from the window was superb. She could see right over the town, to the river and the mountains beyond. The bathroom was almost as big as the bedroom, very stark with old-fashioned black and white tiles on the walls and grubby-looking lino which was peeling back in the corners. But there was a television in the bedroom, tea-making facilities and even a small fridge. She thought she could be quite content here for a week, and it was very cheap.

After she'd unpacked her clothes and put them away, she walked down into the town to explore.

The shops had only just closed and the cafes and restaurants were quiet, as people hadn't yet come out for an evening meal. She saw a shop selling artist's materials which looked as if it had been in the same hands for several decades, and made a mental note to call in there the next day. Most of the shops were the souvenir kind selling china Highland cows, tartan scarves and the like. But there were a couple of art galleries, with good displays of hand-thrown pottery, locally made jewellery and paintings. Eva realized that back in 1969 Pitlochry had probably been far less sophisticated, and yet she was getting a strong sense of what had attracted Flora to stay here.

Back in her room that evening, after fish and chips in a cafe, Eva got out Flora's diaries and found the one which she thought was written while her mother was up here. It was infuriating that the entries weren't dated, because it was impossible to know whether two entries were on consecutive days, or weeks apart.

'Such a long, weary drive,' she read. 'Stopped in M too tired to go any further.'

Was 'M' Manchester?

'Too grim to stay another night. Worse bed I've ever slept in.' That sounded as if she was still in 'M', and the next entry appeared to be on the same day. 'Band playing, I love bagpipes, and the sun is shining. Scenery inspiring, if the B&B wasn't so awful I could stay here for days.'

Eva doubted she'd encountered a pipe band in Manchester, and the views there would not have been inspiring, so she had to be in Scotland. She made a mental note to buy a more detailed map of Scotland in the morning.

'Scotland is bigger than I imagined,' Flora had written next. 'Couldn't drive any further so stopped in P.'

That could be Perth, but Eva felt it was more likely to be Pitlochry, because she didn't mention any further driving after that, only that another bed and breakfast had no hot water.

'Rent for cottage only four pounds a week,' was the next entry. 'A bit primitive, but it's got good vibes. I need that to get me out of these black moods.'

She referred obliquely to depression in several entries after that. The effort required to wash her hair, staying in bed all day, and avoiding someone she called 'D' who she said 'analysed' her. Eva got the idea she was wrestling with depression, wanting to hide away from people, yet knew this wasn't helping her mental state. As Patrick had said, sometimes her entries appeared to be just random thoughts: 'I could just walk into the river at night and let the water embrace me,' was one that sounded very much like the temptation of suicide. But then the very next entry was: 'Watched dragonflies hovering over the water, so beautiful I found myself smiling again.'

She must have warmed to 'D', as she mentioned him or her quite often, usually to say he or she had called, they'd supper, gone to the pub or visited nearby villages together. There was also a 'G'. The latter was clearly someone she did like. Eva thought it was a he, as she mentioned him quite a lot and she went camping, hiking and had dinner with him. In one entry she called 'G' a 'soulmate'.

Could this 'G' be Eva's father?

If he was, Flora wasn't inclined to write about any kind of romantic interludes. She wrote about painting views of the river, taking the garden in hand and long walks in the woods. There was not so much as a hint of a love affair, let alone pregnancy.

Eva's last thoughts as she drifted off to sleep were that if she could find the cottage Flora had stayed in, maybe the present owner would be able to shed more light on her stay here.

She found the cottage the next morning. It was on the far side of the river Tummel, a short distance from the dam for the hydroelectric plant. Even if she hadn't taken the little picture of the cottage painted by Flora with her, she would have recognized it by the old cast-iron latticework forming an arch over the front door.

In the picture the cottage looked charmingly dilapidated, with straggling roses over the arch, the front door in need of a coat of paint, and a weed-strewn path leading from a sagging gate. But it wasn't like that now; it was painted a soft pale pink, the door glossy white, as was the latticework arch. Although it was late in the summer, there was purple clematis scrabbling through the carefully trained rose which still had many pink blooms. New windows, a white picket fence and proliferations of hanging baskets and tubs in the tiny front garden gave the cottage a 'take your snapshots in front of me' look.

Taking a deep breath, Eva lifted the brass lion's-head knocker and rapped on the door.

A pleasant-faced woman who appeared to be in her early forties, wearing an apron that said 'Kiss the Cook', opened the door.

'I'm sorry to disturb you,' Eva said, 'but my mother painted this cottage some twenty-two years ago – I think when she was renting it. She died recently, and I'm trying to find out more about her time in Scotland.'

She showed the painting to the woman.

'Well I never,' she said, taking it in her hands and smiling.

'What a lovely picture. I remember it like that from when I was a girl. How sad you've lost your mother, and you so young.'

The woman had the softest Scottish accent, a smear of flour on her face and more on her apron – all of which suggested she would be very kindly.

'Yes, it was sad.' Eva had no compunction in getting the woman's sympathy. 'Was the cottage owned by your family then?'

'No, my dear,' she said. 'But come on in, you don't want to stand out on the doorstep.'

The woman introduced herself as Janet Mayhew and proceeded to make tea, urging Eva to sit down at the kitchen table.

'I've been baking.' Janet waved her hand at a cooling tray of buns. 'I've just put some shortbread and some flapjacks in the oven. Now what was it you wanted to know?'

The inside of the cottage was as attractive as its outside, with lots of pretty china on a dresser, potted plants on the window sill, and a fat ginger cat lying on the mat by the open back door. But the pine kitchen units looked new, so Eva knew it wouldn't have been the same when Flora lived here.

Eva gave Janet an edited version of her mother's death, finding the diaries and the picture of the cottage. 'Mum didn't talk about her past,' she explained. 'She never even told me she'd been a successful artist, let alone that she lived in Scotland for a while. I want to try to discover why she was here, who with, and just a bit more about her. So do you know who owned the cottage then?'

'Aye, it was the Hamiltons. Old Will Hamilton bought up a few wee cottages in the town after the war. I know this one was rented out to men while they worked on building the dam, but that was finished in the 1950s. When I was a wee

226

schoolgirl there was a crazy old lady lived here. Old Will died sometime around 1967, I think, and his son Gregor took over management of everything.'

'So Gregor Hamilton would have rented this to Mum?' Eva asked, her mind turning to the letter G in the diaries.

'Aye, he would that. My husband bought the cottage from him in 1975.'

'Do you suppose I could meet him?'

'I'm sure he'd like that. He's been in a wheelchair since a climbing accident a few years back. Such a shame, he was always so active. My husband used to climb with him and he goes up to see him at least once a week. Gregor puts a brave face on it, but it can't be pleasant being stuck in a chair all day.'

Eva had a cup of tea with Janet, and tried one of her buns, and they chatted about general things for some time. Eva was very tempted to tell the older woman the whole story, because she was so nice. But she controlled the urge; she didn't want Gregor to get wind of anything that might make him wary of talking to her.

'So how old is Gregor?' she asked. 'It must be hard for his family if he's in a wheelchair.'

'He's around fifty, a bit younger than my hubby. He never married. He's only got his younger sister, Grace, and her family now. They share the big house with him, though Gregor has converted the downstairs rooms for himself.'

Eva liked the way Janet pronounced house 'hoose'. It reminded her of a book she had when she was small that contained a poem about a moose who lived in a 'hoose'. Flora used to put on a Scottish accent when she read it to her. Perhaps she even thought about Pitlochry as she read it.

'I should go now,' Eva said as Janet got her shortbread and flapjacks out of the oven. 'Could you let me have Gregor's address and phone number?'

'I will, but why don't you go up to the house just now?' Janet said. 'He gets lonely, and a nice wee girl like you stopping by will make his day.'

Janet gave her the address and drew a little map. It was only two streets from the hotel, further up the hill.

As it was just on one, Eva didn't go straight there in case Gregor was eating his lunch. She wandered along the main street buying some milk, biscuits, tea bags and fruit to keep in her room. The town was really busy, but it had that leisurely, friendly feeling about it that seaside towns had.

She might have only been there a day, but she really liked it.

Gregor's house was a big gloomy-looking Victorian place, three storeys and double-fronted, with stone steps up to a wide porch. The front door had stained-glass panels and was twice the width of Eva's front door back in London.

The door was opened by a very attractive blonde woman wearing a floaty blue and white dress. She looked about forty, and Eva guessed she must be Grace, Gregor's sister.

'I'm sorry to turn up here uninvited,' Eva began. 'I was talking to Mrs Janet Mayhew about her cottage, and she suggested I come up here to have a chat with Mr Hamilton, because he used to own it.'

'I'm sure Gregor will be delighted to talk to you,' the woman said. 'Do come in, and I'll just check if everything is alright with him. What was your name?'

'Eva Patterson,' she said and watched as the woman tip-tapped in her high heels across the parquet flooring to a door at the back of the house.

The staircase was right at the centre of the hall, very wide with intricately carved banisters and a carved hawk on each of the newel posts. At the half-landing where the staircase turned stood a complete suit of armour, looking as if a man stood inside it, and above it on the wall were two crossed battleaxes. Eva wanted to giggle, thinking that if Gregor was her father, she would have to educate him into going for less intimidating decor.

'Come on in, Eva,' the blonde woman called out from the back of the hall. 'Gregor will be glad to talk to you.'

Eva's first thought on seeing Gregor, sitting in an armchair by the window, was that he couldn't possibly be her father, as his hair was as red as her mother's. She was sure two red-headed people couldn't produce anything but another redhead.

'You'll forgive me if I don't get up,' he said. 'I expect Janet told you that my legs are useless now.'

His voice was beautiful – not just the Scottish accent, but the depth of his voice – and he was a fine-looking man with strong features, reminding her a little of the actor Liam Neeson.

'She told me you had a climbing accident. I'm so sorry,' she said, walking over towards him to shake his hand. 'Thank you for agreeing to see me, it is a bit of an imposition.'

'Not at all.' He smiled warmly and he shook her hand very firmly. 'What man wouldn't like a visit from a bonny lassie? And what can I do for you?'

'My mother died earlier this year and amongst her things I found a painting of Janet's cottage. I believe Mum rented it for a while in 1969. I am trying to find out more about that period in her life. As you would've been her landlord then, I wonder if you remember her. Her name then was Flora Foyle.'

The reaction to her mother's name was instantaneous. A light came into his blue eyes, and his mouth curved into a smile.

'Flora! Aye, I do remember her, Eva, very well. She rented the cottage for going on for a year. Do sit down, I hate looking up at people.'

Eva took the sofa opposite his chair. She explained about the diaries and said she believed her mother was depressed when she was here.

'I have to say that none of us really recognized depression back then,' he said thoughtfully. 'People might refer to someone as "having trouble with their nerves". But as I recall, no one realized it was an illness. As for Flora, I liked her a great deal. She was one of the most interesting women I'd met at that time. With hindsight, perhaps she *was* depressed. But at the time I thought of her as fragile, moody and temperamental. I was aware there was something troubling her, but I never did discover what that was. I never heard from her again after she left here.'

'She mentions someone called "G" in her diaries. Was that you?'

'Probably, I saw quite a lot of her. May I know what she said?'

'All nice things,' Eva assured him. 'Nothing personal, you understand – hiking, camping, dinner with you.'

He smiled. 'Well, that's disappointing. Flora and I used to talk for hours, I told her stuff I'd never told anyone before. I would've quite liked to be reminded of it all again now.'

'She did say you were a soulmate,' Eva said with a smile. 'That's a pretty good thing to say about someone, isn't it?'

'Aye, that it is.' He grinned. 'I'd have preferred "devilishly handsome, irresistible, with the mind of a rocket scientist". But I have found we rarely get what we wish for.'

Eva laughed. She liked his warmth and lack of pomposity; she had expected a man who had lost the use of his legs to be bitter and difficult. 'Did Flora reciprocate with confessions, soul-searching or anything about her past?' she asked. 'You see, she never did to me. I didn't know she'd been a successful artist. I don't even know anything about her parents, or her childhood.'

'She didn't tell you about her parents?'

Eva felt his surprise at this. 'No, nothing. Aside from her growing up in Cornwall.'

'I can't imagine why she kept that back from you – she mentioned them quite often to me. But maybe that was because I admitted that my late father was overbearing and often cruel. Anyway, she told me she felt suffocated by her parents. Her mother had two stillbirths before Flora was born. But instead of being joyful at finally getting a longed-for healthy baby, she spent her whole time fretting about everything. Flora said she didn't like her playing with other children in case she "caught" something from them. She wasn't allowed to go to the beach alone because she might fall on rocks or drown –'

Gregor broke off there, and laughed.

'Flora used to make me laugh with these stories, Eva. It sounded like being kept in a very clean and tidy prison. She said her mother had an obsession with germs, she scrubbed things until her hands were red raw. Nowadays that might be diagnosed as OCD, but back then she was probably thought of as a bit barmy. Anyway, Flora ran away to London when she was sixteen. She said she couldn't take all the rules any more. I did wonder whether that was what made her become so bohemian, so resentful of any kind of authority. I can't believe she didn't tell you that too!'

Eva shook her head. 'Not a word. Did she tell you about Patrick O'Donnell?'

He frowned. 'No. Who was he?'

Eva told him a little about Patrick and the baby they'd lost, and how Eva had found him. 'He was with Mum for a long time. She left him living in the studio she had in London, where I live now. Fancy her never mentioning him!'

'She didn't tell me any of that, except that she had a studio which she was renting out. I assumed that was why she didn't need to find a job here. But I did suspect there had been a broken love affair, and that something bad had happened to her. Sometimes she would stay in bed all day, refusing to answer the door, or be seen roaming around late at night. But though I tried to get to the bottom of it, she just brushed off my questions. She had a thing she used to say . . .' Gregor frowned and hesitated, as if he was trying to recall the exact words, 'it was, "Don't allow yourself to feel jealous or angry about my past or former loves, because who I am now, whatever it is you like about me, is the result of the experiences I had with other people, and the influences they had on me."'

Eva thought about that for a moment. She had never heard Flora say it, but yet her voice was in the words.

'That's actually a profound bit of advice,' she admitted, 'and very typical of Mum. She had a stock of meaningful phrases she could trot out when needed. But I can't help but be cynical and think she said that one to prevent you probing into her past. Can I ask a personal question, Gregor? Were you lovers?'

She knew when he hesitated that they had been.

'"Lovers" suggests a lot more than we shared,' he said eventually. 'I was in love with her, but if she felt the same about me, she never said. We did sleep together, but just for a short while. And I think she regretted it, because it compromised her.'

'Why? She was free, you are an attractive man and must have been a very eligible bachelor.'

'I suspect she had a fear of commitment. I think she

wanted a purely platonic friendship.' He sighed. 'I didn't of course. Do men ever want that?'

'Not often,' Eva replied, and laughed nervously. She was very afraid she was going to discover her mother might have used this kind, attractive and sensitive man as merely a sperm donor.

'Can you remember how long Flora was living in Pitlochry?' she asked. 'You see, her diary has no dates, and I'm trying to fit the pieces of the puzzle together.'

'I know she arrived here in the spring of 1969. She remarked on the primroses coming up in the garden of the cottage, so that would've been late March. She loved gardening, but then you'd know that. She brought the one at the cottage back to life. As for when she left, the last time I saw her was Christmas of 1969. I went away to the Highlands for Hogmanay, and stayed right through till March. She'd gone when I got back. No one could say exactly when she left, as she hadn't been seen around for some time. But I think I only missed her by a few days, because she'd left bread and other stuff in the cottage and it was still reasonably fresh.'

'But I was born on the 26th of April, 1970!' Eva gasped.

Gregor blanched. He looked so shocked that Eva was afraid he might become ill.

'I'm sorry, I shouldn't have blurted that out like that,' she said. 'Didn't you know she was pregnant?'

'But she can't have been.' His voice had risen slightly, and the words came out like a protest. 'I was with her at Christmas and she was as slender as you are. If she was pregnant, someone would've noticed it before she left Pitlochry.'

Eva could hear the anguish in his voice and saw it in his eyes too. She realized in that moment that he really had loved Flora. To be told that she had either got pregnant by someone else while he was seeing her, or she had deliberately deprived him of his own child, was like a dagger through his heart.

'Thick winter clothes can hide a lot,' Eva said. 'I am so sorry, Gregor, to give you such a shock. But it seems to me that you could be my father.'

Chapter Fourteen

'It couldn't be me, Eva. Well, not unless you were very pre-mature,' Gregor said, making a gesture of finality with his hands. 'There was nothing physical between us until late September.'

Eva made a rapid calculation in her head and found that she must have been conceived in July 1969. She had no doubt Gregor was telling the truth; like Patrick, he looked a bit sad that he hadn't gained a grown-up daughter.

'I feel really bad and pretty stupid, assuming that,' she said glumly. 'There must have been another man in her life before you,' she said.

'Don't worry on my account. I'm a big boy now, and it was a long time ago.' He shrugged. 'Obviously there must have been, but she never said anything – not so much as a hint. But he can't have been from around here, as the jungle drums would've been beating loud and clear.'

'She mentions someone with the initial "D" quite a few times. She said something about feeling she was being ana-lysed.'

His expression lightened. 'That would be Dena Deeds,' he said. 'She was just another hippy who turned up here on a quest for salvation, but she claims to have psychic powers these days.'

Eva smiled at his description of the woman. 'She and Flora were friends?'

'That's debatable. They spent a lot of time together when Flora first arrived – understandably, as they were both on their own, and both oddballs. Flora told me she found it comforting to be around someone who had more hang-ups than she did. Back then, Dena was into astrology in a big way. She'd do people's charts and analyse them on the strength of it. I know she told Flora that the only way she would be happy was to "surrender" herself to a man and normality.'

Eva laughed and Gregor joined in. 'Flora and I laughed about that a lot,' he said. 'Dena had no idea what "normal" was, and Flora certainly wasn't the kind to surrender to anyone.'

'Yet she did,' Eva said thoughtfully. 'That's exactly what she did do. She met and married very normal Andrew, gave up painting, surrendered everything she'd been before in order to fit in with his ideals.'

'You are joking?' His red eyebrows shot up in disbelief. 'That's as unlikely as hearing the Pope has signed a pact with the Devil!'

'I'm entirely serious,' Eva insisted. 'Her change of character and lifestyle is all part of this mystery surrounding her.' She went on to explain a little about her mother's life during the years with Andrew.

'So maybe Dena isn't barking, as I've always believed,' he said thoughtfully. 'Maybe you should go and see her? She's got a room above a shop in the main street. She tells fortunes for the tourists.'

Grace came back with a tray of tea and shortbread for them. Gregor introduced them, explaining that Eva was Flora Foyle's daughter.

'Oh aye, I remember her so well.' Grace's face lit up at the memory. 'I adored her, she was so outrageous. She wore a wonderful emerald-green velvet cloak and lacy long dresses. I was at that stage you go through in your teens when you rebel against everything, especially living in a little town in the middle of Scotland. Flora convinced me it was actually pretty cool to live here. I even hoped she and Gregor would get married.'

'That was never on the cards,' Gregor snapped.

Grace made a face at her elder brother and flounced out of the room. Eva sensed that Gregor was prone to putting his sister in her place and that, if they hadn't had company, Grace would have retaliated.

Eva poured the tea. 'I understand Grace lives here with her family,' she said. She thought Gregor was lucky to have a sister who was prepared to look after him, and he ought to be nicer to her.

'Yes, she does. Her husband was here too originally, but they got divorced a couple of years ago. She's got two boys, Cameron and Brett. They are having a holiday in France with their father at the moment. But they'll be back next week.'

There was so much more she wanted to ask Gregor – not just about Flora, but about his life too. But she felt she had intruded enough for one day, and she felt bad at shocking him with the news that Flora had left this town pregnant. How could Flora have kept that from him when she'd called him her soulmate? Why would she go hiking and camping with Gregor when she was already seeing another man? Even worse was that she embarked on an affair with him later, knowing full well she was pregnant.

Eva had been intending to tell Gregor about Flora taking her own life, but she couldn't bring herself to tell him anything else upsetting. So she kept the conversation to lighter subjects, asking him about what places she should visit, and if she could come and see him again in a day or two.

'I'd like that,' he said with a warm smile. 'Why don't you come and have dinner with me tomorrow? Grace will be very relieved to wriggle out of sitting with me all evening. Go and see Dena, then we can laugh about whatever she tells you.'

On the way back to the hotel Eva spotted Dena's sign on a street door next to a gift shop in the main street: 'Madame Dena Psychic. Tarot, Astrology and I Ching'. The sign said 'By appointment only', so as soon as she reached the hotel she rang the number and arranged to see Dena at eleven the following morning.

Back in her room, she felt both excited that she'd discovered so much today, but also sad. She had really liked Gregor and thought it was wonderful that he'd retained his sense of humour even when life had been so cruel to him. But her sadness wasn't so much because of that, or because he wasn't her father, but rather because her mother had been so devious and secretive.

What else was she going to uncover about her?

She really hoped this Dena woman might have something more uplifting to tell her.

At eight that evening, before going down to the Chinese Restaurant to get a takeaway, she rang Phil. He was really pleased to hear from her, and wanted to know every last thing she'd done since leaving London.

She gave him a brief synopsis, and he laughed about her having an appointment with a fortune-teller in the morning.

'I'm not going to tell her who I am,' Eva said. 'Well, not until after she's done her thing. I'm looking forward to it.'

'Well, if she tells you there's a tall dark man coming by train to meet you soon, you'll know she does have powers,' he said. 'I think I should be able to get there on Saturday.'

Eva gave a little squeal of pleasure.

Phil laughed. 'I thought you might do the "I vant to be alone" thing,' he said in a Marlene Dietrich voice.

'Of course I don't. I'm thrilled,' she said eagerly. 'There's a station in Pitlochry,' she added, and gave him the hotel telephone number. She told him to ring and leave a message to say what time the train got in, so she could meet him at the station.

Eva found it hard not to burst into laughter when she arrived for her appointment with Madame Dena the next day. The whole room was festooned with purple and lilac cheap nylon material. It was held in place on the ceiling by a Moroccan-style lantern, then fell in swathes down to what was probably a picture rail, before dropping down to the floor. It bore a passing resemblance to something from *The Arabian Nights*, and was not what anyone would expect in a small Scottish town.

Dena sat at a small card table, and she was wearing an orange and purple Indian-style long jacket with a high collar, with a purple turban around her head fastened with an amber brooch. Maybe if she'd worn Indian-style trousers beneath the jacket she could have carried it off, but she was wearing jeans, and a pair of brown bedroom slippers.

Gregor had said she was in her fifties, but in fairness she didn't look it. Her face was virtually unlined, with a clear complexion. She was almost beautiful, with high cheekbones and very dark eyes like melted chocolate, but her nose was a little too big. A strand of hair escaping from the turban was dyed black, but Eva thought she might have been born with black hair because of the darkness of her eyes.

She beckoned Eva in. 'Come in, my dear, sit down and make yourself comfortable.' Her voice was odd too, as if she'd trained herself to speak in that low, husky way and had ironed out any kind of accent.

Eva sat down and the woman took her hands, holding them very lightly while looking intently at her. Eva felt uncomfortable being under such deep scrutiny and wondered what else she was in for.

'A tarot reading is right for you,' Dena said after a few moments. She picked up the deck of cards and handed them to Eva. 'Shuffle them well,' she ordered. 'And as you shuffle, think hard on the questions you want answers for today.'

Eva shuffled them extremely well, continuing to do so even when Dena looked as if she wanted to take the cards from her. She played along with the game, silently asking who her father was, and where she should take her search from here.

Dena held the stack of cards to her lips for a moment or two, then cut the pile, discarding one half. She did this again and then laid the top card down on the table, face up.

Eva had been expecting traditional tarot pictures, but instead she was startled to see a picture of a snake. It was a beautiful picture, the colours bright as jewels, but she instinctively knew a snake was not a good image.

'Betrayal, dishonesty and guile,' Dena intoned, placing it in the centre of the table. The next card was of an ox, in equally vivid colours. 'Hardworking, strength, lacking in ambition,' she said and placed it to the side of the snake card.

A monkey came next, which meant mischief, speed and agility. There was an eagle, which the woman said represented the enemy. She went on until she had nine different animal cards in a circle around the snake.

She looked hard at Eva again, then spent a few moments staring at the cards before her.

'The snake is at the centre of your anxiety,' she said. 'You have recently experienced the betrayal, dishonesty and guile which he represents.'

Eva hadn't for one moment expected to hear anything she could relate to, and for this woman to get right to the crux of what was on her mind was astounding. But she concentrated on not showing any reaction, maintaining what she hoped was a blank expression.

'This card,' said Dena, pointing to a deer to the right of the snake, 'represents you. The deer's strength is its speed, sharp ears and its ability to sense danger. You are leaning on the strength of the ox and feel braver than is wise, especially when the monkey is creating mischief for you.'

As she named the cards she touched them lightly with her index finger. She had rings on each of her fingers, and her long nails were painted blood red.

'The horse is righteousness, or the law. You will be put in a position where you have to choose between what is right, knowing it will bring trouble for others, or walking away from it, for the sake of those you hold dear. I can see from the other cards that this will be the hardest choice you will ever have to make.

'However, the tiger will strive to protect you, and he is very powerful. Sadly, though, he is no match for the eagle . . .' She paused here, tapping the eagle card as if trying to work out this card's place in the reading. 'I must warn you, this is someone who seeks to hurt you. The eagle has the ability to swoop down and also soar away, suggesting to me that you will not see what is coming.'

Eva thought this was all very dramatic and so slick – Dena had probably used the same patter to dozens of people. While the snake did appear to represent her mother, all the rest could be interpreted by almost anyone to fit problems they might be having.

'Why am I the deer card?' she asked. She saw herself as rather more of a tiger.

'That position on the table is always the one which repre-sents the person asking for a reading. The ox is the man in your life. You are compatible, each one's strengths compen-sating for the other's weakness.'

'Who is the tiger?' Eva asked.

Madame looked hard at her. 'It is normally a father or elder brother. In your case I think it is someone who has, or will choose to, become your protector. The bluebird there,' she said, pointing to a very pretty card she hadn't mentioned before, 'represents happiness, but the position he occupies means you are unable to recognize what, or who, it is that will bring you happiness.'

'And the rabbit?' Eva asked.

'A vulnerable family member,' she said. 'You have, or will have, someone who needs your help.'

Eva assumed that was it, that the reading was over. But Dena took hold of Eva's right hand and studied her palm.

'You have more courage and determination than others credit you with,' she said. 'Practical, and yet sensitive and artistic too. You have in the past strived to fit in with other people, often at the expense of your own needs. '

'Will I marry and have children?' Eva couldn't resist ask-ing that.

'My gift is for guiding,' the woman said sharply. 'I cannot see into the future. But I can see that you have lost someone close to you recently and you have many unanswered ques-tions.'

'How do I get the answers to these questions then?'

'They will come, but not all will be the answers you want.'

Eva felt unable to keep up the pretence any longer. 'I believe you were friends with my mother, back in 1969. Flora Foyle.'

Eva expected the woman to look surprised and delighted.

But instead her face tightened and there was alarm in her dark eyes. 'I see you have inherited her guile,' she said sharply, losing the deep huskiness. 'Did she send you here to try to trip me up?'

'She couldn't send me, she's dead,' Eva said. 'But I suppose I should've introduced myself instead of getting you to do a reading for me. I'm sorry if I've upset you. But you're right, I have a great many unanswered questions, and I had hoped you might know the answers to some of them.'

'I tried to help Flora, but she was cruel to me,' the older woman said, the husky voice turning to a shrill whine. She clutched at the front of her jacket as if distressed. 'Please go now, there is nothing I can tell you that will help you. Just take what I saw in the cards with you.'

Eva got out her purse to pay, but the women waved it away. 'I don't want your money. I don't want anything from you.'

Eva was both bewildered and embarrassed. There was nothing to do but leave. 'I'm staying at Brae Bank, should you have a change of heart,' she said. 'My name is Eva Patterson, and I'm truly sorry I've upset you. That wasn't my intention.'

Eva went for a walk along the river, shaken by the woman's reaction to finding out who she was. What on earth had Flora done to her that was so bad she couldn't put it aside even after more than twenty years?

The whole animal thing was so strange too; she'd never heard of anyone else using such cards.

Was she like a deer? And was Phil her ox? Was Sophie the family member who needed help? Was the law thing something to do with Myles? And this tiger who was going to protect her, who was that? Patrick? But who was the eagle, her enemy? Was that Andrew, or someone she hadn't met yet?

She had always laughed at people who believed the things psychics and fortune-tellers had told them. So why was she rerunning the things Madame Dena had said through her mind, and giving them credence?

It was late afternoon when Eva returned to the hotel. Her feet were throbbing because she'd walked so far, and she wrenched off her trainers and socks and ran a bath.

She had gone over and over what Dena had said during the card reading. However much she wanted to believe the woman was a complete charlatan, she couldn't. Fortune-tellers basically told people the stuff they wanted to hear – that they were going to be successful, that love was around the corner. They wouldn't stay in business for long if they told everyone sad or frightening things.

Was it possible that Dena was the real thing?

Grace was wearing a very elegant black dress when she opened the door to Eva at seven that evening, her blonde hair was pinned up, and she had a chunky glittery bracelet on her wrist. 'He's been champing at the bit, waiting for you to arrive,' she said. 'I hope he's not going to bore you rigid. I noticed he's dug out some old photograph albums.'

'I like looking at old photos,' Eva said. 'And I've been looking forward to seeing him.'

Grace raised one eyebrow as if in disbelief. 'You look lovely tonight. Shame you haven't got someone young and handsome to take you out.'

Eva laughed. She was wearing the turquoise dress she'd worn the night of her twenty-first birthday, and it looked much better now than it had then, because she was sun-tanned. 'You look gorgeous too, and I'm quite happy to stay in with someone old but interesting,' she said.

'I won't tell him you said that,' Grace said. 'Go on in. The starters are on the table and the main course is in the heated trolley. I'm off out to see a friend.'

Gregor did look very pleased to see her. The table in front of a big window overlooking the garden was beautifully laid with a white cloth, blue and white polka-dot napkins, silver cutlery, crystal glasses, candles and an arrangement of blue and white flowers.

'The table looks lovely,' she said.

'I think Grace wanted to make sure you felt welcome. She said I was churlish yesterday.'

'You weren't.'

'I was when she said she hoped I would marry Flora. It wasn't meant to come out the way it did.'

'I think I've learned enough about Flora in the past few weeks to realize she was no saint,' Eva said lightly. 'You can tell me the truth about her, Gregor. That *is* what I want.'

'What did Dena the crackpot tell you?' he asked as he wheeled his chair over to a sideboard and poured them both a glass of wine. 'Tell me about that first, it will get me in the mood.'

Eva sat down on the sofa with her drink and launched into telling him about it.

He was a very good listener, alert, asking the odd question here and there, and clearly taking it all in.

'She's spooked you?' he asked, as she rounded it off with how Dena had told her to leave. 'I'm not exactly surprised, she's an odd fish. Many people are convinced she really does have "powers".'

Eva smiled at the way he made the inverted commas with two fingers. 'Yes, she did spook me. What on earth did Flora do to her?'

'She accused her very publicly of trying to take over her life, claimed that she was a latent lesbian and that she used her supposed "powers" to hide the fact she was a talentless blood-sucking leech.'

Eva's mouth fell open with shock. 'She actually said all that?'

'Yes, exactly that. It was at a Christmas party thrown by some friends of mine. At least twenty people witnessed it. I was appalled! We all were. Dena might be weird, but she didn't warrant that kind of abuse. Dena ran out crying, and I told Flora to leave too. She said some horrible things to me and then stomped off. That was the main reason why I went away for Hogmanay. I realized that she had no feelings for me, and I wasn't going to hang around and be humiliated in the way she'd humiliated Dena.'

'I'm sorry,' Eva said. 'I can't imagine her being like that. She always seemed more of a doormat than a firebrand.'

'Maybe Dena's spells worked then,' he said.

'Spells?'

Gregor chuckled. 'When I got back weeks later to find Flora had left Pitlochry, Dena was in the pub one night, pissed as a newt. She said she'd got even. She claimed she'd used her powers to make Flora suffer.'

'After the way she was with me today, I can almost believe that of her.' Eva smiled. 'Have you seen that room of hers? I half expected the Forty Thieves to jump out from behind the draped walls.'

'I have heard,' he said, pulling a face. 'But the tourists love it. She must be making a small fortune – though what she spends it on is a mystery. She lives in the room behind her office, rarely goes out and she doesn't even have a car.'

'Maybe she sponsors a coven of witches?' Eva giggled.

Gregor grinned. 'Let's have dinner now and talk about something less spooky. Grace is a fantastic cook, and I know she went to a lot of trouble because of you. It's her special goulash. She only makes it for people she likes, so I don't get it much.'

Gregor was such good company. Eva forgot he was in a wheelchair and middle-aged, because he was witty and interesting – and he didn't talk down to her either. He talked about climbing, and how as a young man he'd been in a search-and-rescue team on the mountains. He showed absolutely no bitterness about his accident, but admitted he'd been careless.

'The only thing that makes me mad is people feeling sorry for me,' he said. 'Sometimes people talk to me as if I'm an imbecile because I'm in a wheelchair. I had a good run for my money, I climbed mountains all over the world, saw things most people only ever dream of. All that is still in my head, no one can take it away. Isn't it better to be sitting in a wheelchair looking back on a life that was full of adventure, colour and mind-blowing sights, than to be able-bodied but looking back at a dull life, regretting that you never took a chance?'

'Yes, I suppose it is,' she said.

Both the starter of prawns marinated in garlic, chilli and coriander and the goulash were marvellous, and as they ate Eva confided her two ideas – one to open a guest house, and the other to become an interior designer. 'The idea of having a guest house is only a few days old,' she admitted. 'I'm probably just being silly.'

He asked her how she thought she could finance buying a guest house, and so she explained about the studio Flora had left to her. Now that the horror of her first night there was well in the past, she could tell the story and make it funny. He laughed with her, but then began to ask questions about why she went there in the middle of the night. Before she knew it, she was telling him the whole story – about Flora's death, and Andrew telling her he wasn't her father.

'Oh my goodness, Eva!' he exclaimed. 'And to think I took you for a girl who had never had a moment of insecurity or unhappiness in your whole short life. How wrong could I be?'

'Well, until I found Mum in the bath it hadn't been such a bad one,' she said. 'You probably remember Mum was very fond of saying, "On every life a little rain must fall." Although that seemed like an absolute torrent . . .'

'It is not something any young girl should have to see,' Gregor said. He looked very thoughtful, as if he was mulling over something in the past. 'You know, Eva, I was always afraid that's what Flora would do. When she first came here I half expected to get a call one day to say she'd been found in the river. It wasn't that she was always gloomy or tearful, nothing tangible to point to anything wrong. But there was an aura about her, like a sadness that took her off somewhere else. I did my best to encourage her to talk about whatever it was. That was why I took her out for long hikes. We even camped out in the mountains sometimes. She was lively, chatty, fun to be with, but every now and then I'd see this expression on her face . . .'

He paused. He looked as if he was searching for the right expression to describe it.

'Have you been at a prize-giving when you thought you were going to get one?' he asked Eva.

'No, I was never in line for any prizes,' she said.

'Well, imagine it then. You are sitting there, expecting your name to be called. But someone else's is read out. You have to adjust your face, to hide your disappointment and look glad for the person who has won. Flora often had that kind of look. Dena saw it more than I did. I think she might even have found out the reason for it, and that was why Flora attacked her verbally that Christmas.'

'But you did love her?'

'I certainly thought it was love then,' he sighed. 'When I was away from her I thought about her all the time. Yet when we were together I was frustrated by the distance she kept between us. I used to beg her to tell me what she wanted, but she'd just make a joke of it. Once she said she wanted a golden eagle's tail feather.'

That jolted Eva; she remembered hearing Flora say it to Andrew once. There had been many times when Eva had seen a similar pensive expression on her mother's face to the one Gregor had described. Yet when asked if something was troubling her, she always laughed it off.

'Was she painting while she was here?' Eva asked.

Gregor nodded. 'Yes, and selling some of her work, both here and in Edinburgh. She told me once that a gallery in London wanted to put on an exhibition of her work, but she didn't feel inclined to get back into what she called "the London scene". I did wonder if that was because of someone from her past. She certainly didn't seem to be motivated by making money from her art.'

After they'd finished the meal Gregor wheeled his chair over to a coffee table and picked up a photograph album.

'I dug this out this morning,' he said. 'There aren't many pictures of Flora, she tore up all the ones she didn't like. Until today I hadn't looked at them for about eighteen years, and a lot of good memories came back.'

He turned the pages till he came to one of Flora kneeling beside a stream filling a bowl with water. It was a black and white picture; Flora was wearing jeans and a thick Fair Isle sweater, and her hair was fixed up on the top of her head, the loose curls looking very pretty. She obviously didn't know he was taking the picture, and her face was soft and relaxed.

'She was lovely, wasn't she?' Eva said.

'Yes, very much so. All my best memories of her are like this one. She loved camping, she didn't do that prissy thing most women do about getting wet or mucky. She liked campfires, watching the sunset, hearing owls hoot at night. That day was the one she told me about her parents in Cornwall. I got the idea she felt guilty that she hadn't made the effort to sort out the grievances she had with them before they died. She told me they left her their cottage and some money, and that with hindsight maybe she should've gone to live there and reconciled herself with the past.'

'Maybe, as she had lost a baby then, she finally understood why her mother was so odd with her,' Eva suggested. 'When was this picture taken?'

Gregor slid it out of the plastic folder and looked on the back. 'The 29th of August, 1969. Near Glencoe. I remember we got badly bitten by midges. We went over to the coast the following day to get away from them.'

'So she must have been pregnant then?' Eva said.

'I suppose she must've been, if you were born the following April. But she certainly didn't say or do anything to suggest she was. You'd think someone who had miscarried would be careful with the next pregnancy, but she was climbing up rocks, running and jumping. We walked miles too.'

Gregor turned the pages to a photo of Flora in the garden of the cottage. She was wearing just a man's shirt, with her legs bare and her hair tumbling over her shoulders. She looked so young and beautiful. 'I took that one right after our first night together. That was a few weeks after the trip to Glencoe.'

Eva wanted to ask if he'd noticed she had a tummy. But he was looking at the picture as if reliving the night, and she couldn't bring herself to.

'Do you get on alright with Dena?' she asked instead. 'If you do, could you ring her and talk her round, about me? Try and get her to agree to meet me again?'

He smiled. 'I'll give it a go. That is, if you'll let me look at Flora's diary. I might be able to see something in it that you've missed.'

Eva agreed, and then looked through his album at the pictures of him. She couldn't imagine why her mother hadn't fallen head over heels in love with him, as he looked so tasty in his climbing gear.

She told him a little about Phil and said he was coming up to join her on Saturday. 'May I bring him to meet you? He's not my boyfriend, just a friend, but I'm beginning to think it's time I pushed things on to another level.'

'There's nothing wrong with waiting until you are sure,' Gregor said. 'I feel kind of sorry for kids these days, courtship and wooing seems to have disappeared. In films and on TV girl meets boy and they leap into bed. It seems to be expected now. But where's the romance in that?'

Grace coming in made Eva realize it was getting late, and she said she must go. She left after thanking Gregor for a lovely evening, promising to bring the diary round the next day.

Back at the hotel, on an impulse, Eva rang Ben. She had rung the number he'd given her in Leeds several times, but he'd never been in. But this time he was, and he was very pleased to hear from her.

She told him she was in Scotland, and why. When he'd come to London for the weekend she had intended to show him their mother's diaries. But because he was out most of the time with his friend, she hadn't even told him about them. He was eager to talk to her now, and so she explained how she was trying to find out who her father was, and a great deal more about Flora.

'Dad must know,' he said. 'People tell each other that sort of stuff when they get involved.'

'I'm not going to ask him anything,' she said. 'I don't ever want to speak to him again.'

'I could ask him,' Ben said. 'I don't have to say I'm asking for you, I'll just make out I'm curious.'

Eva agreed it was worth a try, and gave Ben the hotel number in case he managed to get anything useful out of Andrew. She was touched that her brother wanted to help. He said he'd spoken to Sophie just a couple of days ago, and that she seemed more rational and had even admitted she missed Eva.

'She failed the audition for drama college, and she hates Rachel, Dad's girlfriend,' Ben said. 'She said that they'd had a huge row – Rachel said she was selfish and rude and that Dad should cut her down to size.'

'I wouldn't disagree with that,' Eva said. 'But what's she going to do now?'

'Well, she did say she was going to college in September to resit her A levels, so I think she got a wake-up call from somewhere.'

'Next time you speak to her, will you tell her I haven't phoned her because I'm afraid Andrew will answer? Give her my phone number in London, and tell her I'd love her to come for a weekend,' Eva said.

'You should phone her tonight,' Ben said. 'I know she's in, because we spoke earlier. And Dad is staying over at Rachel's. But look, Eva, I'm sorry I didn't treat you very well when I came to London,' he added. 'It was only afterwards that I realized I was out of order.'

'All three of us were pretty screwed up by Mum's death,' Eva said. 'What's important now is to move on and make lives for ourselves. I think Mum would have wanted us to stick together.'

He told her a little about a restaurant he was working in until he started at university, but then Eva's money ran out and she had no more change.

She was halfway up the stairs to her room when she suddenly thought she ought to get more change and phone Sophie right now. It was doubtful she'd gone to bed, as she had always been a night owl.

Armed with more change from the receptionist, Eva phoned The Beeches.

Sophie answered it after only two rings, but her disappointment that it was only her sister was obvious. 'What do you want?' she asked.

'I just wanted to know how you are,' Eva said. 'I miss you. Ben told me you were going back to resit your A levels. I was glad to hear that.'

'Why were you?' Sophie's voice had a hard edge. 'So I have to stay stuck in this house for another year?'

'Of course not. I was thinking of you being able to go to university, like Ben.'

'I don't want to go to university. I want to be an actress.'

Eva wished she hadn't rung, as it was quite clear Sophie was angry about something. She didn't dare say she already knew that she'd failed the audition. 'Maybe you could join a drama group in Cheltenham while you are doing the resits?' she suggested.

'As if I'd want to join one of those crummy amateur things!'

'I can't say anything right to you, so I'd better ring off,' Eva said. 'All I wanted to do was put things right between us. I know it's tough for you there on your own. Ben's got my home phone number if you ever want to talk. And I hope whoever you were expecting to call tonight does ring.'

As Eva went up to her room she didn't feel too badly

about Sophie's reaction to her. She'd caught her at a bad moment, and maybe in a day or two she'd come round. At least Ben had been pleased to hear from her. The day might not have started well with Dena being so peculiar with her, but she'd had a lovely evening with Gregor.

As Flora had been very fond of saying, 'You are stuck with relatives. Thank God we have friends.'

Chapter Fifteen

Eva had just showered and got dressed the following morning, when there was a frantic knock on her room door. She opened it and to her surprise it was Dena. She was wearing a grey tracksuit, her black hair scraped back from her face with one long, untidy plait over her right shoulder, as if she'd just got out of bed. She looked wild-eyed and manic.

'I had to come here,' she said, panting as if she'd run up the stairs. 'I couldn't sleep for worrying about you.'

It was just after eight. Eva had intended to have breakfast and then go for a drive and explore. While what Dena said suggested a change of heart, her appearance, and her arriving so early in the morning, was a little scary.

'There's no need to worry about me,' Eva said. 'But I am very glad you called round because I owe you an apology. I should have told you right off who I was, and I'm sorry it made me seem devious. Come in, I could make us some tea.'

Eva pulled up the bedclothes so Dena could sit down on the bed, then filled the kettle in the bathroom and rinsed out the teacups.

Dena was clenching and unclenching her hands, and looking around the room as if she half expected someone to jump out on her.

'I dare say some people have told you I'm a fraud,' she blurted out. 'But I do see things in the cards, and all I do is pass on what I've seen. I was disturbed by what I saw in yours, and then when you told me who you were I panicked.'

'Gregor Hamilton told me last night that Flora was very

nasty to you, so I understood then why you reacted as you did,' Eva replied, wanting to put the woman at her ease. She somehow doubted that was possible, though. Dena looked as tightly strung as a violin.

'Did Flora commit suicide?' she asked in little more than a whisper.

'What makes you think that?' Eva asked.

'Because I felt it, after you'd gone yesterday. I felt it in here.' She put her hand on her heart, her large dark eyes very troubled.

'Yes, she did, Dena,' Eva replied. 'Back in March.'

'And you came to me hoping to find out why?'

'No, I wouldn't expect you to know, not when you haven't seen her for twenty-odd years. And after what I heard she said to you I wouldn't have blamed you if you hadn't wanted to speak to me.'

'It doesn't matter what happened between Flora and me,' she said. 'But it does matter that you are searching for answers, and that you hoped I might have them.'

Eva thought the woman was frighteningly intense. But she saw no point in beating about the bush, so she explained as concisely as she could the circumstances of Flora's death, and the events which followed it.

The older woman's eyes filled up with tears; they spilled out down her thin cheeks like a waterfall. 'How terrible,' she gasped. 'I am so sorry. I loved her, you know. I knew she didn't feel as strongly about me. I think she felt I was hanging on her coat-tails. But whatever passed between us, it makes me very sad to think she didn't find lasting happiness.'

'There seem to be so many discrepancies between the mother I knew and loved, and the younger Flora Foyle,' Eva explained. 'This is mainly why I came looking for answers. You see, I saw her as kind, maternal, a person who liked solitude, and yet that seems to be almost the exact opposite of how others saw her. I also don't know who my real father is, and I hoped to find out here – where I was presumably conceived. Did you know she was pregnant with me that night she said such awful things to you?'

'No!' Dena exclaimed.

Her stunned expression was evidence she was speaking the truth.

'Well, she was. I was born in April 1970.'

'Why didn't she tell me?' Dena's voice rose to an indignant squeak. 'I can understand most things about her, but not that. She was grieving over the baby she'd lost when I first met her, she cried about it a great deal and I comforted her and tried to help her through her grief. She must have known I would be overjoyed for her that she'd got pregnant again. It's not as if I was the kind to be disapproving that she wasn't married.'

'She was secretive with everyone, even her children. You mustn't take that personally.'

It was obvious Dena was taking it personally; she looked crushed. 'Gregor Hamilton must be your father then,' she said.

'No, Dena, that's part of the mystery. It seems there was nothing between Mum and Gregor until after she had got pregnant. Do you know of any other man she was seeing?'

Dena made a despairing gesture with her hands. 'There wasn't anyone. Well, obviously there must have been. But she never told me, and I never suspected there was anyone else in her life.'

'Did you notice her getting fatter towards that Christmas of 1969? Gregor said she was as slender as she'd always been, but men often don't notice such things – especially if she took to wearing baggy clothes. Did she do that?'

'I don't recall her looking any different. On Christmas Eve she was wearing a long emerald-green velvet dress, with a loose beaded jacket over it. I suppose that might have hidden a bump, but then lots of women don't show until the last couple of months with their first baby. All I recall clearly about that night was the cruel insults . . .' Dena paused, her lower lip trembling as if she was going to cry. 'That was the last time I saw her. I made sure I didn't run into her. I didn't even know she'd left Pitlochry until Gregor told me she'd gone. She left without a word to anyone.'

Eva could hear raw grief in this woman's voice. As it was now two decades later, she wondered what Dena had been like when she found out Flora had left. But she wasn't going to ask; she thought that might open a floodgate of tears.

'Well, do you have any idea who my father could be?' she asked instead.

'None. He can't have been from around here, or I would've known. And anyway, she was scornful of all the local men. Except Gregor, of course, but she treated him badly too. She had such a vicious streak! She reminded me sometimes of a cat playing with a mouse, she thought it was funny to tease and lead men on.'

Eva winced. That wasn't the kind of image she wanted to have of her mother as a young woman.

'You didn't like to hear that, did you?' Dena asked, cocking her head to one side like a bird. 'But it is true. I'll tell you now, you must give up delving into Flora's past. It's better that you remember her as she was, as your mother, because I'm sure you saw the very best side of her.'

'I can't give it up, I need to know,' Eva insisted.

'There was a warning in the cards.' Dena reached forward and grabbed Eva's hands. 'I didn't know who you were then, or what it was about, yet even so it scared me. The snake card can mean many things, none of them good. One interpretation is the sleeping serpent: disturb it at your peril, for once its secrets are uncovered they will not be contained.'

Eva loosened herself from Dena's grip and turned to make the tea. She couldn't make up her mind if the woman was barking mad or just delusional.

'I am not going to give up,' she said firmly as she handed Dena her tea. 'It is important to me to find out about my mother, and hopefully to find my father.'

'Then you must be prepared for more heartache,' Dena said. She put down the tea without drinking it and stood up. 'I sensed something very bad in your cards. I knew this bad thing hadn't been done by you. But once you told me Flora was dead, I knew it was her. I am positive now that whatever it was, it was the reason she took her own life. Trust my instinct, Eva, and leave the sleeping serpent alone. No good will come of prodding it awake, it will put you in danger.'

She was out of the door so fast Eva barely saw her move. Eva just stood there, too stunned to run after her and beg for further explanation.

That same evening there was a knock on her door just as she was settling down to read *Scruples* by Judith Krantz, which she'd found that afternoon in a charity shop. She remembered people raving about it when it came out, but she'd never got around to reading it. She opened her door to find it was the receptionist, who said there was a telephone call for her.

Eva had wanted to lose herself in a book, as she'd felt disturbed all day by what Dena had said. Her rational mind told her the woman was a dramatic crank who got some kind of perverse kick out of giving sinister and even threatening messages. Yet the image of a sleeping serpent was a strong and insidious one, and she couldn't quite shake off the feeling that maybe Dena was on the level.

Assuming the phone call was from her – Phil had said he doubted he'd be able to ring this evening, as he was working late – she ran down the stairs to take the call. She hoped Dena wanted to apologize for how she'd behaved, or had remembered something about Flora. The receptionist said she could take the call on an extension at the end of the reception desk.

But the instant she picked up the phone and heard the familiar deep voice saying, 'Eva, is that you?' she trembled. Andrew was the one person she had never expected to hear from again. Especially here in Scotland.

'Yes, it's Eva. Is there something wrong with Sophie or Ben?' She assumed that would be the only reason he would call her.

'No, at least not aside from you filling Ben's head with foolishness, and pestering Sophie,' he said sharply. 'What is all this about secret diaries of your mother's?'

Eva's stomach turned a nervous somersault. Ben had promised that he would only ask discreet questions of his father. He wasn't supposed to tell him about the diaries, or where she was. She'd only given Ben the phone number of the hotel in case he wanted to ring her back; the last thing she expected was for him to pass it on.

'I found them in the attic at the studio,' she said. 'Why do you call it foolishness for Ben and I to be curious about our mother?'

'It's unhealthy and unnecessary. If she wanted you to know anything more about her, she would have told you.'

'So why did she leave the diaries in a place she knew I would eventually find them?'

'What's in them?' he asked. His voice rasped, as if he hated having to ask.

'That would be telling,' she said lightly. 'Lots of stuff about you, and some of it very worrying,' she lied.

It felt good to get one over on him; she hoped he'd be worrying about it all night.

When he didn't come back with a retort, she knew she'd got him. 'Thanks for ringing, I must get back to the diary. I'm in 1970 now. Tomorrow I'm planning to go and visit some of the places she mentions.'

She put the phone down and turned to the receptionist. 'If Mr Andrew Patterson rings again to speak to me, tell him I've left the hotel.'

As Eva opened the door through to the staircase which led to the guest rooms, she heard the phone ring again. She paused, thinking that it was Andrew again, and looked back to the receptionist, who was answering it.

'Will you hold the line a moment, sir?' she said. Putting her hand over the receiver, she asked Eva if she wanted to speak to a Mr Marsh.

'Oh yes,' Eva assured her, and rushed back to the extension line gleefully.

'That was quick,' Phil said when she answered. 'Were you sitting on the reception desk?'

'A lucky break, I was just nearby,' she said breathlessly. 'Though if I'd known you were going to ring, I would've been camped by the phone.'

'I'm on a quick break so I can't chat, much as I'd like to. Just wanted to say I'm definitely coming on Saturday afternoon, if that's OK with you?'

Eva's heart did a flip with excitement. 'That's marvellous. Of course it's more than OK, I can't wait to see you.'

Phil chuckled. 'Well, that was a good response, because I can't wait to see you too. I think the train gets into Pitlochry at about four thirty in the afternoon. I'll double-check tomorrow and ring and leave a message for you. I'm frantically trying to finish this job, so I've got to go now. I'm looking forward to hearing all your news.'

Eva went back upstairs bubbling with excitement, Andrew and Dena forgotten. Phil had been on her mind a great deal since she'd left London, often imagining erotic scenarios. She had a feeling that their relationship was about to change, and she couldn't be more pleased about it.

As Eva got ready to meet Phil on Saturday afternoon she had butterflies in her stomach. She'd had her hair trimmed and blow-dried that morning, had her legs waxed, painted her toenails, and bought a new set of undies, just in case.

She smiled at herself in the mirror as she put on some lipstick. Her appearance in a blue T-shirt and jeans was perfect – she looked good, but she didn't appear to have tried too hard. She'd also booked a room in the hotel for him. That would save her any potential embarrassment when he first got here.

'Stay cool,' she reminded herself. 'Let him do the chasing.'

An hour later, as she stood on the platform at Pitlochry and saw Phil step out of the train, her resolve to be cool vanished and she ran to hug him. It was only a week since she'd last seen him, but he seemed bigger, more handsome, and his returned hug was as enthusiastic as her own.

'I feel like bursting into "I'd walk a million miles for one of your smiles",' he joked. 'Maybe I ought to write one called "I went four hundred miles on the train to see her again"?'

'It hasn't got the same snappy quality.' She laughed. 'But I'm really glad you came.'

She tucked her hand under his arm as they left the station. 'I've got heaps to tell you,' she said. 'But Gregor invited us up for supper with him tonight. I can phone and cancel if you don't fancy it.'

'I'm easy,' he said. 'As long as we can do what we like the rest of the holiday.'

'I think I'm done here now,' she said. 'I thought we could drive further north tomorrow. Maybe we could go to Glencoe.'

Eva's premonition was right. Even as they walked out of the station to her car, she felt something different between them. He ruffled her hair affectionately, and she tucked her hand through his arm. She wanted to get closer still.

After he'd seen his room, which was on the floor beneath hers, they went out for a walk down by the river, then stopped to have tea and cake in a cafe. She told him about both the visit from Dena and the call from Andrew.

'I want to dismiss Dena as a nutter,' she said. 'But there's something about her that makes me think she really does sense things. As for Andrew, why is he so worried about the diaries? You don't get in a flap about something if you've got nothing to hide, do you?'

'No, you don't,' Phil agreed, and then frowned as if worried. 'You might ask something like "What did she say about me?" but you wouldn't assume she'd written something damaging. It also strikes me as odd that he couldn't stop himself from phoning you, given that you've had no contact with him since you left home. It looks like Ben put the wind up him, asking awkward questions.'

'And what do you think about Dena?'

He shrugged. 'Well, I've never believed in all that tarot and stuff. But she does, and therefore she is convincing. Do you believe your mum did something bad?'

'I don't know. I don't want to believe it. And anyway, what can it be? She murdered someone? She robbed a bank? I can't see her doing either of those. What do we know so far? She was mean to Patrick, abusive to Dena and heartless with Gregor. But I can't see anyone killing themselves over twenty years later because of that.'

'No, but she got pregnant by someone she didn't want anyone to know about. Someone prominent, maybe? And also married? Was it possible she stopped painting because she was afraid he or someone connected with him would be able to trace her through that?'

That hadn't occurred to Eva. She thought about it for a moment. 'But if that was the case, she wasn't likely to go back to live at her old address in London. She could be found straight away there.'

'Umm. We're missing something. I've got a feeling Andrew knows, though, and he's running scared you are going to find out about it.'

Eva changed into a pink dress and high heels when they got back to the hotel. When she met up with Phil in reception, just before seven, he'd shaved and put on a short-sleeved white shirt and navy-blue trousers. As they walked to Gregor's house he took her hand in his. Just the sensation of his warm, big, calloused hand against her far smaller one sent delicious shivers down her spine. She glanced up at him and he grinned. But by then they were at the gate of Gregor's house and she realized she was going to have to spend the whole evening thinking, 'Will he, won't he?'

His grin said that he was totally aware of this.

Grace was joining them for dinner too. Eva introduced Phil to both Gregor and his sister as her 'ox' and everyone laughed. Grace looked him up and down approvingly and added that she thought he looked more like a sleek panther.

'There've been a couple of developments,' Gregor said,

once Grace had poured them all a drink. 'First, I called in a favour from a friend who works at the doctor's. It was completely wrong of me, because it's totally against the law for anyone to divulge anything on medical records, even if they are the records of a patient who no longer uses the practice. But I thought, as Flora is dead, it wouldn't do any harm to know what was on her file. However, this information mustn't go outside this room, as it could cause a lot of trouble for my friend.'

'OK,' she said warily. 'My lips are sealed.'

'He has a finger in every pie in town.' Grace sniggered. 'God only knows what he has found out about me over the years!'

Gregor ignored that remark. 'Guess what I've discovered?'

'Flora had antenatal appointments?' Eva said.

'No, that's just it. There is no mention on her file of pregnancy.'

'As I told Gregor earlier, that doesn't mean much,' Grace said. 'She was a hippy chick, she probably thought it was all so natural she didn't need to.'

'But she'd lost a baby before,' Eva said. 'Would any woman who'd been through that take any chances with the next one?'

'That's what I thought,' Gregor said. 'She got Valium for depression on her first visit, a high dosage. She saw the doc again for a repeat prescription, and also sleeping tablets, but nothing more until January 1970, when she went back for more Valium again. But there was nothing about pregnancy.'

'If she was having a prominent and influential man's baby, he might have paid for her to go private to keep it hush-hush,' Phil suggested.

'That's possible,' Gregor said thoughtfully. 'But I can't imagine Flora doing that, she wasn't the type. She was a socialist through and through, didn't approve of private schools or private doctors and dentists. And surely she shouldn't have been taking antidepressants during pregnancy?'

'Did people think of that kind of thing then?' Eva asked. 'Women used to smoke and drink and no one thought anything of it, so I doubt they worried about antidepressants either.'

'But she would have been booked into a hospital for the birth,' Gregor said. 'Do you know where you were born, Eva?'

'Only that it was in London. Mum never said which hospital, or anything about it.'

'That's odd,' Grace said. 'Women usually talk about their birth experiences, both to their friends and their children. I mean, it's one of the biggest events in any woman's life. My boys aren't interested but on their birthdays I usually say something, even if it's just that I couldn't believe how beautiful they were.'

'My mum often talks about it. She had me at home,' Phil said. 'Dad said she was hanging washing on the line the next day.'

Gregor nodded. 'Our mother had both Grace and me at home too. It was me who called the doctor and midwife when Grace started to come. Mother used to tell Grace how I hung around her crib like a guard dog.'

'And how you nearly dropped me on the stairs!' Grace added. 'He was trying to be helpful, bringing me down when I was crying.'

'Flora wasn't one for talking about things like that.' Eva felt a bit sad that she'd never been told little stories about her birth. But given that Flora wasn't the reminiscing kind, she wanted to defend her. 'She did keep a kind of journal of

266

pictures she drew of me as a baby. It's beautiful – I wish I'd brought it with me now to show you. Ben and Sophie were born in the hospital in Cheltenham. I remember going to see her there with Andrew.'

'Well, let's move on to the other development,' Gregor said. 'I read through the part of the diary that comes after she was here in Pitlochry, where she wrote: "Thought of staying in M but couldn't face another hellhole." I think she's talking about Moffat. I remember she said she stayed there on her way here and thought Moffat was lovely, but the guest house was awful. So I reckon that "C", the place she did stay at, was Carlisle. It's a logical place to stay overnight on the way south, but there's more. That photo you left in the diary, Eva, the one of the row of shops.'

'It's in Carlisle?'

'Yes. I got a magnifying glass on it. I could just make out the name Huggett above one of the empty shops. There was a Huggetts which sold harnesses, saddles and other riding paraphernalia in Carlisle. It closed down years ago, but it was quite well known back in the 1950s and 1960s. My father used to order things from there.'

'How great.' Eva grinned. 'I fancied going to Carlisle anyway, on the way home. I wonder why she took the picture? There's a painting of it back home too.'

'Maybe she took a photo so she could paint the scene later,' Grace said.

'A funny thing to want to paint,' Eva said.

'I thought that too, Eva.' Gregor frowned. 'I only ever saw her drawing or painting beautiful things – views, gardens, trees. And the castle would be the most obvious place to paint in Carlisle. But also, why did she end the diary there? Could something have happened to her there?'

'I thought it was odd the way the diary just ended,' Eva

agreed. 'I know we all kept diaries as kids and they just fizzled out, usually by the end of February, but she'd been keeping hers going for years. Why not carry it on until I was born?'

'I can't help but think that row of shops is a clue,' Gregor said. 'Maybe we should take it to Dena and let her "powers" give us the answer.'

Everyone laughed at that, and it reminded Eva to tell them about Dena's surprise visit to the hotel. 'She was in a right old state,' she said, quickly running though the gist of what was said. 'She maintained Flora killed herself because of something bad that she'd done. She almost had me convinced.'

There was a little discussion on this, and both Gregor and Grace felt that Dena was overexcitable and out of touch with reality.

'Getting back to Carlisle and why Flora stopped writing in her diary,' Phil said, 'maybe she met her man there. That could explain it. Perhaps she was just too engrossed in him to write?'

Gregor and Grace both agreed that was likely.

'But she only went to Carlisle because she'd had a bad experience in Moffat,' Eva argued. 'Why mention another place if you've already arranged an assignation?'

'Ah, now we're getting to the bad thing she did,' Phil said, his eyes twinkling with mischief. 'Maybe the man met her there, told her he was dumping her and she killed him and buried him under the floor in that empty shop. Then she took a photo of the place and later painted it as a kind of memorial to him.'

Eva was drinking her wine but she laughed and spluttered it down her chin.

'Steady on, Eva!' Grace said. 'Sherlock's just cracked the case.'

'So we can all have our dinner now then.' Gregor laughed as he wheeled his chair over to the table. 'And I thought it was we Scots who were supposed to be canny!'

'That was such a good evening,' Phil said as he and Eva walked back rather unsteadily to the hotel.

There had been no more talk about Flora. Gregor told them climbing stories, Grace talked about her time in London in her early twenties, Phil offered up a few hilarious anecdotes about builders, and Eva told them how Phil had stopped the man who tried to rob her. They had all laughed a great deal and drunk too much.

'Gregor is amazing,' Phil said. 'I don't think I'd have much to laugh about if I had to spend the rest of my life in a wheelchair.'

'I wish he had turned out to be my dad,' Eva said. 'But then I like Patrick too. Mum had pretty good taste in men – well, except for Andrew. But I'm not going to talk about that any more, the rest of this holiday is just going to be fun.'

Phil stopped suddenly, put his arms around her and kissed her.

It was the best kiss ever. While he'd taken her by surprise, it was the perfect moment for it. All her nerve endings began tingling, her pulse raced, and she found herself melting into his arms as if that was where she belonged.

'Umm,' he murmured as they came up for air. He was still holding her tightly but covering her upturned face with little kisses. 'This Scottish air seems to be making my heart beat faster.'

'Mine too,' she whispered. 'I think we might need to lie down in a darkened room.'

'Together?' he asked, rubbing his nose against hers.

'If you think you could bear it,' she said.

'I'm brave enough to try,' he said. Then, taking her hand, he began running down the hill, pulling her along, and he didn't stop till they got to the hotel.

'Why the rush?' she asked as they slowed down at the steps up to the door.

'To make sure you don't have time to change your mind,' he said. Putting one hand on either side of her face, he kissed her again tenderly. 'You must know I've wanted you from the first day we met. I'm crazy about you, Eva.'

All the past sadness, hurt and anxiety, and all those years of believing that she was plain and unlovable just faded away at his words. She knew him, he'd been her friend when she most needed one, had helped her, supported and encouraged her. She could trust him, she didn't have a moment's fear that he'd wake up in the morning and regret it.

He kept stopping to kiss her all the way up to her room, red-hot kisses that made her want to pull his shirt out of his jeans and stroke that brown silky body she'd admired so often when he was plastering.

A middle-aged couple came along the landing as they got to the second floor, and looked affronted to see young people canoodling on the stairs. Eva ran the rest of the way to her room giggling.

Everything about the seduction was perfect; there might not have been candlelight and satin sheets, but every kiss and caress was beautiful. He took it slowly, undressing her as if he was unwrapping something fragile and valuable.

'Your hair smells so good,' he murmured into it. 'You've got such a great body too. I want this night to last for ever.'

It did seem to last for ever, waves of pleasure which went on and on, gradually reaching a crescendo of white-hot passion. But Phil showed so much tenderness, and it made Eva cry.

Tod had been an accomplished lover, he knew all the right buttons to push, but experience was nothing compared to being made love to by a man whose heart was truly in it. Phil kissed and stroked every inch of her body, making her moan for more and then plead with him to come inside her. But again and again he only smiled at her while continuing to pleasure her until she came.

When he finally entered her it was the most incredible sensation she had ever known. He was big, and he moved her around, on top of him, to the side, sitting astride him, sitting up and from behind, before finally getting on top of her again, holding her hips as if he never wanted to let go of her. The ecstasy went on and on till they both came together.

They were sticky with perspiration, the sheets were damp and twisted beneath them, but sated at last they clung together as one, whispering endearments.

Later they straightened out the sheets, then curled up together. She loved the way he ran his fingers through her hair, his deep, satisfied sighs. 'You are like a little deer,' he whispered. 'So small and so pretty.'

For the first time in her life she believed that she was. All the unkind things girls at school had said to her didn't matter any more. All those boys who'd been crude and callous to her were forgotten. And she wasn't needy now; she felt she was worth something, and not just because Phil thought so, but because she felt it inside.

'You make me feel brand new,' she whispered in the darkness.

He began to sing the song, out of tune, words wrong, and she shook with laughter.

'Fancy laughing at a man serenading you!' he said with mock indignation. 'And I was just going to suggest I made a cup of tea too.'

He did make tea, and nothing had ever been so refreshing.

'Time to sleep now,' he said, getting back into bed and pulling her into his arms. 'I've dreamed about this for so long, I can hardly believe it's come true.'

They checked out of Brae Bank in the morning after breakfast, giggling because they were sure some of the residents must have heard the squeaky bed during the night.

Before they left to drive to Glencoe, Eva quickly rang Gregor to thank him for dinner and all his help, and said they'd pop back to pick up Flora's diary in a few days.

'I wish you and Phil happiness. You two are made for each other,' he said simply.

'I think you might be right there,' she said. 'Funny that we had to come all this way to discover it.'

Chapter Sixteen

'Will it still be like this when we get back to London?' Phil asked as they drove back to Pitlochry to pick up the diary they'd left with Gregor. He was driving, but kept reaching out for her hand.

Eva knew exactly what he meant. They had been everything to each other in the last few days. The Highlands had been way above their expectations: they'd marvelled at the huge lochs, the mountains and forests, walked for miles, made love in secluded places, stared in wonder at superb views, eaten delicious food, and stayed in cosy guest houses they hadn't really wanted to leave. There had been so much laughter, talking and sharing stories about their pasts, and now Phil was afraid, as she was, that it would disappear when they got back home.

'I hope so,' she said. 'It will if we want it to be, won't it?'

He cast a sideways glance at her and sighed. 'Well, I've got the flat with my brother and you've got your house. Won't that sort of thing get in the way?'

Eva had noticed that Phil was very conventional in many ways. He might be outrageous in that he wanted to make love outdoors, or slide his hand under the table to grope her in a busy restaurant, but from little things he'd said she knew he believed in marriage, being the main breadwinner and sharing everything. He'd also remarked that he thought too many couples rushed into living together without any real thought. But now they'd been together day and night for a few days, and it had been so wonderful. Perhaps he didn't want them to have separate lives?

'It's too soon to worry about that,' she said, squeezing his hand. 'Besides, by the time we get back to London we might both feel we want some space.'

She knew with utter certainty that wasn't going to happen. But at the same time she was wary of jumping into living with him too soon. She needed to get a new job, and they had to look at the practicalities of their separate homes and make decisions that weren't based purely on wanting to sleep in the same bed together every night.

Gregor and Grace both wore slightly bemused expressions when the young couple arrived at their house looking 'loved up', no doubt remembering Eva's claim before Phil arrived in Scotland that he was just a friend.

Grace teased them a little as they all shared a pot of tea, asking about the places they'd stayed at and if it was the Highland air that had given them a certain glow. Eva and Phil hadn't dared look at each other at that point for fear of laughing, because they'd pulled off the road earlier in the day and made love in a wood. They had only just got back on their feet when a man walking his dog had appeared close by, which had sent them into spasms of helpless laughter.

'You're welcome to stay here for a couple of days,' Gregor said. 'Grace and I would love that.'

'That's very kind of you,' Eva said. 'But we're running out of days and there's still so much more we want to see before we go home.'

'You just make sure you keep in touch,' Gregor said as they got up to leave. 'We want to be the first to hear if you find a body in Carlisle. And there will always be room for you here, if you fancy coming up here again.'

Eva bent over his wheelchair to hug him. 'It's been such a

pleasure meeting you. Shame you didn't turn out to be my dad, but a girl can't have everything.'

Grace laughed. 'Well, you've got your ox now, and I think you'll find that is a great deal more exciting than gaining a dad.'

She took Eva to one side. 'Phil's a keeper,' she whispered, 'a lovely man. You hold on to him, and live happily ever after. Take Dena's advice and don't try to wake the sleeping serpent. Let the past go.'

Eva grinned. 'You know, I think you really believe she has "powers".'

Grace frowned. 'Sort of. It might just be coincidence, but she's told me and several friends some rather uncanny things. A year before Gregor's accident she warned him about it. Some would say that there was a fair chance any man who took such risks would eventually come a cropper, but there are many people in this town who pay her a visit whenever they have a problem.'

'Maybe we can go back to Pitlochry next summer?' Phil said as they drove south. 'I'd like to see Gregor and Grace again. You just don't meet people like that in London.'

Eva agreed with that. It wasn't that there was anything wrong with Londoners, but a big city made people harder, more wary and materialistic. Gregor was very open, he was in touch with nature, the elements and the seasons. Grace was more sophisticated than her elder brother, but she too was warm and generous. They both cared about people and valued friendship.

'Whatever Mum got up to in the past, or however devious she was, I'm really glad I've met Gregor and Patrick,' she said thoughtfully. 'They feel almost like family.'

As they'd decided they wanted to spend the rest of their

holiday exploring the Lake District, their plan was to reach Carlisle by mid-afternoon, stay overnight, then push on the next day.

They found a guest house within walking distance of the town centre, left the car there and set off to have a look around. Phil picked up a city map at a Tourist Information kiosk, and after they'd had a look at the castle, they got an outside table in the market square and ordered coffee and cake. Phil got out the map which also listed places of interest.

'We should visit Hadrian's Wall tomorrow,' he said. 'I've always wanted to see that. There's also a covered market close to here.'

'Oh, goody,' she replied. 'Markets in strange towns are always more exciting than the ones back home. Is the street in Mum's picture on that map?'

'Yes, it's here,' he said, pointing to it. 'Botchergate. It's a horrible name. I wonder what it means? Do you think all the city botchers lived there at one time?'

Eva giggled. 'Plumbers who leave leaks, bricklayers whose walls fall down!'

'I didn't imagine it being a main road, but it seems to be,' Phil said. 'We can go down that way on our way back to the guest house, take a look and then cut across through the backstreets to where we are staying. I don't think it's very far.'

'Is there really any point?' she said doubtfully. 'After all, it's not as if we're likely to find out what that place meant to Mum.'

'True. But if you don't bother to go, you'll always wonder about it,' he said.

'You are an amazingly reasonable man,' she said. 'Aren't you sick of me banging on about all this stuff with Mum?'

'No, I'm intrigued.' He grinned. 'Especially as there don't seem to be any secrets to uncover in my family.'

'It could be they are just better at hiding them.'

They had fish and chips in a cafe later, wandered around a little more, then made their way back down Botchergate, intending to find a pub near the guest house afterwards.

Botchergate was a bit rough and dreary; the streets running off it were all terraces of run-down little houses without a tree or plant in sight.

'So that's it,' Eva said as they stood outside The Cranemakers Arms and looked from the pub across the street at the row of shops. It didn't look all that different to the old photograph – except the shops had changed hands since then. There was a fireplace shop on the corner, a newsagent's, a charity shop and what appeared to be a printer's. A man they had spoken to by the new Lanes Shopping Centre in the middle of town had said this area of Carlisle was due for redevelopment too. It badly needed it, especially the houses.

'The fireplace shop is where Huggetts was, and that charity place is the old betting shop,' Phil said. He got out his camera and took a picture. 'Just something for you to tuck away for the next generation to puzzle over,' he added.

'Let's go and have a drink in here?' Eva suggested, looking at the big pub behind them. 'You never know, some old codger might be sitting at the bar ready to tell us about the racy red-haired artist who once called in.'

It was a traditional workingman's pub which had probably changed very little over the years, still with a separate saloon and public bar. They went into the public bar, as there was more likelihood of being able to get into conversation with someone who had lived here twenty-odd years ago. A group of six men wearing navy-blue overalls sat at a table in the corner, a few old men nursing a pint were dotted around, and up at the bar there were two middle-aged men on stools talking to the landlord. He was short but burly, with an impressive moustache.

Phil ordered the drinks and then asked the landlord how many years he'd had the pub.

'Only five as landlord,' he said in a rich Cumbrian accent. 'It was me da's place afore that. It took a lot of effort to lick it back into shape. I was away in the army and he let things slide after me mam died.'

'So you grew up here then?' Eva said eagerly.

The man smiled. 'I certainly did. Got my training in the licensing trade as young as six when I used to fill up crates with empty bottles.'

'You must have seen a lot of changes in Carlisle then over the years?' Phil said. 'Show him the picture, Eva.'

'My mother took this photo, we think in 1970,' Eva said as she took it out of her bag. 'She also did an oil painting of it.'

The landlord took the picture and showed it to the other two men. All three spoke about it eagerly – about who owned the shops at the time of the picture, and before.

'I wonder if you'd remember my mother? She was red-haired, small, slender, called Flora Foyle.' She took one of the pictures of Flora that Gregor had given her out of her bag. 'I think she must've stayed here for a while.'

The landlord looked at the picture and shook his head. 'I'd remember a good-looking lassie like that if I'd seen her, but I was in the army then and only came back occasionally to see the folks.'

The other two men didn't recognize her either. 'Sorry, pet,' one said, 'never seen her. But she looks classy. It were rough around here twenty-odd years ago, can't imagine she'd have lived here.'

'Well, she was an artist, so she might have,' Eva said. 'I can't see why she would want to photograph that row of shops unless she had some connection with it.'

'Back in 1970 there were a lot of people coming around

here taking pictures.' The older of the two customers spoke up. 'Remember! That was when Sue Carling's baby was snatched.'

The three men had a little argument between themselves, the older man insisting it was June 1970 because his own bairn was born around the same time. The younger man said it was 1971, but the landlord said it couldn't have been, as his mother had written to him at Aldershot about it and he'd left there by the summer of 1970. He thought it happened earlier in 1970.

'Did they find the baby?' Eva asked. She felt a tiny pin prick of anxiety at this news.

'No, never,' the landlord said. 'Most think the mother did away with it. Whether she murdered it and buried the body out on the fells, or sold it to someone rich, we'll never know for sure. The police could never prove anything. All we really know is that she left the bairn outside the bookies in its pram. What decent mother leaves a newborn bairn out in the rain while she puts a bet on?'

'Did she have any other children?' Phil asked.

'She had two or three, all taken off her afore that. She ought to have been sterilized years ago, instead of letting her breed like a rabbit with every drunken bum in town. I heard she had another one a few years later, and I think they left that one with her. God knows why, poor kid. She lived in Flower Street then, but she's long gone now. I haven't seen or heard owt else about her for years.'

The sudden arrival of a group of men ended the land-lord's diatribe, and he went to serve them.

Phil picked up their drinks and took them over to a table. 'Whew!' he said. 'Well, I suppose we wanted a bit of local colour.'

When Eva didn't respond, he patted her knee. 'What's up? Did that upset you?'

'It was like a goose ran over my grave,' she admitted. 'That snatched baby couldn't have been me, could it?'

Phil laughed. 'Come on, Eva, of course not. You're letting your imagination run away with you. I bet your mum was told that story while she was here, and being pregnant it played on her mind and that's why she took the picture. Mystery solved!'

In the early hours of the morning Eva lay beside Phil, unable to sleep for the thoughts running around in her head. They had gone on to another pub and drunk far too much. Once back at the guest house they had made love, but Phil fell asleep quickly afterwards.

Phil's explanation of why Flora took the picture was completely logical. Eva could almost see Flora walking along that street, perhaps stopping to ask why the press were there, and being told the sad story. Or she could have read about it in the newspaper, and curiosity made her look for the place where it had happened.

Yet however logical that explanation was, she couldn't help but weigh other facts against it. No one in Pitlochry had known Flora was pregnant, and in the early summer of 1969, when she must have conceived, there appeared to have been no man in her life. It was odd enough that she didn't confide in anyone about her pregnancy, but even more puzzling was the fact that she didn't mention it in her diary. She'd made comments about far more trivial things.

The diary stopped in Carlisle too. If she had been affected so badly by the story of a baby being snatched that she had to photograph the scene and later paint it too, surely she would have written about it?

Eva couldn't believe Flora was capable of stealing a baby, but she had lost her own baby and she was depressed – and it wasn't unheard of for a woman to take a child under those circumstances.

But over and above everything that may or may not add up to a case against Flora, Dena's warning words about sleeping serpents kept ringing in Eva's mind.

What should she do?

Just walk away from Carlisle and try to forget that row of shops and the story behind it?

Or should she dig further and try to find something that would exonerate her mother, for her own peace of mind?

The next morning Phil began talking about going to see Hadrian's Wall almost as soon as he woke.

'It's a beautiful morning,' he said, pulling open the curtains. 'If we go early, we can be in The Lakes by late afternoon with plenty of time to find somewhere nice to stay.'

Eva didn't want to disappoint him, but she'd already made up her mind what she must do.

'I'm sorry, Phil, but I've got to go to the library first and look in their archives to get details of the baby-snatching,' she said.

His face fell. 'No, Eva! That baby can't be you. We don't even know if it was a little girl.'

'I know,' she sighed. 'I'm hoping it will turn out to be a boy. But if it was a girl, then I'm hoping she'll be much older than me, and that Flora arrived here long after she had been taken. But I need certainty. I can't go off trekking around the countryside and dwelling on this.'

Phil went to the window and looked out. He didn't say anything for what seemed ages. Eva was afraid he was cross with her. But then he turned back to her with a resigned expression on his face.

'OK. We'll go to the library and look it up, just to give you peace of mind. Then we're going to Hadrian's Wall, because I know you are going to be laughing again by then and feeling daft that it even crossed your mind that your mother stole you.'

'I do hope so,' she sighed.

As Phil took a shower he felt very concerned about Eva. In the last few months she'd had a lot to deal with, and he thought she was now getting dangerously close to becoming obsessed with her mother's past. He wished that when they arrived in Carlisle he hadn't agreed to find the street, and they hadn't gone into that pub. He doubted the landlord could be certain about the year the baby was taken. Who remembered such details so long after the event?

Aside from Ben – who he thought sounded like a good sort – Eva's whole family left a lot to be desired. Flora appeared to have been a highly strung femme fatale who chewed men up and then spat them out, and finally became a complete self-pitying doormat. Andrew was a louse – Phil didn't know how any man who had brought up a child as his own could suddenly turn on her – and Sophie sounded like she was a totally spoiled brat.

Considering what Eva had been through before he met her, she was remarkably well adjusted, yet even so there were many pointers to her having a very poor self-image. She had told him once that she became a goth because she preferred being considered weird to being pitied for being plain and fat.

He couldn't imagine why she thought that about herself. That first day he saw her, when her bag was snatched, he'd been bowled over by her pretty face, those lovely blue eyes, shiny hair and clear skin. She certainly wasn't fat either. She had a gorgeous body, and if her damned mother hadn't been so wrapped up in herself perhaps she would have noticed that Eva needed encouragement and praise.

It was weeks ago that he realized he had fallen in love with

Eva. It began with him just feeling he wanted to help her because she seemed so vulnerable and scared, but once he discovered that she was plucky, fun and caring, love took over. She was definitely the one he'd been waiting for all these years, yet he'd begun to think such a girl didn't exist.

The reason he didn't try to push her beyond mere friendship was because he felt she needed to get over her mother's death. But he had suggested he join her in Scotland in the hope that something would come of it. And it had, and it was like a dream come true. He had been as intrigued as she was by the mysteries of her mother's past, but it was all getting a bit much now.

He really hoped that digging out the facts on this baby-snatching story today would end all this nonsense and they could get back to where they had been in the Highlands.

Eva sat at a table in the library archives with the folder containing newspapers from 1970 open in front of her. Phil was standing behind her, reading the story of the disappearance of baby Melanie Jane Carling, over Eva's shoulder.

'Only three days old!' he exclaimed. 'But as appalling and heart-breaking as that is, she can't be you. Look, she was born on the 29th of March, 1970.'

When Eva didn't respond he sat down beside her so that he could see her face. She looked stricken. 'What is it?' he asked. 'Surely you weren't hoping it was you?'

'Mum killed herself on the 29th of March,' she said, and her voice shook with emotion.

A cold chill ran down Phil's spine. He wanted to say that it was just coincidence, but he couldn't. 'Come on, Eva, get a grip. Your birthday is in April.'

'That's what it says on my birth certificate, but how do we know if that's correct?'

'The hospital would. And mums keep stuff like wrist tags.'

'She could've told the registrar that I was born at home. Does anyone check that kind of thing?'

Phil had no idea of the drill for registering a birth; all he knew was that he had been registered about three weeks after he was born. 'They must do,' he said, 'or what would stop people registering babies that don't exist and then claiming family allowance and stuff?'

She seemed to rally a bit at that. 'I'm going to ask to photo-copy some of these articles about the case,' she said. 'I'll read them more carefully when I get home.'

Phil thought they'd both already read them quite carefully enough. They knew that the mother had left the pram, which was described as a green carrycot on a wheeled collapsible frame, outside the bookmaker's at approximately 1.45 p.m. The mother claimed she had only been in the shop long enough to put a bet on, no more than five minutes, and she'd come out to find the pram and baby gone.

But a few days after the snatching, the staff in the book-maker's were reported as having said that Sue Carling was a regular customer with a gambling habit, and right up till the baby was born she was in and out of the shop almost every afternoon, often staying to watch the race she'd bet on. The manager couldn't say with any certainty how long she'd stayed that particular day, because there were a lot of people in and out, and he hadn't even known she'd had her baby and certainly didn't know she'd left it outside. The first he knew of it was when she burst back into the shop screaming that the baby had been taken.

There were no pictures of the baby – it was said the mother hadn't got a camera – but the baby's birth weight was 5lbs 6oz.

'She was very small.' Eva looked up at him anxiously. 'A

284

month on she would still have passed for a newborn baby. The mother told the police she was wearing a pink frilly dress, a pink hand-knitted lacy matinee jacket, with matching bonnet and bootees. That is exactly what I found in the box in the attic. I thought at the time they didn't look like clothes Mum would dress a baby in.'

Phil knew he was out of his depth. He didn't know what to do or say.

'I'm scared,' she whispered. 'I really don't want that baby to be me.'

'We'll get the photocopies and then we'll go somewhere away from here so we can talk about it,' he said.

It was a mistake going to see Hadrian's Wall, as Eva was in a world of her own. Phil suggested they stay the night in Wetheral, the village near Carlisle where she had stayed on her way to Scotland, because she'd liked it there. After checking in, they walked down to the river and found a bench to sit on.

'I thought when I stayed here before that it was the kind of place nothing bad ever happens,' Eva said in a small voice. 'What should I do, Phil?'

'We both know the right thing to do is to go to the police,' he said. 'But I don't think you're ready for that yet. And considering this crime happened twenty-one years ago, it's not going to make a scrap of difference if you wait a few more days before you do it.'

'Look,' she said, turning the photocopied pages till she came to the one with a picture of Sue Carling scowling and holding up a fist to the press photographer. 'She looks like one of the Fat Slags in the *Viz* comic. It's hardly surprising there was so little sympathy for her.'

Phil winced, but admired Eva's bluntness. It wasn't just the woman's aggressive stance, or even the very short skirt and tight sweater and hair like a bird's nest, it was more that she looked like everyone's idea of a stereotypical unfit mother. The headline was what she'd screamed at the journalist: 'OK so I like a drink and a f—ing bet, but that don't mean I'm a baby killer.'

Flicking through the photocopied press cuttings, some of which were dated several weeks after the event, it was clear why Sue Carling hadn't got much public sympathy. A spokesperson at the hospital where Melanie had been born said that Sue Carling had discharged herself against their advice, just a few hours before the baby was taken from outside the betting shop. A few days after the event, while scores of local people had joined in the police search for the baby, she'd been photographed buying whiskey at an off-licence. There were reports that she'd got into a fight with a neighbour, been too drunk to do a television appeal for witnesses, and she'd punched a policeman who called on her during the inquiry.

'I don't want a woman like that as my mother,' Eva admitted. 'And if I go to the police, I'll be opening Pandora's box, won't I?'

She was also thinking of how Ben and Sophie would react to having their mother pilloried in the press. She knew it would sever any bond that had ever existed between the three of them. Of course it was right for Sue Carling to be exonerated of any crime, if she should be proved to be her birth mother, and also to have the peace of mind of knowing her baby had been well cared for. But Eva didn't think for one moment she'd want a relationship with this woman, who might latch on to her and become a living nightmare.

'Only if you do turn out to be the missing baby,' Phil reminded her. 'I really can't believe you are. For one thing, everything you know about the young Flora suggests she was quite self-centred. Apart from Dena telling you about her crying over losing her baby there is no other evidence of her dwelling on it. Besides, women who snatch babies because they want one to love are always caught. That kind of impulse surely doesn't go with the cool-headedness needed to successfully pass the baby off as your own?'

'So what should I do?'

'Well, nothing in haste,' he said. 'Maybe we should find a doctor or lawyer for you to talk to first? And what about your stepfather? Wouldn't it be a good idea to talk it over with him?'

'He won't want me stirring anything up that might affect his children,' she said, remembering how snotty Andrew had been when he'd called her in Pitlochry.

'Probably not, but he did bring you up. And I think you owe him the chance to either tell you something which proves Flora gave birth to you or, if he can't, give him some warning of what might possibly lie ahead.'

She heard the understanding in Phil's voice, and when she looked into his eyes she saw the honesty she'd observed the first time she met him. He'd helped her then when she most needed it, and she felt certain he would see this through with her too.

'Will you come with me to see him?' she asked. 'I'm a bit scared of him.'

He took both her hands in his. 'Of course I will. We can go to Cheltenham on the way back to London. You aren't alone any more, Eva. You've got me now.'

There were times during the next few days when Eva thought that the tarot cards must have been spot on when they represented Phil as the ox, in as much as he was patient, calm and reliable. He joked that he was also dim, thick-skinned and likely to charge into things too if the mood struck him. She liked his self-deprecating sense of humour, the fact that he was never boastful, and that he was interested in so many different things, from all kinds of sport to history, current affairs, music and nature. They had travelled on to the Lake District where he bought her a pair of proper walking boots and thick socks, so they could do some serious walking on the fells.

The walks may have been seriously strenuous ones, but Phil made her laugh so much that she barely noticed her aching muscles and even managed to stop dwelling on Sue Carling and her baby.

One afternoon, after lunch in a pub in Grasmere, they had climbed up a steep path to look down on the lake. It had been raining in the morning, but the sun had come out while they were in the pub and everything looked sparkling: white cottages with pretty well-kept gardens, the lush grass and the lake shining like blue glass.

'I feel a Wordsworth moment coming on,' Phil said and stopped to look at the view, shielding his eyes from the sun with his hand.

'I wandered happily with my girl. When all at once my head began to whirl. Was it because my lady was so fair? Or just that I'd eaten a pudding big enough to share?'

Eva giggled. 'I think Wordsworth might turn in his grave at that,' she said.

They sat down on the grass beside the path.

'It's so beautiful here,' Phil said. 'I thought Scotland was fantastic, but this is even better. I think I might give up the idea of seeing the world, and just tour round England.'

'Plastering as you go?' She raised one eyebrow questioningly. 'You could get a van and have a sign painted on it: "Stop me for a plastering job".'

'Not a van, a posh motor caravan,' he said dreamily. 'We'd park up in places like this, and I'd go and do a job while you made the dinner or washed our clothes.'

'Nice daydream,' she said, leaning against his shoulder. 'I won't bring you down to reality by saying how cold it would be in winter or how few people would actually want plastering done by some itinerant man who just knocked on their door.'

'I never used to imagine impractical things until I met you,' he said, putting his arm around her. 'That's what falling in love does to a bloke.'

'You love me?' she asked.

''Fraid so,' he said, kissing her nose. 'I had the idea of getting right to the top of this path, doing the whole romantic bit of taking you in my arms and telling you. But I guess I've blown it.'

Eva felt as if she was melting inside. She'd almost said she loved him several times in the past few days – but she hadn't, for fear of jinxing everything.

'You haven't blown it,' she said, catching hold of his face with both her hands and kissing him. 'I love you too, and nothing in my life has ever felt this good.'

'We'll sort all this stuff with your stepdad – and the police, if they're needed. And even if Ben and Sophie don't want to know you any more, I'll always be there for you.'

No words had ever sounded sweeter to Eva. She felt she had everything she'd ever wanted right here with Phil beside her, and all the beauty of Grasmere and the mountains surrounding it, spread before her. She just wished she had the right words to express what she felt.

Chapter Seventeen

'This is where you grew up?' Phil exclaimed in astonishment as Eva directed him into the drive of The Beeches. Andrew's red BMW was parked up by the front door.

'Well, yes,' she said. 'What sort of house did you imagine then?'

'Something a lot humbler.' He grinned. 'You've never so much as hinted that you lived in a palace.'

Looking at the Georgian house through Phil's eyes she supposed it did look very grand, but she was shocked at how neglected the garden was. The grass had been cut, but the flower beds and the drive were overgrown with weeds.

She was very nervous at seeing Andrew. She'd phoned him two days earlier while still in the Lake District and he had been very chilly. She said she had a dilemma that she needed to discuss with him, and he began to say her dilemmas were of no interest to him. It was only when she said it was to do with her mother in Carlisle that he agreed to see her today at five thirty. Just the fact that Carlisle triggered a response suggested he knew something.

Eva went to the front door and rang the bell; somehow, she knew Andrew would be affronted if she went to the kitchen door. She would have felt easier if Ben and Sophie were there. But Ben was in Leeds, and no doubt Andrew had sent Sophie out.

Andrew looked flushed when he opened the front door. She wondered if he'd been drinking.

'Hello,' she said, and introduced Phil to him.

Andrew looked very hostile. 'Do you think it's appropriate to bring someone else along when we need to talk about family business?' he said in icy tones.

'Yes, I do,' Eva said more firmly than she felt. 'We are an item, and he was with me in Carlisle, so I want him here.'

She shot Phil an 'I told you so' look. He gave a little shrug.

Andrew extended one hand to indicate that they were to go into the sitting room.

As they walked into the hall Eva noticed that Rose must still be coming in to clean, as everything looked much the same as it always had. But when they entered the sitting room she saw straight away that Flora's painting of the Cornish beach had been removed and replaced with a print of a Venetian canal.

Phil sat next to Eva on one sofa, while Andrew took an armchair opposite. 'What is this?' he said without any pre-amble or the offer of a drink.

Eva had rehearsed what she was going to say over and over in her head, but the stony expression on Andrew's face made it hard to get the words out.

'As you so kindly informed me you weren't my father, I wanted to find out who was,' she began. 'You already know about Flora's diaries, and while I was in Pitlochry – where she lived for a year until a short time before my birth – I found out that no one there knew she was pregnant.'

Andrew didn't react to that, so she cut to the chase. 'Flora left both a photograph and a painting of a row of shops in Carlisle. It transpires that on the 1st of April, 1970, a three-day-old baby girl was taken from outside one of the shops in that picture, and has never been found. I think there is a pos-sibility that baby was me.'

'Don't be ridiculous,' Andrew exclaimed.

'I really do hope my fears are ridiculous,' Eva retorted.

'And I'm looking to you for some facts to prove it isn't true. For a start, which hospital was I born in?'

'How do you expect me to know? I hadn't met your mother then,' he said.

'She must have told you, women talk about that kind of thing. Was it in London or somewhere else?'

'I seem to remember her saying it was a home birth.'

'A first baby born at home? I don't think that's even allowed,' Eva said. 'Where? At the studio?'

'Yes, I think so.'

She knew with utter certainty that he was lying. He was sitting on the edge of the sofa, his back hunched, looking down at his knees; even his voice didn't have the conviction he normally spoke with.

'How old was I when you met?' she asked. 'And how did you meet?'

'What's that got to do with anything?'

'I want to understand Mum's frame of mind,' she said wearily. 'Look, I came here because I'm hoping you can re-assure me that she was my birth mother. Unless you can tell me something that will convince me, I'll have to go to the police. They'll soon find out the truth – and once that cat is out of the bag, there's no putting it back.'

He glanced up at her and then looked at Phil, as if weighing them up.

'You were two months old or thereabouts when I met her. I was staying with a friend in a flat just around the corner from the studio. It was a Saturday, and I was having a lunch-time drink sitting outside The Prince of Wales because it was warm and sunny. She was there too, rocking the pram back-wards and forwards to get you to sleep. We got talking, she said she was waiting for a friend. I don't think she was – a friend would have called at the house. My guess was that she was lonely. It can't be much fun being on your own with a young baby.'

'So did she say where my father was? Why she was on her own?'

'She said she'd made the mistake of having an affair with a married man up in Scotland. She'd left there because she didn't want people knowing her business. And anyway, the tenant she'd had in her studio had finally left, so she could move back in.'

'OK.' Eva thought that sounded plausible. 'So how long after that did you move into Pottery Lane with her?'

'A couple of weeks or so later. I was paying rent at my friend's place, but I was spending most of my time with Flora, and it made more sense to help her financially.'

'What did she tell you about my father?'

'Nothing much. It was a brief fling and afterwards she found out she was pregnant.'

'A name?'

'If she did tell me, I don't recall. Surely even you remember how little your mother talked about her past?'

Eva didn't like his scathing tone, but she let that go. 'But if she'd had me all alone, I can't believe she didn't ever talk about that time. Was she coping when you met her? Did she seem calm and serene? What?'

'She was very untidy, stuff everywhere, and she said it had been hard at first. By the time I met her she'd got you in a routine and you were a placid baby. Not that I knew anything about babies back then. But I don't remember you being any trouble. You were always out in the pram in the garden with her. Anyway, I was out at work during the day.'

'If she had a home birth there would've been a midwife,' Phil said. 'And don't health visitors come, and all that?'

'That was all over by the time I came on the scene.' Andrew

shot Phil a look that implied he didn't expect to be questioned by him.

'But surely she spoke about the birth?' Eva asked. 'Women do – if not to you, then to her girlfriends.'

'She made the odd reference to it being an ordeal, but nothing specific,' he said. 'As for girlfriends, there was only really that woman Lauren, who came to the funeral. And she didn't turn up until you were four or five months old. Flora wasn't one for girlfriends.'

'So you haven't got any proof that she actually gave birth to me?' Eva said, trying to push him and get a reaction.

'Have you got any proof that she didn't?' he retorted, and his eyes flashed with anger. 'Why on earth would you want to think otherwise, Eva? Is this your Cinderella complex again? You always did like to make out you were the one no one cared about. Are you so desperate for attention that you like to think you were snatched by a maniac?'

That stung, but Andrew had always been one for cutting remarks.

'Now you are being ridiculous,' she retorted. 'You started this, remember, by telling me you weren't my father. All I wanted was to find out who my real dad was. But as it happened, Mum left diaries, baby clothes, my birth certificate and other things at Pottery Lane, and I believe she left them there for me to find.'

'Flora was one of the most disorganized women I've ever met. If she left things there, it was just because she forgot them – not for anyone, and especially not you.'

'OK then, so why did she take her life on the very day that baby in Carlisle would have been twenty-one?'

'Pure coincidence,' he snapped. 'Really, Eva! Have you based this whole ridiculous idea on something as flimsy as that? You want your head examined.'

'I hope it is pure coincidence,' Eva retorted. 'As I said, I was hoping you'd be able to tell me something which would convince me it was just that. I don't need my head examined at all. I could easily get the proof I need by requesting a simple blood test. But it would be far better for all concerned if I didn't have to go down that road, as it involves talking to the police.'

Andrew's body language changed immediately. He dropped his eyes from hers, rubbed his hands on his thighs and looked nervous.

'I shouldn't have told you I wasn't your real father the way I did. I'm sorry for that,' he said, and his voice was no longer strident. 'But your mother hurt me badly and I was lashing out. You were old enough when Ben and Sophie were born to remember what a good mother she was. Can you possibly imagine her stealing another woman's baby?'

'No, I can't. That's the problem,' Eva said. 'But I do know she lost a baby before me, and that she was depressed because of it. And tell me, why did she stop painting? That's another thing that doesn't make sense to me.'

'There's no mystery about that, it was because she was lazy,' he said. 'Before I met her she had to paint to keep herself, it was the only thing she was good at. But once we bought this house all she wanted to do was be a mother, do some gardening, a bit of cooking and float around in her vintage clothes.'

'I don't call that being lazy,' Eva said with some indignation. 'Three children create a lot of hard work.'

'Yes, maybe, but most women get at least a part-time job once their children are at school. But not Flora, she had enough difficulty getting the breakfast things washed up. A job was beyond her. Stop thinking of her as some kind of mystical heroine, Eva. She was idle, self-centred and perhaps mentally ill. Sadly, I didn't realize the latter – but then, as you must have realized by now, she was very good at hiding things.'

Eva looked at Phil to get his reaction to this.

'What's this got to do with him?' Andrew snapped. 'He didn't know your mother.'

'He knows a lot more now, thanks to the diaries she left. And a whole lot more about you too.'

'What do you mean by that?' Andrew's eyes narrowed.

'I think we ought to go.' Phil got up, reached down for Eva's hand and pulled her up. 'I think Mr Patterson has said all he's got to say.'

Eva wasn't satisfied with what she'd been told; she had hoped to hear Andrew say something tender about Flora. But Phil was right, she wasn't going to get anything more from him. And she probably shouldn't have goaded him by suggesting she had some information about him either.

As she walked out into the hall with Phil right behind her, she saw the Cornish painting leaning against the wall.

She turned back to Andrew. 'What are you going to do with that picture?' she asked.

He shrugged. 'Give it to a charity shop, I suppose,' he said. 'I don't want it here any more.'

'May I take it then?' she asked. 'Sophie or Ben might like it when they get a home of their own.'

'I doubt that, but take it if you want it,' he said brusquely. 'I never liked it.'

Eva thanked him and picked it up. Phil opened the front door and they both walked to the car outside. Andrew came out into the porch and watched them.

'Well, that's it then,' Eva said, looking round at him as she put the picture on the back seat of the car. 'I'll check at the local clinics and hospitals to see if I can find any evidence of where my birth took place. But if I can't find that, then I will have to go to the police.'

Andrew stepped out of the porch towards them. 'Have you for one moment thought of what this will do to Ben and Sophie?' he asked, and there was a plea in his voice.

'Have you thought what all this has done to Eva?' Phil said, moving himself between the pair of them. 'You rejected her at the time she needed a father most. That was shameful. She doesn't want to find out that the woman in Carlisle is her mother, or that Ben and Sophie aren't her true brother and sister. But she has the courage to face up to what Flora may have done, and to try to put it right. She should be admired for that. And if you had any guts, you'd help her.'

Without waiting for a response he opened the car door for Eva so she could get in, then walked round the car and got in himself, started it up and pulled away. Eva looked back to see Andrew just standing there. She didn't know if it was her imagination but he appeared to have shrunk, as if some of the stuffing had been knocked out of him.

'What a bastard!' Phil exclaimed as they pulled out on to the road. 'No wonder you moved out right after the funeral. I felt like decking him for the way he spoke to you.'

Eva didn't respond. She hadn't for one moment expected Andrew to be overjoyed to see her, but she had hoped that he would meet her halfway in resolving the bad feeling between them. He had, after all, known her since she was a tiny baby, and they had both loved Flora. But it was painfully clear he had no feelings for her whatsoever, and perhaps never had.

She sat wrapped in thought all the way to the M5, till they were heading towards London. 'What do you really think, Phil?' she blurted out. 'I'm too close to get any kind of perspective. Does Andrew know something he wasn't telling us?'

'If you mean, does he know Flora snatched you and has concealed it all these years? I doubt it,' Phil sighed. 'Why would anyone do that? But I don't think he married Flora for love. Just the way he talked about moving in with her makes me suspect he had his eye on the main chance.'

'But she was a single mother, all she had was the studio. I bet she was living on benefits.'

'Was she? Gregor said she'd had money coming in from rent, and she sold paintings in Scotland. We don't know if she blew all the inheritance from her parents on the studio either. Andrew had been paying rent, so he wasn't loaded. Yet somehow they managed to buy that big house. Where did the money come from, if not from Flora?'

Eva thought about that for a minute. Then suddenly she remembered something.

'I think you might be right there. They once had a huge row about Mum decorating their bedroom. Andrew hated what she'd done. I heard her screaming at him. She said something to the effect that if it wasn't for her, they wouldn't even have the house. At the time I wondered what she meant by that. But then Andrew was always claiming she talked nonsense, so I forgot about it.'

'Well, there you go!'

Eva turned in her seat to look at Phil. 'You know, it must've been a huge gamble buying it. I was only small when we moved there, so I don't actually remember much. But we kind of camped out in what's now the sitting room. We had a sort of before-and-after photograph album of it – Ben and I were always looking at it. One of the pictures was of daylight coming through holes in the roof. Mum said they used to put saucepans, bowls and buckets down when it rained.

Anyway, they sold off the land at the back to a company that built a small estate of houses, and then all at once there were builders crawling all over our house, doing it up.'

'I wonder if there's a way we could find out how much they paid for it, and how much they got for selling the land? It's not exactly relevant, I know, but it would be good to have the complete picture,' Phil said.

'Someone at school told me they sold the land for a quarter of a million,' Eva said. 'Of course that was just another teenager repeating something she'd heard her parents say, so it might not be true. The same girl said Andrew bribed someone on the Council to make sure planning permission went through. I asked Mum about it, and she just laughed and said, "They say the love of money is the root of all evil, but I think it's jealousy." But with hindsight, it probably was true.'

'I didn't like the defensive way Andrew said that Flora was lazy, as if she was an albatross around his neck,' Phil said. 'She must have been pretty smart to buy the studio in the first place, then rent it out when they moved, plus making sure it remained in her hands. Anyway, what sort of man would expect a mother of three children to go out to work when they lived in a big house like The Beeches?'

'I always thought they were really happy together,' Eva said sadly. 'But with everything that's happened since Mum's death, I can see she can't have trusted him to do the right thing by Sophie and Ben, or she wouldn't have changed her will.'

'The thing we have to ask ourselves,' Phil said, slowly and deliberately, as if still thinking it through, 'is if Flora did snatch you – and I still can't believe she did – would she have admitted it to Andrew? The story would've been in all the nationals and on TV, people would talk about it. And if she did tell him, or he just had suspicions about her, what would make him keep quiet?'

'Well, I'm assuming the answer to that is because he loved her and didn't want to see her go to prison.'

'OK, that's what I would assume too. But now I've met him, and heard his snide comments about her being lazy, I'd be inclined to think it was so that he could control her. An ace card up his sleeve.'

'We're getting a bit ahead of ourselves,' Eva said. 'I agree he's a bastard, and possibly a control freak too. But we've still got no proof Flora didn't give birth to me. And given that she was so secretive, why would she admit to anyone that she'd stolen me? Perhaps the real truth of the matter is more mundane in that Flora married Andrew for security, and he married her because she had the studio. When she killed herself she robbed him of an insurance payout, plus she prevented him having all the assets. And that's why he's so nasty to me.'

'That's a far more pleasing scenario than baby-snatching.' Phil reached out and stroked her thigh affectionately. 'And can I tell you again that I love you? Even if that would make your stepfather think I need my head examining too.'

'It feels good to be home,' Eva said as she and Phil had a glass of wine before going to bed. The studio had felt chilly when they got in, and she'd put the heating on for the first time and drawn the curtains. It felt very cosy and snug now. 'I thought I'd feel dejected and sad that the holiday is over, but I don't – well, except for us not being together all the time, because you've got to go back to work.'

'We've still got tonight,' he said and did a comic thing with one eyebrow, making it go up and down.

Eva giggled and turned towards him on the sofa to hug him. 'We get to christen the bed. We could have a bath together. We could make so much noise that we annoy Nasty Mr Francis next door.'

'It's nice that you are thinking of things like that.'

She knew he meant 'instead of thinking of stolen babies' and she realized that she had mentally put that to one side for now.

'We need to get some advice about that,' she said, snuggling up to him. 'I think I'll talk to Patrick and Gregor and see what they recommend. On top of that, I need to find a job. But meanwhile, Proud and Powerful Prince Phillip, I want your body.'

'Well, extraordinary, elegant, exciting Eva, I am at your disposal.'

He got up from the sofa, reached down to pick her up in his arms and carried her up the stairs.

She squealed as he dropped her on to the bed and then dived on top of her. When she had put the new white-painted iron bed together and made it up with new bed linen, she had wondered if she and Phil would be in it together one day.

'Hmmm,' he sighed as he pulled her T-shirt over her head. 'Should it be a bath first, or later?'

'Later,' she said, unzipping his jeans. 'Much later.'

Phil had to leave early the next morning to go home and get his work clothes and car. Eva didn't wake up when he got out of bed; the first thing she knew was Phil holding out a cup of tea to her.

'I've got to go now,' he said, bending to kiss her. 'I hope I can be back here by six. Have a good time today.'

But just after nine the phone rang, and it was Phil saying he'd got to go to Birmingham at once for a rush job that was likely to last for at least two weeks.

'Sorry, babe, I tried to wriggle out of it, but I couldn't.'

'It's OK,' she said. She was very disappointed; she'd already mentally planned a special dinner at the weekend, but it seemed he'd got to work all the way through the weekend too. But there would be other weekends, and lots of nights when he was working locally. She wasn't going to make him feel bad by sounding miserable. 'Just phone me when you can – and remember, I love you.'

She poured herself a bowl of cereal after she'd put the phone down and began writing a 'to do' list. The puzzle of her birth would have to wait to be solved; she didn't know how to go about checking on hospital and home births. And anyway, she wanted to speak to Patrick about it all first. She decided getting a job had to be her priority for now.

Patrick wasn't at home when she called him. But she did phone Gregor, because she badly needed another perspective on the potential baby-snatching before she did anything else.

'Don't be daft, Eva,' he said when she'd explained everything they had discovered. 'Flora would never have done that.'

'I really hope so,' Eva replied. 'Give me a good reason why she wouldn't do it?'

'Well, surely women desperate for a baby give the game away by hanging over prams and asking to hold babies? Flora was never like that. I don't remember her even talking about babies.'

'Perhaps that was the problem – the fact that she never talked about it?' Eva suggested. 'Or else it's all a wild coincidence. Anyway, I've put it on the back burner for now. I need to get some advice before I take it any further.'

She rang Olive too just for a chat. The last time she'd spoken to her was to say she was going to Scotland for a holiday and that Phil might be joining her there.

Olive was delighted to hear from her, and her first question was about Phil. 'So did he join you? And if so, how did it go?'

'Yes, Phil came. And it was amazing, delicious and I'm so happy,' Eva told her. She didn't want to get into telling her anything about the diaries or Carlisle, as she knew Olive would be too busy for a long phone call. Instead she just told her about the places they'd been and how much fun they'd had. 'But I've got to get a job now I'm back,' she ended up. 'My plan for the day is to start looking.'

'I'm really glad it's working out with Phil,' Olive said. 'I've got to come down to London on business in a couple of weeks' time. Let's meet up and have a real catch-up? Meanwhile, I'll put on my thinking cap about who I know in London that might need someone like you.'

Two days later Eva got back from job hunting to receive a visit from the police. They wanted to clarify a few points in her statement about Myles. It was clear he would be pleading not guilty. And even though the broken wine bottle had his prints all over it, and they had photographic proof of the fingermarks on her neck, they wanted to warn her that his defence would put up a fight.

They didn't really need to draw a picture for her to describe what they meant by that. Eva realized that Myles's looks and bearing would influence the jury, and they were likely to believe she'd led him on or acted provocatively.

As soon as the police had left she started to ask herself if it was really worth going through something that would be so unpleasant, especially if Myles only ended up with a rap over the knuckles.

She knew Phil and Patrick would be horrified if she withdrew her statement. But then they weren't going to be put through the ordeal of being cross-examined.

Eva was still mulling it over in her mind when Serendipity – a shop she loved in Notting Hill that sold all kinds of china, glass and kitchen equipment – rang her and asked if she would like to start working for them immediately. Before she'd left for Scotland she had gone into the shop and said how much she'd like to work there. They didn't have any vacancies then, but they'd taken her telephone number just in case there was one at a later date. A vacancy had occurred now, and it seemed her enthusiasm for the shop had impressed the manager, so he'd rung to offer her a month's trial.

The phone call couldn't have come at a better time. At a stroke it took her mind off Myles, the stolen-baby issue and feeling lonely without Phil.

The manager wanted her to start on the Saturday, just six days after returning from Scotland, and she accepted eagerly.

Right from the first day she took to it like a duck to water. The shop was busy, because they stocked great things at bargain prices. The other staff were fun and friendly, and the customers were all people she could relate to. Unlike the bistro, where she'd had people treat her like an inferior being, at Serendipity the customers were eager to be liked so she would show them the best bargains.

Thursday was her day off. As it was a lovely day, she rushed off to the supermarket first thing to buy food for the weekend. Phil was coming home, and she hurried back to do some cleaning before sitting out in the garden with a book.

Phil didn't ring her as usual in the early evening. Since he'd gone to Birmingham he always rang on his way out from his digs to get an evening meal. But she thought he'd probably had to work late, so he could leave to come home earlier tomorrow. And anyway, she was so busy making a Victoria sponge that it didn't matter to her.

She was in bed by eleven, sitting up painting her fingernails and thinking about what she would cook for dinner the next day and what she would wear to greet Phil. She smiled to herself; she would need to cook something that wouldn't spoil, and wear something that would be easy to take off. She wondered if he would like it if she opened the door to him wearing nothing more than an apron?

Turning out the light, she snuggled down and lay there listening to the sounds of people leaving the pub, calling to one another, and car doors banging. One of her neighbours had stopped her earlier in the week and asked if she didn't think they ought to complain about the noise from the pub. She'd said it didn't bother her. But the truth was, she quite liked that burst of noise which gradually faded away to complete silence. It rounded off the day, just as the sound of the milk float rattling down the road started the new day. She wondered why it was that some people complained about everything.

A noise woke her. She groggily reached out in the dark for the alarm clock, but on seeing it was only three in the morning, she thought it was just a drunk going past the house. But then she heard a sound she didn't recognize – a whooshing noise. And she could smell something too.

Puzzled, she sat upright. It was a few moments before she registered that it was a smell of burning, and that the sound was the crackling of fire.

In panic she jumped out of bed and ran to the door. As she opened it she recoiled in horror when she saw a wall of thick smoke. It was so dense she couldn't even see the banisters of the staircase less than four feet away. But even through the smoke she could see an orangey-red glow coming from down by the front door, and it was making its way towards the stairs.

The house was on fire, and she was trapped.

Shutting the bedroom door to hold the fire at bay until she

could get out, she ran to the windows overlooking the street. They were both locked. She fumbled for the little key on the sill but couldn't find it.

On advice from John, who had installed the windows, she always closed and locked them when she went out for fear of a burglar. She hadn't opened them when she went to bed because moths and daddy long-legs came in, attracted by the light. Frantically she rushed to switch on the light to help her find the key.

She had just got back to the window sill when the light went out; she guessed the fire had burned some wiring and shorted it out. The key had to be on the sill, she always kept it there. But she ran her fingers all along both sills, and it wasn't there.

The smoke was belching in under the door now. She grabbed the duvet and shoved it down to cover the gap. Coughing and spluttering, she went back to the windows and crawled along beneath them feeling with her hands for the key. But she still couldn't find it. Terrified now, she began hammering on the windows with her fists but soon realized that no one was going to hear her. She tried to think of something she could use to break the window.

A chair was first, but when she cracked it against the glass she merely broke off the two front legs. She tried a shoe, but that made no impression at all.

She knew she was going to die. Someone had once told her that smoke killed you before the flames did. And she was choking now – her lungs were filling up with it – and there was nothing heavy enough in the room to break the glass.

In a moment of clarity she remembered what that smell was when she opened the bedroom door. It was petrol. It must have been that bastard Myles who had set the fire – his revenge for her going to the police.

Coughing and wheezing, her lungs feeling as if they were on fire, she fell on to the bed and covered her head with a pillow. She had thought that sometime in the future she and Phil would get married; that they'd have children and have a long and happy life together.

But now she wasn't even going to get a chance to say goodbye to him.

Phil was smiling to himself as he took the Hammersmith turn-off from the M4. He hadn't knocked off work at five o'clock as usual. He knew, if he kept on working, he could finish the job by about one in the morning. And then he could drive home to Eva. The two joiners had teased him about being in love and growing soppy. But it was in their interests for him to finish the plastering early, as it meant they wouldn't be held up in the morning waiting for him to get out of their way.

Eva had given him a key, and he couldn't wait to creep up the stairs and into bed with her. He just hoped she didn't scream, thinking it was an intruder.

As he was about to signal to turn left off Holland Park Avenue, a fire engine with sirens screaming came up behind him, overtook him and turned into Portland Road. Another one followed it, and Phil had pull right over almost on to the pavement.

Even before he turned the corner, he knew the fire was close. The sirens had stopped, but halfway along Portland Road he could smell smoke and see the bright lights from the fire engines. He realized they must be in Pottery Lane.

He parked his car in the first space he saw, got out and ran the rest of the way. As he turned the corner by the pub he saw it was No. 7, the small window beside the front door glowing red with flames. He felt himself go cold with fright.

'My girl's in there!' he yelled at the first fireman he reached. The man had just got out of the fire-engine cab and was unrolling the hose. 'I've got a key, I must go in and get her.'

The fireman caught hold of his shoulders. 'You can't go in, the whole ground floor is alight. Which room will she be in?'

'The front room.' Phil pointed up. 'Get a ladder!'

He was aware that, behind the man he had spoken to, the other firemen were moving quickly into their positions; one hose was already out, and he heard the gushing sound as the water ran into the gutter. The glass in the small window by the front door suddenly exploded, pieces falling out on to the pavement. The men lifted the hose and aimed it through the window. Phil heard sizzling as water hit the flames.

It was then he became aware of how many other people were out in the street. There were dozens of them, huddling in small groups, all wearing dressing gowns or coats over their nightclothes. The police arrived then and started moving people back, away from the fire. One came over to Phil, signalling with his arms for him to go back too.

'My girl's in there,' Phil yelled again over the noise of the engines and the water. 'Please get her out!'

Everything seemed to have gone into agonizing slow motion. He saw the fireman he'd spoken to talking to a colleague, and pointing to the windows upstairs. His colleague spoke to someone else, and it seemed to take for ever before he saw them positioning a ladder.

The first fireman came back to him. 'Is there anything in the house or garage we need to know about. Gas cylinders? Cans of petrol?'

'Her car will be in the garage,' Phil gasped. 'Oh hell, there's not only the petrol in the car, but there's probably paint stripper, white spirit and God knows what else too.'

This news seemed to have a galvanizing effect on the fire crew. The front door was instantly broken down and the hoses played right into the inferno of the hallway.

A ladder was now firmly in place and a fireman with breathing apparatus went up it. Phil was unable to stop himself miming breaking the window, hopping from one foot to the other in agitation. He was vaguely aware that a woman had come to his side – a neighbour, he supposed. She spoke but his focus was on the window and he didn't hear what she said.

She shook his arm to get his attention. 'She'll be alright, they'll get her out,' she said. 'Look, an ambulance is here now.'

At last Phil heard the sound of breaking glass falling into the street. He held his breath as the fireman on the ladder put his mask over his face and climbed in.

'I don't suppose he'd have gone in if the fire was in that room,' the woman said to Phil. 'I called them, you know. I normally curse that I don't sleep well, but I'm glad I was awake tonight. You see, I went out into the backyard, and that's when I smelled the smoke and saw it coming over the gardens. Next door are away – they aren't going to be too happy when they get back and find the smoke has damaged their house, are they?'

Phil wanted her to shut up, even though he knew he should be grateful to her. He wanted to keep his eyes on that window, not look at the woman and make some response. His heart was pounding with fear that Eva was already dead from the smoke. He didn't know what he'd do without her.

At last he saw the fireman at the window with Eva over his shoulder like a sack of coal. At that point there was a loud bang from inside the house and a tongue of flame licked out of the front door and up the front of the house. Two more of the team advanced with another hose, and the sound of spitting and hissing as the water attacked the flames filled the air.

Slowly the fireman came down the ladder with Eva. Phil rushed towards them.

'Steady on,' the fireman said. 'She's alive, but she needs urgent medical treatment.'

The ambulance men came forward with a stretcher and laid Eva down on it, then gave her oxygen as they wheeled her back to the ambulance. She was wearing pink pyjamas with teddy bears on them; in the yellowy glow of the street lighting she looked about twelve.

'Will she be alright?' Phil's words were a plea more than a question. 'I'm her boyfriend. Can I come with you?'

'It's too soon to say,' one of the men replied. 'But sure, you can come with us.'

Chapter Eighteen

Phil felt a surge of emotion as he looked down at Eva in the hospital bed. The smell of smoke still clung to her and she looked so pale, small and vulnerable. Even though he knew she was out of danger now, the terror of the past few hours when he thought he was going to lose her would never leave him.

She had been unconscious on admittance to hospital, and so close to death from the smoke inhalation that they had to put a tube down her throat to give her oxygen. All he had been able to think of as he paced the hospital corridors was that he was to blame. He'd told her the electrical wiring was in good condition, and he'd clearly been mistaken.

When the doctor finally came to tell him she was rallying, Phil wanted to hug the man for saving her. The doctor pointed out that she was very disorientated and nauseous, and she would be plagued by coughing bouts for some time. But he smiled as he said that Eva was a fighter.

While Phil had been waiting for news, two different policemen had called in to see how she was. Phil had admitted that he felt responsible, because he should have recommended she get a qualified electrician to check the wiring in the house. They said that fires started for many reasons and that, until the fire service had made their investigation and discovered what had caused it, he shouldn't go blaming himself.

When he was finally allowed in to see her, the relief of knowing she was going to be alright made him feel almost euphoric.

He took her hand and stroked it. She opened her eyes and looked at him.

He was shocked at how sore her eyes looked, and he bent to kiss her forehead. 'Don't try to speak, your throat must be very raw. I'm going to stay with you, just go back to sleep, you are safe now.'

'Do they know it was Myles?' she croaked out.

'Myles?' he repeated. Then he remembered that it was the name of the man who had assaulted her. 'What makes you think he was responsible?'

'I smelled petrol. Who else would pour that through the letter box?' she said, her voice so hoarse it didn't even sound like hers. 'You must tell the police.'

It hadn't even crossed Phil's mind that the fire had been started deliberately. The possibility that Eva could be right, and that the intention had been to kill her, made his stomach lurch.

Somehow he managed to stay calm, to reassure her that he would deal with it, that everything would be alright and all she needed to do was go to sleep knowing she was in safe hands. But that calmness was just a front – inside, his stomach was contracting with anger. If he could lay his hands on that bastard, he'd tear him limb from limb and take real pleasure in it.

A policeman was waiting out in the corridor, hoping for a few words with Eva. Phil went straight over to him and repeated what she had just said.

The policeman was in his mid-thirties, a pleasant-faced man with brown curly hair. 'One of my colleagues did mention that your girlfriend had been assaulted recently,' he said in a very off-hand manner. 'We will check out Miss Patterson's allegation.'

'The man is out on bail,' Phil tersely reminded him, wondering why he wasn't rushing out of the door now to catch Myles. 'He should be arrested immediately and charged with attempted murder.'

'We will of course question him – should it transpire that the fire was arson,' the policeman said. His tone had more than a touch of 'allow the police to decide what is to be done'. He continued, 'Does Miss Patterson have family we should contact? Will they be able to take her in when she is released from hospital?'

Phil said that he would take care of her and gave the man his address and phone number. He explained that Eva's mother was dead. And although she had a half-brother and half-sister, he didn't think there was any point in contacting them, as their father was not on friendly terms with her.

The policeman nodded, but wrote Andrew Patterson's address down in his notebook anyway. 'We should have the report back from the fire officers shortly,' he said. 'You look as if you could do with some sleep yourself. Go on home for now. And if we need to know anything else, we'll contact you.'

Phil didn't go home. As tired as he was, he felt unable to leave in case Eva needed him. He rang Serendipity in Notting Hill, where Eva had been working, and told them what had happened to her, saying he'd contact them again once she was better. He also phoned his own boss to warn him that he might not be in to work on Monday. He wished he could phone Patrick, Gregor and Olive too, as he felt the need to share what had happened with people who cared for Eva. But she had their numbers, and he didn't even know their addresses to look them up in a directory.

They moved Eva later that morning from intensive care into a medical ward. As the ward sister wouldn't let him sit by her bedside, he had to wait in the visitors' room.

He must have dozed off, as he came to with a start when his name was called.

It was the same curly-haired policeman he'd spoken to earlier. 'I just came to tell you that the fire was started deliberately,' he said, looking grave. 'Forensics have ascertained that rags soaked in petrol were pushed through the letter box. We are doing a house-to-house inquiry in the proximity, in the hope that someone saw something.'

'How likely is that in the middle of the night?' Phil asked. 'It's obvious it was that creep who attacked her.'

'He wasn't at his home when we called there. According to his neighbour, he went on holiday three days ago.'

Phil made a dismissive snort. 'How convenient!'

'We will of course be checking on that,' the policeman said. 'But we'll also be checking around the neighbourhood. I'm going in to speak to Miss Patterson now, to tell her of these developments.'

Phil looked at his watch; it was one forty-five. 'It will be visiting time in another fifteen minutes!'

The policeman gave him a sharp look, as if his visit was far more important. 'I'll come back and tell you when I've finished.'

After the policeman had gone, Phil went to see if he could get a cup of coffee. He was irritated by the policeman's lack of urgency in this case. Would he have pulled his finger out if Eva had died in the fire?

The following evening, while Phil was visiting Eva, the policeman he had spoken to on the previous day came in with another officer.

It was Saturday, so there were more visitors than usual around the other patients' beds. 'Couldn't they have called when there aren't any other visitors to gawp?' Eva whispered to him. 'It makes me feel like I've done something wrong.'

She was much better today. Her face was a yellowy grey, her voice was still very hoarse and her eyes sore, and she also had a headache and the awful cough to deal with. But she had managed to eat some lunch and she was no longer disorientated.

The curly-haired policeman introduced himself as Detective Inspector Turner. He came straight to the point, saying that he had proof Myles was in Cornwall on holiday. It seemed the local police had interviewed him; he had a cast-iron alibi, because he'd been in a restaurant in St Ives on the night of the fire with a group of friends until almost one thirty. Aside from the fact he'd left there very drunk, it was impossible for anyone to reach London by car in just over an hour.

'Maybe he paid someone else to do it?' Phil suggested.

Turner ignored him and just looked at Eva. 'A young woman in Portland Road did see something she thought was suspicious,' he said. 'She was dropped home by taxi about fifteen minutes before the logged call to the emergency services. As the taxi drew up outside her house she saw a man getting out of his car. But when he saw the taxi was stopping, the man ducked down behind his car. It made her nervous, because she thought it was an old boyfriend who has been making a nuisance of himself. However, when she got in she looked out of the window and saw the car was a BMW, and therefore not her old boyfriend's. The man who had hidden was gone. And a little later she heard the car drive off, so she assumed he'd just been calling on one of her neighbours. But she still thought the man's behaviour was suspicious enough to tell us about it. Unfortunately, she couldn't tell us what colour the car was. Do you know anyone with a BMW?'

'Only my stepfather,' Eva said. 'I'm not very good at recognizing cars, unless they're a Mini or a Beetle.'

'My wife is the same,' Turner said. 'But your boyfriend did mention that you weren't on friendly terms with your stepfather. Is this a recent falling out? Could you tell me a little more about your relationship with him?'

Eva began coughing violently. Phil poured her a glass of water and held it for her to sip.

'He was unpleasant to her after her mother died,' Phil said. 'He was angry because she'd been left the house in Pottery Lane.'

'I see,' Turner said. 'When did Miss Patterson last have contact with him?'

'Less than two weeks ago – we called on him on our way back from a holiday in Scotland,' Phil said. 'Eva had things she wanted to ask him about.'

Even through her coughing Phil could see Eva's eyes were imploring him not to say anything further, but he knew he must.

'Eva doesn't want me to tell you this,' he said. 'I understand why, because it's personal, complicated and we could very well be barking up the wrong tree. But whether we are or not, Patterson was rattled by some questions we asked him.'

She caught hold of his hand as if to stop him. Her coughing subsided and she looked scared.

He turned to her and smoothed down her hair. 'I've got to speak out, Eva. You could've died in that fire, and the police need to know all the facts if they are to find the person who started it.'

'We certainly do, Miss Patterson,' Turner said.

Phil was aware that some of the other patients in the ward were watching keenly. He looked back at Turner. 'I'd rather tell you about this down at the police station, not here in a ward full of people with their ears pinned back.'

Turner nodded in agreement.

'Then will you leave first? I'll come down in my van in a few minutes, so no one gets the idea I'm being arrested,' Phil said.

As the two policemen left the ward, Eva clutched his hand even tighter. 'It can't be Andrew,' she said, her eyes welling up. 'He wouldn't do that to me.'

'A year ago you wouldn't have believed he would turn on you the minute your mum was dead,' Phil pointed out. 'While I was away I thought a lot about his reaction when we went to his house. If he had nothing to hide, why was he so defensive? Especially in front of me! So maybe that baby in Carlisle isn't you – I really hope so – but there is something weird about Andrew's attitude towards you. And I think telling Turner about it all is the best way of getting to the bottom of it.'

'But what about Ben and Sophie?' she implored him.

Phil shrugged. 'You can't brush this under the carpet just to save them some grief,' he said. 'And if Andrew did try to kill you, then he deserves whatever comes to him.'

She lay back on the pillows as if defeated.

'I love you, Eva,' he said, leaning over her and kissing her gently. 'I want us to have a happy and secure future together. Whatever your mother did, it isn't your fault, but at the moment it's spoiling your life. What sort of man would I be if I didn't try to make it better for you?'

'I'm afraid I'm spoiling your life,' she whispered hoarsely. 'What will your family think if all this comes out?'

'They'll think the same as me, that it was nothing to do with you. They'll admire you for being brave enough to expose it all. And they will all love you as much as I do.'

He had to go then. He gave her one last kiss and walked away. He didn't dare look round and see her stricken face.

Whatever came of this, he knew it was the right thing to do.

Two hours later Phil was still in an interview room with DI Turner. He had explained everything as well as he could, but as he talked about the diaries and the second-hand information Eva had gleaned on her trip to Scotland, the policeman's incredulous expression made him falter.

'I know it is only supposition that Eva could be the baby snatched in Carlisle,' he said. 'I can't even show you the diary, the picture of the shops or the set of tiny baby clothes – they must all have gone up in flames. But Gregor Hamilton and his sister, Grace, in Pitlochry will confirm the contents of the diary, as they read it. And surely a simple blood test will prove whether the woman in Carlisle is Eva's mother or not . . .' He paused for a moment, aware that this all must sound like a piece of fiction to a policeman. He had to make his case a little stronger if he wanted to be taken seriously.

'Look, I don't want to believe Andrew Patterson tried to kill Eva. Would anyone do that to someone they'd looked after from a small child? But there is something fishy about the man – he couldn't tell us where Eva was born, and I feel certain he did have some sort of hold over Flora. And do you think it's mere coincidence that Flora killed herself on that other baby's twenty-first birthday?'

Turner sighed deeply. 'I don't know what to think. It isn't unknown for a depressed woman, who has lost a baby, to steal one. But in such cases the woman is usually caught very quickly because she isn't capable of all the guile, nerve and planning it would require to get away with it. So maybe Flora was just lucky, and cool-headed enough to drive down south with a new baby without drawing attention to herself. But tell me, why would any man, supposing he did actually know of such a crime, collude in it?'

'The only reason that would make me do it is if I loved the woman so much I couldn't bear to shop her,' Phil said. 'But from what I've seen of Patterson, he isn't a man to allow his heart to rule his head. My gut feeling from the one meeting I've had with him is that he is a control freak. And Flora was the one with money when they met . . .' He paused, looking hard at Turner. 'But whether Flora did or didn't snatch the baby, and whether Andrew knew or not, he's still been very nasty to Eva, he knew where she was living, and a man with a BMW was acting suspiciously near Pottery Lane. So surely that's enough to bring him in for questioning?'

'Oh, we will. And his car will be checked by forensics.'

'Will you also open the case about the missing baby and run some blood tests?'

DI Turner gave him a long, studied look. 'Go home now. Leave it to us.'

On Tuesday, four days after her admittance to hospital, as Eva waited for the doctor to do his rounds and discharge her, she dressed herself in the clothes Phil had brought in for her the night before. He had bought them himself: underwear, jeans, black T-shirt, a grey zip-up jacket and a pair of sandals.

He had seemed embarrassed about them, and apologized for picking such dull things. But Eva didn't care what they were like – she was just impressed that he'd got everything the right size – and was very glad that she'd be going home with him the next day.

Yet when Phil went home after visiting time, all at once the enormity of what had happened hit her.

She didn't think she could ever forget the terror of being trapped in the bedroom with the fire raging downstairs. If it had been an electrical fault, she might just have been able to feel grateful that the firemen had got her out in time. But the thought that someone wanted her dead was going to haunt her for ever.

Aside from almost dying, she'd lost everything she owned. Her clothes didn't matter so much – but photographs, Flora's paintings, and the little things she'd had since she was a child, were all irreplaceable.

Phil had said he thought some things in the bedrooms might be salvageable, as the flames hadn't reached there. But even so, they would be badly smoke-damaged. He found it almost miraculous that the fire hadn't got into the garage. If it had reached the car, the whole house would have gone up – and probably next door too.

All the effort that had gone into making the house nice was for nothing. She tried to tell herself she didn't care, that the place had been full of bad memories: the misery of her first days there, the snootiness of the neighbours and Myles attacking her.

But there were so many good memories there too: Phil taking her home after her bag was snatched, seeing the house come together as Brian and John worked on it, the joy she'd felt at learning to do jobs for herself, meeting Patrick for the first time and planting up the garden. Phil had been the rock she leaned on, the man who made her laugh and restored her faith in men. She had spent so many nights wondering if he would ever make the first move, or whether she'd have to do it.

She'd planned to take their relationship slowly, to savour

what they had between them and just enjoy it. Now she was dependent on Phil, and that wasn't the way she wanted it to be.

Phil had said that the insurance would pay out, that he could get men in to do the house up again and then she could sell it, if she wanted to. She couldn't tell him that she felt she'd been robbed of a period of courtship, that she had intended for them to have separate lives until such time as they were absolutely sure of one another.

On top of all that, she wasn't sure she'd ever feel safe again.

Minutes after the doctor discharged her, Phil arrived to take her back to his place. The doctor had warned her she must take it easy for a while and get plenty of rest. 'No going into smoky pubs or restaurants,' he reminded her. 'Get as much fresh air as you can, and that cough will soon go.'

She felt pretty good, considering what she'd been through. Her eyes and throat were still a bit sore, and the cough was horrible. But when her mind turned to being trapped in the smoke-filled bedroom and the fire that threatened to overwhelm her, she made herself think about Scotland and the Lake District.

'All set then?' Phil asked as he came in. 'If I'm allowed to boast, I think I chose those clothes pretty well. But you'll be able to go shopping yourself in a day or two.'

'You did outstandingly well,' she said, hugging him. 'Everything fits perfectly.'

'We'll stop at Boots on the way home and you can get some toiletries and make-up,' he said. 'That was Mum's input – she turned up at the flat at eight this morning to give it a good clean, and then reminded me that girls need stuff men don't think of. She's dying to meet you, but I said you need a few days to get settled. That was my hint to make sure she left the flat before we got back.'

'Is your brother OK about me coming?'

His brother, Lee, was two years younger than Phil. He had already told her that Lee was messy, played music very loudly and had no respect for his elder brother's possessions.

'He's gone to stay with Mum for a while. It was his suggestion too, so don't feel bad about it.'

Eva's first impression on pulling up outside Phil's home was that it looked very well kept. It was a semi-detached, two-storey ex-council house, his flat being on the ground floor, with a privet hedge and grass in front. It was in a crescent which curved around behind a busy road with a rank of small shops, but there were trees all the way along it. And there was a fenced-off small children's playground in a grassy area further along the crescent.

'The front garden really belongs to the couple upstairs,' he said as he led her through the gate. 'But I cut the grass and trim the hedge for them, because they don't bother.'

Eva saw there was a stone staircase at the side of the house, which presumably was his neighbour's entrance, and beyond that was a fence with a gate to the back garden.

She liked the fact that Phil had painted his front door red – and although she didn't say it, she was already thinking how much nicer it would be with tubs of flowers flanking it.

Inside, the flat was bigger than she had expected. It had a decent-sized sitting room, two bedrooms, a tiny bathroom and a long narrow kitchen with a door at the end leading to the garden.

Phil stood for a moment, and sniffed appreciatively at the smell of lavender furniture polish. 'Mum's done a good job,' he said. 'If you'd seen what it was like a couple of days ago, you'd have run off down the street.'

'No, I wouldn't, I'd have got stuck in to clean it,' she said, and kissed him. 'I'm so glad to be here with you.'

What she really meant was that she was sure she could feel safe with him. His flat might be what she expected from a couple of bachelors – uninspired decor in green and beige, a worn three-piece suite and a stained carpet which was a testimony to many parties – but she was glad to be there. The kitchen was very nice, though: pine units and a sparklingly clean cooker. When she looked out of the back door she found the garden was all paved, with not an empty beer can or overflowing dustbin in sight.

'When the Fire Department gives us the all clear, I'll go and rescue all your tubs of flowers,' Phil said. 'I bet you are thinking how boring it looks out there?'

She laughed; he so often seemed to guess what she was thinking. 'When did men ever think of planting up tubs?' she said. 'I'm just delighted there is somewhere to sit outside. Mind you, autumn is here, it's already chillier.'

'I'm going to make us some tea,' he said. 'Then we'll sit down and make a list of all the things that have to be done. Not that you need to do stuff like contacting the insurance people, your bank, or even going to buy clothes yet. But a list is always a good place to start.'

'Shouldn't you be at work?' she asked, leaning into his chest. 'You've had so much time off because of me.'

'I should go back tomorrow. But only if you'll be OK on your own. Once you'd given me Patrick, Olive and Gregor's phone numbers I did ring them. Patrick wasn't there, so I could only leave a message, but Olive and Gregor have got this number and address and will ring to speak to you. Brian said he'd definitely call round to see you. He was horrified by what had happened, but he did say he could get a team together to sort out the house once the police have finished poring over it.'

'Dear Brian,' she said with affection, 'he's such a nice man. Now, about that tea!'

Chapter Nineteen

Detective Inspector Salway stepped forward as the front door of The Beeches was opened by a tall dark-haired man.

'Mr Andrew Patterson?' Salway inquired.

'Yes, what can I do for you?'

'DI Salway,' he said, and flashed his identity card. Then, half turning towards the other detective who was standing a few feet back, he introduced him as DC Connaught. 'We have a warrant to search your house following the arson attack at your stepdaughter's home in London.'

It was just after eight in the morning – always a good time to catch suspects unawares. Patterson looked ready to go to work; he was clean-shaven, wearing a smart navy-blue suit.

'Arson attack!' Patterson exclaimed. 'Good heavens. How awful! Is Eva alright?'

The two policemen exchanged glances. If they hadn't been told of the issues between this man and his stepdaughter, they might have almost believed his concern for her was real.

'She's recovering now,' Salway said. 'But it was touch and go at first. We're looking at an attempted murder.'

'But why do you need to search my house?' Patterson asked as if bewildered. 'She hasn't lived here for some time, and she took all her belongings with her. Why wasn't I told about this fire before? I would've driven down to London with her brother and sister to see her. Is she still in hospital?'

Salway thought the man was a very cool customer. His reaction to the news of the arson attack was pitched perfectly to make him look entirely innocent: not just his indignation at the fact he hadn't been informed, but his assumption that they wanted to search in order to find clues in the daughter's belongings that might lead them to the arsonist. But Salway had looked at the file from the day when Mrs Patterson died, and it had been noted by WPC Markham that his attitude towards Eva on that occasion had been remarkably lacking in sympathy or support for the girl.

'It was up to her to say who she wanted us to contact,' Salway said. 'I don't believe she included you in that number. The team who will be handling the search will be here any minute. But while they are doing their job we'd like you to accompany us to the station to help us with our inquiries.'

At that moment two police cars turned into the drive. It was only then that Patterson looked nervous. 'I don't want them rampaging around my house while I'm not around,' he said. 'Can't I answer your questions here?'

'We prefer interviews to take place at the station, and my men will take great care not to damage anything,' Salway said. 'Now, if we can have your house and car keys please? They will be returned to you as soon as we've finished.'

Patterson's face darkened. He looked as if he was about to start a protest, but as the other uniformed police got out of their vehicles and began walking towards him, he clearly thought better of it. 'My other daughter Sophie is upstairs. I need to tell her what is happening.'

'One of the women police officers will inform her,' Salway said firmly. 'Now, please come with us.'

Two hours later Salway left the interview room for a breather and to discuss his progress with Wilson, his sergeant.

'Patterson is very calm and controlled,' Salway sighed. 'He claims on that night he left his girlfriend's place at around half eleven and went straight home to bed. He said he heard his daughter come in a bit later – around twelve, he thought. If she corroborates that, he'll be off the hook. Anything found in the house?'

'There was a petrol can in the garage. But then who doesn't have one, if they've got a petrol lawn mower? Nothing suspicious in his car – and nothing that matches the bit of rope used as a wick that they found outside the crime scene.'

Salway was disappointed. But having spent a couple of hours with Patterson, he wasn't really surprised the man had left nothing incriminating for them to find. He was a clever man; even when he was questioned about his relationship with his stepdaughter he managed to remain remarkably convincing that it was Eva who had turned against him. 'We'll get the younger daughter in for questioning too. I want to know where she was that evening, and how she got home.'

'What d'you reckon on this thing about the baby in Carlisle?' Wilson asked. 'Did you bring that up with Patterson?'

'No, I didn't. I was waiting to see if he would mention it. I half expected he might use it to imply his stepdaughter was deranged, but he's a very cool customer. When I asked him what made her go to see him on her way home from Scotland, when relations were a little chilly between them, he said she wanted to ask him some stuff about her birth. He said he couldn't really answer her questions as he hadn't met Flora then. He managed to give a first-class impression of a concerned father, pointing out that she'd lost her mother in the worst possible circumstances, and maybe regretted leaving home so soon afterwards. He even covered all the bases by admitting he could have handled her with more sensitivity and tact, and put that down to his own grief. It would be very easy to believe him.'

'But you don't?'

'No. He's suave, calm and charming, but I sensed the bully

beneath. He's a man who is used to having everything his own way, and he's not a man to cross. I don't know that I believe his wife stole a baby, but I'm sure as hell he did something to that poor woman which made her top herself. And I've got a gut feeling he set that fire, though I can't see a motive for doing it. He doesn't stand to gain anything by it.'

'So what now then?' the sergeant asked.

'Apart from interviewing Sophie Patterson, there's not a lot we can do. The Met will continue to make further inquiries in the locality of the fire, and they'll be checking on all maternity cases on the 26th of April, 1970, when Eva was born – both hospital and home births. I was told they were checking on all doctors in the Holland Park area to find where Flora Foyle and her baby were registered, and to see if there were any checks made by health visitors back in 1970.'

'But what about Patterson?'

Salway shrugged. 'We haven't got anything to hold him with, much less charge him. The daughter Sophie is on her way now. But if she confirms her dad was home in bed that night, we'll have to let him go.'

WPC Markham's most vivid memory of the night of Flora Patterson's suicide had been the coldness Andrew Patterson had shown towards Eva. It had played on her mind for some time afterwards.

On hearing the news that Eva had been the victim of an arson attack in London, that Andrew Patterson wasn't her father, and that he was the prime suspect, she felt justified in many of the thoughts she'd had about the man.

When she was asked to interview Sophie Patterson, she just wished she had studied the younger daughter more closely that night, because all she really remembered about the girl was her hysterics. As she walked into the interview room, where Sophie was waiting with PC Holderness who had brought her in, she was surprised to find the girl had changed a great deal.

Six months ago Sophie had been an innocent, pretty schoolgirl with a clear complexion and shining hair. Now she looked plain tarty: she was wearing far too much make-up, her jeans were so tight she could have been poured into them, and her T-shirt was very low cut, revealing impressive cleavage which Markham felt could only have been achieved with a substantially padded bra. Even her hair was spoiled. She'd had a perm, but it was frizzy rather than the kind of Botticelli curls she'd clearly been aiming for.

She was chewing gum – something Markham hated – and her surly expression and the way she had her arms crossed suggested she no longer had any respect for authority.

'What do you want to talk to me for?' she asked, tossing her hair. 'I haven't done anything.'

'No one suggested you had,' Markham said. 'We just need to know where you were on the evening of last Thursday, the 20th of September.'

'Why?'

'Just answer the question, please.'

'I went to my mate's house,' she said, folding her arms and looking up at the ceiling.

'And her name and address?' Markham asked.

'Louise Randal, 47 Fortworth Road.'

'What time did you go there, and what time was it when you left?'

''Bout seven. Don't know when I left, I never looked at the clock.'

'Roughly will do.

Sophie shrugged. 'Might have been around eleven thirty.'

'And how did you get home?'

'I walked.'

Fortworth Road was some half an hour's walk from The Beeches, and it was unlikely a girl of her age would walk that far so late at night.

'And you were alone?'

'Yes, what of it?'

'Does your father approve of you walking back home so late at night alone?'

Again Sophie shrugged. 'What am I supposed to do, stay in on my own? He doesn't care, he's always out with his bird.'

Despite the girl's belligerent attitude, Markham felt some sympathy for her. She'd lost her mother at a very crucial time. She was neither a child nor yet an adult, and if she was being left to her own devices for long periods, it was hardly surprising she was getting in with bad company.

'So was Louise's mother at home while you were there? We'll need to contact her to verify you were with her daughter.'

Sophie looked panicked then. 'I can't remember,' she said.

Markham knew the houses in Fortworth Road were very ordinary houses – too small for a visitor not to know who else was there.

'You do realize that telling lies to police officers is a serious crime?' Markham pointed out. 'If you weren't really at that address, or you didn't get home until much later than twelve, then it would be far better for you to tell me the truth now, as we'll be checking. So let me ask you again. Where were you that evening?'

Sophie picked at her fingernails. Markham could almost see her weighing up whether the consequences of telling the truth to the police would be greater than the trouble she'd be in with her father when he found out she hadn't been where she'd said she was.

'I was with my boyfriend,' she finally admitted. 'Round at

his place in Gloucester Road. Please don't tell my dad, he'll go mad with me. He doesn't approve of Jake.'

PC Holderness smirked at Markham.

'What time did you get home?' Markham couldn't promise Sophie anything, but she hoped to get at the whole truth before she was forced to admit this to the girl.

Sophie hesitated.

'Tell me the truth, Sophie,' Markham insisted.

'It was almost one,' she admitted reluctantly. 'Jake dropped me home on his motorbike.'

'Was your father home then?'

'Yes, he'd gone to bed. I crept in, so he wouldn't hear me. He asked me the next morning what time I'd got in, and I said it was twelve. Please don't tell him I lied or he'll ground me for ever.'

'Are you sure he was there? Was his car in the drive?'

'It must've been or I wouldn't have worried about him hearing me,' Sophie said. Then she frowned and looked at the policewoman curiously. 'What's this about? Why would you care when I got in anyway?'

'Because, Sophie, someone set fire to your sister Eva's house in London that night. She could've died in the fire.'

Sophie's eyes widened in shock. 'Oh my God!' she exclaimed. 'Is she OK?'

'She is now, but she has been very poorly,' Markham said.

All at once a flash of understanding passed over Sophie's face. 'You think my dad did it! You do, don't you? That's why the police were searching our house. Did she say he did it? The cow! She just wants to get some revenge because he chucked her out of our house.'

'I don't believe Eva has suggested your father was responsible,' Markham said. 'In cases like this we question everyone involved with the victim.'

Markham left Sophie with Holderness and went to report what the girl had said to DI Salway.

His face dropped when she told him. 'Bugger. The chances are she was so drunk or stoned that she just assumed he was there and never thought about his car. And she'll stick to her story no matter what, for fear of him finding out how late she was.'

'She doesn't appear to care much about her sister either. That girl needs to wake up and see where she's heading. We can check on the boyfriend – Jake. But you can bet he cleared off from The Beeches so fast he wouldn't have noticed if there was a double-decker bus parked in the drive.'

'It wouldn't hurt to do a little background check on Patterson, though,' Salway said. 'I've got a hunch he isn't quite what he seems.'

Chapter Twenty

Eva stood at the bottom of the stairs in her house in Pottery Lane, her hand over her mouth in shock at the black walls and charred remains of her furniture and belongings. Everything was ruined; the red sofa was only recognizable because some springs were sticking out of the blackened heap. All that remained of her bookshelves, which she had put up so proudly, were the metal brackets on the wall, the books a soggy mound of ash beneath. The kitchen units were still in place but were burned and distorted, doors hanging open, the contents just so much rubble.

Then there was the smell, as if a hundred people had smoked twenty cigarettes each and then sprayed the room with a toxic mixture of mould, rotting vegetation and some kind of pungent chemical. The floor was still wet from the firemen's hoses, the French doors had buckled, and the glass was broken. There was absolutely nothing left to show that this had once been a bright and pretty room.

Even the garden was a mess, because her tubs of flowers had been knocked over and charred timber and scorched carpet thrown out on top of them. She had told herself over and over again since the fire that everything she had lost could be replaced – and that much was true. But what she couldn't get over was the fact that someone had deliberately set the fire, intending her to die in it.

She was sure that it *was* Andrew, and she sensed the police were convinced too. But without some evidence to prove it, they had no choice but to release him without charge.

It was a horrible feeling, knowing that he was walking around free, probably gloating that he'd been clever enough to cover his tracks. To add to her anxiety, so far the police hadn't been able to confirm or disprove that she was the baby taken in Carlisle. It seemed they hadn't as yet been able to find Sue Carling, or any record of where Flora gave birth to Eva.

But then, as Phil kept pointing out, it was only ten days since Eva had left hospital. And there would be an awful lot of legwork involved in checking London hospitals and doctors' practices, as nothing was on computers twenty-one years ago. As for Sue Carling, it was hardly surprising she'd left Carlisle – no woman would want to stay in a town where people thought she was a baby killer.

Eva wished now she'd never found her mother's diaries, because Dena's prediction about waking the sleeping serpent did appear to have come true. She'd not only put herself in grave danger, but she'd alienated herself from Sophie and Ben too.

She had rung Ben a few days after she got out of hospital, but Andrew had got to him first.

Ben's voice grew harsh and cold as soon as she spoke. 'If you've rung to try to convince me Dad started that fire, don't bother. I can't believe you'd try to destroy him, Sophie and me. You're deranged, Eva. What possible motive would he have to kill you?'

'He didn't like the questions I asked when I stopped by The Beeches on my way home from Scotland,' she said. 'Did he tell you about that?'

'He told me you were talking a load of rubbish, slandering Mum. And you had some lout with you who was backing you up,' he said contemptuously. 'Honestly, Eva, you'd better get a grip. You're heading the same way as Mum – totally unhinged. Sophie and I don't want anything more to do with you.'

'And you, Ben, are heading the same way as your father – cruel and spiteful,' she retorted before banging the phone down and bursting into tears.

Phil had tried to comfort her, saying that the truth would come out eventually, and she must just be patient. But shaken up by her experience in the fire, with everything she owned gone, and still coughing a great deal at night, she found it hard not to sink into self-pity.

There had been brief moments of comfort: she had received flowers and chocolates from Olive and all the staff at Oakley and Smithson; Gregor and Grace had sent a gift box of Scottish biscuits, toffee and cake with a card saying she was welcome to come and stay with them to recuperate. But Patrick was still away, and she wished so much that she could speak to him.

It was Phil's idea to come here today, on Saturday morning, because he felt seeing the damage for herself might help her to move on.

'Your car seems fine. But I'll get one of my mates to check it over and clean it up on Monday night,' Phil said.

Eva could only nod; at the moment she felt she would never laugh again, let alone drive. She wanted it sold. It was just another unwanted memory of Flora and Andrew.

'Why don't we go upstairs and see what we can salvage,' he said, putting his arm around her shoulders. 'Or would you rather leave me to do it at another time?'

'No, I'll do it now. You were right, I needed to see it,' she said, forcing a smile.

The staircase was still intact. The first two steps were badly burned, the carpet almost welded into a solid mass, but they had been told by the Fire Department that they were safe to use.

Phil opened the door to the big bedroom first. Although

it was gloomy, because the windows had been boarded over, the damage here was only from smoke. Eva touched the duvet tentatively, and her fingers came away black. She tried not to think about how she had felt the last time she was in here.

'There's no point in trying to save things like that,' Phil said. 'It would take dozens of washes to get it clean. But your shoes and things in the drawers might be OK.'

He had brought a large suitcase and some bin bags with him. He turned on a big torch, placing it on the dressing table so she could see better.

'Everything smells horrible,' Eva said. She didn't really want to try to salvage anything, but common sense said she must, and as the clothes in the drawers didn't look too bad she scooped them out and put them into a bin bag. A jewellery box that looked OK went into the suitcase, followed by shoes and clothes from the wardrobe.

As she opened the drawer of the bedside cabinet, her spirits were suddenly lifted to find the book of sketches of herself as a baby. 'I'd forgotten I'd put this in here,' she exclaimed in delight. As it had been inside a large envelope, it wasn't even sooty. 'I just wish I'd brought Mum's diaries up here too.'

'This picture doesn't look too bad either,' Phil said, taking down from the wall the Cornish beach scene painted by Flora. 'It will need cleaning by an expert, and the frame looks grotty, but we can always get a new one.'

It was the only one of Flora's paintings that Eva had hung upstairs, and that was purely because it looked right against the turquoise wall. All the others, some on the living-room walls and some still stacked in a box until she decided where to put them, were now just ash. She was thrilled that the beach one was relatively undamaged. If she'd been given a choice of saving just one of Flora's pictures, she would have picked that one.

'The dressing table will be fine with a good clean,' she said. 'I think the bedstead will be too. But where can we store them?'

'I can find room for them in the shed,' Phil said. 'But let's leave them for now. I'll get one of my mates to help me get them out.'

They moved on then to the small bedroom and found that it was just as badly smoke-damaged, but the bedside lamp and a few other items were worth saving.

'That's everything,' Eva said, after rescuing a few toiletries from the bathroom and the ash-covered towels from the airing cupboard. 'I'd like to get the table and chairs from the garden, though, and any tubs that aren't broken.'

Once Phil's van was loaded he locked the padlock on the front door and they drove away.

'The smell has come with us,' Eva said as they drove down Holland Park Avenue.

'It won't stay,' Phil assured her. 'We'll put everything out in the garden, clean up the table and chairs, and wash the clothes. The fresh air and sunshine will make everything as good as new. Or do you mean you are afraid the bad vibes from the house have come too?'

Eva *was* afraid of that, and once again she was astounded at how perceptive Phil could be. He'd been wonderful since she'd come out of hospital. He'd comforted her when she had nightmares, and sat her up and fetched her a drink when she had coughing fits. She had been very down, crying at nothing, yet he'd cooked her meals and listened patiently when she agonized about Andrew, Ben and Sophie. He hadn't snapped at her once, or showed any irritation at her state of mind. But she couldn't expect him to be so tolerant for much longer, and she knew she must pull herself together.

'No, I don't believe bad vibes can travel with possessions,' she said. 'You did the right thing taking me there, it had to be faced. I can move on now.'

Around five o'clock that afternoon, as Eva was emptying the washing machine in the kitchen, there was a ring at the doorbell.

'I'll get it,' she called out, because Phil was in the garden hosing down the table and chairs.

She opened the door, fully expecting it to be one of Phil's friends, only to find it was Patrick, with a huge bouquet of flowers.

'Patrick! What a lovely surprise,' she exclaimed.

'I'm sorry to take so long in coming to see you,' he said. 'I've been in America, and I did try to phone you a couple of times but got no reply. The third time I tried, I got the message that the number was unobtainable. Of course, when I got back and found Phil's message, I understood why. So I've rushed over as soon as I could today, to say how sorry I am and to see how you are.'

'I'm on the mend now,' she said as she kissed him. 'Do come in, Patrick. I'm sorry if Phil's message gave you a shock.'

Over coffee Eva explained everything: the things she'd discovered on the trip to Scotland, going to see Andrew on the way back, and then the fire.

'I know it was Andrew, even if it can't be proved. And why else would he do that unless he'd known all along that Flora had stolen me?'

Patrick looked absolutely stunned. His mouth opened and closed like a goldfish, and he shook his head too – as if finding it hard to believe.

'But the police haven't found Sue Carling yet to test her blood against mine. I kind of made a resolution today that I must put it all behind me. If I don't, I'll go crazy.'

Patrick put his hand on her cheek and smoothed it tenderly, a gesture that said far more than mere words. 'You've been through a terrible ordeal, and all this uncertainty must be a terrible strain on you. But I am cheered in that Phil is taking care of you. Last time I saw you he was just a friend, but it looks to me as if things have moved on there. Am I right?'

Eva smiled, got up and went to the window to beckon Phil to come in. 'Yes, they have. And he's going to be so pleased to meet you. I've told him so much about you.'

Phil came in, drying his hands on a towel. Eva introduced them.

'Patrick!' he exclaimed. 'I glanced through the window but thought you were from the insurance company – that's why I didn't come in. I am so glad to meet you at last.'

'Do I look like an insurance man?' Patrick asked, grinning and shaking Phil's hand. 'I am so pleased that Eva had you to lean on through all this. What a shocking business!'

'It certainly is.' Phil looked grave. 'I don't think I'll ever forget the night of the fire. I was afraid I was going to lose Eva. But she made it! We can replace a house and possessions, but we couldn't have replaced her.'

One of the things that Eva loved most about Phil was his ability to mix with all kinds of people. He didn't try to impress them, or fawn round them; he just had a knack of asking the right questions to get people talking, and he listened. In fact he was a far better listener than a talker, and he made people feel special because he was genuinely interested in what they said.

The way he reacted to Patrick really pleased her. He was as respectful as if the older man was his prospective father-in-law, yet he didn't try to ingratiate himself. They had a brief man-to-man conversation about the damage done to the house, and what would be needed to rebuild it. But he also asked about Patrick's trip to America, and said how much he liked his children's book illustrations.

Later, over a bottle of wine, they told Patrick about the best moments in Scotland and the Lake District. But inevitably they were drawn back to discussing all that they'd discovered about Flora, and Eva's shock when she found out about the snatched baby.

'Tell me honestly, Patrick. Can you believe that of Flora?'

He frowned. 'My first reaction was that it was impossible. But thinking now of how distraught she was when she lost our baby, and how reckless she could be, plus what you've told me about her time in Scotland, I'm no longer quite so positive. While I really hope the police will be able to prove Flora gave birth to you, Eva, this whole business of Andrew setting the fire makes that look unlikely.'

He paused for a moment. 'Eva, you told me that the mother you knew didn't seem to match up to what others told you about her. So it is possible that Andrew may well have used what she'd done to control and manipulate her. But the police seem to be dragging their feet,' he continued. 'How difficult can it be to prove if you are, or are not, this stolen baby?'

'I've rung them twice,' Phil said. 'I think one reason for the delay is because the investigation involves London, Cheltenham and Carlisle police forces.'

'Hmmm,' Patrick frowned. 'I suppose even if they had found the mother in Carlisle straight away, they could hardly rush into telling her about this until they were almost a hundred per cent sure you were her child. Imagine what a shock it would be for her!'

Eva could only nod. From what she knew of Sue Carling from the newspapers of the 1970s, she dreaded being told this woman was her real mother.

'Even if she is your mother, you don't have to meet her,' Patrick said, looking at Eva intently as if reading her mind. 'You are the innocent in all this. And although it might seem too cruel for the woman to be told you are her child but don't wish to see her, that is your prerogative.'

'It would be cruel,' Eva said. 'What sort of person would that make me?'

'An honest one.' He shrugged.

'You have to keep in mind that Sue Carling isn't blameless,' Phil said, reaching out for Eva's hand and squeezing it. 'She cared more about putting on a bet than taking care of her baby's welfare. We read that she'd had other children taken away from her too. I agree with Patrick, you don't have to meet her. You can be nice about it. Write her a letter and say it's all been very distressing but you are an adult now and, on balance, you think that there's nothing to be gained for either of you in taking it further.'

'I don't think I could do that,' Eva said, and her eyes filled up with tears.

'Let's cross that bridge when we come to it,' Phil said soothingly.

'She might not be your mother anyway,' Patrick said. 'But if she is, the police might be able to offer the services of a counsellor to mediate between the two of you. But shall we talk about something more cheerful? Tell me about your plans for the old studio.'

'I couldn't bear to live there again,' Eva said. 'It will have to be renovated, though, so that I can sell it.'

'I'll be getting some of my builder pals to sort it,' Phil said.

'One of my closest friends is an architect,' Patrick said. 'He's drawn up plans for many renovations in that area. I could put him in touch with you, if you like? It might be a good idea to scrap the garage and make an extra room downstairs. You could probably add another room in the attic too. That would get you a much better price when you sell it on.'

Eva thought about that for a moment. 'That makes sense, doesn't it, Phil?'

'It does,' he agreed. 'The garage is a waste of space – too hard to get into, and lots of people buying around there don't even bother with a car. A good architect has ideas that ordinary tradesmen wouldn't think of too. I think you should get Patrick's friend in on it, Eva.'

'Will you give him our number then?' Eva said. 'The insurance is all going through, so we'll have the money to do all the work.'

'I will,' Patrick said. 'My friend's name is Simon Curlew. But there is another thing I wanted to ask you. The man who attacked you, Eva. What's happening about that?'

'He's still on remand, waiting for a court date. But I'm tempted to drop the charges,' she said a little sheepishly. 'In the light of everything else that's happened, I haven't got the stomach for more nasty stuff.'

Patrick raised his eyebrow in surprise.

'I didn't agree when Eva first said that's how she felt,' Phil said. 'But when I think what giving evidence means – the defence lawyers picking holes in Eva's story, maybe even bringing up things about her past – I'm inclined to side with her now. The chances are he'll only get a suspended sentence. Is it worth seeing Eva get upset again just for that?'

'I suppose not,' Patrick said. 'I just don't like the idea of him getting off scot-free.'

'Nor me,' Phil said. 'If it was down to me, I'd like to go round to his house and give him a good kicking. But I'd be arrested immediately, and that won't help Eva.'

Patrick laughed. 'I'm so glad she's got you beside her. You are a man after my own heart. I want you both to know that I will help in any way I can with the house, and giving any evidence about Flora – if that becomes necessary.'

'There is one thing I wanted to ask you about my mother,' Eva said. 'Do you see anything of her in me?'

He looked at her for the longest moment. 'Not physically,' he admitted. 'Well, aside from both being small and blue-eyed. But there is something – whether that's nature or nurture, I couldn't say. You've got the directness I always liked about her, and her inquisitive nature. But you are kinder, Eva, and you also have inner strength. You will come through all this, and she would've been very proud of how you turned out.'

Just before Patrick left, Eva showed him the Cornish painting which had survived the fire.

'I remember that one so well,' he sighed. 'We used to talk all the time about going down to Cornwall to paint – the light is so good there – but sadly we never got around to it. I'm so glad it wasn't burned, it was always my favourite one of hers. It needs a good clean. Would you like me to take it and get it done for you? I have a friend who restores old paintings, and he'll make a lovely job of it. You'll be amazed at how vibrant the colours will come up.'

'That would be lovely,' Eva said. 'I intended to get it reframed too. That gilt frame is horrible and all wrong – it needs something more contemporary.'

'Well, that will be my little housewarming present to you both,' Patrick said with a wide smile. 'I'll take it with me now.'

After Patrick had left Phil looked thoughtful. 'How different things would be for you if either Patrick or Gregor had turned out to be your father.'

Eva felt a shiver run down her spine. It was bad enough thinking Sue Carling might be her real mother. And if that was the case, she certainly didn't want to think who her father might be.

Chapter Twenty-One

'Are you sure this is the place?' PC Clive Avery pulled up the collar of his waterproof coat against the heavy rain and shone his torch at the derelict old cottage tucked down in a dip on the Cumbrian fells. 'I can't imagine anyone choosing to stay out here, especially a young girl.'

The call had come in from a farmer from Caldbeck who had seen a girl out on the fells several times in the past couple of weeks. He didn't think anything of it at first, because the weather was still good – this was, after all, a favourite place for walkers – but when he spotted her early this morning, in the rain, without a waterproof coat, he became suspicious that she was sleeping rough.

Unfortunately, he hadn't got to a phone to call the police until the evening. Now it was pitch dark, raining cats and dogs, and on this part of the northern fells the roads were just dirt tracks to remote farms.

'When Sarge told me we had to check it out I knew where he meant, because I used to come out here camping with my brother,' WPC Sonia Banbury replied. 'We stayed in that cottage one night too when it was tipping down with rain just like tonight. My brother said he'd never take me camping again, because all I did was cry to go home. The cottage was tumbledown then, and that was twelve years ago.'

Avery turned his torch off for a moment and then put it on again. 'Well, there's not a glimmer of light coming from it. So if she's in there, she can't be right in the head,' Avery said. 'Come on then, we'd best go down. I've got water dripping down my neck already.'

Leaving the car headlights on to give them some light, they slithered down a narrow path flanked by rocks on both sides. It was so muddy it looked more like a stream in the light from their torches, and it was hard to get a firm foothold.

'If this is a wild goose chase, or one of Sarge's sick jokes, I'm going to make him pay to get my uniform dry-cleaned,' Sonia said.

The tiny cottage, little more than a hut, was built into the hillside and would have belonged to one of the hardy tenant fell farmers in the last century. Part of the roof had caved in, and if there had ever been glass in the two windows it was gone now.

'My brother tried to light a fire when we stayed here,' Sonia said. 'But the rain came down the chimney and put it out. We were freezing – and that was in July.'

The old door was hanging off its hinges. But when they shone their torches on the ground around it, they could see the earth was well trampled. Avery yanked the door open enough for him to squeeze in, and Sonia followed him.

'Someone's been here alright,' Avery said. In the light of his torch they could see a wooden crate with a saucepan, a tin plate and a mug on it. The side of the cottage where the roof had caved in had nothing in it other than a plastic bucket. But as he shone the torch around the other side, they saw what looked like a mound of old sacks, ancient blankets and bits of carpet. 'No one here now. Maybe the girl was just meeting a boyfriend here, or something.'

Sonia moved closer to the mound of sacks and blankets and shone her torch directly on to it. Seeing a slight movement, she jumped back thinking it was a rat.

'What is it?' Avery asked.

'Something moved there,' she said.

Avery picked up a stick lying on the floor and flicked back the sacks. There was more slight movement and what sounded like a low groan.

Sonia forgot her fear of rats and darted forward to pull the sacks further back, revealing a young girl, her eyes wide and fearful in their torchlight.

'Don't be scared, we're police officers.' Sonia realized the girl was blinded by the torchlight. 'What on earth are you doing here?' She reached out and took the girl's hand in hers; it was very cold, yet she could see beads of sweat on her forehead. 'What's your name, love? Are you hurt?'

'It's Freya,' she whimpered. 'I'm not hurt, but I don't feel very well.'

'Then we'd better take you to the hospital to get you checked out,' Sonia said. 'How long have you been living out here?'

'I think it's about a month,' she said, her voice weak and shaky. 'I lost my job and couldn't pay the rent, so I came here.'

'Freya Carling!' Sergeant Withers exclaimed when Avery and Banbury reported back to him later that night that they'd found the girl and taken her to hospital. 'Is she Sue Carling's daughter?'

Avery and Banbury looked at each other in consternation. In their concern for the girl's health her surname hadn't registered with either of them as being the same as the woman whose whereabouts were currently being sought.

'Sorry, Sarge, we didn't think of that,' Avery said. 'But she isn't fit for questioning right now anyway. They think she's got pneumonia.'

'Well, as soon as she is up for it, she'll have to be asked

where her mother is. The Met are getting very impatient – they think all the police north of Blackpool are useless turnips as it is.'

'She said she lost her job and couldn't pay her rent, so it doesn't sound like her mother is around,' Banbury said. 'She's a sweet little thing, and she strikes me as a kid who hasn't experienced much kindness in her short life.'

'Sue Carling was in trouble right from the age of sixteen. She had two kids taken away from her because of neglect before she had the baby that was allegedly taken from the street,' Withers said. 'But there was another child, born a few years after the baby disappeared. One of the bleeding-heart-brigade social workers got involved at the time, and she wanted Sue to be able to keep the new baby. I'm assuming that child is Freya. I don't recall Sue getting into any more trouble after that, so maybe she did turn over a new leaf – at least for a while.'

'Well, Freya is only seventeen, and I got the idea she's been living on her own for a good while,' Banbury said. 'Maybe her mother isn't around any longer?'

Withers sighed. 'I was always surprised that she didn't leave right after her baby was taken. Most folk believed she killed her and buried her somewhere. Sue certainly never behaved like a grieving mother. Now there's this girl in London who might be that child!'

'And we've got Freya, malnourished and sick in hospital,' Banbury said.

'You found her and clearly made some sort of connection with her, so be at the hospital first thing in the morning and see what you can find out,' Withers said to the WPC.

Two days after seeing Eva, Patrick called on his friend Nathan Cohen with Flora's Cornish painting.

They had been friends since meeting at Goldsmiths Art College in the 1960s. Back then Nathan had ambitions to be an artist too but, as happened to so many of their friends from that era, the need for a real income took over. Nathan was offered work on an art restoration project in Italy and soon found it to be his forte – along with being well paid – and he'd never looked back.

Nathan hadn't aged as well as Patrick. Spending so long bent over old canvases had made his back stooped; he wore thick glasses, and his once thick black hair was now white and sparse. Patrick often dropped into Nathan's home in Primrose Hill to see his old friend. As always when he visited, Nathan's thin lined face broke into a wide and welcoming smile.

'I've brought you work as well as a drink,' Patrick said, waving a bottle of brandy with one hand, the other firmly holding Flora's painting wrapped in brown paper.

'The brandy would've been enough,' Nathan said. 'I've got enough work to last me till they carry me out in a box.'

'Ahh, but I thought you'd like this little job,' Patrick said as he followed Nathan down the passage to his studio at the back of the house.

The house was a beautiful Edwardian semi with three floors and a basement. Rosemary, Nathan's wife, lavished all her time and energy on it now that their four children had all left home. The wood floors were polished, not a speck of dust sullied the lovely antiques, the cushions were always plumped, and the curtains were draped to perfection. But Nathan's studio was a different story, and he claimed he never allowed Rosemary to set foot in it.

Books covered two of the walls – not arranged neatly, but thrust in wherever they would fit. An ancient sofa with stuffing coming out of the arms was by the fireplace. Two large trestle tables were covered in everything from paints, cleaning materials and brushes to piles of papers and works waiting for collection, or to be started. There were several easels by the huge north-facing window, each with paintings on them, and the floor was littered with crates of chemicals, old newspapers, paint-splattered overalls and other equipment.

'I think you'll quite enjoy doing this,' Patrick said once they were sitting on the sofa. 'It's not an Old Master!'

As Patrick removed the brown paper he'd wrapped it in, Nathan grinned. 'Why, it's an Old Mistress!' he exclaimed. 'And it looks like she's been in a fire!'

Patrick chuckled. He had told Nathan about Eva getting in touch with him, and that Flora had died, but little else. 'She has,' he said.

He then proceeded to tell Nathan about the fire in Pottery Lane, though not that it was believed to have been started by her stepfather. Nathan had been to Pottery Lane many times when Patrick and Flora were together. But despite being such good friends, Patrick didn't feel it was right to tell him the whole story. 'I told her I'd ask you to clean it up and reframe it. It's the only one of her mother's paintings that survived the fire, and it's important to her.'

'I'd forgotten what a good painter Flora was,' Nathan said, looking at it carefully. 'Such a shame she gave up, she had a rare talent. It won't be difficult to clean it, though I expect under the ash and soot there's twenty-odd years of grime too. I'll just take the frame off now, so I can assess it better.'

He got up and went over to one of the tables. 'You can make yourself useful and pour me a drink,' he said. 'This is a hideous frame. I suppose her philistine of a husband, thought a gilt frame would make it look more valuable?'

Patrick laughed. He found two dirty glasses and rinsed them out in the sink.

'My God! Flora didn't intend anyone to remove this frame easily!' Nathan exclaimed. 'I'll have to break the frame to get the canvas out.'

'That's no great loss,' Patrick said.

As he was pouring the brandy, he heard the frame crack and then Nathan mutter something. 'What's up? Broken a fingernail?' Patrick joked.

'There're letters or something in the back of it,' Nathan said. Patrick turned to look and saw his friend standing there with a sheaf of foolscap paper in his hands. 'I think Flora must've hidden this. It is her writing, isn't it?'

Patrick moved quickly to take a look. 'Yes, it is.'

He read the first page, which was dated April 1986, the year Eva was sixteen. By the time he'd got to the bottom he felt faint, as if his blood had suddenly been drained away.

'What's up, Pat?' Nathan asked. 'You've gone as white as a sheet.'

Patrick couldn't answer for a moment. Just from the first page he knew that this was Flora's explanation – or perhaps it should be called a confession. He would have to read on to know exactly what it was.

But for the fire it might have stayed hidden behind the painting for a hundred years or more. He had a feeling that was what she had planned – it had been the writing of it which had been important to her. By putting it down on paper she had hoped to find some kind of absolution, rather like making a confession to a priest.

'What is it, Pat? You're scaring me now,' Nathan said.

'I'm scared too,' Patrick said. 'I need to take it home and read it.'

Chapter Twenty-Two

Sonia Banbury looked down at Freya Carling in the hospital bed and felt a rush of pity for the girl. She looked much younger than seventeen: her face was almost as white as the pillowcase, with dark circles beneath her eyes. She had light-brown hair that was straggly and dirty, and she was terribly thin. 'How are you feeling this morning, Freya?' she asked.

'Much better, thank you,' the girl said, but her voice was little more than a husky whisper, and her blue eyes were clouded with anxiety.

'Sister tells me you have a severe chest infection, but the antibiotics they are giving you will soon sort that out. I believe they are going to keep you in until the infection is clear. Now, can you tell me where Sue, your mother, is at the moment?'

'She went to Spain,' Freya whispered. 'But that was nearly two years ago, and I haven't heard from her since.'

'Do you have an address for her?'

'No.' Freya's eyes filled with tears. 'She said she'd send me a ticket to come and join her when she was settled, but I haven't even had one letter or a postcard.'

'Who did she go with?'

'Some man,' Freya sighed. 'She never told me his name. But that's how Mum was – always a man in her life, and they always let her down.'

Sonia sensed this was true by the utter resignation in the girl's voice. She also sensed that Freya had been let down, over and over again, and maybe her mother had taught her she should expect it to always be that way.

'So tell me why you were living in that place on the fells?'

'I had nowhere else to go,' she said.

'You said last night that you lost your job. Where was that, and where did you live then?'

'I worked in the bakery – I'd been there since I left school. Before Mum left for Spain she arranged for me to lodge with Ena Willoughby. She let me live there cheap, as long as I helped her out with the cleaning and stuff.'

Sonia nodded. She knew Ena Willoughby by reputation as a rough loud-mouthed woman who offered long-haul truckers bed and breakfast. Mostly they slept four to a room, and the sheets were rarely changed. It was generally thought that Ena had a price list for her personal services too.

'So what happened at the bakery? Was it Harris's, the one that closed down?'

'Yes, they were going bust,' she said, and her eyes filled with tears again. 'I was happy there, but when I had no wages to pay the rent Ena said I could make more giving the drivers . . .' She paused, clearly unable to say the word 'sex'. 'I didn't want to do that, it's dirty. But she said if I had no money, I couldn't stay there. And then she told me to sod off.'

'You should have gone to talk to a social worker,' Sonia said. 'They would've helped you get benefits and somewhere else to live.'

'Mum said social workers screw up your life,' she said.

Sonia tried hard not to roll her eyes. She'd been a social worker before she joined the police and knew what people like Sue Carling meant by such remarks. Social workers tried to show their clients how to curb their destructive behaviour. But mostly it was a fruitless exercise, as the clients usually ignored all advice. Then when their lives fell apart, or their children went off the rails, they had the cheek to scream from the rooftops that it wasn't their fault. The public were quick to use social workers as the whipping boys for the whole of society; no one these days seemed to believe that people should take responsibility for their own lives.

'That isn't true, Freya; they are there to help people in difficulties. Didn't you have any friends or relatives who could help you?'

The girl shook her head. 'I thought I would be able to find another job straight away, but all I could get was some casual work washing up in a restaurant. You need a deposit to rent a room, and I didn't have that.'

Sonia knew that most people who ended up living on the streets were caught in the same situation. Sadly, the longer they lived rough, the dirtier and less employable they became. But with Freya being so young, if she'd asked for help she would have got it.

'I'll get someone to come and talk to you while you are still in here,' she said. 'No more living rough – you could have died of pneumonia out there. Now, is there anyone who might know where your mother is? A friend maybe?'

Freya gave her a long sad look. 'Mum didn't have no friends, only people she used.'

Sonia was only a child at the time Sue's baby disappeared, and she'd known nothing about the case – not until it was reported that a young woman in London believed she might be that child. Reading through Sue Carling's case notes she couldn't feel any real sympathy for the woman, as she sounded like the mother from hell.

Freya's poignant statement about her mother having no friends said so much about her own mindset. She was a kid who had been brought up in the shadow of her sister's disappearance; she must have heard the gossip that Sue had killed and buried the baby, and always felt like an outcast.

Sonia thought that even the most dysfunctional woman

who had lost a baby for whatever reason would treasure the next one and always keep it in her sight. Yet she'd callously gone to Spain leaving a vulnerable fifteen-year-old in the care of a harridan who was well known for her lack of morality, without any thought as to what might become of her.

And yet, despite the terrible start in life she'd had, Freya seemed a nice kid.

Patrick didn't ring Eva immediately and tell her what had been found in the back of the painting. He needed to read it several times more and think about her reaction first.

The woman who had written the diaries that Eva had found was the Flora he knew. The cryptic style was typical of her – sarcastic, vague, often lacking in feeling. And yet she could be very funny too. It had been no surprise to him that she didn't date any of her entries, because she had never been organized about anything. He hadn't even asked Eva if Flora had written about losing their baby, because he knew she wouldn't have. She had always kept anything that was important to her locked away, inside her.

He could see her now, lying in that bed in Hammersmith Hospital, her red curly hair too vivid against the white pillow and hospital gown. Her mouth was a scarlet gash, because she'd put lipstick on in a desperate attempt to hide her sorrow. But her eyes gave her away – not just the red rims, but the bleakness in her gaze. They had always danced with mischief or sparked with passion. That day they were dead.

Patrick had called an ambulance late on the previous afternoon when the pains started. By the time they reached the hospital she was losing blood. The baby was dead already, but she had to go through with a normal delivery. He wasn't

355

allowed to go into the delivery room with her; he had to wait outside, hearing her screams. But unlike the other expectant fathers pacing the corridor, he knew there would be no joy for him and Flora when the screaming stopped.

It was just on midnight when she haemorrhaged, but he didn't know that until much later. He had to draw his own conclusions as to why the doors of the delivery room suddenly burst open and she was rushed past him on a trolley. All he was told by a young nurse was that Flora had been taken to theatre with 'complications'.

He had never been a man for praying before, but he did that night. He found the hospital chapel and got on his knees, begging God to save her. He was still in there at four in the morning when that same young nurse came to tell him what had happened. She said that he might as well go home, as Flora was now on a ward recovering. He wouldn't be allowed to see her till the visiting hour in the afternoon.

Hospitals weren't like that any more, thank goodness. Maybe if they'd allowed him to be involved, to comfort Flora and to be with her as she came round from the emergency operation, she wouldn't have slipped into that dark hole where no one could reach her.

He remembered going into the little nursery when he got home, looking at the frieze of bears and the second-hand cot they'd bought and painted white. The blankets and sheets were still in their cellophane wrappers, waiting to be opened and the cot made up. He picked up a little white coat he'd bought for the baby, buried his face in it and cried.

When he got to the ward the following afternoon her lipstick told him straight away how her mind was working. They could have the most unholy row and then, when she was tired of it, she'd put new lipstick on and suddenly start talking about something else. He'd admired that attitude once. But that was before he realized that she didn't deal with issues, she just brushed them under the carpet, where they festered, only to rear up again at another time.

'Don't look like that,' she said reproachfully – presumably because his face showed the deep sorrow he felt for her and the baby they'd lost. 'It was an awful thing to happen, but maybe there was a good reason for it.'

Now, in the 1990s, women were encouraged to have counselling; they got to see their baby, and what went wrong was explained to them. But back then there was nothing – no counselling, no advice on how to deal with the grief, and precious little sympathy. Yet although he knew very little about the effects losing a baby could have on a woman, his instinct told him that they should talk about it and cry about it together. But Flora wouldn't do that.

Months later when she turned on him and made his life a misery, he had put it down to her having no heart. But now, as he read this long statement from her yet again, he realized that she'd just grown a thick shell around her heart, thinking she was protecting herself.

Her statement wasn't from the Flora he knew. There were no sharp comments or sarcasm, nothing cryptic that you had to struggle to understand, just an account of what she'd done, and the consequences. But as horrified as he was, he felt he'd found the nicer, kinder sister of the Flora he fell in love with all those years ago.

Although she had written it retrospectively in 1986, and underlined this heavily to make it quite clear, she had begun the confession with the date: 1 April 1970.

Today, sixteen years ago, I took a three-day-old baby girl from outside a betting shop in Carlisle. It was raining and I just wheeled her little pram to two streets away where my car was parked, put the carrycot on the back seat, folded up the wheels and put them in the boot, then drove to London with her.

Patrick brushed away a stray tear and carried on.

I intend to hide this somewhere where it may never be found, and I have asked myself what is the point in writing it if no one ever reads it. But the point to me is that I have a need to put it on paper, and that perhaps when I'm done I might even be brave enough to go to the police and confess.

I went to Scotland after I lost the baby Patrick and I were expecting. I knew I was growing crazier by the minute, being vile to Patrick and anyone else who came near me. I thought if I was entirely alone, in a peaceful place, I could heal myself.

It almost worked. There were good people there – especially Gregor, who made me laugh again and paint. I'm not sure if I really loved him, or just wanted to, but it did feel like love at the time. When I found I was pregnant again, I really thought I could be happy with him for ever. But at only a few weeks, in early December, I miscarried and the craziness came back even worse than before.

I had never told Gregor I was pregnant – perhaps because a sixth sense told me I might lose it. So I did what I always seemed to do in those days: I was nasty to everyone who cared about me, withdrew into myself, and then lay low.

Six months before that, Scotland had looked so beautiful. But after the miscarriage I found it ugly, and all through January and February I holed up in that little house by the river, in bed most of the time, thinking constantly about committing suicide. By late March I knew I had to get away, and I slunk out like a thief in the night without saying good-bye to anyone.

I didn't go straight back to London, even though Patrick had left the studio. I stayed in Edinburgh for a few days, then went down to the Borders – a night here, a night there – and ended up in a village not far from Carlisle. I went into the town one day and I was having a cup of tea in a cafe when a very pregnant woman of about twenty-five came in. She ordered tea and a cake, then found she had no money. The owner of the cafe wouldn't let her have it without any money. I felt sorry for her, because it was very cold, so I paid for her.

She was very rough, her accent was so thick I could barely understand her, but I gathered she didn't want the baby she was carrying. She expected the social workers would take it from her anyway. I left then – I found what she said upsetting, and it played on my mind all that night.

A few days later, on the 1st of April, I decided I would go back to London. My plan was to go into Carlisle first to buy some new canvases, then drive on down as far as the Lakes, stay the night there, then continue on to London the next day. I was told there was an art shop in Botchergate but I couldn't see it, so I parked my car in a side street, and walked up the road a bit to see if I'd missed it. I went into a news-agent's to buy a newspaper, and while I was in there I saw that woman again, passing by the window. She'd had her baby and she was pushing it in a small green pram. I remember thinking that the baby couldn't be more than a few days old, and I couldn't believe she'd brought it out on such a cold, rainy day.

Perhaps it was fate that as I was walking back to my car, I saw the green pram left outside a betting shop. I looked in the pram and saw the tiny baby; she was wearing a pink bonnet and crying. Instinct made me rock the pram. I couldn't see through the shop window, because it was covered over with pictures of racing horses, but I knew the mother was inside, I could hear her voice shouting as she watched a race on the television. I knew I ought to open the betting-shop door and tell the woman off for leaving her baby outside, but suddenly I took the brake off the pram and wheeled it away.

I wasn't thinking clearly at all – although, in my defence, once I'd put the carrycot part of the pram in my car, and folded the wheels and put them in the boot, I did drive back on to Botchergate where I'd found her. I told myself that if the mother was outside, frantic because her baby had been taken, I'd give her back straight away and tell her what a lousy mother she was.

I stayed parked in that street for some time. I saw a couple of men go in to put a bet on, and another three come out. But the mother didn't emerge. So I drove off.

I wish I could claim I was shocked by what I'd done, but all I could think of was that I'd lost two babies, and there was that woman stuck in a smoky betting shop while her brand-new baby was out in the cold and the rain. I felt the baby was meant for me.

I was so calm. On the other side of Carlisle I stopped in a side street and took the waterproof storm apron off the carrycot. There was a bottle of milk made up, wrapped in a nappy to keep it warm, tucked down the bottom, and there was a new tin of baby milk too, which she'd clearly bought earlier. In my mind that was further proof the baby was meant for me.

I got the baby out and fed her there in the car. She took the bottle like a dream, and all I could think while I cuddled her was that I'd rescued her, and no further harm could come to her.

Putting her back in the carrycot, I drove away and didn't stop till I got to Preston. Eva – I decided to call her that – was fast asleep, the rain was pelting down outside, but we were snug and warm.

In Preston I found a branch of Boots that I could park outside, and I bought another couple of bottles and teats, plus other baby essentials, including a packet of disposable nappies, and then drove off again.

Eva slept the whole way to London – I suppose that was the motion of the car – and it was just as well, as I couldn't make up a bottle with powdered milk anywhere without drawing attention to myself. I guessed that Patrick would have left the little nursery at the studio just as it was when I lost our baby; he wasn't the kind to pack it all up without my permission. And knowing it would all be there still – clothes, bedding, even a sterilizer – I felt I was taking our baby home.

Looking back at that day, sixteen years on, I find it astounding that I didn't panic or worry about anything. Eva felt like mine from the moment I held her in my arms. Instinct took over, and I seemed to remember all the advice I'd read in baby books before.

Patrick put down the statement, because his eyes were swimming with tears. He could remember Flora reading baby books all the time when she was pregnant; he used to tease her about it, because he hadn't expected her to become so maternal.

She was right, he hadn't touched the nursery; he couldn't bring himself to. He was also glad he'd cleaned the studio thoroughly before he left it back then. He'd been tempted to trash it just to spite her, but in the end he couldn't.

Reading her words, he got a picture in his mind of her arriving back with the baby. The studio must have been icy after being empty so long, and the enormity of what she'd done must have hit her hard as the practicalities of sterilizing bottles, four-hourly feeds and wet nappies kicked in.

How did she summon up the nerve to go and register the baby's birth? Wasn't she afraid they had some way of checking the baby was really hers? And to put down the date of birth as nearly a month later than it was! That was such a risk, but he supposed she took it because she thought the police might look at all registered births around the time the baby was taken.

But Flora said nothing about any of that. Or perhaps by the time she'd begun her statement all that had faded from her memory, and she could only recall the joy of having a baby to love?

Tomorrow he would have to give this statement to Eva, and he wondered how she would react. At least it was clear that Flora had loved her deeply and that she'd been a far better mother than her real one would ever have been.

Chapter Twenty-Three

Patrick rang Eva on Monday afternoon. Using the excuse that he had a brochure of picture frames for them to choose from, he asked if he could pop round that evening if Phil was home. He didn't want to leave Eva on her own after she'd read Flora's statement.

Phil opened the door when he arrived at seven thirty. 'You didn't have to rush and get the picture done,' he said. 'But Eva will be pleased to have it hanging up again.'

'I'll be there in a sec,' Eva called out from the kitchen. 'Just finished the washing-up, and I'm making a cup of tea.'

They went into the lounge. *Coronation Street* was just starting as Eva came back into the room with the tea on a tray. Patrick thought she looked very pretty in a fluffy turquoise sweater and jeans. Phil had said she hadn't been sleeping well after the fire, and he thought she was depressed, but she looked rested now. He hoped his news wasn't going to set her back.

'Is everything alright, Patrick?' she asked as she put the tray down on the coffee table. 'You look very tense.'

Patrick was tense, but he hadn't thought either of them would notice. If nothing else, her question gave him the perfect opener – something he'd been worried about all day.

'Selecting the right frame is a tough job, but someone's got to do it,' Phil joked.

'Showing you this is an even tougher job,' Patrick said, opening the battered old music case he used as a briefcase and pulling out the folder containing the statement. 'It isn't a catalogue of frames; it's something my friend the art restorer found behind Flora's painting.'

'What is it?' Eva asked. 'You're scaring me, Patrick, with that grim face.'

'It's the answer to all your questions about Flora,' Phil said. 'Written by her. I think you need to sit down and read it.'

Eva frowned in puzzlement. Phil turned off the television, and the pair of them sat down side by side on the sofa. Patrick handed the folder to Eva, then sat back in his chair to watch her reaction as she read it.

Patrick had found it hard to keep a lid on his emotions while he read it – even on the second and third reading it still had the same impact. He didn't know how Eva would take it. She'd already been dealt enough bad cards this year; most girls of her age would have crumbled under the strain.

Yet as disturbing as this statement was, the truth – however unpalatable – was always better than supposition and half-baked theories. He really hoped she would see it that way.

Eva had tears running down her cheeks as she finished the first page and handed it to Phil. But she made no comment and carried on with the second page without once looking up. At the end of the third page, which Patrick knew was the part where she arrived as a baby in Pottery Lane, she handed it to Phil and put the other pages down beside her.

'I know what she did was wrong,' she said to Patrick, the break in her voice even more telling than the tears on her cheeks. 'But why does it sound so right?'

Patrick had asked himself that same question too when he read it. 'Because she gave you the childhood you deserved,' he said. 'Somehow I doubt you'd have fared so well with your birth mother.'

Phil had finished it too. He took Eva's hand in his and for a moment said nothing, clearly overwhelmed by what he'd read. 'Losing the second baby must have tipped her right over the edge,' he said eventually. 'Yet even if she was mentally ill at the time she did it, she sounded rational and calm as she wrote the story.'

'Well, that comes of writing it down sixteen years later. I doubt she'd have presented it so clearly at the time,' Patrick said. 'She doesn't say very much about the first few weeks with you, Eva. I'd say that was because she was overtired, like all new mothers are. But that book of sketches she did of you is proof enough that she held everything together and that the pair of you bonded well.'

'I can't imagine how anyone with no experience of newborn babies could cope alone.' Eva's voice cracked with emotion. 'Especially when the baby isn't your own. I'd be terrified if it wouldn't stop crying.'

'As she makes no comment on that, I think we can surmise that the joy of taking care of you wiped out her depression. The bit that puzzled me most was how she had the nerve to go and register your birth as her child. I think most people would be far too afraid of getting caught out to do that. I didn't know anything about the process of registering, so I made some inquiries. It seems she must've had some prior knowledge, because the only documentation needed is a marriage certificate in order to put the husband's name on the birth certificate. Without that, the section for the father's name is left blank – unless he accompanies the mother. I suppose she rang them and asked what was necessary in advance. It seems a doctor or midwife's signature isn't needed.'

'That's amazing,' Phil said. He shook his head as if in disbelief. 'But then, I suppose the people who set up birth registration never imagined anyone would lay claim to a baby that wasn't theirs?'

'I'm sure that loophole must have been closed by now, since computers have started to take over,' Patrick said. 'Anyway, to move on, the next part is about meeting Andrew – and Flora was clearly back to her normal self by then.'

Eva looked at Phil, then back to Patrick. 'Can you just tell us about it, and we'll read it all later?' she suggested. 'I'm finding it hard to deal with hearing her voice in this. It will be easier if you tell us.'

Patrick knew exactly what Eva meant; there was a rawness in Flora's writing that revealed how painful she found it to open up. Yet she had clearly been determined that she must tell the truth and justify her behaviour to herself – that she hadn't been mad, or bad, but motivated only by wanting to give the baby love and care.

'OK then. But when you read this part yourself, do take it slowly. Try to put yourself in her shoes as a single mum who has been forced to cut herself off from all her old friends because of her guilty secret. She's lonely, and desperately needs someone to talk to. Most new mums I've known can't wait to show off their baby, but she couldn't do that straight away as she had to build a back story for herself that was entirely plausible. You must also remember that back in the early 1970s there was a stigma attached to being an unmarried mother.

'Anyway, to get back to her story, she met Andrew at the end of June. It was a hot day and she'd stopped for a drink at The Prince of Wales and sat outside with the pram –'

'Andrew told us that too,' Eva interrupted.

'I expect he put his own spin on it, and I'm going to tell you how I think it was,' Patrick said. 'I suspect from what she says that she went there on purpose, hoping to get into conversation with other adults and make new friends. She must have been delighted when Andrew began chatting to her. He was, after all, young, handsome and single. Flora pointed out that he was rather serious, but she seemed to see that as a plus. She also liked the fact that he was attentive and seemed very caring – he even asked if he could hold you.

'From then on, it sounds like he really wooed her. He turned up with flowers, toys for you, they had picnics in Kensington Gardens, and cosy dinners together. He moved in with her after just a couple of weeks, and soon asked her to marry him.'

'Did she say she loved him?' Eva asked.

'Not exactly. She speaks of respecting him, that he was the marrying kind. That they were good friends, and that he loved you, but not that he made her weak at the knees! She didn't agree to marry him straight away. That came about when you were almost two, Eva. I got the idea that she thought he was almost too good to be true. Perhaps she wondered why someone as eligible as he was would want a woman with a baby?'

'Lots of men are attracted to the idea of a ready-made family,' Phil said. 'I think it's because they don't have to do the home-building stuff or take any responsibility.'

'I'd never thought of it that way,' Patrick replied. 'There could've been an element of that in Andrew's mind, but I'm afraid I'm more of a cynic. You see, house prices began to rise around the middle of 1970. Later on, in 1972, it got quite crazy. I remember reading about it in Canada. Andrew worked as an estate agent then, and he would've been aware of this happening before the man in the street started to notice.'

'So you think he was a gold-digger?' Phil said.

Patrick grimaced. 'It's difficult for me to imagine any

man not loving Flora just for herself, but I've got a feeling from the moment Andrew got his feet under her table, he saw the main chance. It also transpires – something I never knew – that she had money stashed away. I always thought she used all her inheritance to buy the studio, because that was what she implied. But she says in the statement that she had over twenty thousand pounds in the bank. And that of course explains why she didn't get a job while in Scotland, and how she was supporting herself. With a house and that nest egg she was rich by the standards of the early 1970s.'

Phil let out a low whistle. 'Quite a catch then!'

'So who did Andrew think was my father? Presumably he didn't find out the truth for a while?'

'Like he told you, a brief affair in Scotland,' Patrick said. 'It was a smart explanation on her part, as most men feel less threatened by a casual fling than a serious relationship which might not be quite over. Everything appeared to be fine between them. Flora only admits to a certain wariness just after they got married, when he suggested they should both make a will. She said Andrew sulked for days when she told him she'd already made one, and that she wanted the studio to go to Eva. She said he made a big thing out of her not trusting him. He said that if anything happened to her, then he would take care of you.'

'And we know how well he did that,' Phil said drily.

'Moving on, Flora was all for it when Andrew suggested they move out of London. The studio was getting too small for a growing child. Andrew landed a job in Cheltenham and then they found The Beeches.'

Patrick took the statement and riffled through it. 'I'll read you this section, as it is the bit that explains so much,' he said.

I loved The Beeches as soon as I saw it, even though it was in a terrible state, and I could also see its huge potential. Andrew kept on grousing that we couldn't manage a big mortgage. But I pointed out we'd only need one for a short time, because we could sell the land at the back of the house. That money would not only pay for all the work needed to restore the house, but we could pay off the mortgage too.

Andrew came round to the idea then, but he wanted me to sell the studio for the huge deposit we needed. I wasn't going to do that – I wanted to let it out, and keep it for Eva – so I had to tell him about my inheritance money and use that. But in order to safeguard it, I insisted The Beeches was put in both our names.

Patrick looked from Eva to Phil. 'It wasn't common practice to have property in joint names back then. Men were considered to be the breadwinners and therefore they mostly took the sole responsibility of a mortgage.'

'So was she suspicious of him?' Phil said.

'It doesn't sound like it. But Flora always had a keen sense of the value of money,' Patrick said. 'We were all so poor at college, and she hadn't had much as a child either. All her parents had of value was their home, and she would never have risked or squandered what they left her. I dare say that was why she never told me that she had twenty thousand tucked away along with owning the studio. Perhaps she didn't trust me not to suggest doing something extravagant with it? But what I do get a sense of in this part of her statement is that she is very aware that Andrew wasn't putting anything into the pot. I think that was what made her cautious.'

'He used to boast to our neighbours about his "foresight" in buying The Beeches,' Eva said indignantly. 'Mum never said anything, so it never occurred to me she was the driving force behind it. Why didn't she ever insist on getting the credit for it?'

'The reason for that becomes clear later on,' Patrick said.

'Then carry on reading,' Phil said. 'I'm dying to know more.'

Patrick cleared his throat and continued to read the next section.

We stayed in a lovely hotel in the Cotswolds for a couple of nights while we completed on the house and took possession. Eva stayed with Andrew's parents. I can only put my stupidity in telling Andrew about Eva that night down to the excitement of the move, having a break from Eva, and getting rather drunk. It seemed to me, as we were married and starting a new life in a new town, there shouldn't be any secrets. So out I came with it, while sitting in the hotel garden after dinner with another bottle of wine.

He was shocked of course, but he said he was glad I'd told him. And I felt so relieved, because keeping the secret had been such a huge burden. He said we should change Eva's name to his, and we should tell people we'd got married in 1969 so that everyone would think she was his child. That made me ridiculously happy. I felt that all the sadness and anxiety about the past was over.

The first couple of years at The Beeches were wonderful, so exciting and fulfilling. I found a property developer interested in buying the land at the back, saw to all the legal stuff, and I did all the negotiations with the Council for planning permission for The Beeches. And I found the right tradesmen to do the work. Although I say it myself, I was the brains and the creative force behind it all. Andrew was too staid, unimaginative and often too churlish with people who needed to be won over.

I almost forgot about my art, because along with Eva to look after, there was so much to do in restoring the house and working on the garden. There were banisters and doors to strip, workmen to oversee, materials to be sourced. It was me who repaired and made new pieces of cornice where they were missing, to match the original design. I was really happy too. I felt I'd found my niche in life, creating something beautiful, a 'forever' home.

Andrew was as proud as punch when Ben was born. He was considerate towards me, he was happy at his job and, as we'd paid off the mortgage when we sold the land at the back, there were no money worries either. I thought then that I'd picked the right man to marry and settle down with. We seemed the perfect team.

Everything was fine between us right up till when Sophie was about a year old. We'd more or less finished the work on the house, and the garden was beginning to look beautiful too.

I certainly hadn't expected to get pregnant again so quickly after Ben, and when Sophie was born I was very tired and run-down. I put the changes in Andrew down to that; he often got annoyed when he came home to find toys all over the floor and the kitchen a mess. But then he began picking on Eva about nothing. She was only little and the things he complained about – such as her making crayon marks on the kitchen table, or spilling drinks – were so petty. We had a row about it one night, and he hit me for the first time. His message was clear: 'Don't you dare criticise me, or else.'

What he was implying was of course that he'd tell the police about Eva. I couldn't really believe he meant it, but it frightened me. So I tried harder to keep things in order, just to appease him. But I am what I am: I was never a great housekeeper, and I needed creativity to be happy. He came in once and found me painting in the kitchen, and he went mad. He threw all my paints on the floor, crumpled up the canvas, and said he didn't want a crazy artist for a wife.

That was the start of what was to become my life of walking on eggshells. He would tell me what I was to cook for dinner; he decided which of the neighbours were to become our friends. On occasions if I'd met and liked another mother at the playgroup or school, he would check out who she was, where she lived, and if he didn't approve he said I wasn't to see her. If I ever tried to challenge him about this, he hit me.

Patrick looked at Eva's stricken face. 'You didn't know he hit her?'

She shook her head. 'No, I had no idea. He was like she said – telling her what to cook, who to invite round and stuff – but I thought she did that because that was the way she was.'

'She says he was always careful where he hit her – never her face, or people would know. If she said one wrong thing when they had visitors, she knew he'd punish her as soon as they'd gone,' Patrick explained.

'I saw bruises on her legs and arms sometimes,' Flora exclaimed. 'I believed her when she said she'd banged into something. But after I was about nine or ten Andrew never let me go into their bedroom when she was dressing. He was nasty to me once and said adults needed privacy. But that was so I wouldn't see anything, wasn't it?'

Patrick nodded grimly. 'I would say so. Let me read out another bit where she tries to rationalize it.'

Looking back now, I don't know why I didn't recognize he was a control freak as soon as I met him. Even at the start he liked everything to be his own way. But of course I had the power then. It was my studio and he was living with me. I just thought he was a bit bossy and critical – nothing more – and I could tease him out of it. While we were doing all the work at The Beeches it was impossible to be tidy and organized, and he seemed OK about it then.

But I had no more power once the house was finished, I was just housewife and mother. I've read articles by women who have been beaten and they always say it was a gradual thing. The first time he swears he'll never do it again, and he doesn't for months. But then it happens again and before long every time she puts a foot wrong he lays into her.

Before Andrew began hitting me I thought women who allowed their men to beat them were spineless wretches. I even imagined they got some sort of buzz out of the violence. I couldn't understand why they stayed with the man.

Whoever reads this will probably think I was spineless too. After all, I still owned the studio and I could've gone there. Or they might think my lifestyle, the big house, the garden and a husband who provided for me – along with my inheritance, which I'd sunk into the house – meant more than my own self-respect.

But it wasn't being spineless, or money, that made me stay. I would gladly have lived in one room with the children, just to have peace and happiness. The real stumbling block was that I knew, if I left, he'd make good his threat and tell the police about Eva. That would mean prison for me, and losing all three children. Goodness knows what would've happened to Eva – she'd have probably ended up in care. So I didn't get my paints out again, and I tried very hard to have everything perfect when he came home.

You have to experience living with someone who is

holding a gun to your head to know what it does to you. Bit by bit it erodes your confidence, your personality. The more you try to appease, the more critical they become. You lose your sense of fun, you are always trying to second-guess your abuser. You can't ever let your guard down. I never dared tell people that just about everything in the house was my idea, or that it was me who stripped doors and banisters, repaired damaged cornices and sourced the antiques. I became a nothing. And I missed painting, talking to people who liked to laugh and were fun. I haven't been allowed to make any decisions for years, and I'm not even allowed to have an opinion about anything.

The only thing he hasn't tried to alter is my dress sense. He is sarcastic about me wearing vintage dresses, but he doesn't try to stop me. I know why that is. My clothes are a smokescreen. No one who sees a woman dressed in old velvet and beaded dresses would imagine her husband controlled her. I am quite sure that he confides in people that I am 'arty', highly strung and reclusive, feeding people the idea I'm halfway round the twist. He gains sympathy, and people don't try to get to know me.

It is obvious to me now from spiteful things he says that he's used me right from the start. Moving in with me saved him money; then he wanted to get married, because he saw that as a way to get the studio. Maybe he even knew I had the other money all along. He used my vision and creativity to turn an old wreck of a house into something beautiful and very valuable. Perhaps he hopes, if he keeps pushing and pushing at me, that I will go mad, or kill myself – that way he'll get everything and be free of me.

'Oh God, how could that have gone on and I never knew?' Eva sobbed.

'She made sure you didn't know, because she loved you all so much,' Patrick said. 'I don't think she was even afraid of prison, really – only of losing you three. She says in one part that she sometimes felt it was payback time, for being horrible to both me and Gregor. So she tried so much harder to be a good wife to Andrew.'

'Does she say if she loved him?' Phil asked. 'I mean, in the beginning?'

'She said: "It is true I never felt the same way about him as I did about those two other good men. But I thought that love like that comes to most people only once, and I'd been lucky enough to have it twice already. But I gave Andrew everything I was capable of giving." You can make what you like of that, Phil!'

'I know she hints that Andrew would like her to kill herself, but is there any suggestion that's what she planned to do?' Eva asked in a small voice. 'I was such a pain to her at the time she wrote that. I can't bear to think I added to her misery.'

'No, she doesn't, Eva. She said how glad she was that she had the foresight to put the studio in trust for you, that at least Andrew would never get that. So I think you can put it out of your mind that you were in any way responsible. She was wild herself at the same age, and she understood the need teenagers have to rebel. I think at the time she wrote this she must've had the idea of waiting until all three of you came of age, and then making the break – whatever that cost her.'

There was more – so many unpleasant incidents with Andrew that Flora had described in detail – but Patrick felt Eva had heard more than enough for one night.

Phil had his arm around Eva, because she was distressed and crying. It was an awful lot to take in as it stood, but no doubt Eva would wonder what worse things occurred after Flora had hidden this away. Only a fool would think the abuse and bullying had stopped at that point.

'You must read the rest for yourselves,' Patrick said. 'She could see no way out when she wrote it, you can hear her despair in every word. It was her love for the three of you alone that kept her going, and I feel ashamed that I ever called her selfish.'

'Do we take this to the police?' Phil asked.

'I think you must,' Patrick said, 'because it makes it absolutely clear what kind of man Andrew was. When you read on from the part where I stopped, she goes on to say about him belittling Eva – another new way of hurting Flora. And he was hitting her more and more often for the most trivial of things. She also says she knew he had other women.'

He paused, sighing deeply at the profound sadness of it all. 'I spent all day yesterday reading this over and over again. A part of me didn't want to hand it over to you, because I knew how distressing you would find it. But of course I had to. You needed to know.'

'To think if Eva hadn't asked Andrew for the painting, the statement might never have come to light,' Phil said. 'Why did she put it in there? Why not send it to someone to be opened after her death?'

Patrick rubbed his face with his hands; he felt exhausted now, but he felt a duty to make both these two young people understand Flora's motive. 'I think it was symbolic. He'd taken so much from her – her personality, her self-esteem and also her art. Writing it all down was probably cathartic; she might never have intended it to be found. Yet when I looked at that ghastly frame, which would never in a million years have been her choice, perhaps she smiled to herself knowing that if Andrew ever sold or gave away the painting to spite her, the frame would be ripped off immediately.

'I doubt he had any idea at all what a great artist she was – or he wouldn't have let you have it, Eva. If it went to auction, an art collector would have paid a lot of money for it. There is also a bitter irony in that if he'd let Flora paint, she would've made far more money than he ever did as a salesman.'

He got up then. 'But I must go now and leave you two to discuss this. If you need me, just ring.' He held out his arms, and Eva rushed to them. Patrick held her tightly and stroked her hair with tenderness. 'Look after yourself, little one. I wish I could've brought you something more uplifting that would make you happy. But they do say "the truth shall set you free".'

The next morning Phil got up when his alarm went off and went into the kitchen to make some tea, leaving Eva still sleeping.

After Patrick's visit they had both read the whole statement right through. Eva had been very upset, and she'd woken during the night with a bad dream. Phil didn't really want to leave her today, but he had to. He made two cups and took them back into the bedroom. As he put Eva's down on the bedside table she opened her eyes.

'Morning, handsome,' she said with a forced smile.

Her pluckiness touched him. He knew that this wasn't something she could just brush away and forget about. He bent over to kiss her. 'I don't want to leave you alone today, but I've got to,' he said.

'Stop worrying about me, I'm not that fragile,' she said and sat up to drink her tea.

Phil sat down beside her. 'I'll be home by six,' he said. 'Shall we go out for something to eat?'

'No, I'll cook something nice,' she said. 'I really am alright this morning. I woke up during the night and thought about it all. I feel more clear-headed about it now. I've accepted that I'm a Carling, not a Foyle. I shudder to think who my real father is – I suppose some casual pick-up in the pub or the bookies, as there was no mention of him in the newspaper cuttings we got.'

'He can't have been a bad man to have produced you,' Phil said. 'But all your influences were from Flora. And the way I see it, she made a pretty good job of bringing you up. Until I read that statement I'd imagined her as some self-centred cow, but I was wrong.'

'Why didn't I realize that Andrew was hurting her?' she asked plaintively. 'The policewoman who took my statement after Mum died asked me if they'd had a happy marriage, and I said they had. Did I say that because I believed it? Or just because I chose to forget how sad Mum seemed sometimes? Since then, I've remembered lots of things that were pointers to them being far from OK. At my age surely I should've suspected Andrew was violent. I was a bit afraid of him myself.'

'Does any kid really question whether their parents are happy together?' Phil shrugged. 'I mean, we all grow up thinking whatever our parents do or say is normal. We have no other yardstick to measure it by. Your dad might be grumpy, or your mum a battleaxe, but you just accept that. I had a mate at school whose parents had blazing rows, his mum would throw his dad's clothes out of the window. But they are still together. Flora said how careful she was to keep it from you all. She was afraid of what Andrew might do if she let it slip. Anyway, children – even ones in their twenties – don't examine their parents' behaviour very closely.'

'Do you think that Mum never told us kids anything about

her past, because she was afraid she'd let slip how unhappy she was?'

'Possibly, or she was afraid that if you repeated anything she'd said in Andrew's hearing, she'd get into more bother. But I think it was admirable that she kept quiet about everything that was going on between her and Andrew, so that you three kids would remain secure and happy. That was very noble and loving. If you keep that in mind, perhaps it won't hurt so much.'

'You are loving and noble too,' Eva said, leaning towards him to kiss his cheek.

'And I'll be late for work if I sit here any longer.' He smiled. 'Now go back to sleep. But if you go out later, get that statement photocopied before we show it to the police. We can't risk losing the original.'

He paused as he went to go out of the bedroom door, and looked back at her. 'I love you, Eva. However bad this seems now, together we'll get through it.'

Eva lay in bed for some time after Phil had left, forcing herself to look back with adult eyes at incidents she remembered from her childhood. One which stood out was being dropped home one afternoon by a friend's mother after a sleepover in the summer holidays. She was eleven then, and Sophie and Ben were spending a few days with Granny and Grandpa, Andrew's parents.

She walked in the back door and Flora called out to her from the sitting room. 'I'm in here,' she said. 'I've had a bit of an accident.'

Flora was sitting on the sofa with her legs propped up on a stool. There was a large lint dressing on one of them, but both her legs were covered in what looked like bad grazes. Eva remembered that she was dressed, but that her hair was very tangled, as if she hadn't brushed it that morning. She also looked like she'd been crying.

Eva had been horrified, and asked how it had happened.

'I was really silly,' Flora said. 'I was having a bit of a clear-out in the attic rooms this morning and I tried to carry a heavy box of old tools and things downstairs to put them in the garage. I tripped and fell, and I landed on some of the tools.'

Eva hadn't for one moment thought that it wasn't the truth. Her anxiety was only about how much pain Flora was in and if she should call the doctor, because Andrew was going to be away overnight.

But Flora said it wasn't serious enough to trouble the doctor, and she felt better already now Eva was home.

She quite enjoyed playing nurse that evening, as it wasn't often she got her mum all to herself. She made them both some tea and sandwiches and she supported Flora when she wanted to hobble to the loo and, later in the evening, up to bed.

But looking back at that incident now, where were these tools Flora was supposed to have fallen on? Anyone taking a tumble like that, and finding they could barely walk, would have left them where they'd fallen; they certainly wouldn't have packed them back into the box and got them to the garage! Besides, why would there be tools stored in an attic room anyway?

Clearly the truth of the matter was that there had been a row that morning, and because there were no children in the house Andrew had attacked Flora. Thinking about her injuries now, Eva thought he must have knocked her down and then kicked her hard, again and again. Then he'd calmly gone off to work after warning her she was to tell no one.

Eva's stomach contracted painfully at the thought of Flora being alone and hurt, forcing herself to struggle to dress the worst wound, then inventing a story to cover her injuries.

'You evil bastard,' Eva muttered to herself. 'I'm going to make sure you pay for everything you did to her.'

Chapter Twenty-Four

'Hello, Eva! How are you feeling now?' DI Turner asked as she opened the door to him. He had a WPC with him who she hadn't met before.

'Oh! I'm much better, thank you,' she replied. She was somewhat startled to see Turner, because she'd just been mulling over what she ought to do about Flora's statement.

It was Wednesday morning, pouring with rain. Phil had left for work at seven and would be away for two days. Everything had seemed so clear and straightforward yesterday. She'd gone out and had five photocopies made of the statement. She intended to send one to Gregor – she thought that he deserved an explanation of why Flora had acted as she did towards him – and two would go to Sophie and Ben. But after Phil left this morning she wasn't so sure about anything. Phil didn't think she ought to send copies to Sophie and Ben, at least not yet. He said she mustn't do anything until he got back, when they could discuss it properly.

'May we come in?' Turner asked. He half turned towards the tall and rather severe-looking WPC. 'This is WPC Rose, and we've come to tell you about some new developments.'

'Yes, of course,' Eva said, opening the door wider. 'Do come in. You just took me by surprise.'

She led them into the lounge, picked up a jacket of Phil's from the sofa and quickly removed a plate and mug from the coffee table. 'Let me make you a drink. Tea or coffee?'

Turner and Rose looked at one another as Eva darted into

the kitchen to get them coffee. They could see she was rattled about something and wondered if she was alright.

She returned a few minutes later with coffee and biscuits on a tray.

'You've come to say you can't find any record of my birth here in London?' she said as she put the tray down on the coffee table.

'That's right,' Turner said. 'Not at a hospital, or as a home birth. The first record of you apart from your birth registration is at about three months old when Flora registered you with her doctor in Holland Park for immunization. But you sound as if you already knew that.'

'I did know I wasn't born in London,' she said, handing them their coffee and then sitting down. 'Not about the immunization.'

Turner was confused about how she could know that. 'We've also got the blood test results from Carlisle,' he said.

'And that proves I'm Sue Carling's child?'

Turner glanced at the WPC; neither of them had expected such calm acceptance. The girl seemed almost dazed.

'Well, Eva, I'm afraid I can't give you utter certainty, but it's about 85 per cent probable that she is your mother,' Turner said. 'I did explain to you before that all blood tests tell us is who couldn't be a child's parent. They are not so good at proving parentage. And the Carlisle police were unable to trace Sue Carling –'

'So how did they test her blood then?' Eva interrupted.

'They couldn't. But they were able to test her daughter's. That result, and a close physical resemblance to yourself, does suggest she is your full sister.'

Eva's eyes widened at that. 'I read in the press cuttings that she had two other children before me, but weren't they taken into care?'

'They didn't trace either of them. This girl is younger than you, only seventeen. Her name is Freya. They only found her when they had a report of a young girl sleeping rough out on the fells. She was very sick with a chest infection and malnutrition, and they took her to hospital. On being questioned she said her mother had gone to Spain two years earlier and she hasn't heard from her since.'

Eva frowned. 'She left a girl of fifteen?' she exclaimed. 'My God, it sounds like Flora ought to have been given a medal for taking me away from her! How could anyone leave a girl of that age and go swanning off to Spain?'

'It seems she left her in the care of a friend who ran a bed and breakfast. Freya was already working there on Saturdays and during school holidays, and she had a full-time job lined up in a bakery for when she left school. Of course that doesn't make it right to leave her, she was far too young.'

'Poor kid! Girls of that age can get into serious trouble if they aren't supervised,' Eva exclaimed.

'Quite so, but according to the Carlisle police she's never been in trouble with them. She wasn't sleeping rough for fun or because she got into bad company. She'd lost her job, because the bakery went bust, and the so-called kindly friend of her mother's at the bed and breakfast threw her out when she hadn't got the rent. Anyway, they are keeping her in hospital until she is well again and can be found a new home.'

All at once Eva came out of the dazed state she'd appeared to be in when they first arrived. 'That's absolutely dreadful,' she said, and her eyes filled up with tears. 'She must have felt very alone and frightened. Does she know about me?'

'Only that someone of the right age to be her sister has been making inquiries. She doesn't know your name or address. She did tell the police, though, that her father, and yours too, was Michael Borthwick, a drayman for a local brewery. He died of stomach cancer when Freya was thirteen, but by all accounts he was a good father. He paid her mother maintenance, he visited Freya regularly and took her on little holidays. He had no police record.'

'So we really are full sisters?'

'So Freya said, and the blood test bears that out. The Carlisle police are redoubling their efforts to find her mother.'

Eva didn't respond to that. She just sat there, looking at her hands in her lap, and appeared to have gone back to her earlier dazed state. Knowing what she'd been through recently, Turner wondered if everything was getting too much for her.

'Are you alright, Eva?' WPC Rose ventured. 'I'm sure this has been a great shock to you.'

Eva looked up at her concerned face. 'Perhaps it would be as well if they never found her. There doesn't appear to be anything good about her.'

The policewoman was right about shock. Patrick bringing round Flora's statement had left her reeling. That same night, and all day yesterday, she'd thought about nothing else. She'd told Phil last night that she felt oddly comforted to know the whole truth, but that wasn't strictly true. Flora's statement had been like opening Pandora's box, and the stuff that had come out was just too much to deal with.

Now to get a second shock – to learn that she had a sister – had taken her a step too far. She felt shaky, frightened and very vulnerable, and she knew she really needed time to digest what this meant to her, before she said or did anything further.

Yet at the same time she felt she must tell these two police officers now about Flora's statement. She needed them to understand why Flora took her, and to prove that Andrew had a real motive in trying to kill her.

'Phil, my boyfriend, said I was to wait until he was with me before I showed you this,' she said, and got up to fetch the folder she'd put Flora's statement in. She took out one of the photocopies and handed it to Turner. 'But under the circumstances, I think you need to see it now.'

She explained carefully who had found it, and where. And she confirmed it was Flora's handwriting.

Turner took about ten minutes to read the first few pages. Eva could see he was as affected by it as she and Patrick had been.

'My goodness. This is something I didn't expect,' he said, looking very troubled. 'No wonder you didn't seem that surprised or even interested in our inquiries at the hospitals. I will of course read it all thoroughly when I get back to the station, but there's far too much detail about taking the baby for it not to be true. This must have been overwhelming for you?'

'I seem to reel from one shock to another these days.' Eva shrugged. 'Mostly I wish I'd never found the diaries Phil told you about. I could've been happy in blissful ignorance. A psychic up in Scotland gave me a weird warning about "waking the sleeping serpent", and I really wish I'd heeded that warning, however barmy that sounds. Nothing good has come of me trying to solve the mysteries – even my brother and sister have turned against me for claiming their father set the fire.'

She stopped for a moment, overcome with emotion. The policewoman got up and put a comforting arm around her, urging her to sit down and asking if she'd like a cup of tea.

'A cup of tea won't make this any better,' Eva said sharply. But she did sit down again. 'But what might help me is if you arrest Andrew Patterson.'

'We've already explained we have no evidence against him

to charge him with starting the fire at your house,' Turner said.

'Just read on in that statement and you will see that he knew all about Flora taking me. It was Flora who financed getting the house in Cheltenham, and he used blackmail and physical abuse to keep her in line. He drove her to kill herself too.'

Turner shook his head as if unsure of what to say. 'I will read it, Eva. And believe me, I have every sympathy with you. It must be utterly devastating to discover such unpalatable truths and to realize the people you trusted and believed in were not as you thought.'

'Are you trying to tell me you can't arrest Andrew now?' Eva asked, her voice shaking as she pointed to the statement. 'That proves his motive for trying to kill me.'

Turner looked abashed. 'I'm sorry to disappoint you, Eva,' he said. 'This statement could possibly be used by the prosecution to show the man's character if we had firm evidence to prove he set the fire, but it has no other value in this case. Even then, the defence could argue that Flora had made it all up. A self-confessed baby-snatcher would not be considered a reliable witness alive. And as she's dead, she can't be cross-examined.'

Eva felt as if he'd stuck a pin in her balloon. 'He drove Flora to kill herself, tried to kill me too – and he gets away with it?'

Turner shrugged. 'Unless some firm evidence turns up, I'm afraid so,' he said. 'The police in Cheltenham are still keeping an eye on him and making inquiries. We will pass this on to them too. I know they do think Sophie lied, or was mistaken, when she said he was in bed when she arrived home that night. She might have a change of heart – people do sometimes, when they get a guilty conscience.'

'So it all rests with Sophie to tell the truth?' Eva was incredulous. 'I can't believe I'm hearing this. I know it was Andrew who set the fire. By the time you've read the whole of that statement you won't have any doubt he was capable of it either. Get the police in Cheltenham to talk to Sophie again. Give her a guilty conscience, for God's sake, before he tries to shut me up again!'

'We'll look at this statement and try to see if it gives us another angle to bring him in for questioning,' Turner said soothingly. 'Now, speaking of guilty consciences!' He put his hand in his pocket and brought out an envelope. 'I very nearly forgot this. We had to open it of course, just to check the contents.'

Puzzled, Eva pulled the letter out of the envelope and gasped when she saw it was from Myles.

'Go on, it's alright,' Turner urged her.

Hesitating slightly, Eva started to read.

Dear Eva,

I am writing to apologize for my appalling behaviour towards you. I can offer no good excuse at all; you didn't do anything to deserve it, and I am ashamed of myself. I guess I didn't see the light until I heard about the fire and you being trapped in it. That was such a terrible thing that it made me see it was so very wrong to keep on lying and saying I didn't insult, hurt and frighten you.

You may think this is a cynical attempt to get you to drop the charges against me, but it isn't. I have already changed my plea to guilty, as the police will tell you, and I will accept whatever punishment the court deems appropriate.

I really hope you have fully recovered now and that you can rebuild your house and your life soon.

All my good wishes for your future,
Myles

'Hum!' She sniffed. 'I suppose someone told him I was in two minds about dropping the charges against him!'

'No, they didn't,' Turner said. 'We were told he was changing his plea to guilty some time before we got this. I promise you, I hadn't told anyone that you were considering dropping the charges – because, quite honestly, I didn't think you would. He just came into the police station and asked for this to be passed on to you.'

'He sounds sincere,' Eva said thoughtfully. 'But then men like him always do. There's no way I'm going to decide right now, I need to think about it, and talk to Phil. Meanwhile, he can sweat a bit longer with the court case hanging over him.'

Turner nodded. 'I think that is a wise decision. Now, what would you like to do about this sister of yours?'

Eva realized he wasn't going to be drawn back to the subject of Andrew, and she felt too weary to even try further. 'She's still in hospital in Carlisle?'

'Yes, and I believe she'll be there for a few more days.'

Eva thought about it for a moment. 'I really don't know. Freya and I might have a mother in common, but that's all. And by contacting her I could open up another can of worms.'

WPC Rose had said almost nothing all this time. She reminded Eva of a bird, because she had small dark eyes which kept darting from Eva to Turner.

'I got the Carlisle police to send down a photograph of her for you,' she said. 'It's only a Polaroid, and as it was taken in hospital she looks a bit gaunt. But we thought you might like to see what she looks like.' Rose reached into her tunic pocket and pulled it out. 'We both think she's very like you – her eyes are identical to yours.'

Eva stared at the photograph. 'We'll leave it with you then,'

Turner said, as she seemed to be transfixed by it. 'Thank you for letting me see this statement. I'll be in touch again soon.'

Eva let them out and said her goodbyes. Still holding the photograph, she went into the kitchen and just stood there studying it.

Turner was right, Freya did look like her. She was very thin, and her eyes dominated her face; they were the same blue and the same shape as her own. Her nose looked the same too, though that could just be the angle of the photograph, and her hair was the same light brown that Eva had been born with. It made goosebumps come up all over her to see a younger version of herself. For as long as she could remember she had been compared unfavourably to Sophie and Ben, and in some strange way it was comforting to see her own features mirrored in this photograph.

Yet it was more than the physical resemblance that spoke to her. There was sadness and resignation in Freya's face that said she'd learned long ago not to expect anything good to come to her. Eva had felt that herself at the same age – never quite belonging, feeling second-rate.

She wondered too about their father, Michael Borthwick. She had a mental picture of a big burly man wearing a leather apron, rolling barrels. He was dead, Sue Carling was missing, and Eva didn't think she wanted to know anything more about either of them. But Freya was all alone and sick.

At two that same afternoon the Glasgow train pulled out of King's Cross Station with Eva settled in a window seat.

She felt compelled to visit Freya in Carlisle. It might not be wise, but the more she'd looked at the picture of her sister, the more she felt she must go. Her nerves were jangling, she'd had a recurrence of coughing while on the tube, and she knew Phil wouldn't approve. But the alternative – to sit at home looking at that sad, forlorn face of a young girl with nothing and nobody in her life – was too grim to contemplate.

In an effort to calm her nerves and make the train journey pass more quickly she opened her book, *The Thorn Birds* by Colleen McCullough. It had come out years ago, but it wasn't until she was in Scotland that she had got around to buying it. She had only read a few pages before that copy turned to ashes in the fire, so she'd bought another one from W. H. Smith before boarding the train.

Books had always been her escape from reality ever since she'd learned to read, and she soon found this one so enthralling that she was barely aware the carriage was packed and the rain was pelting down outside. She had a small overnight bag with her, and she intended to find a bed and breakfast when she got to Carlisle. She'd telephoned the hospital to make sure Freya was still there, and had also rung Phil's company so they could tell him where she was. She was afraid he'd be worried, if he rang home tonight and got no reply.

It was only as the train got to Cumbria and the sun came out that she put her book down and looked out of the window. The trees here were already in full autumn colouring and it looked very beautiful, reminding her poignantly of when she saw the same scenery before – on her way to Scotland, full of hope she might find out who her father was.

So much had happened since then: discovering more about Flora, meeting Patrick and Gregor, falling in love with Phil, her terrible suspicions about her own origins, and then the fire. But she did feel that all these things, good and bad, had given her new strengths. She thought she was more confident, less concerned about other people's opinion of her, and she was becoming more self-reliant.

It had been a severe body blow to learn Flora's statement

wouldn't put Andrew behind bars. Not just because he'd tried to kill her, but for all the hurt and humiliation he had heaped on Flora. Maybe a great many people would think Flora deserved the life he gave her. But Eva knew better; she was a good woman.

In another ten minutes she would be arriving in Carlisle, and it was even more unclear to her now why she felt compelled to come all this way.

Was it purely an emotional response, because the girl was young, sick and alone? Did she think that it was some form of restitution or apology for what Flora did? Or because she could hear Dena's voice telling her that one of the tarot cards she'd turned up, the rabbit, was a vulnerable family member who needed her help?

She'd told herself hundreds of times that the whole tarot thing was just ridiculous mumbo-jumbo, but she couldn't deny that some of what Dena had told her appeared to have come true. But what if this was more of the sleeping serpent thing? What if, by seeing this girl she knew very little about, Eva would get drawn into something that she'd regret later?

Her stomach began to churn with anxiety. Freya might be her biological sister, but she'd been brought up in a different world to her, and there was no guarantee that they would like one another. In fact it was quite likely that Freya might resent her, because she'd had a much better childhood.

On arriving at Carlisle Station Eva stood for some time looking at the departure board, wondering if it would be wisest to get the next train back to London. She looked down at the carrier bag in her hand. In it was a basket of toiletries from The Body Shop, which she'd bought for Freya at King's Cross.

'You can't go back without seeing her,' she told herself. 'That's plain cowardice.'

That decided her – she'd come all this way, and it would be pathetic to chicken out now. She didn't have to take on any responsibility for the girl; it was just a visit, and she could make that quite clear.

There was just half an hour before the hospital's evening visiting hour. So, bracing herself, she bought some fruit and chocolate too, then caught a taxi to the hospital.

With butterflies in her stomach, Eva walked down the big ward. Nearly all the other beds were occupied by old ladies, and the sister had told her Freya was at the far end. As she approached the bed, she saw Freya was lying on her side, hunched up with her back to all the visitors coming in. Eva's heart went out to her. It was awful to think that someone so young had no one other than a total stranger to visit her.

'Freya?' Eva said softly as she approached the girl's bed.

She rolled over on to her back to look at whoever was speaking to her. She had been crying, and she quickly put her hand over her eyes to hide the tears.

'I'm Eva, and it looks as though I'm your sister. I've just come up from London.'

Freya took the hand away from her eyes. 'Really!' she said, looking startled. 'You came all that way just to see me?'

Eva didn't speak for a few moments, because it was extraordinary how alike they were: two sets of identical blue eyes, the same full lips and small nose slightly turned up at the tip. Freya was painfully thin, her long hair desperately needed washing and trimming, her nails were bitten down to the quick and her skin was rough and wind-burned from being outside in all weathers. But the similarities between them would be obvious to anyone.

'Yes, just to see you. I was told you weren't very well.' Eva smiled. 'Now, are you going to sit up and talk to me?'

In books long-lost relatives always seemed to have an instant connection, falling upon each other's necks to hug and kiss. But it wasn't like that with Freya. Although she sat up, wiped her eyes and thanked Eva politely for the fruit, chocolate and toiletries, she was very guarded and awkward. Her strong Cumbrian accent was a little hard to understand too.

Eva had to make all the running. Having no idea how much Freya knew about the sister who had been taken long before she was born – or even if the police had made it clear to her why her blood had been tested, and what that meant – she had to explain everything. Then she went on to tell her about finding Flora's diaries after her death, and how she began her own investigation in Scotland to see if she could find out who her real father was.

Freya didn't react to any of it; she just sat up in bed picking at her nails and avoiding looking directly at Eva. It was as if she wasn't interested in any of it.

'There was a photograph of a row of shops in her stuff too, and someone in Scotland told me it was in Carlisle. So I came here looking for that street,' Eva went on. 'Then I found out about the baby that had been taken from there and I suspected that baby was me. I didn't want to think the woman I thought was my mum had done such a terrible thing, but I've found proof now that she did.'

'Is that right?' Freya said with what seemed to Eva complete contempt.

'The police officers who came to tell me about the results of the blood test also said you'd told the Carlisle police we had the same dad. Have we?'

'Yes, but he's dead now. And Mam might as well be. Everyone thought she'd killed you. Did you know that?'

She spoke in an accusatory manner, which unnerved Eva.

'Yes, I did hear that, Freya. That was very unfair of people to claim such a thing without any evidence to back it up. And it must've been very hard for you growing up with that hanging over you?'

'You aren't kidding! It followed me everywhere. Mam didn't help things neither, what with her drinking and carrying on. She's a bad 'un.'

Eva had every sympathy for the girl. But having come so far to see her, she had expected to receive a little warmth, not hostility – after all, she wasn't the one to blame for anything.

'I've had some very bad times too since Flora died,' Eva said in an attempt to make Freya see she wasn't the only one who had suffered. 'The man I always thought was my dad tried to burn my house down with me in it, because I found out about all this. His children, the people I thought were my brother and sister, hate me too because of it. But what I'm trying to say, Freya, is I know you've had it tough ever since your birth, and I had a cushy childhood compared with you. But we've both got to move on now.'

'Easy for you to say.' Freya looked Eva up and down in contempt. 'You're posh. You've got nice clothes, pretty hair and somewhere to live. I've got nothing. Even my clothes are all gone, and they weren't much to start with. Mam sent me to stay with that witch Ena, even though she knew what it was like there. Ena wanted me to have sex with old dirty truckers. She hit me when I refused! She asked what I had to be so stuck up about. Said that my mam would do it with anyone for just a tot of whiskey. She threw me out then, and I had nowhere to go.'

Freya's voice was gradually becoming shriller with anger. The old lady in the next bed and her younger visitor were both turning their heads to see what was going on.

'Do you know what it's like to sleep on the ground in a

place that don't even have a roof?' Freya went on. Her eyes were flashing, and red angry spots were appearing on her white face. 'Do you know what it's like to not be able to have a bath? To sleep in your clothes because it's so cold, and be so hungry that you hang around picnic places hoping someone will leave some food behind? You were the lucky one that got taken away by someone that wanted a bairn. All I've ever known was people sneering at me.'

Eva was shocked at the angry tirade. She was embarrassed that people were looking at them, and her instinct was to rush out of the ward and get the first train back to London. But she knew, if she did that, it would just prove to Freya that no one in the world cared whether she lived or died.

'Don't take it out on me, Freya,' she said sternly. 'I didn't have to come all this way, remember. I came because I thought it was the right thing to do, because we are sisters. I know you've had a bad time, and I am sorry about that, but don't think I'm so soft that I'll take you being rude to me.'

Freya looked quite shocked.

'Right, I'm leaving now. I've got to find somewhere to stay tonight. I'll be back tomorrow to see you again, before I go home. And I hope by then you can be polite.'

Eva walked away and didn't look back, but as she got to the ward door the sister stopped her. She was a big woman with a very red face. 'How was Freya with you? I can see you are related, you are so much alike.'

Eva guessed that the sister knew about the Carling family history and perhaps even suspected Eva was the stolen baby. But Eva felt far too despondent to admit that.

'Yes, we are related, but I've never met her before today,' Eva said. 'I don't know that I did the right thing coming here. She's so angry, and I might have just made her feel worse.'

The ward sister put her hand on Eva's shoulder. 'That

poor girl has been dragged up. God only knows what terrible times she's lived through. She doesn't feel she can moan or complain to any of us – I dare say she's afraid, if she did, we'd discharge her. But you arriving with gifts probably made her think that she could unburden herself. She's only a child still, and one that has had precious little love and care.'

'I know that.' Eva felt she just might burst into tears herself. 'I just don't know what to say, or what to do to make her feel better. She seems to resent me.'

'I'm sure she doesn't really. From what I've seen of her, she's grateful for just the tiniest bit of kindness. But I'll have a word with her later, tell her that you only wanted to help her. And please come back tomorrow. Don't worry about visiting hours, morning or afternoon will be fine with me, I know you've come a long way.'

Eva looked at the nurse's kindly face and felt a bit ashamed. 'If I buy her some clothes and bring them in, will she see that as me being a do-gooder and throw them back in my face?'

'If she does, I'd be tempted to smack her bottom!' the sister said with a smile. 'The trouble is, she's got so far down she can't see the way to climb up. New clothes will be the first rung on the ladder. And I know the social worker is looking out for a home for her right now.'

Eva spent a restless night in a guest house she found close to the hospital. All she felt for Freya was pity; she was sad, because she had wanted to feel something more than that. But maybe Freya thought she had come to visit her like she was some exhibit in a zoo. And then she was going to walk away, having satisfied her curiosity.

She did intend to go back again – she hadn't come all this way to give up at the first hurdle. But what then? Even if Freya was different in the morning and they did strike up some sort of bond, she couldn't go home afterwards and forget about her. That would be too cruel, and would confirm the girl's opinion that no one cared about her. But it was 300 miles from London; she couldn't keep coming up here. She needed to get back to work, and there was Phil to think of too. Neither could she suggest that Freya came to London. If she did that, she'd have to be responsible for her.

That was what really scared her – feeling she owed Freya. But why should she feel that way?

After a shower the next morning and a big breakfast, Eva was feeling a little more positive. She'd made a list of things Freya would need when she left hospital, including a warm coat. Buying clothes for her might be seen as patronizing, but Freya needed them desperately. And besides, it was the only concrete thing Eva could do to help.

She packed up her overnight bag, paid her bill and then walked into the town centre to the Lanes Shopping Centre.

It suddenly occurred to her as she was walking that Flora must have come back here at some time and taken the photograph of the row of shops where she'd snatched Eva. She couldn't have taken the photo on the day she stole the baby – no one would do such a thing, even if they had a camera on them. Did she feel it was a necessary bit of evidence to back up what she'd done? Was it some compulsion to return to the scene of the crime? Or maybe she even hoped to discover if the hue and cry was still going on here? Eva doubted that the papers in London would have reported much on a crime that took place so far away.

She'd learned so much about Flora in the past few months, yet there was still so much she didn't know.

In Chelsea Girl she first ascertained if whatever she bought could be changed if she'd got the wrong size. Finding that it could, she picked out a black quilted coat with a furry lining and some size eight jeans in a short leg length. She'd noticed a lot of girls in Carlisle were still wearing ski pants – a trend that had died a death in London. But she wasn't going to encourage that awful fashion statement. She bought two different sweaters – one pale blue with an appliquéd satin bow on the front, and the other a leopard print in fuchsia pink and black. With a pair of plain black trousers and a couple of T-shirts, she felt she'd made up a basic wardrobe that was both practical and pretty.

Next she went to Marks and Spencer to buy underwear. She got two bras, in different cup sizes – to be sure one fitted her properly, and then she would change the other one – and four pairs of knickers that matched the bras. Some pyjamas, slippers and a warm turquoise scarf with matching gloves, plus a card with three different-coloured scrunchies to tie her hair back, and a hair brush completed the shop.

At a charity shop she found a small black suitcase with wheels for just £2 and packed everything into it.

Finally, she just needed to buy some shoes. She chose trainers, after noticing that was the only thing anyone of Freya's age was wearing, and bought size four – that was what she wore, and the chances were her sister had the same size feet.

Then, pulling the suitcase along behind her with her own bag over the handles, she walked to the hospital.

It was scary that she'd written out cheques for nearly £200. The money she'd got from her mother was nearly all gone now, and she had to find the payments for the £1,000 loan she'd got for the central heating and new bathroom. She had no money coming in, but the bank did know about the fire now. She supposed they'd let her be overdrawn until the insurance for the house paid out.

Besides, Freya's need was greater than hers — and she knew what it was like to lose all her clothes.

Freya was sitting up in bed reading a magazine. She was still wearing a hospital nightgown, but she'd washed her hair and it was fluffed out around her face making her look a great deal prettier.

'I'm sorry I was nasty yesterday,' she blurted out, looking embarrassed.

'That's OK,' Eva said. 'It must have given you quite a shock, me turning up out of the blue.'

'Yes. And you look so nice and you speak so posh,' she said. 'People like you don't usually speak to me, so I said stuff that was just in my head.'

'You are going to look nice soon too,' Eva said. 'I've bought you some new clothes and packed them all in this case so nothing gets lost. Would you like to see them?'

Freya's face tightened. 'You didn't have to do that,' she said.

Eva realized that Freya didn't like to be thought of as a charity case, so she tucked the suitcase beneath the bed. 'No, I didn't have to, but I wanted to. I'll leave them for you to look at later. I hope they are the right size, but the receipts are in the case. You can change them, if they're wrong.'

The start of the visit wasn't much better than the one on the previous day. Eva was at a loss what to say to the girl. When she said anything about her own life, she felt she sounded smug. And yet when she asked Freya any questions, the girl's face tightened again as if she thought Eva was prying.

'Will you tell me about our mother?' she asked after a somewhat stilted conversation about the television programmes they liked.

'What's to tell? Mam never stood a chance – her mam was a drunk, and so was her da. And one minute they were skelping her backside and the next sending her off to chaff.'

Eva didn't understand; she had to ask what those words meant. 'Skelping' meant slapping and 'chaff' meant stealing, Freya told her.

'They took Mam's other bairns away cos she didn't care for them right. But then when she met our da, I think she settled down a bit – well, she always said she did. But then when you was stolen, she went right off the rails again. Da loved her, you know. He used to say, "Freya, I wish I could walk away from her, but I can't, she's only got to flash those lovely eyes at me and I forgive her."'

'Did she look after you properly when you were a baby?'

'Well, me da said she tried. He was staying with her then, that's why they never took me away from her too. But when I was about four she met some new bloke and she was up and off with her old tricks and chucked Da out. He did his best for me, the only one that ever did. See, everyone but Da thought she'd killed you. He always stuck up for her. I know she weren't a good mam, but with everyone whispering about her all the time, that got her down and she'd turn to the drink.'

Freya had painted an even uglier picture than Eva had imagined. But she was touched that Freya seemed to understand why their mother was like she was.

'Did she ever tell you how she felt about losing me?' Eva so much wanted to hear that Sue Carling had normal maternal feelings.

'She was always goin' on about folk believing she'd killed you. Every time she got drunk she'd bring that up. "As if I could do that," she'd say. But she knew she were a bad mam. She said she couldn't help it cos no one had ever taught her how to be a good one. She used to talk about how she thought your life was. She'd show me pictures in magazines of rich people's houses and say she knew you lived somewhere like that. I think that's the way she made it alright in her head, like she gave you away to a nice woman with lots of money. Was your mam nice?'

'Yes, she was,' Eva said. 'She was wrong to do what she did, and she had to live with the guilt of it. But she loved me and I loved her.'

'Then you was bloody lucky,' Freya said, her mouth turning down. 'It's like one of those kids' storybooks where one gets brought up like a princess and the other in a slum.'

'Not quite, Freya. My life hasn't been a bed of roses either.' Eva felt she had to drive that point home. 'Flora committed suicide, and my stepfather, brother and sister have all turned against me. But one thing I've learned is that you have to live with the cards you've been dealt, and make something out of that.'

'You tell me what I can make with my cards,' Freya said.

'I don't know you well enough yet,' Eva said. 'But I'm sure there is something you are good at that you can use. Now, will you me tell about our dad?'

Freya's face softened then. 'He was that handsome, lovely thick hair and eyes like ours. He could've had his pick of women really. He wasn't too tall, but big muscles and very strong. He liked to go walking up on the fells, he knew the names of all the birds, and he liked animals an' all. He used to take me up there with him. When I was little he used to carry me on his shoulders and sing to me as we walked. That's how I knew that old house was there, where the police found me. I had a couple of pictures of him that I took up there with me. Maybe when I get out of here I'll go up and see if they are still there.'

'He sounds lovely,' Eva replied. 'I'm sorry I won't ever get to know him. The man I thought was my father turned out to be a real snake.'

It was only at that point that Freya began to look interested, so Eva told her about finding Flora dead in the bath. 'She did it on what was my real twenty-first birthday. But now I know her whole story and how that man was with her, I understand why she felt she couldn't go on. He told me he wasn't my father just a week after she died. He couldn't wait to get me out of the house. I hate him now.'

'I don't hate our mam,' Freya volunteered. 'She was useless, selfish and weak, but when she weren't drinking and that, she could be right funny. Sister, the big lady with the red face, she said to me last night. 'Now look here, Freya, you can lie here and feel sorry for yourself if you like, and nothing will change for you. But if you put on a brave face, let the social worker find you a new home and a job, you could do well.'

Eva had to smile, as Freya had copied the sister's way of speaking exactly. 'And what did you think of that?'

'She were right. Now I know Mam didn't kill you and bury you out on the fells, I can put two fingers up to all them what said she did. I'll be OK, I'll get them to stick me on some training course, or whatever. I reckon I'm more like me dad than me mam. I can stick at things.'

'You are a brave girl,' Eva said. Tears came into her eyes, because she knew she could never be that stoic.

'Don't you start crying over me,' Freya said sharply. 'A year from now I'll be doing alright. I can promise you that.'

'Is that a firm promise?' Eva asked. She couldn't look after this girl – and what's more, she realized Freya wouldn't want her to. To make some sort of pledge that they'd look each other up a year from now seemed the ideal solution.

'Yeah, OK. But you've got to promise me you'll stop harping on about the past and make a future for yourself too. We ain't so different, Eva, even if you do speak posh and wear nice clothes.'

Eva wrote down her address and telephone number, and also took out twenty pounds from her purse. 'That's just so you've got a little something to fall back on until they arrange benefits for you. I wish I could give you a bit more, but that's all I've got. Let me know how you get on and your address once you're settled.'

Freya looked hard at her and smirked. 'You're a bit of an old woman,' she said. 'I don't mean that nasty, like, cos you've been right kind coming here and getting me clothes and stuff. But I don't need another mam. And from what you've told me about your'n, you don't neither.'

Eva put her arms around Freya then and hugged her tightly. She was so thin, she felt like a child, and that brought more tears to her eyes. 'Get well soon, and be good,' she said. 'Old woman or not, I'm pleased to be your sister.'

She had to break away then, hurrying down the ward and struggling not to break down. When she got to the ward door, she glanced back and saw Freya was already out of bed and opening the case of clothes.

Chapter Twenty-Five

March 1992

Eva came out of Holland Park tube station deep in gloomy thoughts. Not only had she had an awful row this morning with Phil, but it was the 29th of March, the first anniversary of the day Flora died. She'd been having flashbacks about it all day. She wondered whether Sophie and Ben were thinking about it too, and what their reactions had been to the copies of Flora's statement she'd sent them a couple of weeks ago.

But the way she felt today was not just a one-off mood, brought about by the row or the anniversary. Since Christmas she'd felt herself gradually sinking deeper into a black hole of depression, and she couldn't find a way of climbing out of it.

She had tried to pull herself together, asked herself a thousand times what she had to be depressed about. She wasn't broke, homeless, sick or hungry. She had people who cared about her, and Phil would probably lay down his life for her. But telling herself these things didn't make her feel any better. She might be able to function adequately on a day-to-day basis – indeed, at work no one even realized there was anything wrong. But she knew there was, and she was scared it would escalate into something much worse.

Looking back she could see it was the fire that started the ball rolling. What with the revelations from Flora, and the police unable to prove it was Andrew, she was left feeling shaky.

On top of that, after returning from her trip to Carlisle last autumn she had been left with anxiety about Freya. She had waited in vain for a phone call or letter from her, telling her where she was and what she was doing, but there was nothing. As she had no idea how to contact Freya, all she could do was just wait and hope she would surface again.

Phil was not all that sympathetic; he took the line that it was probably for the best. He felt that Freya was too damaged to value a relationship with the sister who she possibly felt had ruined her own life. Although that might be the case, Eva still couldn't help feeling hurt and disappointed.

Fortunately Horace, the owner of Serendipity – the shop she'd just started working at before the fire – rang her around the same time to say her job was still open for her if she wanted to come back. She had accepted gratefully; aside from desperately needing to earn some money, she needed something to divert her.

It felt good to be back at work, and it lifted her spirits to be part of a friendly team. There was exciting new stock coming in daily for Christmas, and each day the shop became busier, leaving her little time for dwelling on unresolved problems.

Soon after starting back at Serendipity she also got the go-ahead from the insurance company to renovate the house in Pottery Lane. Phil took over as project manager, getting Patrick's architect friend to draw up plans, and took on builders he knew to do the work.

All through November and December she and Phil saw far less of each other: she was working more hours in the run-up to Christmas, Phil had his own work, and in the evenings and weekends he helped out at Pottery Lane.

She had sent off a copy of the statement to Gregor, and

he'd rung her as soon as he'd read it. Like Patrick he had found it very upsetting, especially the revelation that Flora had been pregnant by him and had later miscarried without ever telling him.

'I would've made sure she had the best of care,' he said, his voice breaking. 'Maybe she would still have lost the baby, but at least I would've been there to comfort her.'

Eva didn't really know what to say; to bring up that Flora had withdrawn into herself when she lost Patrick's baby too, seemed like rubbing salt in the wound.

'If only I hadn't gone off after the Christmas party,' he admitted. 'If I'd stayed, she might have told me about it. And even if she hadn't, I might have realized something was badly wrong and got her some help. I can't bear to think she spent the rest of her life being blackmailed into subservience. I know taking you was wrong, but she didn't deserve such a terrible punishment.'

Eva felt very sorry for Gregor, yet at the same time she began to feel resentment that Gregor and Patrick seemed to be centring on their part in the story and forgetting about what it had done to her. When she brought this up with Phil one night it caused their first real row.

'How can you even think that?' he said. 'Both of them really care about you, they would gladly have accepted you as their daughter, and they both feel for you. Of course they talk about their part in it, that's just human nature.'

'Don't take that line with me,' she said angrily. 'I was only trying to explain how I felt. I've lost my entire family through this, but it couldn't be helped that Flora miscarried.'

'They both loved her,' he said in an exasperated voice. 'It isn't all about you, Eva.'

She should of course have agreed that he was right, because she knew he was. But instead of admitting it, she brought up other minor grievances – such as his mother saying, 'She should forget all about it.'

'As if I can,' she screamed at him like a madwoman. 'It's there in my head every day. It's not like a light switch I can turn off.'

Phil's mother tended to rub her up the wrong way. It wasn't that she was nasty; she was just blunt, and she had very fixed opinions about almost everything. Eva felt she wished her son had found himself someone without emotional baggage. Mrs Marsh wasn't worldly enough to understand such a phrase, her words would be that she wanted someone 'normal' for him.

That time, Phil just put his arms around her and cuddled her until she cried, said she was sorry and they made it up. But it was still at the back of her head. She tried to explain to Olive how she felt when her old boss came down to London and they met up for a meal. But Olive wasn't overly sympathetic either. She just said that Eva had to look at the positive things in her life, like Phil, and stop torturing herself with all the negatives ones.

Eva held off sending Sophie and Ben copies of Flora's statement because of Phil. He believed they were too young to cope with such distressing information, and it was likely to send them both off the rails. She hadn't agreed with him, and that caused another argument just before Christmas.

'You just want to punish them for not believing Andrew tried to kill you,' Phil flung at her. 'But we haven't any proof he set the fire, have we? I agree that they ought to see that statement, but not now when Ben is in his first year at uni, and Sophie is already a mess. Give them a chance to grow up a bit.'

That was the point when this awful sinking feeling began. She was convinced that Ben and Sophie ought to know what their father was capable of, so they would be wary of him. And she couldn't believe Phil would interpret her anxiety for them as spite.

Myles was due to be in court in early December, and she was dreading having to attend – even though the police had said it would be all over in minutes, as he'd pleaded guilty. Three days before the hearing she told the police she was dropping the charges against him. She thought it would be a relief, but instead she just felt cowardly and weak.

Phil's mother looked like she was sucking lemons when Eva told her what she'd done. She didn't actually comment, but Eva got the idea Mrs Marsh thought she'd done it out of guilt because she had led Myles on.

When Christmas arrived – the first one away from her brother and sister – she felt very sad. She went through the motions of decorating the flat, buying Christmas food and presents for Phil and his family, but her thoughts kept turning back to previous Christmases. There wasn't even a card from Freya, let alone one from Ben and Sophie, and though Phil claimed to understand how she felt, he kept pointing out that his family were hers now. But they weren't her family. They had no shared history, and his mother's tactless remark – that from what she'd heard about Ben, Sophie and Freya, Eva was better off without them – really hurt.

In mid-January, when the shop grew quiet after the New Year Sale, that was when she began to find it increasingly difficult to cope. It was cold and wet. The washing machine broke down and flooded the kitchen one day, and that made her hysterical. The noise of traffic seemed too loud, the crowded tube in the morning made her panic, she craved solitude. But then when Phil went away to work, she felt frightened on her own.

She knew Phil was growing irritated by her moods and

negativity, but she just couldn't snap out of it. She felt so tired, she would go to bed the minute she'd eaten the evening meal. And on her day off during the week, and Sunday, all she did was lie on the sofa watching television. She didn't want to make love any more; she gave up caring about her appearance, and even keeping the flat clean and tidy. She wasn't interested in how the work at her house was going. If she was truthful, she didn't really care about anything.

To be fair to Phil, he did try to jolly her along by saying they could go on a holiday to somewhere warm once the house was finished and up for sale. But she couldn't even find the will or enthusiasm to collect any brochures from a travel agent. She got the idea into her head that the only thing which would make her feel better was if Ben and Sophie knew the truth about their father. Finally, two weeks ago, without telling Phil, she sent the statements off to them.

Because she was afraid Andrew might intercept Sophie's copy, she sent them both to Ben at his address in Leeds with a covering letter asking him to give Sophie hers, and explained where the original was found. She said how sorry she was that they had to find out the truth about their father like this. But even though she knew now she had no blood tie to them, she would always think of them as her brother and sister. Her final message to them was that they mustn't allow Andrew to get them to sign over their half of the house to him. She pointed out that Flora had gone through hell to keep them all secure. Therefore, it was their duty to ensure they honoured her wishes.

Yet almost as soon as she'd posted the envelope, she'd panicked and wished she hadn't. Was her motive just spite, as Phil had suggested, because they hadn't believed their father had set the fire? What she had perceived as an act of a loving and concerned elder sister suddenly seemed callous and irresponsible. As her anxiety grew, she withdrew even further from Phil, and she sensed he was reaching the end of his tether.

The row they'd had this morning had brought all this home to her. He'd brought her a cup of tea, as he always did. But as he went to put it down on the bedside table he tripped and spilled the tea everywhere – on the bedclothes and the carpet.

'I tripped on those magazines,' he said accusingly, pointing to the ones on the floor. 'Do you have to leave them there?'

She said something about him leaving dirty clothes on the bathroom floor for her to pick up. And suddenly they were off, shouting at one another.

'I haven't seen you pick up anything for weeks,' he yelled. 'Yours or mine. We haven't had a decent meal, a night out, or even a real conversation either. It's like living with a zombie.'

'You're never here to do any of those things,' she screamed at him. 'What's the point in me cooking something nice if you don't come back until after nine at night?'

He exploded with rage then. 'Anyone would think I was down the pub. I'm at your house, doing it up, for God's sake! I don't know what to do with you any more. I thought we were going to be a team, but it's more like I'm your handyman, here to sort everything for you so you can lie around feeling sorry for yourself. My friends have all worked their socks off to get the house finished, and you can't even be bothered to go along there and see their work!' he roared at her. 'Every time I go there I have to make excuses for you, but I'm running out of them now. I wouldn't blame them at all if they thought you were just a money-grabbing bitch who can't wait to pocket the huge profit you'll make when you sell it. What happened to the girl who used to help, who made the guys tea and cared about someone other than herself?

My mum takes more interest in the house than you do. She even came over on the bus last Sunday to bring some stew for our lunch. She was shocked that you weren't there helping.'

'I bet that made her day!' Eva shouted back. 'It would confirm what she's always thought – that I'm not good enough for you. I bet she even told you that you should get half the money from the sale of the house, and get shot of me.'

The moment those words came out of her mouth, Eva knew she'd gone too far. She had never seen Phil look so angry.

His face darkened and his eyes flashed dangerously. 'You bitch!' he exclaimed. 'You must think that's what I'm after, or you wouldn't have said it. As for my mother, she was just being kind bringing some stew round. She wanted to please both of us, to help in the house if she could. If you can twist that into something nasty, then I think it's time we called it a day.'

He left then, slamming the front door so hard it was a wonder the glass panel didn't break.

It was fear that he would leave her that had made her decide to meet him this afternoon at Pottery Lane. Any other man would have taken himself off to the pub, or even to his mother's for Sunday lunch and some sympathy, and he certainly wouldn't have gone back to Pottery Lane after the things she'd said to him. But Phil would never leave a job half done, however angry he was with her.

Besides, Brian would be there, completing the new kitchen. And aside from trying to make amends with Phil, she needed to thank Brian for all the work he'd done on her behalf.

She did feel very ashamed that she hadn't been there since early January, but she just hadn't been able to face the cold, the mess and seeing Brian too. She knew a perceptive man like him would notice she wasn't herself and would ask Phil about it.

She hated the idea of them discussing her state of mind.

As she opened the door at No. 7, she was greeted by loud male laughter. To her surprise Patrick was there with Brian and Phil, and she felt a momentary stab of jealousy that they were all happy when she was feeling miserable.

'Eva!' Patrick came towards her, smiling a warm welcome. 'I only came round on the off chance Phil might be here, and now I've got you both. How lucky is that?'

Patrick had gone to America for a month in late January. Although he had phoned a couple of times since he got back, they hadn't met up.

'Hi, Patrick,' she replied, trying hard to smile. 'Lovely to see you.'

'Isn't this great now?' Patrick waved his arms at the room. 'It's so huge.'

When Eva had last seen it the men were knocking down the wall to the garage. There were metal posts holding up the ceiling, with debris everywhere, so it was difficult to imagine what it would be like when it was finished. Patrick was right; it was a huge room now, with two windows either side of the front door. The kitchen area was at the front by the left-hand window. A new, very attractive pine staircase came down against the opposite wall, leaving what seemed a vast open living space with new, even larger patio doors on to the garden.

She knew she should be jubilant that it looked so wonderful, yet she couldn't feel anything but resentment that Phil had become far more interested in the project than he was in her.

'The kitchen looks beautiful, Brian,' Eva said, knowing

she must say something positive or risk alienating him too. It was pale golden beech and it had everything: integrated fridge-freezer, washing and dishwashing machines, even a cooker hood that was ducted to the outside. 'I don't believe this one fell off a lorry!'

'Sadly not.' He grinned. 'If you want the top price for this place you have to put in top-quality fittings. But why are you looking so sad today? Come here and have a hug.'

'I'm not sad, just tired,' she lied and let him hug her.

Phil hadn't greeted her at all. She could hardly blame him after the things she'd said that morning. To imply he was after her money was appalling, and completely untrue. Phil hadn't got a mercenary bone in his body.

She looked across at him fixing door handles on the cupboards; he hadn't even looked round to acknowledge she was there. But she couldn't apologize to him in front of Brian and Patrick – and she was afraid he wouldn't accept her apology either.

'You've done a wonderful job here, Brian,' she said, hoping he hadn't noticed how Phil was with her. 'I'm sorry I haven't been round lately, you must think I'm really unappreciative.'

'I expect you find being on your feet all day in the shop tiring,' Brian replied, his voice oozing sympathy. 'By the time you get out of there, I expect the last thing you want is a walk down cold dark streets to come here.'

She saw Phil looking at her, his expression so cold it sent shivers down her spine.

'I'll just go and look upstairs,' she said. 'Then I'll make you all some tea. The staircase is fabulous, by the way.'

She scooted up the stairs to inspect the newly painted bedrooms. Phil never normally brooded about things she said in the heat of the moment. But then she'd never said anything as bad as that before. She would have to try to make it up to him tonight.

Phil had skimmed all the walls and ceilings upstairs and painted the walls cream. There were smart new doors too. When she went into the small bedroom she felt a stab of sorrow to see the teddy bear frieze had gone, even though she knew it had to go. She looked out of the window to see that the garden was bare aside from one lone daffodil. She remembered that she'd intended to plant at least a hundred when they got back from Scotland last year. That was just another bitter reminder that the fire had spoiled all their plans.

'Brian was right, you do look sad.' Patrick's voice came from behind her. 'What's wrong?'

Eva wheeled round. 'It was a year ago today that Mum died. And today is my real birthday,' she said. 'I didn't think it would affect me, but I woke up thinking about it this morning.'

He came up to her and put his arms around her. 'I didn't know it was today. But maybe I tuned into something, because I felt I had to come here this afternoon, and I hoped you'd be here. What's wrong between you and Phil? And don't say "nothing" – I could sense the bad vibes.'

She leaned into his shoulder, very tempted to admit she felt like an overwound clock and that the spring was likely to break any minute. But she couldn't admit such things – not to him, or anyone.

'We had a row this morning. I said something very hurtful. I seem to get sad about everything these days, even because the bears have gone from this room. And that's really silly, as they were covered in soot.'

Patrick stroked her hair gently. 'When you have a baby I'll paint more bears for him or her.'

'I've done something really stupid,' she whispered. 'Phil will be even angrier with me if I tell him, because he said I wasn't to do it.'

'You've ordered a Rolls-Royce?' he joked. 'Or is it just a very expensive dress?'

'I wish it was something like that,' she sighed. 'You see, I sent the statements off to Ben and Sophie. Phil said they were too young. I'm panicking now in case he's right.'

'Oh dear,' Patrick sighed, but he rubbed her back soothingly. 'Last time I saw you both, Phil told me his views on that. My opinion was that there would never be an ideal time to read something so dreadful but that, young as they were, they had every right to see something their mother had written. I even told Phil that I felt the longer you kept it back, the more likely they were to hold that against you.'

'So you don't think it was wrong of me?'

'I do think you ought to have told Phil what you intended to do. You are a couple, and such things should be shared. But Phil is a good and compassionate man, not an ogre. He's not going to bite your head off for doing something you thought was right.'

'But what if Ben and Sophie do go off the rails because of it? I'll always feel responsible for that.'

'You must stop feeling responsible for everyone,' he chided her, putting his finger under her chin to lift it and look at her face. 'You might be officially twenty-two today, but that's very young to be a mother hen. You need to have some fun, splash out a bit, be rash and bold.'

'I seem to have forgotten how to have fun,' she admitted.

'Then you must remember,' he said. 'Now, patch it up with Phil this evening. This place is nearly ready to sell, and you must make plans then together about what comes next. You look pale and listless, I think you need a holiday. That will get you focused again.'

Phil had said similar things to her many times, but she always took it as criticism. Coming from Patrick it sounded caring, and she wondered why she got things so mixed up.

'I'd better go down and make the tea,' she said. 'Don't let on I've said anything to you, will you? I'll apologize to Phil on the way home. I'll make it right between us again.'

'Good girl. I'd like to take you both out for lunch next Sunday. There's a lovely place down by the river in Chiswick. I want you dressed up all pretty and a big smile on your face.'

An hour later, as Phil drove Eva towards home, she turned to him. 'I'm really sorry, Phil. I've been awful to you for weeks now, but what I said today was unforgivable. I don't know why I said it. I think I must be losing my mind.'

He was clenching his jaw. When he didn't answer immediately, she was afraid he'd say that an apology was useless now and he'd had enough.

But after a few moments he glanced round at her and reached out to take her hand. 'I'm sorry too that I didn't insist you talk to an expert on these things after the fire. It's pretty obvious to me now that the fire and finding out the truth about Flora unhinged you a bit.'

'You mean you think I really am losing my mind?' she asked in horror.

Phil chuckled. 'Not as in needing a straitjacket or a spell in the funny farm, but I'd say you were clinically depressed. That is of course unless it's because you wish you weren't with me?'

'Of course not,' she exclaimed. 'You are the one good thing in my life.'

'There's more than one good thing in your life,' he reproved her. 'You have lots of people who care about you. You've got

417

a job you like, and enough money coming to set you up for ever. You are young and pretty, you've got plenty of love to give, and the world is at your feet. You've got to find a way of seeing that, and give up dwelling on the past.'

She didn't reply immediately, just sat there looking at her lap. 'In my defence can I just say that it's a year today since Flora died? And it's my real birthday.'

'Is that a defence or an excuse?' he said. 'But Happy Birthday anyway, and let's go out tonight to celebrate it with a slap-up meal?'

A warm feeling ran through her. One of the things she loved most about Phil was the way he didn't sulk or bear grudges.

'That would be lovely,' she replied. 'But then you are always lovely. I've got the day off tomorrow. I'll make an appointment at the doctor's, get my hair done and clean the flat up. I think if I try to think positive, I can prevent a trip to the funny farm.'

He squeezed her thigh. 'Get some holiday brochures. And ring some estate agents to get them to value the house. That should keep you from moping!'

At half past eight they were sitting at a table in the Italian restaurant they both loved in Chiswick. Phil had ordered a taxi, so he could drink. They started on a bottle of wine while they looked at the menu.

It was good to be out in a busy place surrounded by other people enjoying themselves, and Eva found herself sitting back and relaxing in a way she hadn't done for a very long time.

They talked about places where they'd like to live: Phil said he thought a village in Buckinghamshire would be good, while Eva said she fancied living by the sea. But they both agreed, if they had to stay in London, Chiswick would be ideal. It felt like a village, it had the river and it was easy to get out into the countryside from there on the M4.

While they were eating their main course the music began – a duo playing guitars, who made their way through the tables singing Italian songs.

'We should go to Italy for our holiday,' Eva suggested. Patrick talked about it often, and had made her want to see Florence and Rome.

'I don't mind where we go, as long as it's warm and the food's good,' Phil said with a smile. 'And I can make love to you to the sound of waves breaking on the shore.'

They were both quite tiddly when the taxi came to take them home, and Eva nestled happily into Phil's arms in the back seat.

'This is what's important,' Phil whispered to her. 'Just you and me, and a night of love ahead of us. We've got it all, babe. Don't let's fight any more.'

Chapter Twenty-Six

As Phil and Eva were sitting in the restaurant being sere-
naded, Ben and Sophie were driving down from Leeds to
The Beeches.

Ben had asked Sophie up for a long weekend with him in
Leeds. Once he'd read the statement Eva had sent him, he
knew he must let Sophie read it while they were together.

The large manila envelope addressed to him had been
posted to his old flat, and it hadn't been redirected to the
halls of residence where he'd been living since October. It
was pure chance that he happened to call round there to see
a friend and saw it lying in a pile of other mail on the hall
table.

He thought it was only junk mail, as his name and address
were typed. Once he opened it and found it was from Eva,
he knew why she'd typed it – she was afraid he wouldn't open
anything with her handwriting on.

She was right about that; he would have binned it
unopened. And as he began to read the contents, he wished
that was just what he'd done. He only read the first page as
he stood in that grubby draughty hallway. Instead of going
into his old flat, he had to rush out to his car, drive away and
find somewhere away from other people to read the rest.

He had believed, until he read Flora's statement, that the
worst experience he would ever have in his life was his mother
killing herself. That still haunted him; he had once described
it to a friend as like having some sort of growth inside him.
A benign one – he knew it wouldn't grow or kill him – but it

was just there, something he felt compelled to prod at, and feel the ache. And it would never go away.

But as he read his mother's story that ache he'd learned to live with grew into real pain.

He had fully understood why Eva wanted to discover who her father was, and he was as intrigued as she was about their mother's time in Scotland. But he hadn't for one moment believed her insane idea that Flora could have taken some other woman's baby and brought it up as her own.

Then there was her conviction that his father had tried to burn her house down with her in it! That was so far-fetched, it was laughable. Yet neither he nor Sophie had laughed, because they'd seen what it had done to their father being taken off for questioning like a criminal and having The Beeches searched. Hadn't they all suffered enough in just one year?

Ben had always taken Eva's part in the past. He knew his dad had hurt her badly, and she must have felt totally isolated when she rushed off to live in London. When he visited her there she didn't tell him what had gone wrong with Tod, but he'd guessed the guy had dumped her. She'd had a struggle to make the house habitable, and she was only working part time as a waitress. He knew too that he had disappointed her that weekend by going off to see some friends.

But none of that was a good enough reason to blame their father for the fire, and Ben had felt he must distance himself from Eva. He was inclined to agree with his father's opinion that she'd got mixed up with a rough crowd again. Possibly she'd been taking drugs too, which would account for her paranoia. He thought it was likely that when she went off to Scotland with the man she later took to The Beeches, who-ever she'd been keeping company with till then didn't like it, and he torched the house when she returned.

Yet as Ben read his mother's words, hearing her voice as if she was talking to him, he wasn't quite so sure he was right to dismiss Eva's claims. He totally believed that his mother had stolen Eva; no one would make up something like that, and he knew women sometimes got very low after losing their own child. But he didn't want to accept that his father was a bully and a blackmailer.

At the first reading he thought his mother had lied about his father's behaviour to justify herself, but by the second reading incidents that he hadn't understood at the time came back to him.

One which stood out in his mind most clearly was when he was about eight. He woke in the night to hear banging and shouting. He got out of bed and went downstairs, and through the open door to the sitting room he saw his father struggling with his mother. She was crying, the coffee table was turned over and there was broken china on the floor.

He was frightened and he ran back to his room. His father came after him, and he made a joke of it, saying Mummy had tripped over the coffee table and was upset because she broke a vase she really liked. The next day he'd asked his mother about it, but she said exactly the same as his father. She even said she was silly to make such a fuss about a broken vase. But that didn't explain the big bruise on her arm or the fact that she was limping. He looked in the bin too – there were broken cups and glasses, but no pieces of a vase.

There were so many other times too when he had a feeling something was badly wrong. He had memories of Mum with puffy eyes, of her shouting to him to help clear up before Dad got home, and of her looking scared. There were the long silences and tense atmosphere when Dad was home, with Mum scurrying around to appease him with drinks or cake. Ben had always wondered why she never stood up for herself when his father laid down the law about what he wanted. Or why she would laugh, dance and sing with her children when their father was out, but was always so quiet and subservient when he was in.

When Ben was about sixteen, he remembered helping her to prepare vegetables for a dinner party. He asked who was coming, and she'd told him. But she sighed as she said the names and wrinkled her nose, the way she always did when she didn't like something. He asked why they'd been invited, if she didn't like them.

And she'd replied, 'It doesn't matter how I feel, your father wants them here.'

Ben had come downstairs to the kitchen later that evening. He could hear the guests talking in the dining room, but to his surprise his mother was in the kitchen, just standing there, staring into space. He sensed something was wrong and asked what it was.

She smiled at him, and cupped his face in her hands. 'Just escaping from the boredom,' she said. 'They are the most tedious bunch of right-wing morons I've ever met.'

'Can't you pretend you've got a headache and go to bed?' Ben suggested.

'No, I can't. I'll have to go back in there and be nice. But this is our little secret. Don't you say anything to Dad.'

With hindsight many other incidents took on a different hue. He had heard his father speaking on the phone and been puzzled that he wasn't using his usual brisk tone – often he waved Ben out of the room. Were those other women he was speaking too? How many of the nights away from home were really work?

Ben was absolutely certain by the third reading that his mother had been entirely truthful. There was such clarity, no flowery adjectives, no attempts to pull at heart strings, just a plain statement of facts. And the financial transactions could be checked. He even felt her deep fear that she would lose her children, if she went against her husband.

Yet despite Ben's disgust at how his father had entrapped her, and his growing conviction that his mother was driven to suicide, he still couldn't really believe his father had tried to kill Eva. Why would he? He didn't know of this statement's existence then. He hadn't met Flora at the time she took Eva, and so he could never be charged with being an accessory. So why would he take such a huge risk?

'What are you thinking about?'

Sophie's question startled Ben out of his reverie.

'Same old stuff,' he said, glancing sideways at her.

Their weekend together had started out badly. Sophie had arrived looking like a tart in a very short leopard-print skirt, a black lace shirt that left nothing to the imagination and boots with four-inch heels. She had expected a weekend of wild student parties, and she sulked when he explained he had got her a guest room in the halls with no question of her taking anyone back to her room. He had intended to wait until Sunday before showing her the statement. He wanted her to have one day of shopping and chatting and drinking with his friends before he had to break the news. But she showed him up on Friday night – not just by the way she was dressed, but by being rude to some of the girls Ben liked, being too full on with two of his friends, and guzzling down drink like she had a death wish.

He had to almost carry her up to the guest room, and she threw up on the landing before he could get her into the room. It took him about half an hour to clear it up, and at one point she came out of the room again and shouted that he was a drag because he wouldn't take her clubbing.

She looked pale and shaky the next day when he met her

in the refectory for breakfast, but she was still eyeing up his friends and kept going on about wanting to go to a club that night. She didn't even apologize for showing him up the previous evening and expecting him to clear up her vomit. Then when she asked why he was being such a bore, he lost patience with her and blurted out about the package he'd got from Eva.

That did bring her round quickly. He told her the main facts of Flora's statement and suggested they go somewhere quiet where she could read it herself and they could talk about it.

He took her to a cafe he knew that had a room at the back with old-fashioned booth seats. And there, armed with coffee, Sophie read it.

As she read, Ben watched her. He despaired over the way she'd changed since their mother had died: the tarty appearance and the rough people she was hanging around with. And her general belligerence was awful. She was eighteen now, and she kept saying she was old enough to do what she wanted. But Ben felt she was as lost as Eva had been when she went through her goth period.

'I can't believe it!' she said at one point, her eyes full of tears. 'Could Dad really be that evil?'

They talked through many aspects of the story, both bewildered that the home they had always thought of as happy, had only been that way because their mother made it seem so.

'There is one thing,' Sophie said. 'Dad did suggest a few weeks ago that we both see a solicitor and sign our half of the house over to him. He made it sound like a really good idea. The Beeches is too big, and we could both have smaller places of our own. He said now was the time for us to get a foot on the property ladder and have our independence.'

'I hope you didn't agree,' Ben said. 'Eva warned us about that.'

Sophie shrugged. 'I wouldn't take any notice of Eva. Dad's the one that knows about property, not her.'

'If he was on the level, he'd just put it on the market in all our names, then the solicitors would divide up the proceeds when it was sold. Do it his way and he could walk off with the lot.'

'He wouldn't do that, he loves us.'

'He said he loved Mum too, but he didn't mind hitting and blackmailing her.'

'We can't be certain this is true,' Sophie said desperately, pushing the statement away from her in defiance. 'Mum could've written it when she was upset about something, and exaggerated.'

'You don't really believe that,' Ben said. 'No one writes a pack of lies and then hides it. Nor do they change their will just a few weeks before they kill themselves unless they don't trust the person they were previously intending to leave everything to.'

'But you surely don't believe Dad tried to kill Eva?'

Ben shrugged. 'How could he have done? You said he was home that night.' He sensed Sophie squirm. 'You were speaking the truth, weren't you?'

'Yes, of course,' she said.

But she dropped her eyes, and Ben knew she was lying.

They had spent the rest of the day mooching about Leeds, half-heartedly looking in the shops and trying to put aside the question of what they were going to do about this statement. Ben felt they should confront their father with it; Sophie wasn't so sure, because she was still living at home with him.

After having a meal they went to the cinema to see *Final Analysis* with Richard Gere and Kim Basinger. One of Ben's friends in Leeds had asked them both to his parents' house for Sunday lunch. On the way home to the halls of residence, Ben asked Sophie if she wouldn't mind toning down her appearance for the day. Predictably, she was offended and stalked off to bed. Yet to Ben's surprise, this morning she was dressed in jeans and a sweater, with very little make-up.

It turned out to be a good day. Mr and Mrs Price, Rod's parents, welcomed them warmly to their rambling and comfortable house in Bramhope and fed them an enormous roast dinner – the first one Sophie and Ben had eaten for weeks. Rod's two sisters and his brother were there too, and after lunch they all lay around in the drawing room watching television and chatting. Ben couldn't help but compare it with past Sundays with his parents. He didn't ever remember all of them relaxing together like this. Sophie enjoyed it too; on the way home she said she wished she came from a family like the Prices. But at least a nice day with them had stopped her thinking about what would happen when she got home.

Perhaps it was because they were both scared of going back to Cheltenham that Ben took Sophie to see all his favourite places in Leeds before going back to the halls, and then said there was no point in leaving till after the early-evening traffic had cleared. Now it was gone nine and they were just leaving the motorway. In ten minutes they would be back at The Beeches.

'How are you going to start?' Sophie asked.

When Ben looked at her he could see she was biting her lip with nerves. 'I don't really know,' he admitted. 'I'll have to wing it. But if he gets nasty with us, we'll leave, right? We'll get a bed and breakfast, or something.'

'It's not going to come to that,' she said, but there was alarm in her voice.

The gates to the drive were open, the light above the porch

illuminating their father's car parked close to it, and there was a light on in both the hall and the sitting room.

'He's going to be surprised to see you,' Sophie said as they parked up. 'He was expecting me to come back on the train.'

Ben was suddenly very scared. He had only been slapped by his father a couple of times in his life, so he had nothing really to fear. But then he'd never before tried to stand up to him.

Sophie opened the door with her key. As they walked in she called out, 'It's me, Dad. Ben's come back with me.'

Andrew appeared in the hall within seconds, wearing a dark-red pullover and grey slacks. 'Well, this is a nice surprise,' he exclaimed with a wide smile. 'Good to see you, son. Needed to see your old man?'

'Yes, Dad,' Ben said somewhat sheepishly. 'I had some things I wanted to talk over with you.'

Sophie shot him a 'not straight off' kind of look.

'Need some cash, I suppose?' Andrew said. 'You kids need to learn to live within your means. I'm not a bottomless pit of money.'

'I'll make us some tea,' Sophie said, dropping her holdall in the hall and darting into the kitchen. 'Do you want one, Dad?' she called out.

'No, I've got a whiskey,' he said, moving to go back into the sitting room. 'Get your tea and come in here to sit down, it's chilly in the kitchen.'

Ben hung his coat up on the peg in the hall, took Flora's statement from the pocket and went to the kitchen.

Sophie made a face when she saw it in his hand.

'I have to,' he said.

'What's that you've got there?' Andrew asked as Ben sat down in an armchair, a mug of tea in one hand and the statement in the other. 'A list of your debts?'

'No, it's something Mum wrote six years ago. You know the Cornish picture you let Eva have? Well, it survived the fire, and when she took off the frame, this was tucked behind the canvas.'

There was a momentary tightening of Andrew's face, but he was quick to control it. 'Oh, I see. Not satisfied with blaming me for the fire, now she's trying a different tack.'

'What makes you think it's something bad?' Ben asked. 'My first thought would've been that it was a love letter, or something along those lines.'

'Are you trying to be clever with me, son? Has that little witch been getting to you?'

'Dad, I'm telling you Eva found this behind a painting – or rather, an art restorer did. It isn't a fake, it's Mum's handwriting. And Eva sent Sophie and me a copy because she thought we had a right to see it. Why are you being so defensive? Do you know what's in it?'

'I can tell by the way you look that it's upset you. I know too that your mother could be a conniving bitch. All I did was put two and two together. So what is she claiming?'

'She has written about stealing Eva,' Ben said. 'Exactly how, why and when.'

Andrew didn't come back with a retort immediately. 'She really did that?' he said after a few moments.

'You know perfectly well that she did,' Ben said scornfully. 'Maybe not at first, but she told you when you were buying this house. Don't lie about it now, it's all in here.'

'OK, so she did tell me. What was I supposed to do? Go to the police and get her arrested? She was my wife, for God's sake. She'd had Eva for nigh on two years. And from what she said about the real mother, Eva was better off with us.'

'I can understand you not wanting to shop her, if you loved her,' Ben said. 'But what excuse are you going to offer

429

for the blackmail, the bullying and for hitting her? That disgusts me.'

'How dare you say such things to me!' Andrew got to his feet and moved threateningly towards Ben, his hand clenched in a fist. 'Your mother was a pathetic, neurotic woman with an overactive imagination. What blackmail? When was I supposed to have hit her? God Almighty, Ben, she killed herself. Doesn't that tell you that she was loopy? You don't know what I had to put up with.'

'Stop it, Dad!' Sophie yelled out from the doorway. 'Don't you dare hit Ben. I've read it too, and I believe it.' She moved to stand next to Ben's armchair in a gesture of support for her brother.

Andrew looked at Sophie, his face darkening. 'You too? She's poisoned your mind against me? You would rather take the word of a woman who didn't even care enough about her children's feelings to live and sort herself out, rather than the man who has fed and clothed you all these years? I've worked my fingers to the bone to buy and restore this house, I even took on her kid. Your mother was an idle, selfish woman who thought of no one but herself.'

'But you didn't work your fingers to the bone to buy and restore this house,' Ben said, jumping up and getting between Andrew and Sophie. 'It was Mum's money that secured it. She came up with the plan to sell the land. She made you a rich man, and you treated her like the housekeeper. That's why she killed herself. You pushed her into a corner she couldn't get out of – she knew you would grass her up for taking Eva, she would be sent to prison and she'd lose all three of us kids.'

Andrew gave an angry hollow laugh. 'Is that what crap she's told you? Well, look here, any woman who snatches another woman's baby is mad. And if she gets away with it she's clever too. This is her final bit of revenge, leaving a pack of lies behind to make me look like the villain of the piece.'

'Why didn't you let her paint?' Sophie burst out, squaring up to her father. 'Why didn't you let her have her own friends? And how come you had another woman lined up before she was even buried? And why did you ask me to sign over my share of the house?'

Andrew's fist shot out at Sophie before Ben could prevent it. It made a loud crunch as it connected with her cheek, and Sophie screamed.

Enraged, Ben swung a punch at Andrew, and knocked him back on to the sofa. 'You bastard,' he hissed at him. 'You have just proved everything Mum said about you.'

'I'm sorry, Sophie,' Andrew pleaded. 'You just made me so mad, accusing me of all those things.'

Sophie had blood dripping out of her mouth; she was holding her cheek and looking at her father in horror. 'You did try to kill Eva, didn't you? You weren't here that night. I told the police you were, because I came home late and I didn't want you to know. How stupid am I? To think I was angry with Eva for saying such things about you. I'm going to call the police right now and tell them the truth.'

As she walked towards the phone in the hall, Andrew leapt off the sofa to stop her. Ben went to hit him again, but Andrew parried the blow, kneed Ben in the groin, then grabbed his shoulder and punched him so hard in the face that he fell to the floor writhing in agony.

'Run, Sophie!' Ben managed to call out before he felt the room swirl around him and blackness descend.

Chapter Twenty-Seven

Ben came to, and the moment he felt the broken tooth in his mouth he remembered Sophie and hauled himself to his feet. The room was spinning but a cold blast of air told him the front door was open. Fear for his sister overrode his giddiness and the pain, and made him move.

As he got to the front door he saw movement on the lawn down by the gate. It was too dark down there and too far away to make out whether it was Andrew or Sophie. Seeing the heavy cast-iron boot scraper in the porch, he picked it up and ran with it across the gravel on to the lawn.

Once out of the pool of light from the porch lamp he saw it was both of them. Andrew had Sophie pinned down on the ground. Her legs were moving but she was silent.

'Get off her, you bastard,' Ben roared out, and forced himself to run even faster despite the pain in his groin. Andrew looked round at him for a second, his face just a white blur. Ben could now see he was pressing Sophie down by the throat.

Ben hurled himself forward, brandishing the boot scraper. Once he was close enough, he brought it down hard on Andrew's head. He heard a scrunching noise and a gasp, and Andrew toppled to one side, across Sophie's body.

Pushing him off his sister, Ben picked Sophie up in his arms and ran back to the house with her. He kicked the front door shut, laid her on the hall floor and grabbed the phone to dial 999. While he told the operator the address he knelt beside Sophie. He could see the vivid marks on her neck where Andrew had tried to throttle her, and he knew that if he hadn't got there when he did, she would have died.

'My father tried to kill my sister,' he cried out to the operator, tears running down his face and his heart thumping like a steam hammer. 'He's in the front garden. I hit him very hard, he might even be dead. But I had to stop him.'

He heard the operator repeat his address. She calmly asked him how Sophie was, and where she was now. 'I brought her into the house. She's breathing, but only just. Please get here quickly.'

The phone ringing at eight in the morning woke Eva. She had stirred when Phil got up for work at seven, but went back to sleep again as she had the day off. She thought the call was going to be from Horace, to ask if she could come into work because someone hadn't turned up. But instead it was DI Turner.

'I'm sorry to call you so early in the morning,' he said. 'I wanted to catch you before you left for work. There was an incident last night at your old home in Cheltenham, and I felt you should know about it. I don't want to discuss it over the phone. Can I pop round now?'

'Yes, of course. But what do you mean by "incident"?' she asked, suddenly wide awake. 'Is it Sophie? Is she in trouble?'

'I'll explain when I get there,' he said. 'Ten minutes at most.'

Eva rushed to wash and dress. Her mind was working overtime, imagining what could have happened.

What had Sophie done? And why were the police informing her?

Half an hour later, Eva was crying. Turner was trying to convince her that the 'incident' in Cheltenham, which he'd explained, was not her fault.

Sophie had been admitted to hospital. Turner said she was in no danger now, but she was traumatized and needed to be kept under observation. Ben had been checked out at the hospital too. Although he was battered and sore, he was just relieved Sophie was alive.

Andrew, however, was in a coma. The heavy object Ben had hit him with may have caused permanent brain damage, but as yet the medical staff were unable to say more.

Turner hadn't yet got the full story. The information that had been passed on to him from the local police was that Ben and Sophie had a row with their father. Sophie said she was going to tell the police he wasn't home the night of the fire in London. Ben tried to stop his father from strangling her by hitting him with a heavy object, and it was he who then called the emergency services.

'How can it be your fault?' Turner asked.

'I sent Ben copies of Flora's statement,' she sobbed out. 'He must have done what I asked and showed it to Sophie. Then they went together to have it out with Andrew. I should've warned them not to be too hasty. But I never thought Andrew would hurt them. Why did he?'

'Well, that's for the Cheltenham police to uncover,' Turner said. 'But it seems to me that your stepfather is trying to keep the lid on something more. This isn't just about not blowing the whistle on Flora snatching you, or even abusing her.'

Eva was too upset to be intrigued by his remarks. 'What will Sophie do now? She can't live there on her own, she's too young.'

Turner put one hand on her shoulder to comfort her. 'She is eighteen, Eva, not a child. But I tell you what, I'll phone Cheltenham Police Station and ask them to get your brother to ring you.'

Eva spent the day like a coiled spring, waiting and hoping

that the phone would ring. She wanted to go to Cheltenham. But not knowing what kind of reception she'd get, she didn't dare. She hoped Phil would phone her, so she could tell him about it. But she was frightened of that too, because he'd said she shouldn't give them Flora's statement.

Last night had been almost like the old times. When they got home from the restaurant in Chiswick they'd made love for the first time in weeks, and Phil had been so loving and tender. She actually believed that she'd turned a corner and things would get back to how they'd been before the fire.

To fill the time she spring-cleaned the flat – even the windows and inside the oven. Then, at four o'clock, finally the phone rang.

She rushed to snatch it up. It was Ben.

'Oh, thank goodness! I've been worrying all day,' she blurted out. 'I'm blaming myself for sending you that stuff. How are you? Were you badly hurt? And what about Sophie?'

'One thing at a time,' he said, and he sounded bone weary. 'I've got a real shiner, a broken tooth and I feel sore where he kneed me in the groin. But I'll live. Sophie was bad last night. Dad tried to strangle her, and if I hadn't hit him over the head he would've killed her –'

He broke off for a moment, and Eva guessed he had been overcome by the memory. 'It's OK, Ben. Take your time,' she said. Then she sat down, because her legs were shaking.

'I've never been so scared. It was awful. Dad punched and kneed me first, and I think I was knocked out for a moment. When I came to, I ran out into the front garden and saw he had Sophie pinned down on the grass with his hands around her neck.'

'Oh, Ben, how terrible for you both! Can you tell me more about how Sophie is now?'

'She was unconscious when the ambulance arrived, but

she came round in hospital. Her neck looks terrible, but she's recovering now.'

'Thank God for that!' Eva exclaimed. She wanted to know every detail – where they were in the house, what was said, everything – but she knew it wasn't appropriate now. 'I never thought he'd harm either of you.'

'Until this happened I couldn't really believe he'd hurt anyone. But he was like a savage animal,' Ben said, his voice thickening with emotion. 'Anyway, I've been with Sophie all afternoon. She's doing her drama queen act of course. But she's entitled to, after what she went through. She said to phone you. And one of the police officers said it too.'

'She wanted you to phone me?'

'Yes. Well, I wanted to anyway. We both need you, Eva.'

He was crying as he spoke. Although Eva thought it was the nicest thing he'd ever said to her, her eyes welled up; she couldn't delight in it under these circumstances. 'I'll be there, Ben. I'll leave right now. I sold my car, so I'll have to come by train. Where will you be? At the hospital?'

'Yes, I don't want to go back to the house on my own.'

'No, of course not. I should be there by nine.'

'Bye then, and I'm sorry.'

'Sorry for what?' she asked.

'For not believing you. We both are.'

'None of that matters now. You know I've always loved you both.'

Eva had flung a few things in an overnight bag, and was just writing a note for Phil when he came in.

'Leaving me?' he said, half serious.

She explained in a hurry.

'Bloody hell!' he exclaimed. 'Those poor kids.'

'It's all my fault. You were right – I shouldn't have sent them that statement. That's what's done this. And I didn't even tell you I'd sent it. But I've got to go. Ben can't be on his own overnight.'

He put his arms around her and held her tight for a moment. 'I'll drive you there in the van. I can't stay with you – I've got a rush job at work – but at least we can talk on the way.'

She leaned into his chest, finding comfort in his calm manner. 'I expected you to say "I told you so",' she whispered. 'I should've listened to you. But thank you for not saying it, and for being so nice.'

He lifted her face up and kissed her nose. 'Things are bad enough without me adding to them. Now, if you'll make me a couple of sandwiches while I have a quick shower and change, we can leave in ten minutes.'

They arrived at the hospital just after eight. Visiting time was over, and Ben was in the waiting room. He looked terrible; his eye was closed over and very swollen.

Eva introduced him to Phil.

'I'd have liked to meet you under better circumstances,' Phil said. 'But if there's anything I can do to help you and Sophie, just ask. Now, let me take you home. You look dead on your feet. And when Sophie is discharged, if she wants to come and stay with Eva and me, she'll be very welcome – as you will be too.'

'Thanks, Phil.' Ben tried to smile, but his eyes were brimming with tears. 'I'm really glad Eva's got you in her corner.'

The police had been at The Beeches all day, making a thorough search of the place, and a couple of them were still there when they got back. They left shortly afterwards, but told Ben to stay out of his father's study and bedroom, also the sitting room and dining room, until they'd completed their investigation.

Phil stayed only long enough for a cup of tea, and to see Ben into bed. As he kissed Eva goodbye he advised her to sleep in the other bed in Ben's room. 'He might have a nightmare tonight. And if he does, you'll be right there. Tell him how proud you are of him for defending Sophie and saving her life. I expect his feelings are very mixed up – no boy ever expects to have to fight off his own father.'

She looked up at his face, which was wreathed in concern, and felt bad that she'd been nothing but trouble for him. He was such a good man; he deserved better than a girlfriend who lurched from one crisis to another.

After he'd gone she went back upstairs to check on Ben. He was fast asleep already – which was hardly surprising after not getting any sleep the previous night, and so much stress today. The other bed was already made up; it was a little reminder of when Ben and Sophie were small. They used to share the room then, because they didn't like being alone, and the bed had remained in here. She tucked the duvet around him more firmly and lightly kissed his forehead, struggling not to cry.

She went back down to the kitchen and washed up the cups. It felt very strange, being back in the house – creepily strange. So quiet, so large and empty. She had never imagined that it would feel so alien, and even hostile, when it had been her home for as long as she could remember.

Ben had said on the way home from hospital that he'd refused to see Andrew in intensive care, and he was never going to. It was obvious that he was freaked out by opposing feelings. On the one hand, he was shocked to find he was capable of hitting someone so hard, and felt an enormous amount of guilt that Andrew might never recover. Yet on the other hand, he was also still full of anger that the father he had loved and looked up to had tried to kill Sophie.

Phil had been good with him; when they got back to the house he gave Ben a man hug and said he'd done the right thing. 'Your instinct was right – to save your sister at any cost. Don't be ashamed of that, because it was very brave. It won't be easy to come to terms with what your dad has done, but you've got nothing to feel guilty about.'

But what was going to become of Sophie now? Ben would of course go back to university. He'd spent enough time away from home during the last year to adjust to taking care of himself. But Sophie needed a real home with supervision; after this attack she was likely to feel very insecure for some time.

And what would happen to this house? If Andrew remained in a coma, would it even be possible to sell it without his agreement? If he recovered and was sent to prison – what then? Sophie couldn't live in the house alone or take care of it. But without anyone living there, it would fall into disrepair.

'No point in worrying about that now,' she said aloud, and her words seemed to echo eerily. The house didn't look as well cared for now as it had when Flora was alive: she saw there were fingermarks on the cupboards and doors, the skirting boards were dirty, and when she opened the oven she saw it hadn't been cleaned for a long time, perhaps not since she last cleaned it.

Was Rose still coming in? If she was, she wasn't doing a very good job.

She wandered back into the hall. Sophie's holdall from the weekend in Leeds was still there, and there were traces of grass on the carpet that had perhaps been brought in on Ben's shoes when he carried Sophie back in here.

Looking up at the skylight above the stairs, she could see stars in the night sky. She remembered she used to sit on the stairs as a little girl and look up at them, imagining angels lived on them. She thought Flora must have told her that.

'If you're there, Mum,' she whispered, 'help us through this. I don't know what to do.'

When Eva walked into the hospital ward to see Sophie the next day, she wasn't prepared for the rush of emotion she experienced on seeing her sister after such a long time.

In a hospital gown, and wearing no make-up, she looked closer to fifteen than eighteen. When she saw Eva she held out her arms, like a small child wanting to be picked up.

Eva ran the last few yards to her and hugged her tight to her chest. 'Poor baby,' she murmured. 'I'm so sorry.'

'It's me who should be apologizing to you,' Sophie sobbed, her voice very hoarse and strained. 'I've been so horrible to you. I can't believe that you would come to see me after that.'

'I didn't stop loving you,' Eva said, and she gently moved back from Sophie and dried her eyes with a tissue. 'How are you feeling?'

She could see the vivid fingermarks on her sister's neck. The amount of pressure Andrew must have used left no doubt that he really was trying to kill her.

'It's hard to swallow. I've only had drinks so far,' Sophie said, catching hold of Eva's hand and holding it as if she wasn't going to let go. 'But the worse thing is, I thought Dad really loved me. How could he do it?'

'I don't know,' Eva admitted. 'But people do all kinds of things when they feel threatened – even to those they love.'

'I've been knocking off college, hanging around with dodgy people and staying out half the night,' she croaked out to Eva. 'If I hadn't lied to the police about Dad being home that night, it wouldn't have come to this.'

'But you've told the truth now,' Eva said. 'And you haven't

440

done anything worse than I did when I was younger – with far more reason, because you'd lost your mum.'

'Everyone will be talking about us. First Mum and now Dad. I don't want to be in Cheltenham any longer,' she said as tears trickled down her cheeks.

'You don't have to stay here,' Eva said. 'You can go anywhere you want. But first you have to recover from this, and I'll help you.'

'Is Dad going to die?' Sophie asked.

'I don't think so. But he's got to have an operation in the next day or two. The doctor told me he's got a good chance of recovery, but I doubt he'll come out of it the way he was before.'

She didn't tell Sophie that perhaps it would be better if he didn't recover completely, because with two charges of attempted murder hanging over him, he was likely to face a long prison sentence. But she knew there was something more. The police were going through the house with a fine-tooth comb, delving into all his business interests, as if they knew something more about him.

'Do you mean he'll be paralysed.' Sophie's eyes went wide with horror.

'I doubt that,' Eva said. 'But stop worrying about him. You need to think about yourself and what you want to do now.'

'I'm so glad you're here, Eva,' Sophie said, and tears filled her eyes again.

'Ben's here too. He decided to wait outside to give us time to talk alone. Shall I go and get him now?'

Sophie just nodded, and tears streamed down her face.

In the next few days, in between visiting Sophie in hospital and supporting Ben as he struggled to make sense of everything, Eva tried to do what Flora would have done. She cleaned the kitchen properly, tidied Sophie's bedroom ready for when she came home, made a couple of cakes and cooked enough food so that, after she and Ben had eaten some, the rest could go into the freezer for the future.

But however upbeat Eva tried to be, she could see no easy solution to anything. The Beeches, the on-going police investigation, Sophie and herself – it was all too much to get her head around. She'd rung Horace at Serendipity, and he'd been very understanding. But he wouldn't keep her job open for her indefinitely.

If she took Sophie back to London, it would soon become difficult living in Phil's flat; it wasn't as if the two of them had always been close. Eva didn't want to stay one minute longer in The Beeches than she had to. But she couldn't leave Sophie on her own in the house – or anywhere – until she knew she really was alright.

Ben had made a tentative suggestion that if The Beeches was sold, he could buy a place in Leeds and Sophie could stay there with him. But did he mean that? And how long before the house could be sold, if at all?

The police finished searching the house and took away many files and folders from Andrew's study. Eva wondered what they were hoping to find. What had Andrew's accounts and work files got to do with his attack on her, or on Sophie? When she asked one of the officers, he just said it was routine. But she didn't really believe that.

At the end of the week Sophie was allowed to come home. Phil drove up on Friday night with more clothes for Eva. Despite all the problems they had, nothing seemed quite so bad once he was there: the weather suddenly turned a bit warmer, the bruising on Sophie's neck was less vivid and she could eat solid food again. The swelling around Ben's eye had gone down a little, and it had turned from red to black – but with a purple tinge, which Phil said meant it was beginning to fade. Phil was good at making them all feel more optimistic and safe. He cut the grass, checked all was well with the swimming pool, and even changed the locks on the front and back doors as a precaution in case Andrew had given a key to anyone.

It was Phil's suggestion that Ben go back to Leeds on Sunday. He pointed out that Ben shouldn't miss any more lectures, and Sophie was fine with Eva. The plan was that both Ben and Phil would come back the following weekend, and during the week Eva would go to the solicitor to find out the legal position about the house. Phil hoped that by the time he and Ben returned, the police would have decided if Ben was going to be charged with anything. And there might also be news from the hospital about Andrew.

'Ring an insurance company and get cover to drive Andrew's car, so you aren't cut off,' he said to Eva as he was getting ready to leave for London on Sunday night. She had been using taxis and she couldn't go on doing that. 'Sophie will be OK in a week or two. Try not to agonize over what to do with her, because she's a bright girl and I bet she'll come up with a plan of her own. I'm going to ring Patrick tonight when I get back, to tell him what's happened. I wouldn't be surprised if he didn't come up here to hold your hand. Now, will you promise me you'll stop worrying about your job and being away from me? The job isn't really important, and we've got the rest of our lives to be together.'

Eva clung to him. She knew he was right about everything, but she didn't want him to go. She was scared to be alone with Sophie in a house that held so many bad memories.

Several journalists had phoned to ask questions, and a few of the neighbours had called too. They pretended to be concerned, offering help, but she knew full well they were just digging for dirt. She hated being under all this scrutiny, feeling she had to be responsible for everyone. The awful tiredness and the desire to be alone that she'd been experiencing in London were coming back again. They had disappeared in the rush to Cheltenham, and stayed away while she was looking after Ben, but now the feelings were here again, creeping up on her, and she was afraid she wouldn't be able to cope.

But she didn't say any of that to Phil. He'd said today that he was proud of the way she'd taken care of Ben and Sophie, and how she'd instantly forgiven them for doubting her. She didn't want him to think she was the kind of person who crumbled under pressure.

Chapter Twenty-Eight

Eva opened the front door of The Beeches on Tuesday morning to a policewoman accompanied by a tall dark-haired man in plain clothes.

'Hello, Eva,' the woman said. 'I'm WPC Markham. We've met before, but I don't expect you'll remember.'

'I do,' Eva replied. 'You were here the night I found Mum. You were very kind. Do come in. I expect you want to talk to Sophie.'

Markham introduced the man with her as Detective Inspector Fellows. 'We'd actually like to speak to both of you – and Ben too, if he's here. How is Sophie now?'

'Ben's gone back to Leeds,' Eva said over her shoulder as she led them to the sitting room. 'We thought it was best he didn't miss any more lectures. He'll be down again at the weekend. Sophie's a bit down in the dumps, but no one would expect her to be anything else after just a week.'

'And you? It must be difficult for you to come back here after all that's happened?' Markham said as she sat down on the sofa.

'I don't like being here much, but I've got no choice. Sophie needs me right now. But let me call her, she's doing something upstairs.'

Eva returned moments later with Sophie and introduced her sister to both the officers.

Markham hadn't been on duty the Sunday night when Patterson attacked his daughter. The last time she'd seen Sophie was when she interviewed her after the fire in London, when she had looked very tarty and was astoundingly belligerent. But now she looked more like the young girl she'd met a year ago: no make-up, hair in two plaits, wearing a pink tracksuit, and with a thin scarf hiding the bruises on her neck. But she was very pale and she looked frightened.

'Hello, Sophie,' Markham said. 'You've had an awful time of it. I wish we could tell you and Eva that it's almost over, but I'm afraid these things take time.'

'I made a statement in hospital about lying when I said my dad was here the night Eva's house was set alight,' Sophie blurted out, perching on the edge of the sofa and wringing her hands with nerves. 'Am I in trouble for that?'

'No, we aren't here about that. Detective Inspector Fellows wants to tell you both about some new developments.'

Sandra Markham liked and looked up to Ian Fellows. He was in his late forties and dedicated to his job. He had superb insight into the criminal mind, was sensitive with victims and easy on the eye – six foot tall, with sparkling blue eyes and a physique a man half his age would envy.

'Well, Eva and Sophie,' he said, looking from one to the other, 'you've both had a rough time of it and I don't want to make it any worse for you. But as you are probably aware, we've been checking your father's financial affairs since he attacked Sophie. We found some discrepancies, which led to an audit at Portwall Papers, his employers. I'm sorry to tell you this, but we have found evidence of fraud and embezzlement.'

Eva gasped, and Sophie looked at her as if expecting an explanation of what this meant.

'I won't go into all the details, as it's very complicated, but basically your father has been supplementing his salary in various fraudulent ways for some years. He's been very clever. The system he used might not have been discovered for years, if we hadn't been looking for reasons why a normally calm family man had reacted so violently towards his daughter.'

'But fraud, embezzlement?' Eva repeated in puzzlement. 'How can that be connected to what he did to Sophie?'

'Let me take you back and make things a little clearer,' Fellows said patiently. 'When our colleagues in London were investigating the fire there, Andrew appeared to have no motive for an arson attack. I know that you believed it had to be because of Flora taking you as a baby, but Andrew had no involvement in that – he hadn't even met Flora when that took place.'

'My boyfriend thought that,' Eva nodded. 'He said it didn't add up.'

'Well, Eva, we still have no real proof that Andrew did start the fire. But thanks to Sophie admitting he wasn't at home that night, and his reaction to her telling him that she was going to the police, we are fairly certain he was responsible. As for his motive – well, we think it was because of the old diaries you found. This is only supposition, but it is possible that when you told him about them he was afraid there was something damaging about him in them.'

'But there wasn't. The diaries ended before Flora even met him,' Eva said.

'Did you imply to him there was something?'

Eva thought about it. 'Well, yes. He phoned me when I was up in Scotland and he was nasty, so I hinted I'd read something about him. I just wanted to wind him up, because he'd been mean to me.'

'That clearly touched a nerve,' Fellows said. 'You see, the Met made some inquiries about him after the fire. They found that, a year or so after he met your mother, he came under suspicion of malpractice with the company of estate agents he worked for in London. It was believed, though it couldn't be proved, that he was taking bribes from people wishing to buy a property. He ensured the owners of the property never knew that there were other potential buyers offering a higher sum. It is well nigh impossible to get evidence of such transactions – the person who offered the bribe isn't going to admit it, and it would have been cash, no paper trail. But he acquired a very expensive car at that time, for which there was no hire-purchase transaction. Nor was there any other evidence of how he paid for it. And he left the company he worked for, and London, in something of a hurry.

'We have no way of knowing whether your mother knew about this – somehow, I doubt it. But he might have thought she had her suspicions and had written about it in her diary. When you said you were going to the police about your birth, he was afraid you'd be showing the diary to us.'

'But if it couldn't be proved twenty years ago, how could it be proved now?' Eva asked.

'Quite so,' Fellows agreed. 'But his alarm was caused by realizing that if the suspicions about him back then came to light, we would be likely to probe into his more current affairs. As we did.'

'So it wasn't ever about me being snatched then?'

'We very much doubt it. We think his plan was just to destroy any evidence that might be in your house.'

'And me with it!'

Fellows pursed his lips. 'That appears likely. Though he may have thought you weren't in the house.'

'Was it a lot of money he took?' Sophie asked in a small voice.

'Over the years, yes.'

'But how?' Eva was puzzled. 'Portwall is a paper company,

448

and he was a sales manager. He didn't handle cash, as far as I know.'

'But he did deal with setting up subsidiary companies, factories, finding new suppliers. And he made deals on Portwall's behalf all around the world. Portwall have some proof now that he took backhanders for negotiating deals, just as he did in the London estate agency. That sort of practice is common enough in many quarters, and not actually illegal, but a principled company like Portwall would never countenance it. But what is totally illegal is diverting funds owed to Portwall into an account he set up for himself.'

Eva didn't really understand how that would work. But then she knew very little about business, and it hardly mattered to her anyway.

Fellows nodded knowingly, as if he understood what she was thinking.

'We think it took him a few years working for Portwall, and gaining their trust, before he took the plunge and began his scams. We found a ledger kept by Flora while they were renovating The Beeches. Bribes, by their very nature, have to be kept secret by both parties and cannot go through a bank account. We found nothing to suggest that Andrew had been paying workmen in cash at that time. Flora had itemized the cost of materials, labour charges and so on, and also cheques issued from their bank account to pay for this. It all balanced with invoices that show who the payments were made to.

'But several years later, although your parents appeared on the face of it to be living within their earned income, we discovered they had acquired a number of expensive items, which did not show up as having been paid for through a bank account or with a hire-purchase agreement. We found the bank account into which he'd been diverting money from Portwall, and also a fake passport. We also found proof that he was disposing of the cash he took in bribes by paying workmen in cash, buying antiques and gambling.'

'How can you prove that?' Eva asked.

'A tip-off led us to a builder who admitted he was paid several thousand for a job he did for Andrew, and signed a statement to that effect. He's no longer working and has cancer. But we think he was anxious to cooperate because he was appalled by Flora's death and Andrew's attack on Sophie.'

Eva didn't know what to say. She found it hard to believe that the man who had always lectured her about being honest should be so dishonest himself.

'However, I don't expect you two girls need chapter and verse about the fraud. I think that you probably want to know what made a measured, methodical man turn into someone capable of being a potential killer?'

Eva nodded.

'We believe it was the aftermath of your mother's death,' Fellows said, looking from Eva to Sophie. 'First, he found he couldn't claim on her life insurance, and it was a very large sum he lost out on. Then, to find she'd made a new will leaving her half of the house to his children, and not to him, must have enraged him. We also discovered in our investigation that he took out a mortgage on this house ten years earlier, forging Flora's signature to do it.

'If the house had been left to him, he could have sold it, paid off that mortgage and bought somewhere smaller. Maybe he'd even intended to go straight. But that was no longer possible. He couldn't sell the house without Ben and Sophie's agreement. He must also have been very scared that it would come to light that he'd forged the mortgage document.'

'But why did he take out the mortgage?' Eva asked. 'He had a good job, and if he was doing all this other stuff on the side, what did he need more money for?'

Fellows shrugged. 'He had become accustomed to living beyond his regular salary. We discovered he stayed in very grand hotels when he was away from home, entertained lavishly, and he frequently gambled in casinos. All in all, he appeared to see himself as something of a playboy.'

'What an idiot!' Eva exclaimed.

'Don't say that,' Sophie retorted. 'We all had a nice life, and he paid for it – like the swimming pool, for instance.'

At the mention of the swimming pool Eva was sharply taken back to five years earlier. She had come downstairs late at night to get some hot milk, because she couldn't sleep, and had stopped short in the hall. She could hear Andrew talking on the phone in the kitchen.

'I'll collect the cash from you in Paris,' she heard him say. 'Next Monday OK for you? Usual place?'

There was a brief silence as he listened to the person he was speaking to. 'Don't worry, it won't be going in any bank. I've got plans for it.'

Andrew laughed at something the caller said. 'Don't worry on that score. I don't tell her anything,' he said. 'She'll just think I got a bonus.'

Afraid she would be caught eavesdropping, Eva went back upstairs. She was a bit puzzled by what she'd overheard, but it didn't really mean anything to her. The only reason she remembered it now was because of what Fellows had said. A few days after hearing her father speaking on the phone she saw architect's drawings on the dining-room table.

She asked Flora what they were for.

'Your dad has been going on about converting the old stables into an indoor swimming pool for years,' she replied. 'Those are plans for it.'

Eva was thrilled and started asking excited questions about when it would be done, how big it would be, and things like that. But she immediately saw by Flora's expression that her mother didn't share her excitement.

In fact Flora looked very worried about it. 'I can't tell you anything, darling. It's probably only one of your dad's pipe dreams that won't happen anyway. We haven't got that kind of money.'

Eva almost reassured Flora that Andrew had got the money. But she stopped herself, because she thought he might be intending to surprise them all with it. It certainly never occurred to her then that he was doing something shady.

But she *was* a little confused by Flora's continuing anxiety when the work began. She spent a lot of time watching the men with diggers doing the excavation work, and Eva overheard her questioning Andrew more than once about where the money was coming from to pay for it. Andrew had been very flippant about it, at least in Eva's hearing, but back then she fully believed that her dad was utterly reliable, straight as a die, and if he said he could afford it, then he could.

In fact her view was that Flora was just being a wet blanket.

But now that she knew the other side of Andrew, she wondered if Flora had guessed he was getting the money dishonestly. Was that another worry for her?

'Was that builder who made the statement the same one who put in the swimming pool?' she asked.

'Yes, Eva. He was – it was the last big job he did before he became ill. He kept a record of all the payments he received, and the dates.'

'What's going to happen to my dad?' Sophie asked in a small voice.

'We're told they are operating on him tomorrow,' Fellows replied.

'But what if he dies?' Sophie asked, her eyes wide and frightened.

'His surgeon is optimistic he will recover, Sophie.'

'But if he does, you'll send him to prison for years and years,' she said accusingly.

'Let's just cross that bridge when we come to it, shall we?' Fellows said.

Sophie fled from the room, sobbing.

'Oh dear, I did my best to be tactful,' he said.

'Sophie's very confused and easily upset at the moment,' Eva said. 'She was always a daddy's girl and it's very hard for her to hear he wasn't the man she thought he was. But can you tell me what will happen to him, if he does pull through?'

'He'll be charged with two counts of attempted murder, and fraud. I'd say he's likely to get something between ten and fifteen years in prison.'

Eva thought about this for a moment. 'But what will happen to this house then? And if he should die, will Ben be charged with manslaughter?'

'First, I would advise Ben to see a solicitor as soon as possible to apply for Power of Attorney – that way, he can deal with the sale of the house and any other assets. As to whether he will be charged . . .' Fellows paused, looking to Markham as if unsure of how to proceed.

'Well, Eva,' Markham continued for him, 'should Andrew not recover from the operation, I'm afraid that unfortunately Ben will be charged with murder or manslaughter. That is the law. I know that sounds grossly unfair, given the circumstances, but let me assure you now that he is certain to be acquitted at the trial. Self-defence, or defence of another, is a complete defence, and no jury would find him guilty when they hear what happened. But Ben will need a solicitor to act for him.'

This all sounded like a nightmare to Eva. 'So what happens if Andrew survives the operation but doesn't recover enough to stand trial?' she asked.

'I can't answer that now, Eva,' Fellows replied. 'Let's wait until he's had the operation. We'll be off now, and let you go and comfort Sophie.'

'I hate that expression, "Let's cross that bridge when we come to it", don't you?' Sophie said when Eva found her up in her bedroom after the police had gone. 'It means people haven't got a clue about anything.'

'No, it doesn't,' she said, sitting down on the bed beside Sophie. 'It means that sometimes things sort themselves out while you wait. So there's no point in getting into a state about something that might not happen.'

'I suppose you mean if he dies, then they don't have to do anything. What about if Dad does survive but he's a vegetable?'

'I suppose he would have to be moved to a hospital that specializes in head injuries.' Eva stroked her sister's hair in an effort to comfort her. 'But he'll still get the kind of care other injured people get. They won't ill-treat him because he's done something bad.'

'It will be better for him if he does die during the operation, won't it?'

Eva didn't know how to respond to that. It was true Andrew's future looked grim – either stuck in a hospital for the rest of his life, or a very long spell in prison. She thought he richly deserved it. But it was different for Sophie; until she read her mother's statement she'd only seen his good side. She was probably clinging to the idea that, if she hadn't said she was going to tell the police he wasn't home on the night of the fire, none of this would have happened.

'But it won't be better for Ben. He'd be charged with murder, or manslaughter,' Eva said. She wasn't going to remind Sophie now that, but for her father, she might still have a mother.

'Can you still love someone who isn't the person you thought they were?' Sophie asked, and she began to sob.

Eva drew her sister into her arms. She realized she'd underestimated the effects of shock. Sophie had seemed almost her old self by Sunday afternoon when Ben left for Leeds. But perhaps the reality of what had happened to her hadn't quite kicked in then. Now, along with knowing her father had tried to kill her, she had to deal with hearing he was a thief.

'Remember Mum used to say she didn't always like us, but she'd always love us?' Eva said. 'I think that covers your dad too. He's turned everything upside down for you. But it will get better, I promise. Look at me! I thought I was going to die in that fire, but I got over it.'

'I used to think I was an OK person.' Sophie sobbed into Eva's shoulder. 'I was popular at school, top of the class most of the time, I had a nice home and parents that all my friends envied, and I thought I was going places. But it was all fake. I'm not anything, I'm going nowhere. People will always whisper about me. "Remember her? Mum topped herself, dad tried to kill her, and he stole money from the company he worked for." I can hear them saying it, Eva. I can't bear it.'

Eva's stomach turned over in sympathy. She had never thought of herself as a person who was going places, but she certainly knew what it felt like to be ashamed and second-rate.

'You are still an OK person, Sophie,' she replied, holding her sister tightly and rocking her. 'You are still clever and pretty, and all that stuff about our family will be forgotten in a while. You don't have to stay in Cheltenham. You can start again in London, Leeds, anywhere you fancy. Just like I did.'

'I'm not strong like you,' she whimpered. 'I go to pieces.'

'Ben and I won't let you go to pieces. We'll hold you up until you are strong enough to stand on your own. I promise you.'

'Why are you so kind to me?' Sophie asked. 'I've been vile to you. I didn't stick up for you when Dad was nasty to you. I didn't try to see you after you left here. I didn't even care that you'd nearly died in a fire.'

'If you can admit that now, there's nothing much wrong with you,' Eva said. 'You were too young to lose your mum, you were all mixed up. I was vile when I was your age too. Can you remember me in all my goth stuff? How embarrassing was that!'

'You were never vile – not to me.' Sophie sniffed. 'And I thought your goth stuff was pretty cool.'

Eva chuckled. 'Now that is worrying! So why don't we go and get a couple of videos to watch tonight, and make ourselves something nice to eat?'

Sophie moped about all day, and finally went off to bed about ten. Eva rang Ben then to tell him about the police visit. She hadn't wanted to talk about it in Sophie's hearing, as she was so disturbed by it all. She ran through about the operation Andrew was to have at the hospital, and then the news of the fraud.

Ben was horrified and incredulous to hear about the fraud. 'But he was always going on about being honourable and stuff,' he said. 'What a bastard! After everything he's done, I hope he does bloody well die.'

'I think I'd rather he gets well enough to stand trial and go to prison,' Eva said. 'Then he'll have years of reflecting on what he's done to us all. Besides, I don't want you to be charged with murder – even if the police do say you'll be acquitted. However much public sympathy there would be for you, it would still follow you around afterwards like a bad smell. Journalists are sniffing around here still. They can't write anything much about it until after his trial, but you can bet they're collecting up dirt even now in readiness.'

The story about 'an incident' at The Beeches had hit the local paper within twenty-four hours. It described the Pattersons as a troubled family, and went on to report that Andrew Patterson was believed to have attacked both his two younger children. In defending themselves, their father had received serious injuries and was in hospital in a critical condition. A footnote was added about Flora's suicide a year earlier and an arson attack on Eva's home in London. As yet there was no mention of Eva's parentage, or about Andrew's fraudulent transactions, so presumably the police had kept a lid on that. But the reporters were bound to discover it soon, and when they did, they were going to have a field day. It wouldn't only be the local papers then, but also the nationals – and even television. It didn't bear thinking about what that would be like for all of them.

'You must stop Sophie talking to any of her friends about it,' Ben said. 'No one would have known much about Mum's suicide, if it hadn't been for her.'

'To tell the truth, I'd be happy to see her phoning someone, or talking to a friend,' Eva admitted. 'But she doesn't want anyone near her. It's like all the stuffing has been knocked out of her. She's totally mixed up about her dad, one minute hating him, and then crying because she loves him. I really don't know what to do or say.'

'I just hate him now,' Ben spat out. 'To think I used to look up to him! All those lies he's told, making out he was so perfect when he was stealing from his company. The stuff he did to Mum, and to you. Just dying on the operating table is far too good for him. I want to see him suffer.'

'I don't like to hear you saying things like that,' Eva reproved him. 'You've always been such a peacemaker and so understanding. Don't let this change you, Ben.'

'Of course it will change me, none of us is the same any more,' he said.

'But we must all try to calm down. I know it's terrible, but ranting doesn't make it any better.'

'I bet you're glad you aren't his daughter now,' Ben said, and it sounded as if he was crying. 'At least you can distance yourself from it.'

'No, I can't, Ben,' she said. 'Because I love you and Sophie. And I will never distance myself from either of you.'

She asked him a few questions about how he was coping, and how he'd explained away his black eye to his fellow students. Then she told him he must get a solicitor's advice about both Power of Attorney and in case any charges were brought against him.

'You are being marvellous. I don't know what Sophie and I would've done without you,' he suddenly blurted out. 'I'm so very sorry that I was nasty to you about the fire. I should've trusted you.'

'Forget it, Ben,' she said. 'I have. The main thing now is to stick together and get through this.'

Olive came round on Wednesday afternoon. She looked anxious when Eva opened the door to her. 'Tell me to go away if you can't face visitors,' she said.

'I'm really pleased to see you,' Eva said, and meant it. Olive's straightforward manner and often blunt advice was just what she needed. 'Come on in.'

'I see a couple of vultures are still hanging on outside.' Olive looked back at the two journalists hovering down by the gate. 'I hoped they'd stop me to ask me something. If they had, they'd have got a piece of my mind. But I suppose they took one look at my face and guessed what I was about.'

'They've been such a pain,' Eva sighed. 'They try to waylay us if we go out, and they keeping phoning. I'd disconnect the phone, but someone important might want us.'

Olive handed Eva a carrier bag with some clothes in it. 'I don't suppose you care much what you look like now,' she said. 'But there's a dress, some trousers and a couple of tops. When things are bad for me, I always feel a bit better if I'm wearing something nice. I hoped it might work for you too.'

'That was very thoughtful,' Eva said, peeping in the bag. 'You've got such good taste. I'm sure I'll love them when I try them on later. Thank you so much. Coffee?'

Over coffee Eva explained the recent events. Olive only knew what had been in the local paper, and that had been somewhat vague. 'I should have phoned you. But to be honest, I've found it a struggle just to keep things together here.'

'I can imagine,' Olive said in sympathy. 'You've been through hell, Eva, and it sounds as if it will get worse before it gets better. But if there's anything I can do, anything at all, even if you just need someone to let off steam to, I'll be there. You can just leave a message on the answerphone for me, and I'll pop round. Everyone at work is thinking of you and your family. They all want to express their sympathy and affection too, but I did tell them you need privacy just now.'

'I really appreciate their kindness. Do tell them that I miss them all, and think of them a lot. Perhaps one day when this is all over I can pop in and say hello to them . . .' She paused then, overcome by emotion.

Olive put her arm around her and gave her a rather awkward

hug. 'I'm no good at this kind of thing,' she said. 'I think when I was designed they left out the "ability to demonstrate affection" bit. But I think you know that I care about you, Eva. So forgive me for my shortcomings.'

Eva gave her a watery smile. 'You've proved to me many times that you've got a big heart,' she said. 'If you started to get soppy with me now, I'd find it scary.'

Olive planted a kiss on her forehead and said she had to get back to work. 'Keep in touch, even if it's only a couple of words on a postcard or the briefest phone call. And ask that man of yours to feed you up. You are getting very skinny.'

On Thursday afternoon Patrick arrived with all his customary warmth and strength – and a couple of bags of food from Marks and Spencer. 'I'm betting you haven't felt much like shopping or even eating much, so I got a few treats.'

He had rung Eva the previous night to suggest he came, and it was just what Eva needed. She was finding it hard to cope with Sophie's mood swings. One minute it seemed like she was on the mend – calm, rational and even talking about the future – the next she was crying, full of self-recrimination and convinced that her whole life was ruined for ever. In the blink of an eye she veered from hating her father to feeling sorry for him. Whatever Eva said was wrong, and there were moments when Eva wanted to run away and hide from it all. She was all talked out, her sympathy was drying up, and although she might have promised Sophie she would support her till she was strong enough to stand alone, she could feel herself buckling under the strain.

Ben had telephoned in the morning to say that Andrew's surgeon had contacted him. He reported that the operation had gone well and he was cautiously optimistic for his father's recovery. He was being kept in a coma for now, to aid the healing process, and when they did bring him out of it they would assess if he had any permanent brain damage.

It was typical of Ben to say little about his own plight. He was happy to tell her he had been busy finding a solicitor to help him sort out the Power of Attorney. And he had also arranged an introductory meeting with another solicitor in the same practice who handled criminal cases. But he dismissed Eva's anxious questions about how he felt. All he would say was that immersing himself in his studies worked for him, and that he wasn't allowing himself to look further ahead than a week at a time.

Eva felt that meant he was avoiding thinking about himself. And because of that, she didn't burden him with her anxiety about Sophie. She could share that with Patrick – she knew his advice would be sound, and Sophie would probably feel more secure with an older man around.

As Eva had expected, Patrick was brilliant with Sophie, hugging her and telling her she was beautiful. He struck just the right note, saying he hoped she would look upon him as an uncle, and that he'd come to offer support – not to judge or criticize.

It was good to have him there that evening. They had a meal together, and Sophie came out of herself a bit, telling Patrick how she'd wanted to be an actress but had been turned down at an audition for drama school.

'Even if you aren't made of the right stuff to be an actress, there are other jobs going in the film, TV and theatre world,' he said encouragingly. 'Backstage work, costumes, make-up, all sorts. I could make some inquiries for you – I know lots of people in that world.'

Later, they watched television together. Sophie lay with her head on Eva's lap and seemed much calmer. When Eva went up to tuck her into bed at the end of the evening she said how nice Patrick was, and that she was sorry she was being such a drip. Eva's response was to laugh and say she thought Sophie was entitled to be a drip for a while.

Eva stayed up with Patrick for some time, talking over all that had happened. She felt she had to tell him how shaky she was feeling.

'I'd been sinking into a black hole since Christmas,' she admitted. 'The stuff about the fire was dragging me down, and then Freya. And I was being horrible to Phil. That Sunday I saw you at Pottery Lane I resolved to pull myself together, and that same evening I went out with Phil for a meal and we had a good time. So I felt things were taking a turn for the better. But that very night all this happened. I didn't have time to consider how I felt at the time, I just had to come and take over, but now –' She broke off in a flood of tears.

He hugged her to his chest and let her cry. 'I'm not surprised,' he said. 'You've had one hell of a year – even the strongest will in the world would crack with it. But why don't you try to tell me about what's worrying you most now?'

'I can't see an end to it, that's the biggest thing. This house, Sophie, Ben – I can't just walk away from it all when I've had enough. Sophie needs looking after, and I want to do the right thing by her. But how long can I reasonably go on doing that for?' She leaned back, moving away from him. 'And I keep worrying about stupid stuff – like what we'll do with all the furniture when the house is sold, and how to pay the bills when they come in. It's really scary.'

'OK. First, there's no point in worrying now about what will happen when the house is sold. That might be a year on – or it might never be sold. As for the bills, that's a far more sensible thing to worry about. And the solution is to sell something from the house to pay them and to buy food for you and Sophie. Is there any of your mother's jewellery around? Any antiques? That china cabinet there, for example.' He pointed to a walnut bow-fronted cabinet in the alcove by the fire. 'If I'm not much mistaken, that's Queen Anne and must be worth at least six or seven hundred pounds. That would solve any immediate money worries. But you should talk to Ben about it at the weekend and get his approval. Likewise, you should have a chat with him about Sophie too. Maybe she has a friend in Cheltenham she could move in with in a few weeks' time, or perhaps she could go to Leeds and share a flat with Ben? But I suspect that there's something else the matter. Whatever it was that got you down in London isn't resolved. Can you tell me about that?'

'I just kept wanting to be alone,' she said in little more than a whisper, because she felt ashamed to voice it. 'I didn't want people around me. I was even pushing Phil away, and I love him. I used to daydream of being in a little house miles from anywhere, with absolutely no one asking anything of me. Isn't that crazy?'

'Not at all. I've felt like that many times in my life,' he admitted.

'You have?' She was astonished.

Patrick smiled and stroked her cheek affectionately. 'Oh yes. The time I remember best of all was after Flora left me, and I was living in Pottery Lane. People kept coming round, trying to jolly me along. But all I wanted to do was to be utterly alone, in silence. It got so bad I didn't answer the door or the phone. I used to go for a walk at night, so I wouldn't run into anyone. People who were concerned about me tried to get me to go to yoga, on blind dates, adopt weird diets, or take tranquillizers. They didn't understand that I was OK alone, that people were the problem. Mostly I found they only really came round, under the guise of sympathy, so they could tell me their own troubles. I felt like I was a crutch to half the world.'

'How did you get over it?' Eva asked.

'I went to Canada.'

Eva was surprised that was his solution. 'That's why you went there? I thought you had a job lined up?'

'I had had a tentative offer of work there, but nothing definite. I was just running away from everyone and everything. I didn't go to the people I knew there – not for a while. I travelled around, looking at the breathtaking scenery. And the beauty of the mountains, forests and lakes cured me.'

'That sounds wonderful. But I can't run away, though,' she said, pulling a glum face. 'Sophie and Ben need me to be strong for them, to stay here keeping things together.'

'No, you can't, not now. But later, when this is all over – and it will be over, you must believe that. Then you can go somewhere peaceful. But I'll pass on a tip to you that always works for me. When you are feeling stressed and worn out, lie back in a chair or in bed and picture a turquoise sea, palm trees and a white sandy beach with no one on it but you. Listen for the sound of gulls and the waves lapping, feel the hot sun burning into your skin. Just keep that image in your head, and let your mind float off there.'

'I'll try it tonight,' she said. 'Thank you, Patrick. I feel a bit better just for talking about how I feel.'

The following morning it felt like summer was almost here, with warm sunshine and a clear blue sky. Over breakfast Patrick suggested they go out to the Cotswolds for a walk and to have some lunch in a pub. Sophie seemed very listless and distant, but she didn't put up any resistance to the plan.

They arrived back soon after four – all with flushed faces

from the sun and feeling tired, because they'd walked a long way. Sophie had remained distant, hardly speaking at all, and she'd only picked at her lunch in The Swan.

'I'm going up to have a bath and a lie-down,' she said, as soon as they got in. 'Thank you, Patrick, for today. It was lovely.'

Eva raised her eyebrows to Patrick. When Sophie had gone upstairs she remarked that Sophie didn't often remember to thank anyone for anything.

'Did any of us at that age?' he said. 'Sometimes it takes tragedy and disaster to make us see what we've got.'

Eva and Patrick took cups of tea out to the conservatory. The sun was shining in there and it was really warm. Patrick dropped off to sleep after just a few minutes, and soon Eva reclined her chair and followed suit.

She woke feeling cold, and saw that the sun had sunk down behind the garden wall. As she got up, Patrick woke and looked at his watch. 'Heavens, it's nearly seven,' he exclaimed. 'I must be turning into an old man, nodding off even in stimulating company.'

Eva giggled. 'I've only just woken up too. Must have been the wine at lunchtime.'

They went into the kitchen and Eva put the kettle on. Patrick was looking in the fridge and suggesting he make a prawn salad for them.

'I'll just go and see what Sophie's doing,' Eva said.

Upstairs, Sophie's bedroom door was open. But she wasn't in there. The duvet was crumpled, though, as if she'd just got up.

Eva went to the bathroom next door. 'Do you fancy some prawn salad for tea?' she shouted out at the closed door.

There was no response.

Eva tried the handle, but the door was locked. She hammered on the door with her fists and shouted more loudly.

Patrick came running up the stairs. 'What's the matter?'

'She's in there with the door locked, but she's not answering,' Eva said in alarm. A cold feeling of dread and déjà vu was creeping over her. 'You don't think –?' She couldn't bring herself to finish the sentence.

Patrick banged on the door. 'Sophie! Answer me! You're scaring us,' he yelled.

When there was still no reply he told Eva to stand back. Putting his shoulder to the door, he forced it open. The door frame creaked and splintered and the door gave way.

'Oh God!' he exclaimed, then pushed Eva back. 'Don't look,' he said.

But it was too late. She'd seen Sophie lying there in a copycat death of her mother's – the bathwater red with blood, her eyes wide open, staring sightlessly, and a similar knife dropped on the floor.

'No!' Eva screamed. 'Not Sophie too!'

Chapter Twenty-Nine

Ben increased the pressure on Eva's hand as the curtains closed around Sophie's coffin and they heard the faint whirr of machinery rolling it away to the incinerator. Eva was blinded by her tears, but she knew Ben was crying too.

Phil, on the other side of her, put his arm around her. But there could be no real comfort for her and Ben today. It made no difference that the sun was shining, slanting in through the chapel windows of the crematorium and playing on the flowers, the polished wooden floor and the faces of all those who had come to pay their last respects to Sophie. Eva felt cold – as if it was midwinter, not a glorious spring day.

'I can't live with this' was all Sophie had written in a note left by her bed. When Eva had been told about this by the police, she'd felt much the same.

It was like being caught up in an avalanche: first the shock of the impact, then the desperate struggle to the surface to deal with everything. If it hadn't been for Patrick – and Phil too when he arrived late the same night – Eva felt she would have crumbled completely. As it was, she was barely holding it together.

She'd scarcely heard a word of the service, because her mind kept turning to what more she might have done to prevent Sophie thinking that suicide was the answer. Ben, Phil, Patrick – even the police – all said she had done everything possible, but she still kept asking herself why she hadn't taken Sophie to see a doctor when it was obvious that she needed professional help. And how could she have slept peacefully in the conservatory while her sister was preparing for, and taking, her last breath?

But Sophie had always been a drama queen. And because of that, Eva had imagined that if she ever had suicidal thoughts, she would have announced them loudly. The knife was new; she must have bought it on the one occasion she went out for a walk alone. She'd picked her moment to do it when Eva wasn't alone, which was uncharacteristically thoughtful. But alone, or with Patrick, the moment of finding her dead was just as terrible and devastating. A young life had been wiped out because Sophie was unable to bear the shame of what her parents had done.

Andrew was at the back of the chapel in a wheelchair, handcuffed to a police officer. He'd been told of his daughter's death shortly after he was brought out of the drug-induced coma he'd been kept in since his operation. Ben said he looked wizened and very old, but Eva had refused to even glance at him. She hoped he'd be in that wheelchair for the rest of his useless life. He was to blame for everything.

They weren't inviting the ten or twelve people – mostly friends of Sophie's – who had turned up today back to the house afterwards, because neither she nor Ben could face their inevitable questions. Ben had said he'd rather have a little wake with Phil, Patrick and Eva, because they'd been the only people who had helped since the night Andrew attacked Sophie. The neighbours and many of Sophie's friends hadn't even rung or written a card to show they cared.

The whole story was out now. The headline on the front page of the local paper two days after Sophie died, was 'The Sins of the Father', and the story of how Andrew Patterson allegedly drove first his wife to suicide, then assaulted his son and attempted to kill his daughter, who subsequently took her own life in a carbon copy of her mother's death, was sensationalized for the maximum effect.

It was clear that someone in the police must have leaked the story, because it was all there – albeit using the word 'allegedly' in front of everything. The fraud and the arson attack on Eva's house were dredged up, and the fact that Ben had attacked his father to defend Sophie, which had left Andrew with brain damage. There was a picture of The Beeches, taken with the wrought-iron gates closed, and they'd used that image to suggest that neighbours never really knew what went on behind closed doors. A quote from one of them, who chose not to give his or her name, was: 'We always wondered how Patterson could afford his millionaire lifestyle.' They had published a photograph of Andrew and Flora – one taken at a black-tie dinner and dance a few years earlier – and that too implied that the Pattersons lived a glamorous life.

The only part of the story which hadn't come out in the local press, even when the nationals picked it up, was about Eva being a stolen baby. That in its way was so juicy that, when it did leak, it was likely to cause mayhem. Eva didn't know what she should do about it. Even now, ten days after Sophie's death, reporters with cameras were still hanging around The Beeches. She'd had to take the phone off the hook, because it rang so often.

Phil had said last night that she mustn't let bitterness take over. He didn't really understand that she wasn't bitter. Just empty. What she wanted now was to see Ben go back to his studies, and then she craved being entirely alone. Ben seemed to understand what she meant; he'd admitted that he felt much the same way as her, only his way of dealing with it was to immerse himself in books.

The final prayers were over. As 'Nessun Dorma' sung by Pavarotti began to play, Ben and Eva looked at one another and tried to smile. It was the most unlikely record for Sophie to love – she was more of a Madonna and Kylie Minogue fan. But she had loved it, playing it over and over again. As Pavarotti's voice soared, Eva hoped he was letting her sister's spirit free.

She could remember one day, shortly after Sophie was born, when she'd heard her crying in her crib. Although Eva was only four, she'd gone into the nursery and picked her up to cuddle her. Flora had laughed when she found them together, but after giving her a warning that babies were far more fragile than dolls, and she wasn't to do it again, they'd sat together on the nursing chair. Flora let her continue to hold Sophie, and she said she wanted Eva to always be a good big sister and to love Sophie. She said she hoped they'd always be best friends as well as sisters.

Eva knew she had always loved her, even when she didn't like the way she behaved. She felt now as if a chunk of her heart had been ripped out, and she couldn't possibly imagine a time when it wouldn't continue to hurt.

Six weeks after the funeral, Phil arrived home early from work one day. He was holding in his hand the details of a house that was for sale. 'I've found this dream house in Chiswick,' he said, grinning like a Cheshire Cat. 'It's everything we want. I looked at it this morning and knew it was the right one. I said I'd bring you round at five. I'll just have a shower and change. After we've seen it I thought we could go to the Italian place we like.'

He disappeared into the bathroom, seemingly unaware that she hadn't grabbed the details with any enthusiasm. She glanced at the leaflet, and then tossed it aside. She couldn't cope with viewing a house right now, and she felt angry that he expected her to.

He was back in ten minutes, buttoning up a clean blue shirt. 'What do you think?' he asked. 'Isn't it great? Those photos show how it really is too – high ceilings, all the original cornices, fireplaces and doors. It needs a new kitchen, but that's no problem. You are going to love the garden.'

'I don't want to see it,' she said quietly. 'I've had enough stress and upheaval to last me a lifetime. I've got nothing left in me to cope with moving.'

'You what?' he exclaimed. 'Oh come on, Eva. You can't carry on sitting in here day after day doing nothing. You'll love this place, I know you will. And if we don't act quickly, we'll lose it.'

'You call all I've been doing nothing?' she said, her voice rising. 'I've hardly had a minute to myself since Sophie died. Aren't I entitled to sit about for a while?'

Since the funeral she'd had so much to do that she hadn't even been able to consider going back to her old job. But keeping busy hadn't made the desire to be alone go away. In truth, the only times she'd felt anywhere near being happy again was when she *had* been alone, giving the house at Pottery Lane a final clean and sprucing up the garden before putting it on the market, and sorting out things at The Beeches.

She hadn't expected to feel so sad about Pottery Lane being sold. Once she'd cleaned it all and polished the windows till they gleamed, she had sunk down on to the floor and cried. It wasn't just because it was so beautiful now, but because it was her legacy from Flora. On some deep level she could understand how Flora must have felt arriving back there with a tiny baby. And whether it was right or wrong to take another woman's baby, Flora had loved her, and to Eva she would always be her dearly beloved mother, whatever the rights or wrongs were.

The terror of the night of the fire wasn't something she wanted to be reminded of, but she did want to hold on to the memory of Phil taking her back to the studio after the handbag snatching, all those lovely moments with him when they were just friends, and the bliss of their lovemaking after they came back from Scotland.

She'd become a grown-up there, met Patrick for the first time, learned so many practical skills from Brian. Flora had always claimed the dead looked down from heaven and watched over those they loved. She hoped that was true, and that Flora would understand why she was selling it now.

Phil's words interrupted her thoughts and brought her rudely back to the present. 'I was only trying to help you to move forward,' he said. 'This place is dreary and shabby, and a few pots on a patio don't make a garden. You'd be much happier with a project. Interior design is your thing, isn't it?'

'I thought it was, until my house was set on fire,' she snapped at him. 'All that effort, planning and hard work went up in smoke. I can't think about doing another house yet.'

'What can you think about then?' he asked, his voice dripping with sarcasm. 'You certainly aren't thinking about us any more. Since you wound up things in Cheltenham I've hardly had a word out of you. You're either sitting there staring into space, or you've got your nose stuck in a book. You show no enthusiasm for anything – not food, lovemaking, going to the pictures, or even your appearance. I get the impression that you'd rather be anywhere but here with me.'

'For God's sake!' she yelled at him. 'Just because you saw a house today that you like, it doesn't mean I've got to jump up and down with joy. Just leave me alone, can't you!'

His face darkened. 'Did you want me to leave you alone after that guy assaulted you, or after the fire? Did I leave you alone when all the stuff happened with Andrew and poor Sophie? I've bent over backwards to help, I've felt for you every step of the way. But it's all over now. And now you say you want me to leave you alone? How could you?'

'You don't understand. I'm sick of being talked at, of being expected to snap out of it. I don't want people asking me how I am, what I'm going to do next. I can't answer those questions, because I never get the peace and silence I need to find the answers,' she snarled at him.

'Fuck off! I was only trying to help,' he roared back at her. 'If you bloody well thought of someone else but yourself for five minutes, maybe you'd see that.'

He turned and stomped off into the bedroom, leaving Eva shocked that he'd sworn and shouted at her. But she felt unable to go after him to apologize, so she just stayed on the sofa, rigid with tension.

He came back a few minutes later. He'd taken off the smart shirt and trousers and had changed into jeans and a T-shirt.

'We can't go on like this, Eva,' he said. His voice was so sad, it made her feel even worse than when he shouted at her. 'You won't let me in. It's like living with a domesticated robot. I've done everything I can to try to help. But the Eva I fell in love with has gone, and somehow I don't think she's ever going to come back.'

'I want that Eva to come back,' she said wearily. 'But it's like I'm dead inside. The kindest thing I could do for you would be to clear off, and leave you to find happiness with someone else.'

'I don't want anyone else,' he said firmly. 'But I can't live like this either. The sale of the studio will be completed next week. You've dealt with everything you can at The Beeches, Ben's got the Power of Attorney now, and he's off the hook with the police. It's time you decided what it is you want for yourself. I had hoped that it would be a house for us to share, and to get married. But I know that isn't what you want, and so I'm waiting to hear an alternative.'

He picked up the keys to his van.

'Where are you going?' she asked.

'To see a man about a plastering job,' he said. 'Think on what I've said while I'm gone. I want an answer when I get back.'

He left then. No door slamming, that wasn't his way.

Eva sat there on the sofa, not even able to cry.

Andrew hadn't made a complete recovery. But he was considered well enough to stand trial, which was set for September. Meanwhile, he was being held on remand in Gloucester Prison. When Ben went to see him there, he likened his condition to that of a stroke victim. His speech was slurred, his memory was affected, and he had lost some of the movement in his right arm and leg.

Ben said it was difficult to know whether it was the shock of Sophie's suicide or his injuries that had changed Andrew, but he had broken down and made a complete confession to the police. He was intending to plead guilty, and he insisted that all charges were to be dropped against Ben. He also agreed to give him the Power of Attorney.

'He's like a pathetic little old man,' Ben had said after the visit. 'He's shrunk in every possible way. I wanted to hate him, but the dad I knew isn't there any more to hate. Maybe it's just as well his mind isn't as sharp as it used to be – the

474

way he is now, he can deal with the boredom and the lack of freedom in prison.'

Ben didn't intend to see his father again once he was sentenced. 'There's no point, Eva. I can't forgive him, so I'm going to airbrush him out of my life. I'll take the money due to me from The Beeches when it's sold, for Mum's sake, but only my share. I'm going to instruct the solicitor to give Dad's share to Portwall, to try to make restitution for what he took from them.'

Eva could hardly believe that her little brother could be so grown-up and honourable.

She had stayed in a bed and breakfast near The Beeches while she sorted everything out and organized an auction room to take all the furniture and household goods.

It was so strange going through drawers and cupboards, finding things – such as the board games she, Sophie and Ben had played with. She found an emerald-green scarf of Flora's in Sophie's room, and guessed her sister had been taking it to bed with her. She found herself rubbing it against her cheek and sobbing for both of them. Reminders were everywhere: old dressing-up clothes in the attic rooms, dolls and teddy bears packed away in an old suitcase, Ben's collection of Matchbox cars, and a bracelet of Flora's tucked down the side of a chair.

It was easy to stuff Andrew's clothes into bin bags for the charity shop; nothing of his brought on pangs of sorrow. Yet the pastry cutters in the kitchen, the secateurs in the potting shed, and a half-used pot of face cream of Flora's made her dissolve into tears.

Olive came round to give her a hand on two consecutive days, and it was good to have her there. She was practical – not given to analysing, or offering advice unasked. What she brought to the table was common sense, a listening ear, and a knowledge of which items were valuable enough to sell in an auction and which would only be fit for a charity shop. Anyone else would have picked over things and asked about them. But Olive didn't; she just packed them into the appropriate boxes and didn't allow Eva to get sentimental about anything.

'Take a few things of Sophie's and Flora's, if you like,' she said casually. 'But let everything else go. Possessions can become like chains – especially ones that act as unwelcome reminders.'

When Olive left on the last day, she hugged Eva. 'I know you think you'll never get over all this. But you will. Write down how you feel each day. In a little while you'll have a day when you feel happy, and you'll suddenly realize you haven't thought about Sophie or your mum for a few hours. That will be the start of better times. And believe me, it will come sooner than you think.'

Once everything had been taken away by the auctioneers, and the clothes and oddments had gone to a charity shop, Eva had cleaned the house from top to bottom. Then she locked the doors and took the keys to the estate agent who would be handling the sale for Ben.

Eva had hired a car for the week she was at The Beeches. As she drove out for the last time, she stopped at the gates, got out and took one last look back at the house. It was beautiful, and even though the flower beds around the lawn were now choked with weeds, she could still imagine Flora kneeling on the grass, at her happiest with dirty hands, tending her flowers. Eva hoped whoever bought it would be happy. She certainly wouldn't want to buy a place where so much tragedy had taken place.

Later she drove round to Crail Road. The house looked just the same, though the tenant in her old room had stuck plastic sunflowers on the window. She wondered if Tod still lived there, and if he ever did enrol on his counselling course. He was bound to have read about what happened to her family in the papers; she wondered how he had reacted to the story.

Now as she sat in the flat, Phil's last words ringing in her ears, she remembered that Tod had said she was needy. She had been then, but she wasn't now. Needy people didn't want to be alone.

It crossed her mind that her real sister, Freya, could be feeling the same as her, and maybe that was why she'd never got in touch again. It was just as well she hadn't – Eva knew she hadn't got anything left inside her for anyone else.

So what was she to do? Patrick had said shortly after Sophie's funeral that she ought to get away, right out of England and far away from all the bad memories. He'd meant with Phil of course – but even if Phil was free to go, that wouldn't work. They'd just be taking the same problem with them.

But what if she was to go alone – go to Paris, to Rome and Florence? See all those works of art Patrick often talked about, and find out if being completely cut off from everything and everyone was what she really wanted?

She considered that for a few minutes, but just the thought of having to get tickets, then pack and get on the right plane all seemed far too hard. Yet imagining herself walking around Florence, seeing the wonders of the Uffizi Gallery or the Pitti Palace, was a lovely daydream. Patrick had once said that he'd like to take her there and show her all his favourite paintings and sculptures, but that wasn't likely to happen, he was always too busy.

Would it really be that hard to pack and get tickets? Why was she being so pathetic?

Just thinking of doing it gave her a twinge of hope.

But would Phil go along with it? If she went, would he say that was the end?

Did she want it to be the end?

She put her head in her hands. She felt like that famous and hideous picture called 'The Scream'. Was she speeding towards a mental breakdown? How did she think she was going to cope in Europe with just schoolgirl French and a smattering of Italian, if she couldn't cope here?

Yet that in itself was an attraction – if she didn't know the language, she couldn't be drawn into conversations. Without talking, perhaps she could nurse her inner self back to what it once was?

One thing was very clear to her. If she stayed here, she was never going to recover. She would carry on doing what she'd done for months now – pretending she was fine, and dying a little more inside every day. Sooner or later, Phil would lose patience and ask her to go. He was getting nothing out of this relationship now, other than having his clothes washed and his meals cooked. A few months down the line she might be too apathetic to even do that.

He deserved better.

Phil came back soon after eleven, and she knew he'd been in a pub by the smell of cigarettes clinging to him. He went to the fridge and got himself a can of beer.

'Well,' he said as he sat down opposite her, 'your time is up. What are you going to do?'

His directness was one of the things she loved about him. He said he liked it in other people, now she was going to test him.

'I'm going to Europe,' she said.

'For ever?'

'No, just for as long as it takes to find myself again. Sorry, that sounds like one of those dippy-hippy sayings.'

'And am I supposed to sit and wait for you?'

'I wouldn't have the cheek to ask that of you,' she said, hanging her head. 'But you're right, we can't go on like this. I've leaned on you long enough, Phil. It's time I learned to stand alone.'

He leaned back on the sofa and put his hands on his head. The gesture was one of bewilderment.

'I love you,' he said, his voice cracking. 'Right now I wish I didn't, because then I could show you the door, and I could pick up the life I had before I met you. We've been through so much together, Eva. This isn't how it should end.'

'I don't want it to be the end. But until I'm mended inside I'm no good to you.'

'And how will going off to Europe "mend you"?' he said with more than a touch of sarcasm.

'I don't know if it will. But I know if I stay here, feeling the way I do now, I'll end up in a loony bin.'

'Then go. I don't want that for you,' he said, getting to his feet. 'But do it quickly. Send me a postcard from Paris.' He walked off to the bedroom and shut the door.

An hour later, Eva was still sitting in the same place and crying. She knew she'd hurt Phil really badly, and her whole being wanted to go and cuddle him and say she wasn't going anywhere. But she couldn't do that. All she could guarantee was that she would wear him down with her silences, her distance, and it would poison his life. He really did deserve better.

She slept in his brother's old room, and she woke in the morning to hear the familiar sound of Phil making tea and his sandwiches for the day. He didn't bring her tea as he usually did. She was glad, because she didn't think she could bear to go if she saw his face one more time.

Lying there, she waited for him to open the front door and then close it behind him. She half expected him to shout out something nasty, or at least bang the door shut. But, considerate as always, he did it quietly with no last bitter remark or an order to leave her key behind when she left.

She heard the van starting up and then pulling away. Tears rolled down her cheeks, because she knew he was hurting. It was tempting to pull the duvet over her again and cry into the pillow. But having told him she was leaving, she had to.

It was after twelve when she left, with a medium-sized wheeled suitcase. She'd packed the rest of her more wintry clothes into a bin liner and tucked it tidily under the spare bed. She would go directly to the airport and buy a ticket to Paris there. She'd telephoned her solicitor and her bank to tell them she was going away and would be in touch with a forwarding address.

Her bank and solicitor had Phil's bank details, and she'd asked them to pay £10,000 into his account when the funds from the sale of Pottery Lane were cleared. She knew he would have refused it if she'd given him the money, but she owed it to him for all the work he'd done on the house. She intended to write to Ben, Patrick, Olive and Gregor later today and explain.

Finally she cleaned the flat, and last of all wrote Phil a letter. She had so much she wanted and needed to say, but couldn't put it into words. So in the end all she wrote was a simple note.

Phil,

I loved you, I still love you and I always will. But I can't make you happy until I've learned how be happy again. I wish more than anything that all this awful stuff hadn't happened, because it's made me someone I don't want to be.

You have been the very best person in my whole life. I'll never forget how you loved and supported me through everything.

My love always,
Eva

PS: I've left some clothes under your brother's bed. I'll understand if you throw them out.

Chapter Thirty

Sorrento

Eva leaned on the rail of the swimming deck, looking down at the waves washing over the rocks some four feet below. The afternoon sun was beating down on to her bare shoulders, and the decking was too hot to stand on without her flip-flops.

She was alone. All the other guests at The Royal Hotel had retreated up the steps to the shade of the gardens. She could hear their laughter, the clink of glasses and the soft murmur of the same tape of Italian songs she'd heard almost continually for the whole time she'd been staying here.

The ferry to Capri was just leaving the harbour. As the swell from the boat reached the rocks beneath her, she saw the fish. About ten of them, four or five inches long – sardines perhaps, because there was a flash of silver as the waves tossed them from one rock pool to another. They appeared to be trying to get back into the open sea. But each time they almost made it, another wave came and they were tossed back.

She watched them for some time, almost mesmerized by the futility of their exertions, and it suddenly occurred to her that she had a great deal in common with them. In the last eighteen months she'd been tossed around by outside forces, reeling from one disaster to another, and she had lost all sense of direction.

Just like the sardines in the rock pools with their desire to reach the open sea, she'd left England almost four months ago in the belief that if she just removed herself from her past life, she'd save her sanity and find the ability to be happy again.

She had no fears for her sanity any more. She'd had times in Paris when she'd been terrified and felt totally isolated. She missed Phil so badly, it was like an open wound – just a glimpse of a man who looked a bit like him made her heart race. Yet however scared she was at first in Paris, and however much she wanted to stay in her room in the pension in St Germain and hide away, she made herself leave there each morning. First, she would have a coffee and a pastry at a sidewalk cafe while she people-watched, then later she went off to explore whichever area of Paris she'd decided on the night before. She found her way around the Métro, walked for miles, visited all the well-known tourist attractions – and most of the less well-known ones too – and went to parts of the city most people avoided.

Loneliness, she had finally decided, wasn't the same as being alone. She did feel lonely sometimes – usually when she saw something that made her laugh, and she wished Phil was there to laugh with her. She felt lonely too eating on her own; food was something which was always better shared. But mostly she found being alone almost a guilty pleasure, because she didn't have to consider anyone else's feelings or tastes.

She had felt very sorry for herself when she first left England, and the only way she could stop the self-pity was by reminding herself how fortunate she was that she had money in the bank from the sale of the studio. She still lived frugally, because she had never been a spendthrift, but it was a safe feeling knowing she had that big cushion of money behind her.

Maybe it was Olive's suggestion that she write down her feelings, or Flora's influence, which made her start a diary. Certainly the irritation she'd felt at Flora's lack of real information in her diaries made her record not just what she did each day, what she saw, but also how she felt.

Looking back at her diary almost a month after she started, she noticed that on her tenth day in Paris she'd written that the noise of the traffic didn't seem so loud any more, that she didn't mind being jostled by crowds, and that sometimes she even wanted to talk to people. That was the point when she began to feel better.

Writing had been her saviour, she was convinced of that. It filled time, it both soothed and kept her mind sharp, and it became a reason to explore further, just for the joy of writing about it. One day, while scribbling away in a cafe, an English couple asked her if she was writing a book. They said they'd seen her there before, always writing. Without even stopping to think, she said she was. Knowing she was never going to see them again, she told them a fictionalized story about herself, and it was the most liberating thing she'd ever done. To invent a different past for herself meant she could be anyone – rid herself of old scars, emotional baggage and a bad self-image.

It was only once she was in bed that night, with the hum of traffic coming through the open windows on a warm breeze, that it occurred to her that maybe the story she'd told that couple – that she'd taken a sabbatical from her job in an advertising agency in order to write – could well be turned into a book. She found herself excited by the idea of inventing a heroine who was all the things she was not: someone beautiful and brave, who had amazing adventures as she travelled from city to city.

From Paris she took a train, first to Florence and then to

Rome. As she assimilated the turbulent histories of both those ancient cities and found herself wowed by their magnificent works of art, she wrote about what she saw and experienced. Paige, her heroine, did things Eva wouldn't dare: she had an affair with a snake-hipped, doe-eyed waiter who stole her money, she accepted invitations to the homes of total strangers, and she drove a Vespa around the narrow streets in tiny shorts and a cropped top. There really were handsome waiters who made eyes at Eva, but her heart was still with Phil.

Sitting in pavement cafes or strolling past expensive shops, she'd observed the elegance and self-assurance of Italians, marvelling that the design of their clothes, shoes, lighting and furniture were all so much more stylish than their English counterparts. She wrote that Paige was asked to design an interior for a splendid old palazzo Eva had seen on the banks of the Arno near the Ponte Vecchio, and she delighted in writing about the fabrics, the colours and the beautiful antique furniture she would put into it. Paige let herself be seduced by the owner of the palazzo – wild, steamy sex that made Eva feel even more wistful about Phil.

However exciting Rome was, in July it became too hot and crowded for even Paige to enjoy it. So Eva caught the train to Naples and arrived here in Sorrento, on the Amalfi coast, to find that it was a different Italy, one of natural beauty that man had no hand in.

The clear sapphire sea, the terrifying hairpin bends along cliff roads with sheer drops to rocks hundreds of feet below, the slower pace and the air, heavy with the scent of lemons, enchanted her. She had no fear of walking about here after dark, although it was often more tempting to just sit on the balcony of her room in The Royal Hotel to watch the sun set over the sea. The comfort, friendliness and location of the hotel were seductive too. She could catch a train to see the wonders of Pompeii. The ferry went to Capri, Positano and Amalfi, and the staff greeted her each morning like an old friend or a member of the family. And they seemed glad she stayed on and on.

The warm sun, the profusion of flowers and trees, and the writing of her story all worked more magic on her. When she looked in the mirror she saw an attractive girl of twenty-two, small, curvy, with pretty blue eyes and shapely legs – not the plump and plain girl she had once believed she was. So the blonde streaks in her hair weren't natural, but her suntan was; she looked good in her turquoise bikini, and not a day passed without men smiling or whistling at her.

She wasn't really aware that she'd started to enjoy other people's company. It crept up on her. First it was Sadie from Essex, who sat next to her going across on the ferry to Capri. She was a student, alone too, and they talked about the thriller Sadie had with her, one Eva had also read, found on the hotel bookshelf. By the time they got to Capri it seemed natural to explore the pretty little town together. When they arrived back in Sorrento harbour in the early evening, they went to a bar and got drunk together, talking and laughing as if they'd known each other for years. Eva told Sadie about how she had first lied and said she was writing a book, and now it seemed she really was. They talked about the plot and made up more and more ridiculous adventures for her heroine. It was such a fun evening, but Sadie was going back to Naples to meet up with a couple of friends the next day. When they parted, Eva was genuinely sad.

From then on there were many more people to spend a day, an afternoon or an evening with. Each one had a story: betrayal, divorce, sickness, trouble with parents, or a love affair that had ended badly. And this made her realize she wasn't unique; everyone got a share of misery in their lives. Eva found she had no need to tell anyone about herself. She listened and sympathized and let them believe she was intrepid, independent and nothing had given her a moment of heartache.

From almost the first day in Sorrento, she'd thought this was the place she could stay for ever. She'd begun to learn Italian, she'd asked about getting work and even buying property here. She'd been convinced that there was no reason to go back to England.

Until she noticed those fish.

The fish were better off in the rock pools than out in the sea. They could bask in the warmer water; they couldn't be eaten by bigger fish or be scooped up by fishermen. Yet they were striving to get back into deeper water because, dangerous or not, they knew that's where they belonged.

So where did she belong? Was it in England? There was only really Ben, Patrick, Gregor and Olive there. Freya had no interest in her, and she'd burned her bridges with Phil. The half a dozen postcards she'd sent him had been intended to tell him she was thinking of him all the time, but maybe he found them insulting – as if she was thumbing her nose at him. She thought he must have got a new girlfriend by now. That stung – she didn't want to picture him with someone else – but after leaving the way she did, she couldn't expect anything else.

As for her brother and friends, they would be perfectly content with only a letter or phone call from her once in a while, if she was really happy and settled here. Ben and Patrick would come and visit her too. Maybe even Olive would.

She *was* happy now. She woke every morning feeling good about herself, and felt she'd dealt with all the hurt of the past. Her sorrow at losing Flora and Sophie would never go away completely. And in some strange way she was glad she had that sore place inside her; it was evidence of their importance to her.

But loving Phil was quite different from loving Flora or Sophie. He was there in her heart and mind every day, a raw place that would not heal. Each meal she ate, every beautiful view she saw, she wished she was sharing it with him. At night in bed she pictured his face: those soft brown eyes, the way his lips curled up at the corners like a smile even when he was serious. She heard his laughter, whispered words of love, and she remembered how the lovemaking had been before all this other stuff got in the way.

Lots of men had tried to chat her up – in France, Rome, Florence and here – but she had no interest in any of them. Phil had been special, and no one else would ever make her feel the way she did about him.

But she had blown it with him. She had to accept that and just be grateful that he'd been there for her when she most needed love. He'd also let her go without bitter words or nasty accusations. A man like Phil only came along once in a lifetime.

Tears sprang up in her eyes at the thought of what she'd thrown away. She might have got used to loss in her life: Flora and Sophie, and all those precious things that were destroyed in the fire. She could accept that now. But there would always be deep regret at losing Phil.

She moved away from the rail and walked towards the ladder at the end of the deck to go for a swim. As she descended each rung of the steel ladder and the cold water crept up her sun-baked skin, she gasped at the exquisite torture. Her hands moved down the smooth steel of the rail till they met the waterline, where the metal became covered in slimy green weed.

She let herself flop into the sea and then swam away from the ladder, out to where a string of buoys prevented boats coming close to shore. She turned on to her back and floated, enjoying the sensation of the cool water caressing her hot scalp.

Above, wisps of cloud like candyfloss drifted across the periwinkle-blue sky. And as always when she swam here, she looked back at the land and marvelled at the hotels and houses built right on to the edge of the cliffs. She wondered how the builders had the confidence to believe the cliffs would hold them safe.

Perhaps that was the exact reason why she needed to go back to England? To prove to herself that all the bad things in her past life really were over, that nothing else was going to crumble beneath her, and more importantly too that she was now capable of forging a career and a real life for herself, that she wasn't dependent on anyone but herself.

Eva turned over, took a deep breath and dived down into the water, staying down and swimming fast until her lungs felt as if they were about to burst.

As she surfaced, she heard someone shout. Treading water, she wiped the water from her eyes and saw a man on the decking, waving. She looked around, but there was no one else in the water, so she swam back to the ladder to see what he wanted.

Climbing up, to her astonishment she saw that the man was Patrick.

'I can't believe it!' she exclaimed breathlessly as she scrambled up the last few steps. 'How did you know I was here?'

'Elementary, my dear Watson,' he said with a wide smile. 'The last postcard you sent was a picture of this hotel. I rang yesterday to check you were still here.'

'I'm too wet to hug you,' she said, but she took his hand in both of hers. He looked handsome in a white short-sleeved shirt and pale-blue slacks, still with his ponytail, but his face and arms were brown and healthy-looking. 'What a wonderful surprise.'

'It's equally wonderful to see you looking so fit and well,' he said. 'But it's too hot for me down here. Shall we go up into the garden and have a drink or two?'

'What made you come?' she asked later, when she'd dried herself off, put on a sarong and they were sitting at a table under the trees.

'To get you to come home,' he said simply. 'Phil showed me the postcards you'd sent him a couple of nights ago when I called round to see him. I got the distinct impression from them that you might be miles away but he was still in your heart.'

'He is,' she admitted. 'But I've hurt him too badly to hope that he still feels anything for me.'

'He isn't one to wear his heart on his sleeve or to cry into his beer, but I know he wants you back.'

'He does?'

Patrick put one of his hands over hers and smiled at her. 'Yes, he does, Eva. He understood why you felt you had to run away. He said if he'd been through what you had, he'd have done the same.'

A delicious bubbly feeling coursed through her veins. 'Does he know you've come here?'

'No, I came on an impulse after I phoned the hotel. But the fact you were still here might have meant you had someone new, so I wasn't going to give Phil false hope. Have you got someone?'

She shook her head. 'There hasn't been anyone at all. I don't want anyone but Phil.'

Patrick grinned. 'Well then, this wasn't a wild goose chase. I suggest we ring the airport and get you on a plane home tomorrow.'

The waiter came, and Patrick ordered a bottle of wine. Then he beamed at her. 'You mentioned on my card you were writing a book. Tell me about it?'

Eva laughed. 'Oh, it's just rubbish really – certainly not publishable. But writing it brought back my sanity. It is very cathartic. While pretending something is fiction you can write stuff that really happened, and make it so it doesn't hurt any more.'

He put a hand over hers. 'I can see that you are mended. You look fabulous – shining eyes, glowing skin – the way you looked the first time I met you. You know that I went through something like this after Flora left me. I know how it feels. I cured myself too by going to Canada. I wrote very bad poetry, I drank far too much, and then gradually that black mist lifted. But no one ever captured my heart again the way Flora did. I sometimes think if I'd gone up to Scotland and found her, that maybe . . .' He paused. 'I think she sent me that picture of the cottage because she wanted me to come. But I was too hurt and weak to do that. Phil is stronger than that. I dare say some people would say he let you walk all over him. But in my opinion only the strongest of men can let their woman go, and trust that love doesn't die, and she'll come back.'

Eva's eyes filled with tears.

'Don't cry,' Patrick said reprovingly. 'Today is for celebrating. I've got something else to tell you too. Your sister, Freya, is in London. I met her with Phil a few weeks ago.'

'Really!' she exclaimed. 'Where? What is she doing? Is she OK?'

'Don't look so anxious, she's a great little thing. As

491

forthright as you! It seems after you saw her in hospital she had a spell working in a nursing home, which she hated. She didn't ring you, because she was too proud. She said she was determined to get a good job before she contacted you again. She said something about you both making a promise to see each other in a year's time.'

'Yes, we did.' Eva grinned. 'But I'd forgotten about that. So tell me, what is she doing?'

'Well, she managed, in her words, to "blag her way on to a training course in computers". That was in Newcastle. She lived in some seedy digs while she did it, passed with flying colours and got taken on by one of the big computer companies in London. I can't remember the name of it.'

'That is amazing!' Eva felt jubilant to hear such unexpected good news.

'She's a tough cookie and fiercely independent. She's sharing a flat in Hammersmith with three other girls, and she only rang to speak to you at Phil's after she'd settled in. Phil was a bit worried about meeting her alone, having to explain how things had been for you and such like. So he took me along. I really liked her – she's an awful lot like you.'

'Did she say anything about her mother . . . ?' Eva paused. 'Our mother,' she added.

'She still hasn't surfaced, and Freya said quite bluntly that part of her reason for coming south was so that Sue will never be able to find her. Both Phil and I thought that was both wise and brave.'

'What did Phil tell her about me being away?'

'The truth, Eva. That everything got too much for you after Sophie died, and that you were cracking up. Do you know what she said?'

'No, tell me.'

'"I felt so ashamed that I didn't tell her how glad I was she came to see me. She bought me such lovely clothes, and it was she who really motivated me to get myself together. I have to be on my own too when I'm troubled. I bet even now she's wanting to come back, but afraid you don't want her any more.'"

Eva's eyes filled up again.

'Don't,' Patrick said, and with his thumb he wiped her tears away. 'Celebration time now. So let's drink to the future? You've got a good man, a sister and a brother to go back to. And I'll be around too.'

Eva had felt she was happy earlier, but now she felt she just might burst with it. Life, it seemed, was offering her a second chance. She was going to take it with both hands.

'To new beginnings,' she said, raising her glass.

Patrick clinked with his glass. 'And happy ever after, like my Mr Bear books,' he grinned.

It was two thirty in the afternoon on the following day when Patrick put Eva's case into a taxi to take her to Naples airport. He was going to stay on at the hotel for a few days, then go to Rome to meet up with an old friend. He said he had rung Phil that morning, as soon as he'd arranged her flight home.

'I'm scared,' Eva admitted and leaned against his chest.

Scared didn't really cover it. She was in a state of elation mixed with terror.

'Flying's nothing these days,' he said, even though he knew perfectly well it wasn't the flight home she was afraid of. He hugged her tightly. 'Everything will be fine, trust me. Now clear off so I can get back to some sunbathing.'

She laughed then.

'That's better,' he said approvingly. 'And I forgot to say you look so gorgeous that if there're any newshounds at

Heathrow waiting for someone famous to appear, they'll think you're a celebrity and they'll start snapping you.'

Eva was actually very pleased with her appearance. She'd bought the short pink silk dress and matching jacket a couple of weeks earlier in Capri, and had been shocked at how many lira it cost. But it had been irresistible; it fitted her perfectly, and it made her feel like a film star. She'd had her hair cut and blow-dried that morning, and she looked so different from the pale miserable girl who had left London four months ago. She could hardly believe it.

'Off with you.' Patrick kissed her on both cheeks and nudged her into the taxi. 'When I get back, we'll all go out to dinner.'

Eva's suitcase was one of the first to come round on the carousel. As she moved forward to take it off, a man lifted it for her and smiled at her. 'A pretty woman shouldn't have to haul her own case,' he said.

She blushed and thought how odd it was that she got compliments like that all the time now. But perhaps it was only her tan.

Clutching the bag with duty-free Bacardi and some Chanel aftershave for Phil in one hand, and dragging her case along behind her, she passed through Customs and out into the arrivals hall, wondering whether she could justify the expense of a taxi rather than a bus or the tube. There were lots of people waiting, many of them holding up cards with a name on, and a young man just ahead of her was nearly knocked over by a girl who ran towards him full tilt to welcome him with a hug.

Then she saw Phil.

He was just leaning on the rail, smiling at her, and he'd clearly spotted her some time before she'd seen him. Her legs turned to rubber, her heart began to race. How like him to surprise her.

He looked far more handsome than she remembered. He was very tanned, wearing a pale-blue short-sleeved shirt and smart grey trousers. His smile lit up the whole arrivals hall.

She tried to run to him, but the wheels on the case seemed to stick to the floor. But suddenly he was there in front of her, arms open wide, and she forgot the case and threw herself into his embrace.

'Welcome home, babe,' he said. 'You look amazing.'

He caught hold of the case with one hand, and her with the other. Then he drew her over to a spot near a wall, and kissed her.

Nothing in the world was so good. Warm, loving and full of promise for the night ahead. It sent tingles all through her body to her toes.

'Patrick said you'd be waiting at home,' she said when he finally released her.

'Do you really think I could wait there?' He laughed. 'Anyway, we aren't going home tonight. My brother is back there again. So I've got us a room in a posh hotel in Bayswater. We've got a lot of time to make up for.'

Eva felt unable to speak as they walked to the car park. There was so much she wanted to say, to ask, but the words just wouldn't form.

As they approached the van she remembered the day they first met, and how he'd helped her into the front seat and put the seat belt around her as if she was a child. And all at once she knew what she had to say.

'Will you forgive me?'

He stopped, reaching out to tuck a strand of hair behind her ear, and his hand lingered on her cheek. 'For what?' He smiled. 'For just having a wobbly attack and needing space?

You don't need forgiveness for that. I'm your lover, not your keeper.'

She put her arms around him and stood on tiptoe to kiss him. Just the word 'lover' sent delicious thrills down her spine.

It was going to be a night to remember.

Acknowledgements

A huge thank you to Emma Housby, who sent me so much material about Carlisle in the 1970s. It never ceases to amaze me that people I have never met will rally round to help with research just because they are kind.

Also, thank you to everyone in Carlisle who offered me snippets of information while I was there researching. You are wonderful.

Last, but not least, a virtual hug to everyone at Carlisle Central Library for the warm welcome you gave me. I love your city and its friendly people so much.

LESLEY
PEARSE

To discover more about Lesley
sign up to her newsletter by visiting

www.lesleypearse.co.uk

and follow Lesley on

twitter 🐦 **@LesleyPearse**

About LESLEY

Lesley Pearse is one of the UK's best-loved novelists, with fans across the globe and book sales of nearly four million copies to date.

A true storyteller and a master of gripping plots that keep the reader hooked from beginning to end, Lesley introduces readers to unforgettable characters about whom it is impossible not to care. There is no easily defined genre or formula to her books: some, like *Rosie* and *Secrets*, are family sagas; *Till We Meet Again*, *A Lesser Evil* and *Faith* are crime novels; and others, such as *Never Look Back*, *Gypsy* and *Hope*, are historical adventures.

Remember Me is based on an astounding true story about Mary Broad, who was convicted of highway robbery and transported on the first fleet to Australia, where a penal colony was to be founded. The story of the appalling hardships that faced the prisoners there, and how courageously determined Mary was to gain a better life for her children is one you are unlikely to forget.

Passionately emotive *Trust Me* is also set in Australia, and deals with the true-life scandal of the thousands of British children who were sent there in the post-war period to be systematically neglected, and in many cases, abused.

Stolen is a thriller – the first of her books with a contemporary setting – and was a Number One bestseller in 2010.

Painstaking research is one of Lesley's hallmarks: first, Lesley reads widely on the subject matter, and then she goes to the place she has chosen as a setting. Once there, digging up local history, the story begins, whether it is about the convicts in Australia, the condition for soldiers in the Crimean war, the hardships facing gold miners in the Klondike or the sheer jaw-dropping courage of the pioneers who forged their way across America in covered wagons.

History is one of Lesley's passions, and mixed with her vivid imagination and her keen insight into how people might behave in dangerous, tragic and unusual situations, she is soon able to weave a plot with many dramatic twists and turns.

'It wasn't until I was working on my sixth or seventh book that I became aware that my heroines had all had to overcome and rise above emotional damage inflicted upon them as children,' Lesley points out. 'I have had to do this myself; my childhood certainly wasn't a bed of roses, but the more people I get to know, the more I find that most have some kind of trauma in their past. Perhaps this is why so many of us like to read about triumph over adversity.'

Lesley's colourful past has been a very useful point of reference in her writing. Whether she is writing about a grim post-war orphanage, about a child that doesn't quite fit in at school, about adoption, about a girl who leaves home too young and too ill-equipped to cope adequately, about poverty or about the pain of first love, she knows how it feels first hand. For as Lesley laughingly puts it, 'my life has had more ups and downs than a well bucket.'

But the sadness and difficulties in Lesley's life are in the past now. With three grown-up daughters, two much-loved grandsons and a brand new granddaughter, Lesley feels her life is wonderful.

'I live in a pretty cottage in Somerset, with my two dogs, Maisie and Lotte, and my garden there is an all-consuming passion,' she says. 'I feel I am truly blessed to wake up each morning with such lovely choices: writing or gardening. They fit so well together; if I'm stuck for an idea in the latest book, I go out and dead head the flowers, mow the lawn or weed. Often I'm writing until two or three in the morning as there is complete silence then and no distractions like the phone to disturb my concentration. I have to confess to wasting a lot of time on Twitter though. It's like having a whole bunch of invisible friends, many of whom are writers too; we comment on things, tell each other what we are up to, and sometimes we get into conversations that are so funny I'm sitting there howling with laughter.'

Friends are very important to Lesley; some of them go right back to ones she met at school and as a teenager. They are her life blood and she likes nothing better than a girl's day out with shopping and lunch with a group of them.

'I love it too when I'm invited to give a talk: sometimes these are in libraries, where I get to meet my readers; sometimes as a guest speaker at lunches or dinner for a charity function. A writer needs to have direct contact with people – without it they would be working in a vacuum. The feedback you get is so valuable: which book they liked best, or least, and why. It's also a great opportunity to get out of my wellies and old gardening clothes, dress up in something glamorous and visit a part of the country I haven't been to before.'

Lesley's storytelling abilities are even more evident when she is speaking to a group of people, for she can trawl through her past and tell hilarious anecdotes that have her audiences in fits of laughter. One suspects she might have made a good actress if she hadn't drifted from job to job as a young girl. As it was, she was a nanny, a bunny girl, a dressmaker and spent many years in promotion work, along with more mundane temping office work. 'The companies I was sent to as a temp were usually glad to see the back of me,' she laughs. 'I was a distraction because I talked to everyone and I was an appalling typist. Funny that I now type fast and accurately. But back then I was always tempted to put my own words into letters I was asked to write, just to pep up the dullness of them.'

Lesley is also the president of the Bath and West Wiltshire branch of the NSPCC – a charity very close to her heart because of physical and mental abuse meted out to her as a child.

'In my ideal world all children would be wanted, valued, loved and cared for,' Lesley says. 'I know from research that child abuse is an evil as old as time, and the only way to stamp it out is through education. I wish every school would put parenting on the curriculum and drum into teenagers the importance of taking care to ensure that they are in stable, loving relationships before embarking on having a baby.'

The Books

Georgia

Raped by her foster-father, fifteen-year-old
Georgia runs away from home to the seedy
back streets of Soho . . .

Tara

Anne changes her name to Tara to forget
her shocking past – but can she really become
someone else?

Charity

Charity Stratton's bleak life is changed
for ever when her parents die in a fire.
Alone and pregnant, she runs away to London . . .

Ellie

Eastender Ellie and spoilt Bonny set off
to make a living on the stage. Can their
friendship survive sacrifice and ambition?

Camellia

Orphaned Camellia discovers that the past
she has always been so sure of has been built
on lies. Can she bear to uncover the truth
about herself?

Rosie

Rosie is a girl without a mother, with a past full of trouble. But could the man who ruined her family also save Rosie?

Charlie

Charlie helplessly watches her mother being senselessly attacked. What secrets have her parents kept from her?

Never Look Back

An act of charity sends flower girl Matilda on a trip to the New World and a new life . . .

Trust me

Dulcie Taylor and her sister are sent to an orphanage and then to Australia. Is their love strong enough to keep them together?

Father Unknown

Daisy Buchan is left a scrapbook with details about her real mother. But should she go and find her?

ill We Meet Again

usan and Beth were childhood friends.
ow Susan is accused of murder, and Beth
nds she must defend her.

emember Me

ary Broad is transported to Australia as a
onvict and encounters both cruelty and passion.
an she make a life for herself so far from home?

crets

dele Talbot escapes a children's home to find
er grandmother – but soon her unhappy
other is on her trail . . .

Lesser Evil

ristol, the 1960s, and young Fifi Brown defies her
arents to marry a man they think is beneath her.

ope

omerset, 1836, and baby Hope is cast out
om a world of privilege as proof of her
other's adultery . . .

ith

otland, 1995, and Laura Brannigan is in prison
r a murder she claims she didn't commit.

read more
www.penguin.co.uk

Gypsy

Liverpool, 1893, and after tragedy strikes the Bolton family, Beth and her brother Sam embark on a dangerous journey to find their fortune in America.

Stolen

A beautiful young woman is discovered half-drowned on a Sussex beach. Where has she come from? Why can't she remember who she is – or what happened?

Belle

London, 1910, and the beautiful and innocent Belle Reilly is cruelly snatched from her home and sold to a brothel in New Orleans where she begins her life as a courtesan. Can Belle ever find her way home?

The Promise

When Belle Reilly's husband Jimmy enlists and heads for the trenches of northern France, she knows she cannot stand idly by awaiting his return. Volunteering to help the wounded, Belle is posted to France as a Red Cross ambulance driver. There, tragedy brings her face to face with Etienne, a man from her past who she'd never quite forgotten.